THE SYSTEMS VIEW OF LIFE

A Unifying Vision

Over the past 30 years, a new systemic conception of life has emerged at the forefront of science. New emphasis has been given to complexity, networks, and patterns of organization, leading to a novel kind of "systemic" thinking.

This volume integrates the ideas, models, and theories underlying the systems view of life into a single coherent framework. Taking a broad sweep through history and across scientific disciplines, the authors examine the appearance of key concepts such as autopoiesis, dissipative structures, social networks, and a systemic understanding of evolution. The implications of the systems view of life for healthcare, management, and our global ecological and economic crises are also discussed.

Written primarily for undergraduates, it is also essential reading for graduate students and researchers interested in understanding the new systemic conception of life and its implications for a broad range of professions – from economics and politics to medicine, psychology, and law.

FRITJOF CAPRA is a Founding Director of the Center for Ecoliteracy in Berkeley, California, and serves on the faculty of Schumacher College (UK). He is a physicist and systems theorist, and has been engaged in a systematic examination of the philosophical and social implications of contemporary science for the past 35 years. He is also the best-selling author of *The Tao of Physics* (1975) and *The Web of Life* (1996).

PIER LUIGI LUISI is Professor of Biochemistry at the University of Rome 3. He started his career at the Swiss Federal Institute of Technology in Zurich, Switzerland (ETHZ), where he became full Professor of chemistry and initiated the interdisciplinary Cortona Weeks. His main research focuses on the experimental, theoretical, and philosophical aspects of the origin of life and the self-organization of synthetic and natural systems. He is also the author of *The Emergence of Life* (2006) and *Mind and Life* (2008).

THE SYSTEMS VIEW OF LIFE

A Unifying Vision

FRITJOF CAPRA
Formerly of the Lawrence Berkeley National Laboratory, California, USA

PIER LUIGI LUISI
University of Rome 3, Italy

CAMBRIDGE
UNIVERSITY PRESS

University Printing House, Cambridge CB2 8BS, United Kingdom

Cambridge University Press is part of the University of Cambridge.

It furthers the University's mission by disseminating knowledge in the pursuit of education, learning and research at the highest international levels of excellence.

www.cambridge.org
Information on this title: www.cambridge.org/9781107011366

First published 2014
3rd printing 2014

Printed in the United States of America by Sheridan Books Inc.

A catalogue record for this publication is available from the British Library

Library of Congress Cataloguing in Publication data
Capra, Fritjof.
The systems view of life : a unifying vision / Fritjof Capra, formerly of the Lawrence Berkeley National Laboratory, CA, USA, Pier Luigi Luisi, University of Rome 3, Italy.
pages cm
Includes bibliographical references and index.
ISBN 978-1-107-01136-6 (hardback)
1. Science – Philosophy. 2. Science – Social aspects. I. Luisi, P. L. II. Title.
Q175.C2455 2014
304.201 – dc23 2013034908

ISBN 978-1-107-01136-6 Hardback

Additional resources for this publication at www.cambridge.org/9781107011366

To the memory of
Francisco Varela (1946–2001),
who introduced us to each other and who inspired both of us
with his systemic vision and spiritual orientation

Contents

Preface

As the twenty-first century unfolds, it is becoming more and more evident that the major problems of our time – energy, the environment, climate change, food security, financial security – cannot be understood in isolation. They are systemic problems, which means that they are all interconnected and interdependent. Ultimately, these problems must be seen as just different facets of one single crisis, which is largely a crisis of perception. It derives from the fact that most people in our modern society, and especially our large social institutions, subscribe to the concepts of an outdated worldview, a perception of reality inadequate for dealing with our overpopulated, globally interconnected world.

There *are* solutions to the major problems of our time; some of them even simple. But they require a radical shift in our perceptions, our thinking, our values. And, indeed, we are now at the beginning of such a fundamental change of worldview in science and society, a change of paradigms as radical as the Copernican revolution. Unfortunately, this realization has not yet dawned on most of our political leaders, who are unable to "connect the dots," to use a popular phrase. They fail to see how the major problems of our time are all interrelated. Moreover, they refuse to recognize how their so-called solutions affect future generations. From the systemic point of view, the only viable solutions are those that are sustainable. As we discuss in this book, a sustainable society must be designed in such a way that its ways of life, businesses, economy, physical structures, and technologies do not interfere with nature's inherent ability to sustain life.

Over the past thirty years it has become clear that a full understanding of these issues requires nothing less than a radically new conception of life. And indeed, such a new understanding of life is now emerging. At the forefront of contemporary science, we no longer see the universe as a machine composed of elementary building blocks. We have discovered that the material world, ultimately, is a network of inseparable patterns of relationships; that the planet as a whole is a living, self-regulating system. The view of the human body as a machine and of the mind as a separate entity is being replaced by one that sees not only the brain, but also the immune system, the bodily tissues, and even each cell as a living, cognitive system. Evolution is no longer seen as a competitive struggle for existence, but rather as a cooperative dance in which creativity and the constant emergence of novelty are the driving forces. And with the new emphasis on complexity, networks, and patterns of organization, a new science of qualities is slowly emerging.

This new conception of life involves a new kind of thinking – thinking in terms of relationships, patterns, and context. In science, this way of thinking is known as "systemic thinking," or "systems thinking"; hence, the understanding of life that is informed by it is often identified by the phrase we have chosen for the title of this book: the systems view of life.

The new scientific understanding of life encompasses many concepts and ideas that are being developed by outstanding researchers and their teams around the world. With the present book, we want to offer an interdisciplinary text that integrates these ideas, models, and theories into a single coherent framework. We present a unified systemic vision that includes and integrates life's biological, cognitive, social, and ecological dimensions; and we also discuss the philosophical, spiritual, and political implications of our unified view of life.

We believe that such an integrated view is urgently needed today to deal with our global ecological crisis and protect the continuation and flourishing of life on Earth. It will therefore be critical for present and future generations of young researchers and graduate students to understand the new systemic conception of life and its implications for a broad range of professions – from economics, management, and politics to medicine, psychology, and law. In addition, our book will be useful for undergraduate students in the life sciences and the humanities.

In the following chapters, we take a broad sweep through the history of ideas and across scientific disciplines. Beginning with the Renaissance and the Scientific Revolution, our historical account includes the evolution of Cartesian mechanism from the seventeenth to the twentieth centuries, the rise of systems thinking, the development of complexity theory, recent discoveries at the forefront of biology, the emergence of the new conception of life at the turn of this century, and its economic, ecological, political, and spiritual implications.

The reader will notice that our text includes not only numerous references to the literature, but also an abundance of cross-references to chapters and sections in this book. There is a good reason for this abundance of references. A central characteristic of the systems view of life is its nonlinearity: all living systems are complex – i.e., highly nonlinear – networks; and there are countless interconnections between the biological, cognitive, social, and ecological dimensions of life. Thus, a conceptual framework integrating these multiple dimensions is bound to reflect life's inherent nonlinearity. In our struggle to communicate such a complex network of concepts and ideas within the linear constraints of written language, we felt that it would help to interconnect the text by a network of cross-references. Our hope is that the reader will find that, like the web of life, this book itself is also a whole that is more than the sum of its parts.

FRITJOF CAPRA, *Berkeley*
PIER LUIGI LUISI, *Rome*

Acknowledgments

The synthesis of concepts and ideas we present in this book took three decades to mature. During this time, we were fortunate to be able to discuss most of the underlying scientific models and theories with their authors and with other scientists working in those fields, as well as with each other. Many of our insights and ideas originated and were further refined in those intellectual encounters.

We are especially grateful

- to Humberto Maturana for many stimulating conversations about autopoiesis, cognition, and consciousness;
- to the late Francisco Varela for illuminating discussions and inspiring collaborations over two decades on a wide variety of topics in cognitive science;
- to the late Lynn Margulis for inspiring dialogues about microbiology, symbiogenesis, and Gaia theory;
- to Helmut Milz for many clarifying discussions of medicine and the systems view of health; and
- to Brother David Steindl-Rast for enlightening conversations over three decades about spirituality, art, religion, and ethics.

Fritjof Capra also would like to express his gratitude

- to the late Ilya Prigogine for inspiring conversations about his theory of dissipative structures;
- to the late Brian Goodwin for challenging discussions over many years about complexity theory, cellular biology, and evolution;
- to Manuel Castells for a series of stimulating, systematic discussions of fundamental concepts in social theory, of technology and culture, and of the complexities of globalization; and for his critical reading of parts of our manuscript;
- to Margaret Wheatley for inspiring dialogues over several years about complexity and self-organization in living systems and human organizations;
- to Hazel Henderson and Jerry Mander, for challenging discussions since the 1970s about sustainability, technology, and the global economy;

- to Miguel Altieri for enlightening tutorials about the theory and practice of agroecology and organic farming; and to Vandana Shiva for numerous inspiring conversations about science, philosophy, ecology, community, and the Southern perspective on globalization;
- to Terry Irwin, Amory Lovins, and Gunter Pauli for many informative conversations about ecodesign;
- to the late Ernest Callenbach for reading portions of the manuscript and offering many critical comments.

Pier Luigi Luisi would like to convey his thanks, in particular,

- to Michel Bitbol (MD, then PhD in quantum physics, and now professor of philosophy at CREA (Centre de Recherche en Épistémologie Appliquée), Paris, where he has worked with Francisco Varela), Matthieu Ricard (Tibetan monk, one of the main figures in the entourage of the Dalai Lama, who started as a PhD student in molecular biology and is still a lover of science), and Franco Bertossa (director of the ASIA center in Bologna) for stimulating discussions about life and consciousness;
- to Paul Davies, Stuart Kauffman, Denis Noble, and Paolo Saraceno for wide-ranging discussions on the subjects of their books; and, last but not least,
- to his students and younger coworkers for their continuous questioning, which obliged him to study more and come up with unexpected answers; special thanks are due, among many, to Matteo Allegretti, Luisa Damiano, Rachel Faiella, Francesca Ferri, Michele Lucantoni, and Pasquale Stano.

Both of us are greatly indebted to Angelo Merante for producing numerous technical drawings, and to Julia Ponsonby for three beautiful line drawings in Chapters 5, 8, and 16. Last but not least, we are grateful to our editor Katrina Halliday at Cambridge University Press for her enthusiastic support during the writing of this book, and to Ilaria Tassistro for seeing the manuscript through the publishing process.

Introduction: paradigms in science and society

Questions about the origin, nature, and meaning of life are as old as humanity itself. Indeed, they lie at the very roots of philosophy and religion. The earliest school of Greek philosophy, known as the Milesian school, made no distinction between animate and inanimate, nor between spirit and matter. Later on, the Greeks called those early philosophers "hylozoists," or "those who think that matter is alive."

The ancient Chinese philosophers believed that the ultimate reality, which underlies and unifies the multiple phenomena we observe, is intrinsically dynamic. They called it *Tao* – the way, or process, of the universe. For the Taoist sages all things, whether animate or inanimate, were embedded in the continuous flow and change of the *Tao*. The belief that everything in the universe is imbued with life has also been characteristic of indigenous spiritual traditions throughout the ages. In monotheistic religions, by contrast, the origin of life is associated with a divine creator.

In this book, we shall approach the age-old questions of the origin and nature of life from the perspective of modern science. We shall see that even within that much narrower context the distinction between living and nonliving matter is often problematic and somewhat arbitrary. Nevertheless, modern science has shown that the vast majority of living organisms exhibit fundamental characteristics that are strikingly different from those of nonliving matter.

To fully appreciate both the achievements and limitations of the new scientific conception of life – the subject of this book – it will be useful first to clarify the nature and limitations of science itself. The modern word "science" is derived from the Latin *scientia*, which means "knowledge," a meaning that was retained throughout the Middle Ages, the Renaissance, and the era of the Scientific Revolution. What we call "science" today was known as "natural philosophy" in those earlier epochs. For example, the full title of the *Principia*, Isaac Newton's famous work, published in 1687, which became the foundation of science in subsequent centuries, was *Philosophiae naturalis principia mathematica* ("The Mathematical Principles of Natural Philosophy").

The modern meaning of science is that of an organized body of knowledge acquired through a particular method known as the scientific method. This modern understanding evolved gradually during the eighteenth and nineteenth centuries. The characteristics of the scientific method were fully recognized only in the twentieth century and are still frequently misunderstood, especially by nonscientists.

The scientific method

The scientific method represents a particular way of gaining knowledge about natural and social phenomena, which can be summarized as occurring in several stages.

First, it involves the systematic observation of the phenomena being studied and the recording of these observations as evidence, or scientific data. In some sciences, such as physics, chemistry, and biology, the systematic observation includes controlled experiments; in others, such as astronomy or paleontology, this is not possible.

Next, scientists attempt to interconnect the data in a coherent way, free of internal contradictions. The resulting representation is known as a scientific model. Whenever possible, we try to formulate our models in mathematical language, because of the precision and internal consistency inherent in mathematics. However, in many cases, especially in the social sciences, such attempts have been problematic, as they tend to confine the scientific models to such a narrow range that they lose much of their usefulness. Thus we have come to realize over the last few decades that neither mathematical formulations nor quantitative results are essential components of the scientific method.

Last, the theoretical model is tested by further observations and, if possible, additional experiments. If the model is found to be consistent with all the results of these tests, and especially if it is capable of predicting the results of new experiments, it eventually becomes accepted as a scientific theory. The process of subjecting scientific ideas and models to repeated tests is a collective enterprise of the community of scientists, and the acceptance of the model as a theory is done by tacit or explicit consensus in that community.

In practice, these stages are not neatly separated and do not always occur in the same order. For example, a scientist may formulate a preliminary generalization, or hypothesis, based on intuition, or initial empirical data. When subsequent observations contradict the hypothesis, he or she may try to modify the hypothesis without giving it up completely. But if the empirical evidence continues to contradict the hypothesis or the scientific model, the scientist is forced to discard it in favor of a new hypothesis or model, which is then subjected to further tests. Even an accepted theory may eventually be overthrown when contradictory evidence comes to light. This method of basing all models and theories firmly on empirical evidence is the very essence of the scientific approach.

Crucial to the contemporary understanding of science is the realization that all scientific models and theories are limited and approximate (as we discuss more fully in Chapter 4). Twentieth-century science has shown repeatedly that all natural phenomena are ultimately interconnected, and that their essential properties, in fact, derive from their relationships to other things. Hence, in order to explain any one of them completely, we would have to understand all the others, and that is obviously impossible.

What makes the scientific enterprise feasible is the realization that, although science can never provide complete and definitive explanations, limited and approximate scientific knowledge is possible. This may sound frustrating, but for many scientists the fact that we *can* formulate approximate models and theories to describe an endless web of interconnected phenomena, and that we are able to systematically improve our models or approximations over time, is a source of confidence and strength. As the great biochemist Louis Pasteur (quoted by Capra, 1982) put it:

> Science advances through tentative answers to a series of more and more subtle questions which reach deeper and deeper into the essence of natural phenomena.

Scientific and social paradigms

During the first half of the twentieth century, philosophers and historians of science generally believed that progress in science was a smooth process in which scientific models and theories were continually refined and replaced by new and more accurate versions, as their approximations were improved in successive steps. This view of continuous progress was radically challenged by the physicist and philosopher of science Thomas Kuhn (1962) in his influential book, *The Structure of Scientific Revolutions*.

Kuhn argued that, while continuous progress is indeed characteristic of long periods of "normal science," these periods are interrupted by periods of "revolutionary science" in which not only a scientific theory but also the entire conceptual framework in which it is embedded undergoes radical change. To describe this underlying framework, Kuhn introduced the concept of a scientific "paradigm," which he defined as a constellation of achievements – concepts, values, techniques, etc. – shared by a scientific community and used by that community to define legitimate problems and solutions. Changes of paradigms, according to Kuhn, occur in discontinuous, revolutionary breaks called "paradigm shifts."

Kuhn's work has had an enormous impact on the philosophy of science, as well as on the social sciences. Perhaps the most important aspect of his definition of a scientific paradigm is the fact that it includes not only concepts and techniques but also values. According to Kuhn, values are not peripheral to science, nor to its applications to technology, but constitute their very basis and driving force.

During the Scientific Revolution in the seventeenth century, values were separated from facts (as we discuss in Chapter 1), and ever since that time scientists have tended to believe that scientific facts are independent of what we do and are therefore independent of our values. Kuhn exposed the fallacy of that belief by showing that scientific facts emerge out of an entire constellation of human perceptions, values, and actions – out of a paradigm – from which they cannot be separated. Although much of our detailed research may not depend explicitly on our value system, the larger paradigm within which this research is pursued will never be value-free. As scientists, therefore, we are responsible for our research not only intellectually but also morally.

During the past decades, the concepts of "paradigm" and "paradigm shift" have been used increasingly also in the social sciences, as social scientists realized that many characteristics of paradigm shifts can be observed also in the larger social arena. To analyze those broader social and cultural transformations, Capra (1996, p. 6) generalized Kuhn's definition of a scientific paradigm to that of a social paradigm, defining it as "a constellation of concepts, values, perceptions, and practices shared by a community, which forms a particular vision of reality that is the basis of the way the community organizes itself."

The emerging new scientific conception of life, which we summarized in our Preface, can be seen as part of a broader paradigm shift from a mechanistic to a holistic and ecological worldview. At its very core we find a shift of metaphors that is now becoming ever more apparent, as discussed by Capra (2002) – a change from seeing the world as a machine to understanding it as a network.

During the twentieth century, the change from the mechanistic to the ecological paradigm proceeded in different forms and at different speeds in various scientific fields. It has not been a steady change, but has involved scientific revolutions, backlashes, and pendulum swings. A chaotic pendulum in the sense of chaos theory (discussed in Chapter 6) – oscillations that almost repeat themselves but not quite, seemingly random and yet forming a complex, highly organized pattern – would perhaps be the most appropriate contemporary metaphor.

The basic tension is one between the parts and the whole. The emphasis on the parts has been called mechanistic, reductionist, or atomistic; the emphasis on the whole, holistic, organismic, or ecological. In twentieth-century science, the holistic perspective has become known as "systemic" and the way of thinking it implies as "systems thinking," as we have mentioned.

In biology, the tension between mechanism and holism has been a recurring theme throughout its history. At the dawn of Western philosophy and science, the Pythagoreans distinguished "number," or pattern, from substance, or matter, viewing it as something which limits matter and gives it shape. The argument was: do you ask what it is made of – earth, fire, water, etc. – or do you askwhat its *pattern* is?

Ever since early Greek philosophy, there has been this tension between substance and pattern. Aristotle, the first biologist in the Western tradition, distinguished between four causes as interdependent sources of all phenomena: the material cause, the formal cause, the efficient cause, and the final cause. The first two causes refer to the two perspectives of substance and pattern which, following Aristotle, we shall call the perspective of matter and the perspective of form.

The study of matter begins with the question, "What is it made of?" This leads to the notions of fundamental elements, building blocks; to measuring and quantifying. The study of form asks, "What is the pattern?" And that leads to the notions of order, organization, and relationships. Instead of quantity, it involves quality; instead of measuring, it involves mapping.

These are two very different lines of investigation that have been in competition with one another throughout our scientific and philosophical tradition. For most of the time, the study of matter – of quantities and constituents – has dominated. But every now and then the study of form – of patterns and relationships – came to the fore.

Pendulum swings between mechanism and holism: from antiquity to the modern era

Let us now very briefly follow the swings of this chaotic pendulum between mechanism and holism through the history of biology. For the ancient Greek philosophers, the world was a *kosmos*, an ordered and harmonious structure. From its beginnings in the sixth century BC, Greek philosophy and science understood the order of the cosmos to be that of a living organism rather than a mechanical system. This meant for them that all its parts had an innate purpose to contribute to the harmonious functioning of the whole, and that objects moved naturally toward their proper places in the universe. Such an explanation of natural phenomena in terms of their goals, or purposes, is known as teleology, from the Greek *telos* ("purpose"). It permeated virtually all of Greek philosophy and science.

The view of the cosmos as an organism also implied for the Greeks that its general properties are reflected in each of its parts. This analogy between macrocosm and microcosm, and in particular between the Earth and the human body, was articulated most eloquently by Plato in his *Timaeus* in the fourth century BC, but it can also be found in the teachings of the Pythagoreans and other earlier schools. Over time, the idea acquired the authority of common knowledge, and this continued throughout the Middle Ages and the Renaissance.

In early Greek philosophy, the ultimate moving force and source of all life was identified with the soul, and its principal metaphor was that of the breath of life.

Indeed, the root meaning of both the Greek *psyche* and the Latin *anima* is "breath." Closely associated with that moving force, the breath of life that leaves the body at death, was the idea of knowing. For the early Greek philosophers, the soul was both the source of movement and life, *and* that which perceives and knows. Because of the fundamental analogy between microcosm and macrocosm, the individual soul was thought to be part of the force that moves the entire universe, and accordingly the knowing of an individual was seen as part of a universal process of knowing. Plato called it the *anima mundi*, the "world soul."

As far as the composition of matter was concerned, Empedocles (fifth century BC) claimed that the material world was composed of varying combinations of the four elements – earth, water, air, and fire. When left to themselves, the elements would settle into concentric spheres with the Earth at the center, surrounded successively by the spheres of water, air, and fire (or light). Further outside were the spheres of the planets and beyond them was the sphere of the stars.

Half a century after Empedocles, an alternative theory of matter was proposed by Democritus, who taught that all material objects were composed of atoms of numerous shapes and sizes, and that all observable qualities derived from the particular combinations of atoms inside the objects. His theory was so antithetical to the traditional teleological views of matter that it was pushed into the background, where it remained throughout the Middle Ages and the Renaissance. It would only surface again in the seventeenth century, with the rise of Newtonian physics.

The teachings of Democritus (460–340 BC) were expanded by Epicurus (341–270 BC), also an atomist, who restated that everything that occurs is the result of the recombination of atoms, and that there is no purpose behind their motions, nor any design of the gods. Epicurus had a great follower in the first century BC in the Roman poet Lucretius, whose poem *De Rerum Natura* is a remarkable exposition of the science of his time, also with a strong atheist flavor.

For the history of science in the subsequent centuries, the most important Greek philosopher was Aristotle (fourth century BC). He was the first philosopher to write systematic, professorial treatises about the main branches of learning of his time. He synthesized and organized the entire scientific knowledge of antiquity in a scheme that would remain the foundation of Western science for 2,000 years.

Aristotle's treatises were the foundation of philosophical and scientific thought in the Middle Ages and the Renaissance. Christian medieval philosophers, unlike their Arab counterparts, did not use Aristotle's texts as a basis for their own independent research, but instead evaluated them from the perspective of Christian theology. Indeed, most of them were theologians, and their practice of combining philosophy – including natural philosophy, or science – with theology became known as scholasticism.

The leading figure in this movement to weave the philosophy of Aristotle into the Christian teachings was Thomas Aquinas (1225–1274), one of the towering intellects of the Middle Ages. Aquinas taught that there could be no conflict between faith and reason, because the two books on which they were based – the Bible and the "book of nature" – were both authored by God. He produced a vast body of precise, detailed, and systematic philosophical writings, in which he integrated Aristotle's encyclopedic works and medieval Christian theology into a seamless whole.

The dark side of this fusion of science and theology was that any contradiction by future scientists would necessarily have to be seen as heresy. In this way, Thomas Aquinas enshrined in his writings the potential for conflicts between science and religion – which reached a dramatic climax with the trial of Galileo, and have continued to the present day.

Between the Middle Ages and the modern era lies the Renaissance, a period stretching from the beginning of the fifteenth to the end of the sixteenth century. It was a period of intense explorations – of ancient intellectual ideas and of new geographical regions of the Earth. The intellectual climate of the Renaissance was decisively shaped by the philosophical and literary movement of humanism, which made the capabilities of the human individual its central concern. This was a fundamental shift from the medieval dogma of understanding human nature from a religious point of view. The Renaissance offered a more secular outlook, with heightened focus on the individual human intellect.

The new spirit of humanism expressed itself through a strong emphasis on classical studies. During the Middle Ages, much of Greek philosophy, and science had been forgotten in Western Europe, while the classical texts were translated and examined by Arab scholars. Their rediscovery and translation into Latin from Greek and Arabic greatly extended the intellectual frontiers of the European humanists. Scholars and artists were exposed to the great diversity of Greek and Roman philosophical ideas that encouraged individual critical thought and prepared the ground for the gradual emergence of a rational, scientific frame of mind.

According to Capra (2007), modern scientific thought did not emerge with Galileo, as is usually stated by historians of science, but with Leonardo da Vinci (1452–1519). One hundred years before Galileo and Francis Bacon, Leonardo single-handedly developed a new empirical approach, involving the systematic observation of nature, reasoning, and mathematics – in other words, the main characteristics of the scientific method. But his science was radically different from the mechanistic science that would emerge 200 years later. It was a science of organic forms, of qualities, of processes of transformation.

Leonardo's approach to scientific knowledge was visual; it was the approach of the painter. He asserted repeatedly that painting involves the study of natural forms, and he emphasized the intimate connection between the artistic representation of those forms and the intellectual understanding of their intrinsic nature and underlying principles. Thus he created a unique synthesis of art and science, unequalled by any artist before him or since.

Many aspects of Leonardo's science are still Aristotelian, but what makes it sound so modern to us today is that his forms are living forms, continually shaped and transformed by underlying processes. Throughout his life he studied, drew, and painted the rocks and strata of the Earth, shaped by erosion; the growth of plants, shaped by their metabolism; and the anatomy of the animal body in motion.

Leonardo did not pursue science and engineering to dominate nature, as Francis Bacon would advocate a century later, but always tried to learn from her as much as possible. He was in awe of the beauty he saw in the complexity of natural forms, patterns, and processes, and aware that nature's ingenuity was far superior to human design. Accordingly, he often used natural processes and structures as models for his designs. This attitude of seeing nature as a model and mentor is now advanced again, 500 years after Leonardo, in the practice of ecological design (see Section 18.4).

Leonardo's scientific work was virtually unknown during his lifetime, and his manuscripts remained hidden for over two centuries after his death in 1519. Thus his pioneering discoveries and ideas had no direct influence on the further development of science. Eventually, they were all rediscovered by other scientists, often hundreds of years later.

A century after Leonardo's science of qualities and living forms, the pendulum swung in the other direction – toward quantities and a mechanistic conception of nature. In the sixteenth and seventeenth centuries the medieval worldview, based on Aristotelian philosophy and Christian theology, changed radically. The notion of an organic, living, and spiritual universe was replaced by that of the world as a machine, and the world-machine became the dominant metaphor of the modern era until the late twentieth century when it began to be replaced by the metaphor of the network.

The rise of the mechanistic worldview was brought about by revolutionary changes in physics and astronomy, culminating in the achievements of Copernicus, Kepler, Galileo, Bacon, Descartes, and Newton. Because of the crucial role of science in bringing about these far-reaching changes, historians have called the sixteenth and seventeenth centuries the age of the Scientific Revolution.

Galileo Galilei (1564–1642) postulated that, in order to be effective in describing nature mathematically, scientists should restrict themselves to studying those properties of material bodies – shapes, numbers, and movement – which could be measured and quantified. Other properties, like color, sound, taste, or smell, were merely subjective mental projections which should be excluded from the domain of science.

Galileo's strategy of directing the scientist's attention to the quantifiable properties of matter proved extremely successful in physics, but it also exacted a heavy toll. During the centuries after Galileo, the focus on quantities was extended from the study of matter to all natural and social phenomena within the framework of the mechanistic worldview of Cartesian-Newtonian science. By excluding colors, sound, taste, touch, and smell – let alone more complex qualities, such as beauty, health, or ethical sensibility – the emphasis on quantification prevented scientists for several centuries from understanding many essential properties of life.

While Galileo devised ingenious experiments in Italy, in England Francis Bacon (1561–1626) set forth the empirical method of science explicitly, as Leonardo da Vinci had done a century before him. Bacon formulated a clear theory of the inductive procedure – to make experiments and to draw conclusions from them, to be tested by further experiments – and he became extremely influential by vigorously advocating the new method.

The shift from the organic to the mechanistic worldview was initiated by one of the towering figures of the seventeenth century, René Descartes (1596–1650). Descartes, or Cartesius (his Latinized name), is usually regarded as the founder of modern philosophy, and he was also a brilliant mathematician and a very influential scientist. Descartes based his view of nature on the fundamental division between two independent and separate realms – that of mind and that of matter. The material universe, including living organisms, was a machine for him, which could in principle be understood completely by analyzing it in terms of its smallest parts.

The conceptual framework created by Galileo and Descartes – the world as a perfect machine governed by exact mathematical laws – was completed triumphantly by Isaac Newton (1642–1727), whose grand synthesis, Newtonian mechanics, was the crowning achievement of seventeenth-century science. In biology, the greatest success of Descartes' mechanistic model was its application to the phenomenon of blood circulation by William Harvey, a contemporary of Descartes. Physiologists of that time also tried to describe other bodily functions, such as digestion, in mechanistic terms, but these attempts were bound to fail because of the chemical nature of the processes, which was not yet understood.

With the development of chemistry in the eighteenth century, the simplistic mechanical models of living organisms were largely abandoned, but the essence of the Cartesian idea survived. Animals were still viewed as machines, albeit much more complicated ones than mechanical clockworks, since they involved complex chemical processes. Accordingly, Cartesian mechanism was expressed in the dogma that the laws of biology can ultimately be reduced to those of physics and chemistry.

Mechanism and holism in modern biology

The first strong opposition to the mechanistic Cartesian paradigm came from the Romantic movement in art, literature, and philosophy in the late eighteenth and early nineteenth centuries. William Blake (1757–1827), the great mystical poet and painter who exerted a strong influence on English Romanticism, was a passionate critic of Newton. He summarized his critique in the celebrated lines (quoted by Capra, 1996):

> May God us keep
> From single vision and Newton's sleep.

In Germany, Romantic poets and philosophers concentrated on the nature of organic form, as Leonardo da Vinci had done 300 years earlier. Johann Wolfgang von Goethe (1749–1832), the central figure in this movement, was among the first to use the term "morphology" for the study of biological form from a dynamic, developmental point of view. He conceived of form as a pattern of relationships within an organized whole – a conception which is at the forefront of systems thinking today.

The Romantic view of nature as "one great harmonious whole," as Goethe put it, led some scientists of that period to extend their search for wholeness to the entire planet and see the Earth as an integrated whole, a living being. In doing so, they revived an ancient tradition that had flourished throughout the Middle Ages and the Renaissance, until the medieval outlook was replaced by the Cartesian image of the world as a machine. In other words, the view of the Earth as a living being had been dormant for only a relatively brief period.

More recently, the idea of a living planet was formulated in modern scientific language as the so-called Gaia theory. The views of the living Earth developed by Leonardo da Vinci in the fifteenth century and by the Romantic scientists in the eighteenth contain some key elements of our contemporary Gaia theory.

At the turn of the eighteenth to the nineteenth century, the influence of the Romantic movement was so strong that the primary concern of biologists was the problem of biological form, and questions of material composition were secondary. This was especially true for the great French schools of comparative anatomy, or morphology, pioneered by Georges Cuvier (1769–1832), who created a system of zoological classification based on similarities of structural relations.

During the second half of the nineteenth century, the pendulum swung back to mechanism, when the newly perfected microscope led to many remarkable advances in biology. The nineteenth century is best known for the emergence of evolutionary thought, but it also saw the formulation of cell theory, the beginning of modern embryology, the rise of microbiology, and the discovery of the laws of heredity. These new discoveries grounded biology firmly in physics and chemistry, and scientists renewed their efforts to search for physico-chemical explanations of life.

When Rudolf Virchow (1821–1902) formulated cell theory in its modern form, the focus of biologists shifted from organisms to cells. Biological functions, rather than reflecting the organization of the organism as a whole, were now seen as the results of interactions at the cellular level. Research in microbiology was dominated by Louis Pasteur (1822–1895), who was able to establish the role of bacteria in certain chemical processes, thus laying the foundations of biochemistry. Moreover, Pasteur demonstrated that there is a definite correlation between microorganisms and disease.

As the new science of biochemistry progressed, it established the firm belief among biologists that all properties and functions of living organisms would eventually be explained in terms of chemical and physical laws. Indeed, cell biology made enormous progress in understanding the structures and functions of many of the cell's subunits. However, it advanced very little in understanding the coordinating activities that integrate those phenomena into the functioning of the cell as a whole. At the turn of the nineteenth century, the awareness of this lack of understanding triggered the next wave of opposition to the mechanistic conception of life, the school known as organismic biology, or "organicism."

During the early twentieth century, organismic biologists took up the problem of biological form with new enthusiasm, elaborating and refining many of the key insights of Aristotle, Goethe, and Cuvier. Their extensive reflections helped to give birth to a new way of thinking – "systems thinking" – in terms of connectedness, relationships, and context. According to the systems view, an organism, or living system, is an integrated whole whose essential properties cannot be reduced to those of its parts. They arise from the interactions and relationships between the parts.

When organismic biologists in Germany explored the concept of organic form, they engaged in dialogues with psychologists from the very beginning. The philosopher Christian von Ehrenfels (1859–1932) used the German word *Gestalt*, meaning "organic form," to describe an irreducible perceptual pattern, which sparked the school of Gestalt psychology. To characterize a Gestalt, Ehrenfels coined the celebrated phrase, "The whole is more than the sum of its parts," which would become the catchphrase of systems thinking later on.

While organismic biologists encountered irreducible wholeness in organisms, and Gestalt psychologists in perception, ecologists encountered it in their studies of animal and plant communities. The new science of ecology emerged out of organismic biology during the late nineteenth century, when biologists began to study communities of organisms.

In the 1920s, ecologists introduced the concepts of food chains and food cycles, which were subsequently expanded to the contemporary concept of food webs. In addition, they developed the notion of the ecosystem, which, by its very name, fostered a systems approach to ecology.

By the end of the 1930s, most of the key criteria of systems thinking had been formulated by organismic biologists, Gestalt psychologists, and ecologists (see Section 4.3 below). The 1940s saw the formulation of actual systems theories. This means that systemic concepts were integrated into coherent theoretical frameworks describing the principles of organization of living systems. These first theories, which we may call the "classical systems theories," include, in particular, general systems theory and cybernetics. As we discuss in Chapter 5, general systems theory was developed by a single scientist, the biologist Ludwig von Bertalanffy, while the theory of cybernetics was the result of a multidisciplinary collaboration between mathematicians, neuroscientists, social scientists, and engineers – a group that became known collectively as the cyberneticists.

During the 1950s and 1960s, systems thinking had a strong influence on engineering and management, where systemic concepts – including those of cybernetics – were applied to solve practical problems. Yet, paradoxically, the influence of the systems approach in biology was almost negligible during that time.

The 1950s was the decade of the spectacular triumph of genetics, the elucidation of the physical structure of DNA and of the genetic code. For several decades, this triumphal success totally eclipsed the systems view of life. Once again, the pendulum swung back to mechanism.

The achievements of genetics brought about a significant shift in biological research, a new perspective which still dominates our academic institutions today. Whereas cells were regarded as the basic building blocks of living organisms during the nineteenth century, the attention shifted from cells to molecules toward the middle of the twentieth century, when geneticists began to explore the molecular structure of the gene.

Advancing to ever smaller levels in their explorations of the phenomena of life, biologists found that the characteristics of all living organisms – from bacteria to humans – were encoded in their chromosomes in the same chemical substance, using the same code script.

This triumph of molecular biology resulted in the widespread belief that all biological functions can be explained in terms of molecular structures and mechanisms. At the same time, the problems that resist the mechanistic approach of molecular biology became ever more apparent. While biologists knew the precise structure of a few genes, they knew very little of the ways in which genes communicate and cooperate in the development of an organism. In other words, molecular biologists realized that they knew the alphabet of the genetic code but had almost no idea of its syntax.

By the mid 1970s, the limitations of the molecular approach to the understanding of life were evident. However, biologists saw little else on the horizon. The eclipse of systems thinking from pure science had become so complete that it was not considered a viable alternative. In fact, systems theory began to be seen as an intellectual failure in several critical essays. One reason for this harsh assessment was that Ludwig von Bertalanffy (1968) had announced in a rather grandiose manner that his goal was to develop general systems theory into "a mathematical discipline, in itself purely formal but applicable to the various empirical sciences." He could never achieve this ambitious goal because in his time no mathematical techniques were available to deal with the enormous complexity of living systems. Bertalanffy recognized that the patterns of organization characteristic of life are generated by the simultaneous interactions of a large number of variables, but he lacked the means to describe the emergence of those patterns mathematically. Technically speaking, the mathematics of his time was limited to linear equations, which are inappropriate to describe the highly nonlinear nature of living systems.

The cyberneticists did concentrate on nonlinear phenomena like feedback loops and neural networks, and they had the beginnings of a corresponding nonlinear mathematics, but the real breakthrough came several decades later with the formulation of complexity theory, technically known as "nonlinear dynamics," in the 1960s and 1970s (see

Chapter 6). The decisive advance was due to the development of powerful, high-speed computers, which allowed scientists and mathematicians for the first time to model the nonlinear interconnectedness characteristic of living systems, and to solve the corresponding nonlinear equations.

During the 1980s and 1990s, complexity theory generated great excitement in the scientific community. In biology, systems thinking and the organic conception of life reappeared on the scene, and the strong interest in nonlinear phenomena generated a whole series of new and powerful theoretical models that have dramatically increased our understanding of many key characteristics of life. From these models the outlines of a coherent theory of living systems, together with the proper mathematical language, are now emerging. This emerging theory – the systems view of life – is the subject of this book.

Deep ecology

The new scientific understanding of life at all levels of living systems – organisms, social systems, and ecosystems – is based on a perception of reality that has profound implications not only for science and philosophy, but also for politics, business, healthcare, education, and many other areas of everyday life. It is therefore appropriate to end our Introduction with a brief discussion of the social and cultural context of the new conception of life.

As we have mentioned, the *Zeitgeist* ("spirit of the age") of the early twenty-first century is being shaped by a profound change of paradigms, characterized by a shift of metaphors from the world as a machine to the world as a network. The new paradigm may be called a holistic worldview, seeing the world as an integrated whole rather than a dissociated collection of parts. It may also be called an ecological view, if the term "ecological" is used in a much broader and deeper sense than usual. Deep ecological awareness recognizes the fundamental interdependence of all phenomena and the fact that, as individuals and societies, we are all embedded in (and ultimately dependent on) the cyclical processes of nature.

The sense in which we use the term "ecological" is associated with a specific philosophical school, founded in the early 1970s by the Norwegian philosopher Arne Naess (1912–2009) with the distinction between "shallow" and "deep" ecology (see Devall and Sessions, 1985). Since then, this distinction has been widely accepted as a very useful term for referring to a major division within contemporary environmental thought.

Shallow ecology is anthropocentric, or human-centered. It views humans as above or outside of nature, and as the source of all value, and ascribes only instrumental, or "use," value to nature. Deep ecology does not separate humans – or anything else – from the natural environment. It does see the world not as a collection of isolated objects but as a network of phenomena that are fundamentally interconnected and interdependent. Deep ecology recognizes the intrinsic value of all living beings and views humans as just one particular strand in the web of life.

Ultimately, deep ecological awareness is spiritual awareness. When the concept of the human spirit is understood as the mode of consciousness in which the individual feels a sense of belonging, of connectedness, to the cosmos as a whole, it becomes clear that ecological awareness is spiritual in its deepest essence. Hence, the emerging new vision of reality, based on deep ecological awareness, is consistent with the so-called "perennial philosophy" of spiritual traditions, as we discuss in Chapter 13.

There is another way in which Arne Naess characterized deep ecology. "The essence of deep ecology," he wrote, "is to ask deeper questions" (quoted by Devall and Sessions, 1985, p. 74). This is also the essence of a paradigm shift. We need to be prepared to question every single aspect of the old paradigm. Eventually, we will not need to abandon all our old concepts and ideas, but before we know that, we need to be willing to question everything. So, deep ecology asks profound questions about the very foundations of our modern, scientific, industrial, growth-oriented, materialistic worldview and way of life. It questions this entire paradigm from an ecological perspective: from the perspective of our relationships to one another, to future generations, and to the web of life of which we are part.

In our brief summary of the emerging systems view of life in the Preface, we have emphasized shifts in perceptions and ways of thinking. However, the broader paradigm shift also involves corresponding changes of values. And here it is interesting to note a striking connection between the changes of thinking and of values. Both of them may be seen as shifts from self-assertion to integration. These two tendencies – the self-assertive and the integrative – are both essential aspects of all living systems, as we discuss in Chapter 4 (Section 4.1.2). Neither of them is intrinsically good or bad. What is good, or healthy, is a dynamic balance; what is bad, or unhealthy, is imbalance – overemphasis on one tendency and neglect of the other. When we look at our modern industrial culture, we see that we have overemphasized the self-assertive and neglected the integrative tendencies. This is apparent both in our thinking and in our values. It is very instructive to put these opposite tendencies side by side.

thinking		values	
self-assertive	integrative	self-assertive	integrative
rational	intuitive	expansion	conservation
analysis	synthesis	competition	cooperation
reductionist	holistic	quantity	quality
linear	nonlinear	domination	partnership

When we look at this table, we notice that the self-assertive values – competition, expansion, domination – are generally associated with men. Indeed, in patriarchal societies they are not only favored but also given economic rewards and political power. This is one of the reasons why the shift to a more balanced value system is so difficult for most people, and especially for most men.

Power, in the sense of domination over others, is excessive self-assertion. The social structure in which it is exerted most effectively is the hierarchy. Indeed, our political, military, and corporate structures are hierarchically ordered, with men generally occupying the upper levels and women the lower. Most of these men, and also quite a few women, have come to see their position in the hierarchy as part of their identity, and thus the shift to a different system of values generates existential fears in them.

However, there is another kind of power, one that is more appropriate for the new paradigm – power as empowerment of others. The ideal structure for exerting this kind of power is not the hierarchy but the network, the central metaphor of the ecological paradigm. In a social network, people are empowered by being connected to the network. Power as empowerment means facilitating this connectedness. The network hubs with the richest connections become centers of power. They connect large numbers of people to the network and are therefore sought out as authorities in various fields. Their authority allows these centers to empower people by connecting more of the network to itself.

The question of values is crucial to deep ecology. In fact, it is its defining characteristic. Whereas the mechanistic paradigm is based on anthropocentric (human-centered) values, deep ecology is grounded in ecocentric (Earth-centered) values. It is a worldview that acknowledges the inherent value of nonhuman life, recognizing that all living beings are members of ecological communities, bound together in networks of interdependencies. When this deep ecological perception becomes part of our daily awareness, a radically new system of ethics emerges.

Such a deep ecological ethic is urgently needed today, especially in science, since most of what scientists do is not life-furthering and life-preserving but life-destroying. With physicists designing weapons systems of mass destruction, chemists contaminating the global environment, biologists releasing new and unknown types of microorganisms without knowing the consequences, psychologists and other scientists torturing animals in the name of scientific progress – with all these activities going on, it seems most urgent to introduce "eco-ethical" standards into science.

Within the context of deep ecology, the view that values are inherent in all of living nature is based on the spiritual experience that nature and the self are one. This expansion of the self all the way to the identification with nature is the proper grounding of ecological ethics, as Arne Naess clearly recognized:

> Care flows naturally if the "self" is widened and deepened so that protection of free Nature is felt and conceived as protection of ourselves . . . Just as we need no morals to make us breathe . . . [so] if your "self" in the wide sense embraces another being, you need no moral exhortation to show care . . . You care for yourself without feeling any moral pressure to do it.
>
> *(quoted by Fox, 1990, p. 217)*

What this implies, according to the eco-philosopher Warwick Fox (1990), is that the connection between an ecological perception of the world and corresponding behavior is not a logical but a psychological connection. Logic does not lead us from the fact that we are an integral part of the web of life to certain norms of how we should live. However, if we have the deep ecological experience of being part of the web of life, then we *will* (as opposed to *should*) be inclined to care for all of living nature. Indeed, we can scarcely refrain from responding in this way.

By calling the emerging new vision of reality "ecological" in the sense of deep ecology, we emphasize that life is at its very center. This is an important issue for science, because in the mechanistic paradigm physics has been the model and source of metaphors for all other sciences. "All philosophy is like a tree," wrote Descartes (quoted by Vrooman, 1970, p. 189). "The roots are metaphysics, the trunk is physics, and the branches are all the other sciences."

The systems view of life has overcome this Cartesian metaphor. Physics, together with chemistry, is essential to understand the behavior of the molecules in living cells, but it is not sufficient to describe their self-organizing patterns and processes. At the level of living systems, physics has thus lost its role as the science providing the most fundamental description of reality. This is still not generally recognized today. Scientists as well as nonscientists frequently retain the popular belief that "if you really want to know the ultimate explanation, you have to ask a physicist," which is clearly a Cartesian fallacy. The paradigm shift in science, at its deepest level, involves a perceptual shift from physics to the life sciences.

I

The mechanistic worldview

1

The Newtonian world-machine

To appreciate the revolutionary nature of the systems view of life, it is useful to examine in some detail the history, principal characteristics, and widespread influence of the mechanistic paradigm, which it is destined to replace. This is the purpose of our first three chapters, in which we discuss the origin and rise of Cartesian-Newtonian science during the Scientific Revolution (Chapter 1), as well as its impact on both the life sciences (Chapter 2) and the social sciences (Chapter 3).

The worldview and value system that lie at the basis of the modern industrial age were formulated in their essential outlines in the sixteenth and seventeenth centuries. Between 1500 and 1700, there was a dramatic shift in the way people in Europe pictured the world and in their whole way of thinking. The new mentality and new perception of the cosmos gave our Western civilization the features that are characteristic of the modern era. They became the basis of the paradigm that has dominated our culture for the past 300 years and is now changing.

Before 1500, the dominant worldview in European civilization, as well as in most other civilizations, was organic. People lived in small, cohesive communities and experienced nature in terms of personal relationships, characterized by the interdependence of spiritual and material concerns and the subordination of individual needs to those of the community.

The scientific framework of this organic worldview rested on two authorities – Aristotle and the Church. In the thirteenth century, Thomas Aquinas had combined Aristotle's comprehensive system of nature with Christian theology and ethics, and, in doing so, had established the framework that remained unquestioned throughout the Middle Ages. The nature of medieval science was very different from that of our contemporary science. It was based on both reason and faith, and its main goal was to understand the meaning and significance of things, rather than prediction and control. Medieval scientists, looking for the purposes underlying various natural phenomena, considered questions relating to God, the human soul, and ethics to be of the highest significance.

During the sixteenth and seventeenth centuries, the medieval outlook changed radically. The notion of an organic, living, and spiritual universe was replaced by that of the world as a machine, and the mechanistic conception of reality became the basis of the modern worldview. This development was brought about by revolutionary changes in physics and astronomy, culminating in the achievements of Copernicus, Galileo, and Newton.

Figure 1.1 Galileo Galilei (1564–1642). iStockphoto.com/© Georgios Kollidas.

Seventeenth-century science was based on the new empirical method of inquiry advocated forcefully by Francis Bacon, and it included the mathematical description of nature and analytic method of reasoning conceived by the genius of Descartes.

1.1 The Scientific Revolution

The Scientific Revolution began with Nicolaus Copernicus (1473–1543), who overthrew the geocentric view of Ptolemy and the Bible that had been accepted dogma for more than a thousand years. After Copernicus, the Earth was no longer the center of the universe but merely one of many planets circling a minor star at the edge of the galaxy, and humanity was robbed of its proud position as the center of God's creation. Copernicus was fully aware that his view would deeply offend the religious consciousness of his time. He delayed the publication of his epochal book, *De revolutionibus orbium coelestium* ("On the Revolution of the Celestial Spheres"), until 1543, the year of his death, and even then he presented the heliocentric view merely as a hypothesis.

Copernicus was followed by Johannes Kepler (1571–1630), a scientist and mystic who searched for the harmony of the spheres and was able, through painstaking work with astronomical tables, to formulate his celebrated empirical laws of planetary motion, which gave further support to the Copernican system. But the real change in scientific opinion was brought about by Galileo Galilei (Figure 1.1), who was already famous for discovering the laws of falling bodies when he turned his attention to astronomy. Directing the newly invented telescope to the skies and applying his extraordinary gift for scientific observation to celestial phenomena, Galileo was able to discredit the old cosmology beyond any doubt and to establish the Copernican hypothesis as a valid scientific theory.

1.1.1 Galileo: mathematical description of nature

The role of Galileo in the Scientific Revolution goes far beyond his achievements in astronomy, although these are most widely known because of his clash with the Church. After Leonardo da Vinci, Galileo was the first to combine scientific experimentation with the use of mathematical language, and is therefore generally considered the father of modern science.

To make it possible for scientists to describe nature mathematically, Galileo postulated, as we have mentioned, that they should restrict themselves to studying only those properties of material bodies – shapes, numbers, and movement – that can be measured and quantified. Other properties, like color, taste, or smell, are merely subjective and should be excluded from the domain of science. In the centuries after Galileo this became a very successful strategy throughout modern science, but we also had to pay a heavy price. As the psychiatrist R.D. Laing (quoted by Capra, 1988, p. 133) put it emphatically,

> Galileo's program offers us a dead world: Out go sight, sound, taste, touch, and smell, and along with them have since gone esthetic and ethical sensibility, values, quality, soul, consciousness, spirit. Experience as such is cast out of the realm of scientific discourse. Hardly anything has changed our world more during the past four hundred years than Galileo's audacious program. We had to destroy the world in theory before we could destroy it in practice.

1.1.2 Bacon: domination of nature

Galileo's empirical approach was formalized and advocated with great vigor by his contemporary Francis Bacon (Figure 1.2), who boldly attacked traditional schools of thought and developed a veritable passion for scientific experimentation. The "Baconian spirit," as it was called, profoundly changed the nature and purpose of the scientific quest. From the time of the ancients, the goals of natural philosophy had been wisdom, understanding the natural order, and living in harmony with it. Science was pursued "for the glory of God." In the seventeenth century, this attitude changed dramatically.

As the organic view of nature was replaced by the metaphor of the world as a machine, the goal of science became knowledge that can be used to dominate and control nature.

The ancient concept of the Earth as nurturing mother was radically transformed in Bacon's writings, and it disappeared completely as the Scientific Revolution proceeded to replace the organic view of nature with the metaphor of the world as a machine. This shift, which was to become of overwhelming importance for the further development of Western civilization, was initiated and completed by two towering figures of the seventeenth century, Descartes and Newton.

Figure 1.2 Francis Bacon (1596–1650). iStockphoto.com/© Georgios Kollidas.

1.1.3 Descartes: the mechanistic view of the world

René Descartes (Figure 1.3) was not only the first modern philosopher but also a brilliant mathematician and scientist, whose philosophical outlook was profoundly affected by the new physics and astronomy. He did not accept any traditional knowledge but set out to build a whole new system of thought. According to the philosopher and mathematician Bertrand Russell (1961, p. 542), "This had not happened since Aristotle, and is a sign of the new self-confidence that resulted from the progress of science. There is a freshness about his work that is not to be found in any eminent previous philosopher since Plato."

Cartesian certainty

At the very core of Cartesian philosophy and of the worldview derived from it lies the belief in the certainty of scientific knowledge; and it was here, at the very outset, that Descartes went wrong. As we have discussed in the Introduction, twentieth-century science has shown very clearly that there can be no absolute scientific truth, that all our concepts and theories are necessarily limited and approximate.

Cartesian certainty is mathematical in its essential nature. Descartes believed that the key to the universe was its mathematical structure, and in his mind science was synonymous with mathematics. Like Galileo, Descartes believed that the language of nature was mathematics, and his desire to describe nature in mathematical terms led him to his most celebrated discovery. By applying numerical relations to geometrical figures, he was able

Figure 1.3 René Descartes (1596–1650). iStockphoto.com/© Georgios Kollidas.

to correlate algebra and geometry and, in doing so, founded a new branch of mathematics, now known as analytic geometry. This made it possible to represent geometrical curves by algebraic equations, whose solutions he studied in a systematic way. His new method allowed Descartes to apply a very general type of mathematical analysis to the study of moving bodies, in accordance with his grand scheme of reducing all physical phenomena to exact mathematical relationships. Thus he could say, with great pride, "My entire physics is nothing other than geometry" (quoted by Vrooman, 1970, p. 120).

Descartes' genius was that of a mathematician, and this is apparent also in his philosophy. To carry out his plan of building a complete and exact natural science, he developed a new method of reasoning which he presented in his most famous book, *Discourse on Method* (Descartes, 2006/1637). Although this text has become one of the great philosophical classics, its original purpose was not to teach philosophy but to serve as an introduction to science. Descartes' method was designed to reach scientific truth, as is evident from the book's full title, *A Discourse on the Method of Correctly Conducting One's Reason and Seeking Truth in the Sciences*.

The analytic method

The crux of Descartes' method is radical doubt. He doubts everything he can manage to doubt – all traditional knowledge, the impressions of his senses, and even the fact that he has a body – until he reaches one thing he cannot doubt, the existence of himself as a thinker. Thus he arrives at his celebrated statement, "*Cogito, ergo sum*" ("I think, and therefore I exist"). From this Descartes deduces that the essence of human nature lies in thought, and that all the things we conceive clearly and distinctly are true. Descartes' method is analytic. It consists in breaking up thoughts and problems into pieces and in arranging these in their logical order. This analytic method of reasoning is probably Descartes' greatest contribution to science. It has become an essential characteristic of modern scientific thought and has proven extremely useful in the development of scientific theories and the realization of complex technological projects. It was Descartes' method that made it possible for NASA to put a man on the Moon. On the other hand, overemphasis on the Cartesian method has led to the fragmentation that is characteristic of both our general thinking and our academic disciplines, and to the widespread attitude of reductionism in science – the belief that all aspects of complex phenomena can be understood by reducing them to their smallest constituent parts. (As we have discussed, no scientific description of natural phenomena can be completely accurate and exhaustive. In other words, all scientific theories are reductionist in the sense that they need to reduce the phenomena described to a manageable number of characteristics. However, science does not need not be reductionist in the Cartesian sense of reducing phenomena to their smallest constituents.)

Division between mind and matter

Descartes' *cogito*, as it has come to be called, made mind more certain for him than matter and led him to the conclusion that the two were separate and fundamentally different. The Cartesian division between mind and matter has had a profound effect on Western thought. It has taught us to be aware of ourselves as isolated egos existing "inside" our bodies; it has led us to set a higher value on mental than manual work; it has enabled huge industries to sell products – especially to women – that would make us owners of the "ideal body"; it has kept doctors from seriously considering the psychological dimensions of illness, and psychotherapists from dealing with their patients' bodies.

In the life sciences, the Cartesian division has led to endless confusion about the relation between mind and body, which has begun to be clarified only very recently by decisive advances in cognitive science (see Chapter 12). In physics, it has made it extremely difficult for the founders of quantum theory to interpret their observations of atomic phenomena (see Chapter 4). According to Werner Heisenberg (1958, p. 81), who struggled with the problem for many years, "This partition has penetrated deeply into the human mind during the three centuries following Descartes and it will take a long time for it to be replaced by a really different attitude toward the problem of reality."

Descartes based his whole view of nature on this fundamental division between two independent and separate realms; that of mind, or *res cogitans* (the "thinking thing"), and that of matter, or *res extensa* (the "extended thing"). Both mind and matter were creations of God, who represented their common point of reference, being the source of the exact natural order and of the light of reason that enabled the human mind to recognize this order. For Descartes, the existence of God was essential to his scientific philosophy, but in subsequent centuries scientists omitted any explicit reference to God while developing their theories according to the Cartesian division, the humanities concentrating on the *res cogitans* and the natural sciences on the *res extensa*.

Nature as a machine

To Descartes the material universe was a machine and nothing but a machine. There was no purpose, life, or spirituality in matter. Nature worked according to mechanical laws, and everything in the material world could be explained in terms of the arrangement and movement of its parts. This mechanical picture of nature became the dominant paradigm of science in the period following Descartes. It guided all scientific observation and the formulation of all theories of natural phenomena until twentieth-century physics brought about a radical change. The whole elaboration of mechanistic science in the seventeenth, eighteenth, and nineteenth centuries, including Newton's grand synthesis, was but the development of the Cartesian idea. Descartes gave scientific thought its general framework – the view of nature as a perfect machine, governed by exact mathematical laws.

The drastic change in the image of nature from organism to machine had a strong effect on people's attitudes toward the natural environment. The organic worldview of the Middle Ages had implied a value system conducive to ecologically minded behavior. In the words of Carolyn Merchant (1980, p. 3),

> The image of the earth as a living organism and nurturing mother served as a cultural constraint restricting the actions of human beings. One does not readily slay a mother, dig into her entrails for gold, or mutilate her body . . . As long as the earth was considered to be alive and sensitive, it could be considered a breach of human ethical behavior to carry out destructive acts against it.

These cultural constraints disappeared as the mechanization of science took place. The Cartesian view of the universe as a mechanical system provided a "scientific" sanction for the manipulation and exploitation of nature that became typical of modern civilization.

Descartes vigorously promoted his mechanistic view of the world in which all natural phenomena were reduced to the motions and mutual contacts of small material particles. The force of gravity, in particular, was explained by Descartes in terms of a series of impacts of tiny particles contained in subtle material fluids that permeated all space (see Bertoloni-Meli, 2006). This theory was highly influential throughout most of the seventeenth century, until Newton replaced it with his conception of gravity as a fundamental force of attraction between all matter.

Mechanistic view of living organisms

In his attempt to build a complete natural science, Descartes extended his mechanistic view of matter to living organisms. Plants and animals were considered simply machines; human beings were inhabited by a rational soul, but as far as the human body was concerned, it was indistinguishable from an animal-machine. Descartes explained at great length how the motions and various biological functions of the body could be reduced to mechanical operations, in order to show that living organisms were nothing but automata.

Descartes' view of living organisms had a decisive influence on the development of the life sciences. The careful description of the mechanisms that make up living organisms became the major task of biologists, physicians, and psychologists during the subsequent 300 years. The Cartesian approach has been very successful, especially in biology, but it has also limited the directions of scientific research. The problem has beeen that many scientists, encouraged by their success in treating living organisms as machines, tended to believe that they are *nothing but* machines. The adverse consequences of this reductionist fallacy have become especially apparent in medicine, where the adherence to the Cartesian model of the human body as a clockwork has prevented doctors from understanding many of today's major illnesses, as we discuss in Chapter 2.

Although the severe limitations of the Cartesian worldview have now become apparent in all the sciences, Descartes' general method of approaching intellectual problems and his clarity of thought remain immensely valuable. As the political philosopher Montesquieu (1689–1755) put it brilliantly, "Descartes has taught those who came after him how to discover his own errors" (quoted by Vrooman, 1970, p. 258).

1.1.4 Newton's synthesis

Descartes created the conceptual framework for seventeenth-century science, but his view of nature as a perfect machine, governed by exact mathematical laws, had to remain a vision during his lifetime. He could not do more than sketch the outlines of his theory of natural phenomena. The man who realized the Cartesian dream and completed the Scientific Revolution was Isaac Newton (Figure 1.4), born in England in the year of Galileo's death, 1642.

Newton developed a comprehensive mathematical formulation of the mechanistic view of nature, and thus accomplished a grand synthesis of the works of Copernicus and Kepler, Bacon, Galileo, and Descartes. Newtonian physics, the crowning achievement of seventeenth-century science, provided a consistent mathematical theory of the world that remained the solid foundation of scientific thought well into the twentieth century. Newton's grasp of mathematics was far more powerful than that of his contemporaries. He invented a completely new method, known today as differential calculus, to describe the motion of solid bodies; a method that went far beyond the mathematical techniques of Galileo and Descartes (as we discuss in more detail in Chapter 6). This tremendous intellectual achievement has been praised by Einstein (1931) as "perhaps the greatest advance in thought that a single individual was ever privileged to make."

Kepler had derived empirical laws of planetary motion by studying astronomical tables, and Galileo had performed ingenious experiments to discover the laws of falling bodies. Newton combined these two discoveries by formulating general laws of motion governing all objects in the solar system, from stones to planets. According to the well-known legend, the decisive insight occurred to Newton in a sudden flash of inspiration when he saw an apple fall from a tree. He realized that the apple was pulled toward the Earth by the same

Figure 1.4 Isaac Newton (1642–1727). iStockphoto.com/© Georgios Kollidas.

force that pulled the planets toward the Sun, and thus found the key to his grand synthesis. He then used his new mathematical method to formulate the exact laws of motion for all bodies under the influence of the force of gravity. The significance of these laws lay in their universal application. They were found to be valid throughout the solar system and thus seemed to confirm the Cartesian view of nature. The Newtonian universe was, indeed, one huge mechanical system, operating according to exact mathematical laws.

The Principia

Newton (1999/1687) presented his theory of the world in his magnum opus, *Mathematical Principles of Natural Philosophy*. The *Principia*, as the work is usually called for short after its Latin title, comprises a comprehensive system of definitions, propositions, and proofs, which scientists regarded as the correct description of nature for more than 200 years. It also contains an explicit discussion of Newton's experimental method (quoted by Randall, 1976, p. 263), which he saw as a systematic procedure whereby the mathematical description is based, at every step, on critical evaluation of experimental evidence:

> Whatever is not deduced from the phenomena is to be called hypothesis, and hypotheses, whether metaphysical or physical, whether of occult qualities or mechanical, have no place in experimental philosophy. In this philosophy, particular propositions are inferred from the phenomena, and afterwards rendered general by induction.

Before Newton there had been two opposing trends in seventeenth-century science; the empirical, inductive method represented by Bacon and the rational, deductive method represented by Descartes. In the *Principia*, Newton introduced the proper mixture of both methods, emphasizing that neither experiments without systematic interpretation nor deduction from first principles without experimental evidence, will lead to a reliable theory. Going beyond Bacon in his systematic experimentation and beyond Descartes in his mathematical analysis, Newton unified the two trends and developed the methodology upon which natural science has been based ever since.

1.2 Newtonian physics

The stage of the Newtonian universe, on which all physical phenomena took place, was the three-dimensional space of classical Euclidean geometry. It was an absolute space, an empty container that was independent of the physical phenomena occurring in it. In Newton's own words (written in a special *Scholium on Absolute Space and Time*, attached to the *Principia*), "Absolute space, of its own nature, without reference to anything external, always remains homogeneous and immovable." All changes in the physical world were described in terms of a separate dimension, time, which again was absolute, having no connection with the material world and flowing smoothly from the past through the present to the future. "Absolute, true, and mathematical time," wrote Newton, "in and of itself and of its own nature, flows uniformly without reference to anything external."

The elements of the Newtonian world that moved in this absolute space and absolute time were material particles; small, solid, and indestructible objects out of which all matter was made. The Newtonian model of matter was atomistic, but it differed from the modern notion of atoms in that the Newtonian particles were all thought to be made of the same material substance. Newton assumed matter to be homogeneous. He explained the difference between one type of matter and another not in terms of atoms of different weights or densities but in terms of more or less dense packing of atoms. The basic building blocks of matter could be of different sizes but consisted of the same "stuff," and the total amount of material substance in an object was given by the object's mass.

The motion of the particles was caused by the force of gravity, which, in Newton's view, acted instantaneously over a distance. This conception was criticized by many of Newton's contemporaries, who were shocked by the idea that a force of attraction should act at a distance without being transmitted by any medium. The definitive solution of this vexing problem had to wait until the development of the field concept by Faraday and Maxwell in the nineteenth century (see Section 1.2.3) and of Einstein's theory of gravity in the twentieth (see Section 4.2.10).

For Newton, the material particles and the forces between them were indeed created by God and thus not subject to further analysis. In his second major scientific work, the *Opticks*, first published in 1704, Newton (1952/1730, Query 31) gave a clear picture of how he imagined God's creation of the material world:

It seems probable to me that God in the beginning formed matter in solid, massy, hard, impenetrable, movable particles, of such sizes and figures, and with such other proportions, and in such proportion to space, as most conducted to the end for which he formed them; and that these primitive particles being solids, are incomparably harder than any porous bodies compounded of them; even so very hard, as never to wear or break in pieces; no ordinary power being able to divide what God himself made one in the first creation."

In Newtonian mechanics, all physical phenomena are reduced to the motion of these material particles, caused by their mutual attraction – that is, by the force of gravity. The effect of this force on a particle or any other material object is described mathematically by Newton's equations of motion. These were considered fixed laws according to which material objects moved, and were thought to account for all changes observed in the physical world. In the Newtonian view, God created in the beginning the material particles, the forces between them, and the fundamental laws of motion. In this way the whole universe was set in motion, and it has continued to run ever since, like a machine, governed by immutable laws. The mechanistic view of nature is thus closely related to a rigorous determinism, with the giant cosmic machine completely causal and determinate. All that happened had a definite cause and gave rise to a definite effect, and the future of any part of the system could – in principle – be predicted with absolute certainty if its state at any time was known in all details.

Even though the Newtonian worldview was based on laws that ultimately were of divine origin, the physical phenomena themselves were not thought to be divine in any sense. In subsequent centuries, science made it more and more difficult to believe in a creator God, and thus the divine disappeared completely from the scientific worldview, leaving behind a spiritual vacuum that became characteristic of the mainstream of modern culture.

The philosophical basis of this secularization of nature was the Cartesian division between mind and matter. As a consequence of this division, the world was believed to be a mechanical system that could be described objectively, without ever mentioning the human observer. In particular, human values were separated from scientific facts, and scientists henceforth tended to believe that scientific facts are independent of our values. Such an objective description of nature became the ideal of all science, an ideal that was maintained until the twentieth century when the fallacy of the belief in a value-free science was exposed, as we have discussed.

1.2.1 Success of Newtonian mechanics

In the eighteenth and nineteenth centuries, Newtonian mechanics was applied with tremendous success to a variety of phenomena. The Newtonian theory was able to explain the motion of the planets, moons, and comets down to the smallest details, as well as the flow of the tides and various other phenomena related to gravity. Newton's mathematical system of the world established itself quickly as the correct theory of reality and generated enormous enthusiasm among scientists and the lay public alike. The picture of the world as a perfect machine, which had been introduced by Descartes, was now considered a proven fact, and Newton became its symbol. During the last twenty years of his life, Sir Isaac Newton reigned in eighteenth-century London as the most famous man of his time, the great white-haired sage of the Scientific Revolution. Accounts of this period of Newton's life sound quite familiar to us because of our memories and photographs of Albert Einstein, who played a very similar role in the twentieth century.

Encouraged by the brilliant success of Newtonian mechanics in astronomy, physicists extended it to the continuous motion of fluids and the vibrations of elastic bodies, and again it worked. Finally, even the theory of heat could be reduced to mechanics when it was realized that heat was the energy generated by a complicated "jiggling" motion of atoms and molecules. Thus many thermal phenomena, such as the evaporation of a liquid, or the temperature and pressure of a gas, could be understood quite well from a purely mechanistic point of view.

The study of the physical behavior of gases led John Dalton (1766–1844) to the formulation of his celebrated atomic hypothesis, probably the most important step in the history of chemistry. Using Dalton's hypothesis, chemists of the nineteenth century developed a precise atomic theory of chemistry which paved the way for the conceptual unification of physics and chemistry in the twentieth century.

Thus Newtonian mechanics was extended far beyond the description of macroscopic bodies. The behaviors of solids, liquids, and gases, including the phenomena of heat and sound, were explained successfully in terms of the motion of elementary material particles. For the scientists of the eighteenth and nineteenth centuries this tremendous success of the mechanistic model confirmed their belief that the universe was indeed a huge mechanical system, running according to the Newtonian laws of motion, and that Newton's mechanics was the ultimate theory of natural phenomena.

With the firm establishment of the mechanistic worldview in the eighteenth century, physics naturally became the basis of all the sciences. Indeed, if the world is really a machine, the best way to find out how it works is to turn to Newtonian mechanics. It was thus an inevitable consequence of the Cartesian worldview that the sciences of the eighteenth and nineteenth centuries modeled themselves after physics. Descartes himself had sketched the outlines of a mechanistic approach to the life sciences (see Chapter 2). The thinkers of the eighteenth century carried this program further by applying the principles of Newtonian mechanics to the sciences of human nature and human society (see Chapter 3).

1.2.2 Limitations of the Newtonian model

As a result of extending the mechanistic approach to the life sciences and the social sciences, the Newtonian world-machine became a much more complex and subtle structure. At the same time, new discoveries and new ways of thinking made the limitations of the Newtonian model apparent and prepared the way for the scientific revolutions of the twentieth century.

Electromagnetism

One of these nineteenth-century developments was the discovery and investigation of electric and magnetic phenomena that involved a new type of force and could not be described appropriately by the mechanistic model. The important step was taken by Michael Faraday (1791–1867) and completed by James Clerk Maxwell (1831–1879) – the former one of the greatest experimenters in the history of science, the latter a brilliant theorist. Faraday and Maxwell not only studied the effects of the electric and magnetic forces but also made the forces themselves the primary objects of their investigation. By replacing the concept of a force with the much subtler concept of a field, they were the first to go beyond Newtonian physics, showing that fields had their own reality and could be studied without any reference to material bodies. This theory, called electrodynamics, culminated in the realization that light is in fact a rapidly alternating electromagnetic field traveling through space in the form of waves.

In spite of these far-reaching changes, Newtonian mechanics still held its position as the basis of all physics. Maxwell himself tried to explain his results in mechanical terms, interpreting the fields as states of mechanical stress in a very light, all-pervasive medium, called ether, and electromagnetic waves as elastic waves of this ether. However, he used several mechanical interpretations of his theory at the same time and apparently took none of them really seriously, knowing intuitively that the fundamental entities in his theory were the fields and not the mechanical models. It remained for Einstein to clearly recognize this fact in the twentieth century, when he declared that no ether existed, and that electromagnetic fields were physical entities in their own right, which could travel through empty space and could not be explained mechanically.

Evolutionary thought

While electromagnetism dethroned Newtonian mechanics as the ultimate theory of natural phenomena, a new trend of thinking arose that went beyond the image of the Newtonian world-machine – a trend that was to dominate not only the nineteenth century but also all future scientific thought. It involved the idea of evolution; of gradual change, growth, and development. The notion of evolution arose in geology, where careful studies of fossils led scientists to the idea that the present state of the Earth was the result of continuous development caused by the actions of natural forces over immense periods of time. But geologists were not the only ones who thought in those terms. The theory of the solar system proposed by Kant (1724–1804) and Laplace (1749–1827) was based on developmental, or evolutionary thinking; evolutionary concepts were crucial to the political philosophies of Hegel (1770–1831) and Engels (1820–1895); poets and philosophers alike, throughout the nineteenth century, were deeply concerned with the problem of becoming.

These ideas formed the intellectual background to the most precise and most far-reaching formulation of evolutionary thought – the theory of the evolution of species in biology. Ever since antiquity, natural philosophers had entertained the idea of a “great chain of being.” This chain, however, was conceived as a static hierarchy, starting with God at the top and descending through angels, human beings, and animals to ever lower forms of life. The number of species was fixed; it had not changed since the day of their creation.

Lamarck and Darwin

The decisive change came with Jean-Baptiste Lamarck (1744–1829) at the beginning of the nineteenth century – a change that was so dramatic that Gregory Bateson (1972, p. 427), one of the deepest and broadest thinkers of the late twentieth century, compared it to the Copernican revolution:

> Lamarck, probably the greatest biologist in history, turned that ladder of explanation upside down. He was the man who said it starts with the infusoria and that there were changes leading up to man. His turning the taxonomy upside down is one of the most astonishing feats that has ever happened. It was the equivalent in biology of the Copernican revolution in astronomy.

Lamarck was the first to propose a coherent theory of evolution, according to which all living beings have evolved from earlier, simpler forms under pressure of their environment. Although the details of the Lamarckian theory had to be abandoned later on, it was nevertheless the first important step.

Several decades later, Charles Darwin (1809–1882) presented an overwhelming mass of evidence in favor of biological evolution, establishing the phenomenon for scientists beyond any doubt. He also proposed an explanation, based on the concepts of chance variation and natural selection that were to remain the cornerstones of modern evolutionary thought (as we discuss in detail in Chapter 9). Darwin's monumental *Origin of Species*, published in 1859, synthesized the ideas of previous thinkers and has shaped all subsequent biological thought. Its role in the life sciences was similar to that of Newton's *Principia* in physics and that of astronomy two centuries earlier.

The discovery of evolution in biology forced scientists to abandon the Cartesian conception of the world as a machine that had emerged fully constructed from the hands of its creator. Instead, the universe had to be pictured as an evolving and ever-changing system in which complex structures developed from simpler forms. While this new way of thinking was elaborated in the life sciences, evolutionary concepts also emerged in physics. However, whereas in biology evolution meant a movement toward increasing order and complexity, in physics it came to mean just the opposite – a movement toward increasing disorder.

Thermodynamics

The application of Newtonian mechanics to the study of thermal phenomena, which involved treating liquids and gases as complicated mechanical systems, led physicists to the formulation of a new branch of science, thermodynamics. The first great achievement of this new science was the discovery of one of the most fundamental laws of physics, the law of the conservation of energy. It states that the total energy involved in a process is always conserved. It may change its form in the most complicated way – for example, from electrical energy to the energy of motion and energy of heat – but none of it is lost. This law, which physicists discovered in their study of steam engines and other heat-producing machines, is also known as the first law of thermodynamics.

It was followed by the second law of thermodynamics, that of the dissipation of energy. While the total energy involved in a process is always constant, the amount of useful energy is diminishing, dissipating into heat, friction, and so on. The second law was formulated first by Sadi Carnot (1796–1832) in terms of the technology of thermal engines, but was soon recognized to be of much broader significance. It introduced into physics the idea of irreversible processes, of an "arrow of time," as it came to be called. According to the second law, there is a certain trend in physical phenomena from order to disorder. Mechanical energy is always dissipated into heat that cannot be completely recovered. "You can scramble an egg," as physics teachers like to put it, "but you cannot unscramble it."

According to the second law, any isolated physical system will proceed spontaneously in the direction of ever-increasing disorder. To express this direction in the evolution of physical systems in precise mathematical form, physicists introduced a new quantity called "entropy," which measures the degree of disorder, and hence the degree of evolution of a physical system. According to classical thermodynamics, the entropy, or disorder, of the universe as a whole keeps increasing. The entire world-machine is running down and will eventually grind to a halt.

This grim picture of cosmic evolution is evidently in sharp contrast to the evolutionary idea held by biologists. At the end of the nineteenth century, the Newtonian image of the universe as a perfectly running machine had been supplemented by two diametrically opposed views of evolutionary change – that of a living world unfolding toward increasing order and complexity, and that of an engine running down, a world of ever-increasing disorder. Who was right, Darwin or Carnot?

It would take another hundred years to resolve the contradiction between the two theories of evolution developed in the nineteenth century (see Chapter 8). What would become clear is that the mechanistic conception of matter as a system of small billiard balls in random motion, which lies at the basis of thermodynamics, is far too simplistic to understand the evolution of life.

1.3 Concluding remarks

In this chapter we discussed the rise of Cartesian-Newtonian science during the Scientific Revolution, which would have a profound impact on Western culture during the subsequent 300 years. As we mentioned in the Introduction, there existed alternative, holistic views of reality during that era, those of the Renaissance and the Romantic movement being perhaps the most powerful ones. But the *Zeitgeist* of the Scientific Revolution defined the modern era for three centuries.

At the end of the nineteenth century, Newtonian mechanics had lost its role as the fundamental theory of natural phenomena. Maxwell's electrodynamics and Darwin's theory of evolution involved concepts that clearly went beyond the Newtonian model and indicated that the universe was far more complex than Descartes and Newton had imagined. Nevertheless, the basic ideas underlying Newtonian physics, though insufficient to explain all natural phenomena, were still believed to be correct. The first three decades of the twentieth century changed this situation radically, as we discuss in Chapter 4. Two new theories of physics, relativity theory and quantum theory, shattered all the principal concepts of the Cartesian worldview and Newtonian mechanics. The notion of absolute space and time, the elementary solid particles, the fundamental material substance, the strictly causal nature of physical phenomena, and the objective description of nature – none of these concepts could be extended to the new domains into which physics was advancing.

2
The mechanistic view of life

Descartes' uncompromising image of living organisms as mechanical systems established a clear conceptual framework for future research in biology, but he himself did not spend much time on physiological observations, leaving it to his followers to work out the details of the mechanistic view of life.

A comment on terminology is perhaps in order here. In this book we use the terms "Cartesian," "mechanistic," and "reductionist" interchangeably. All three terms refer to the scientific paradigm formulated by René Descartes in the seventeenth century (see Section 1.1.3), in which the material universe is seen as a machine and nothing but a machine.

In Descartes' mechanistic conception of the world, all of nature works according to mechanical laws, and everything in the material world can be explained in terms of the arrangement and movements of its parts. This implies that one should be able to understand all aspects of complex structures – plants, animals, or the human body – by reducing them to their smallest constituent parts. This philosophical position is known as Cartesian reductionism.

The fallacy of the reductionist view lies in the fact that, while there is nothing wrong in saying that the *structures* of all living organisms are composed of smaller parts, and ultimately of molecules, this does not imply that their *properties* can be explained in terms of molecules alone.

As we discuss in Section 4.3, the essential properties of a living system are emergent properties – properties that are not found in any of the parts but emerge at the level of the system as a whole. These emergent properties arise from specific patterns of organization – that is, from configurations of ordered relationships among the parts. This is the central insight of the systems view of life.

2.1 Early mechanical models of living organisms

In the seventeenth century, the first to be successful in applying the Cartesian approach was Giovanni Borelli (1608–1679), a student of Galileo, who managed to explain some basic aspects of muscle action in mechanistic terms. But the great triumph of seventeenth-century physiology came when William Harvey (1578–1657) applied the mechanistic model to the phenomenon of blood circulation and solved what had been the most fundamental and difficult problem in physiology since ancient times. Harvey's treatise *De motu cordis* ("On the Movement of the Heart"), published in 1628, gave a lucid description of all that could be known of the blood system in terms of anatomy and hydraulics without the aid of a microscope. It represented the crowning achievement of mechanistic physiology and was praised as such with great enthusiasm by Descartes himself.

Inspired by Harvey's success, the physiologists of his time tried to apply the mechanistic model to describe other bodily functions, such as digestion and tissue metabolism, but these attempts were dismal failures. The phenomena they tried to explain – often with the help of grotesque mechanical analogies – involve chemical and electromagnetic processes that were unknown at the time and could not be modeled in mechanical terms.

2.1.1 Cartesian reductionism

The situation changed considerably in the eighteenth century, which saw a series of important discoveries in chemistry, including the discovery of oxygen and the formulation of the modern theory of combustion by Antoine Lavoisier (1743–1794), the "father of modern chemistry." Lavoisier also demonstrated that respiration is a special form of oxidation and thus confirmed the relevance of chemical processes to the functioning of living organisms. At the end of the eighteenth century a further dimension was added to physiology when Luigi Galvani (1737–1798) demonstrated that the transmission of nerve impulses was associated with an electric current. This discovery led Alessandro Volta (1745–1827) to the study of electricity, which became the source of two new sciences, neurophysiology and electrodynamics.

These developments raised physiology to a new level of sophistication. The simplistic mechanical models of living organisms were abandoned, but the essence of the Cartesian idea survived. Animals were still machines, although they were much more complicated than mechanical clockworks, as they involved chemical and electrical phenomena. Thus biology ceased to be Cartesian in the sense of Descartes' strictly mechanical image of living organisms, but it remained Cartesian in the wider sense of attempting to reduce all aspects of living organisms to the physical and chemical interactions of their smallest constituents.

2.2 From cells to molecules

In the nineteenth century, the mechanistic view of life progressed further, due to remarkable advances in many areas of biology, including the formulation of cell theory, the beginning of modern embryology, the rise of microbiology, and the discovery of the laws of heredity. Biology was now firmly grounded in physics and chemistry, and scientists devoted all their efforts to the search for physical and chemical explanations of life.

2.2.1 Cell theory

One of the most powerful generalizations in all of biology was the recognition that all animals and plants are composed of cells. It marked a decisive turn in biologists' understanding of body structure, inheritance, development, evolution, and many other characteristics of life. The term "cell" was coined by the physicist and naturalist Robert Hooke in the seventeenth century to describe various minute structures he saw through the newly invented microscope, but the development of a proper cell theory was a slow and gradual process that involved the work of many researchers and culminated in the nineteenth century with the formulation of modern cell theory by Robert Virchow (1821–1902). This achievement gave a new meaning to the Cartesian paradigm. Biologists thought that they had definitely found the fundamental units of life. From now on, all functions of living organisms had to be understood in terms of the interactions between cellular building blocks, rather than as reflecting the organization of the organism as a whole.

Understanding the structure and functioning of cells involves a problem that has become characteristic of all modern biology. The organization of a cell has often been compared to that of a factory, where different parts are manufactured at different sites, stored in intermediate facilities, and transported to assembly plants to be combined into finished products, which are either used by the cell itself or exported to other cells. Cell biology has made enormous progress in understanding the structures and functions of many of the cell's subunits, but it still has revealed very little about the coordinating activities that integrate those operations into the functioning of the cell as a whole. Biologists have come to realize that cells are living systems in their own right, and that the integrating activities of these living systems – especially the balancing of their interdependent metabolic pathways and cycles – cannot be understood within a reductionist framework.

2.2.2 Microbiology

The invention of the microscope in the seventeenth century had opened up a new dimension for biology, but the instrument was not fully exploited until the nineteenth century, when various technical problems with the old lens system were finally solved. The newly perfected microscope generated an entire new field of research, microbiology, which revealed an unsuspected richness and complexity of living organisms of microscopic dimensions. Research in this field was pioneered by the genius of Louis Pasteur (1822–1895), whose penetrating insights and clear formulations made a lasting impact on chemistry, biology, and medicine.

With the use of ingenious experimental techniques, Pasteur was able to clarify a question that had agitated biologists throughout the eighteenth century, the question of the origin of life. Since ancient times it had been a common belief that life, at least in its lower forms, could arise spontaneously from nonliving matter. During the seventeenth and eighteenth centuries that idea – known as "spontaneous generation" – was questioned, but the arguments could not be settled until Pasteur demonstrated conclusively that any microorganisms that developed under suitable conditions came from other microorganisms. It was Pasteur who brought to light the immense variety of the organic world at the level of the very small. In particular, he was able to establish the role of bacteria in chemical processes like fermentation, thus helping to lay the foundations of the new science of biochemistry.

After twenty years of research on bacteria, Pasteur turned to the study of diseases in higher animals and achieved another major advance – the demonstration of a definite correlation between "germs" (bacteria) and disease. Pasteur's discovery led to a simplistic "germ theory of disease," in which bacteria were seen as the only cause of disease. This reductionist view eclipsed an alternative theory that had been taught a few decades earlier by Claude Bernard (1813–1878), a celebrated physician who is generally considered the founder of modern physiology. Bernard insisted on the close and intimate relation between an organism and its environment, and was the first to point out that there was also a *milieu intérieur*, an internal environment in which the organs and tissues of the organism live. Bernard observed that in a healthy organism this internal environment remains essentially constant, even when the external environment fluctuates considerably. His concept of the constancy of the internal environment foreshadowed the important notion of homeostasis, developed by the neurologist Walter Cannon in the 1920s.

2.2.3 Darwin and Mendel

While the advances in cell theory and microbiology supported the mechanistic view of life, biology's main contribution to the history of ideas in the nineteenth century was Darwin's theory of evolution, which forced scientists to abandon the Newtonian image of the world as a machine that had remained unchanged since the time of its creation. As we discuss in detail in Chapter 9, Darwin's discovery that all forms of life have descended from a common ancestor by a long process of modifications over billions of years introduced a radical shift in biological thought – a change of perspective from being to becoming. Moreover, by realizing that all living organisms are related by common ancestry, the Darwinian conception of life was utterly holistic and systemic: a vast planetary network of living beings interlinked in space and time.

Although Darwin's twin concepts of chance variation – now known as random mutation – and natural selection would remain essential elements of modern evolutionary theory (as we discuss in Chapter 9), it soon became clear that chance variations, as envisaged by Darwin, could never explain the emergence of new characteristics in the evolution of species. Darwin shared with his contemporaries the assumption that the biological characteristics of an individual represented a "blend" of those of its parents, with both parents contributing more or less equal parts to the mixture. This meant that an offspring of a parent with a useful chance variation would inherit only 50% of the new characteristic, and would be able to pass on only 25% of it to the next generation. Thus the new characteristic would be diluted rapidly, with very little chance of establishing itself through natural selection. Darwin himself recognized that this was a serious flaw in his theory for which he had no remedy.

Ironically, the solution to Darwin's problem was discovered by the Austrian monk and scientist Gregor Mendel (1822–1884) only a few years after the publication of the Darwinian theory, but it was ignored until the rediscovery of Mendel's work at the turn of the twentieth century. From his careful experiments with garden peas (see Section 9.2), Mendel deduced that there were "units of heredity" – later to be called genes – that did not blend in the process of reproduction, but were transmitted from generation to generation without changing their identity. With this discovery it could be assumed that random mutations would not disappear within a few generations but would be preserved, to be either reinforced or eliminated by natural selection.

Mendel's discovery not only played a decisive role in establishing the Darwinian theory of evolution but also opened up a whole new field of research – the study of heredity through the investigation of the chemical and physical properties of genes. The biologist William Bateson (1861–1926), a fervent advocate and popularizer of Mendel's work, named this new field "genetics" at the beginning of the twentieth century and introduced many of the terms now used by geneticists. He also named his youngest son Gregory, in Mendel's honor.

2.3 The century of the gene

In the twentieth century, genetics became the most active area in biological research and provided a strong reinforcement of the Cartesian approach to living organisms. It became clear quite early on that the material of heredity lay in the chromosomes, those threadlike bodies that are present in the nucleus of every cell. Soon thereafter it was recognized that the genes occupy specific positions within the chromosomes; to be more precise, they are arranged along the chromosomes in linear order. With these discoveries geneticists believed that they had now pinned down the "atoms of heredity" and proceeded to explain the biological characteristics of living organisms in terms of their elementary units, the genes.

This new perspective brought about a significant shift in biological research – a shift that may well turn out to be the last step in the reductionist approach to the phenomenon of life, leading to its greatest triumph and, at the same time, to its demise. Whereas cells were regarded as the basic building blocks of living organisms during the nineteenth century, the attention shifted from cells to molecules toward the middle of the twentieth when geneticists began to explore the molecular structure of the gene. Their research culminated in the elucidation of the physical structure of DNA – the genetic component of chromosomes – which stands as one of the greatest achievements of twentieth-century science. This triumph of molecular biology led biologists to believe that all biological functions can be explained in terms of molecular structures and mechanisms.

2.3.1 Genes and enzymes

During the first half of the twentieth century it became clear that the essential constituents of all living cells – the proteins and nucleic acids (DNA and RNA) – were highly complex, chainlike structures containing thousands of atoms. The investigation of the chemical properties and exact three-dimensional structure of these large-chain molecules became the principal task of molecular biology.

The first important step toward a molecular genetics came with the discovery that cells contain agents, called enzymes, that can catalyze (i.e., mediate) specific chemical reactions. During the first half of the twentieth century biochemists managed to specify most of the chemical reactions that occur in cells, and found out that the most important of these reactions are essentially the same in all living organisms. Each of them depends crucially on the presence of a specific enzyme, and thus the study of enzymes became of primary importance.

During the 1940s geneticists had another decisive insight when they discovered that the primary function of genes was to control the production, or "synthesis," of enzymes. With this discovery the broad outlines of the hereditary process emerged: genes determine hereditary traits by directing the synthesis of enzymes, which in turn mediate the chemical reactions corresponding to those traits.

Although these discoveries represented major advances in understanding heredity, the nature of the gene remained unknown during this period. Geneticists ignored its chemical structure and were unable to explain how it managed to carry out its essential functions: the synthesis of enzymes, its own faithful replication in the process of cell division, and the sudden permanent changes known as mutations. As far as the enzymes were concerned, it was known that they were proteins, but their precise chemical structure was unknown and so, as a consequence, was the process by which enzymes catalyze chemical reactions.

2.3.2 Schrödinger's **What Is Life?**

This situation changed radically over the next two decades, which brought the major breakthrough in modern genetics, often referred to as the breaking of the genetic code: the discovery of the precise chemical structure of genes and enzymes, of the molecular mechanisms of protein synthesis, and of the mechanisms of gene replication and mutation. A crucial element in the breaking of the genetic code was the fact that physicists moved into biology. Max Delbrück, Francis Crick, Maurice Wilkins, and several of the other protagonists had backgrounds in physics before they joined the biochemists and geneticists in their study of heredity. These scientists brought with them a new perspective and new methods that thoroughly transformed genetic research.

The main reason for these scientists to leave physics and turn to genetics was a short book entitled *What Is Life?*, published in 1944 by the famous quantum physicist Erwin Schrödinger (1887–1961). The fascination of Schrödinger's book came from the clear and compelling way in which he treated the gene not as an abstract unit but as a concrete physical substance, advancing definite hypotheses about its molecular structure that stimulated scientists to think about genetics in a new way. Schrödinger was the first to suggest that the gene could be viewed as an information carrier whose physical structure corresponds to a succession of elements in a hereditary code script. His enthusiasm convinced physicists, biochemists, and geneticists that a new frontier of science had opened where great discoveries were imminent. From now on these scientists began to refer to themselves as "molecular biologists."

2.3.3 The structure of DNA

The basic structure of the biological molecules was discovered in the early 1950s through the confluence of three powerful methods of observation: chemical analysis, electron microscopy, and X-ray crystallography. The first breakthrough came when the chemist Linus Pauling determined the structure of the protein molecule. Proteins were known to be long-chain molecules, consisting of a sequence of different compounds, known as amino acids, linked end-to-end. Pauling showed that proteins need to have a stable, folded, three-dimensional structure, coiled in a right-handed helix (known as the alpha-helix), and that the rest of the structure is determined by the exact linear sequence of amino acids along this helical path. Further studies of the protein molecule showed how the specific structure of enzymes allows them to bind the molecules whose chemical reactions they promote.

Pauling's great achievement inspired the geneticists James Watson and Francis Crick to concentrate all their efforts on elucidating the structure of DNA, which by then had been recognized as the genetic material in the chromosomes. After two years of strenuous work (1951–3), vividly described in Watson's (1968) personal account, *The Double Helix*, they were rewarded with tremendous success. Using the X-ray data of the biophysicists Rosalind Franklin and Maurice Wilkins, Watson and Crick were able to determine the precise architecture of DNA, now called the Watson–Crick structure. This is the now well-known double helix made up of intertwined, structurally complementary strands (see Section 9.3).

2.3.4 Breaking the genetic code

It took another decade to understand the basic mechanism through which the DNA carries out its two fundamental functions of self-replication and protein synthesis. This research, again led by Watson and Crick, revealed explicitly how genetic information is encoded in the chromosomes. These twin achievements – the discovery of the DNA structure and the unraveling of the genetic code – have been hailed as the greatest scientific discovery of the twentieth century. Advancing to ever smaller levels in their exploration of biological life, molecular biologists found that the characteristics of all living organisms – from bacteria to humans – were encoded in their chromosomes in the same chemical substance, using the same code script. After two decades of intensive research, the precise details of this code had been unraveled. Biologists had discovered the alphabet of a truly universal language of life.

2.3.5 *Genetic determinism*

With this momentous discovery, the link between genes and biological traits seemed compellingly simple and elegant: Genes specify the enzymes that catalyze all cellular processes. Thus genes determine biological traits and behavior, each gene corresponding to a specific enzyme. This explanation has been called the "central dogma of molecular biology" by Francis Crick. It describes a linear causal chain from DNA to RNA, to proteins (enzymes), and to biological traits.

In reality, however, the linear chain described by the central dogma is far too simplistic to describe the actual processes involved in the synthesis of proteins. And the discrepancy between the theoretical framework and biological reality is even greater when the linear sequence is shortened to its two endpoints, DNA and traits, so that the central dogma is turned into the statement, "Genes determine behavior." This view is known as genetic determinism.

During the decades following the discovery of the genetic code, genetic determinism became the dominant paradigm in molecular biology, generating a host of powerful metaphors. DNA was often referred to as the organism's genetic "program" or "blueprint," and the genetic code as the universal "language of life." The exclusive focus on genes implied by these metaphors largely eclipsed the organism from the biologists' view. Living organisms tended to be viewed simply as collections of genes, subject to random mutations and selective forces in the environment over which they have no control.

Further research in genetics, however, forced molecular biologists to realize that the elegant principle of "one gene – one protein" had to be abandoned, and in spite of many brilliant achievements the awareness that we may need to go beyond genes to really understand genetic phenomena is now increasing. Indeed, some geneticists are even speculating that we may be forced to abandon the concept of the gene altogether, as we discuss in Section 9.6.

2.4 Mechanistic medicine

Throughout the history of Western science the development of biology has gone hand in hand with that of medicine. Naturally then, the mechanistic view of life, once firmly established in biology, also dominated the attitudes of physicians toward health and illness. The influence of the Cartesian paradigm on medical thought resulted in the so-called biomedical model, which constitutes the conceptual foundation of modern scientific medicine. For Descartes, a healthy person was like a well-made clock in perfect mechanical condition, and a sick person like a clock whose parts were not functioning properly. The principal characteristics of the biomedical model, as well as many aspects of current medical practice, can be traced back to this Cartesian imagery (see Capra, 1982).

Following the Cartesian approach, medical science has largely limited itself to the attempt to understand the biological mechanisms involved in injuries to various parts of the body. These mechanisms are studied from the point of view of cellular and molecular biology, leaving out all influences of nonbiological circumstances on biological processes. Out of the large network of phenomena that influence health, the biomedical approach studies only a few physiological aspects. Knowledge of these aspects is, of course, very useful, but they represent only a small part of the story. Medical practice based on such a limited approach is not very effective in promoting and maintaining good health. This will not change until medical science relates its study of the biological aspects of illness to the general physical and psychological condition of the human organism and its environment.

The conceptual problem at the center of contemporary healthcare is the confusion between disease processes and disease origins. Instead of asking why an illness occurs and trying to remove the conditions that led to it, medical researchers try to understand the mechanisms through which the disease operates, so that they can then interfere with them. These mechanisms, rather than the true origins, are seen as the causes of disease in current medical thinking.

In the process of reducing illness to disease, the attention of physicians has moved away from the patient as a whole person. By concentrating on smaller and smaller fragments of the body – shifting its perspective from the study of bodily organs and their functions to that of cells and, finally, to the study of molecules – modern medicine often loses sight of the human being, and having reduced health to mechanical functioning, it is no longer able to deal with the phenomenon of healing. Over the past four decades, the dissatisfaction with the mechanistic approach to health and healthcare has grown rapidly both among healthcare professionals and the general public. At the same time, the emerging systems view of life has given rise to a corresponding systems view of health, as we discuss in Chapter 15, while health consciousness among the general population has increased dramatically in many countries. The growing awareness of the power and the responsibility of individuals to maintain themselves in good health has expressed itself in increased attention to healthy nutrition, exercise, yoga and other "mind–body" practices, as well as in the rising popularity of a wide range of alternative therapies.

In the late 1970s and early 1980s, the leading catchphrases of this broad popular movement were "holistic healthcare," "holistic medicine," and "wellness," and in the subsequent decades the phrase "integrative medicine" established itself as the unifying term. We shall argue in Chapter 15 that, in our view, integrative medicine represents the conscientious application of the systems view of life to health and healing.

2.5 Concluding remarks

The spectacular success of molecular biology in the field of genetics – the discovery of the structure of DNA and the "breaking of the genetic code" – has been hailed as the greatest achievement in biology since Darwin's theory of evolution. Indeed, at the close of the twentieth century the biologist and science historian Evelyn Fox Keller wrote a review of genetics titled *The Century of the Gene*. However, Keller gave this phrase a double meaning. The main point of her brilliant evaluation of genetics is the observation that the most recent advances in this field are now forcing molecular biologists to question many of the fundamental concepts on which their whole enterprise was originally based. Thus Keller comes to the conclusion:

> Even though the message has yet to reach the popular press, to an increasingly large number of workers at the forefront of contemporary research, it seems evident that the primacy of the gene as the core explanatory concept of biological structure and function is more a feature of the twentieth century than it will be of the twenty-first.
>
> (*Keller, 2000, p. 9*)

3
Mechanistic social thought

3.1 Birth of the social sciences

While Descartes himself has sketched the outlines of a mechanistic approach to physics, biology, and medicine, the thinkers of the eighteenth century carried this program further by applying the principles of Newtonian mechanics to the study of human nature and human society. In doing so, they created a new branch of science, which they called "social science" (later changed to the plural "social sciences" to denote a variety of disciplines outside the natural sciences). This new science generated great enthusiasm, and some of its proponents even claimed to have discovered a "social physics."

The Newtonian theory of the universe and the belief in the rational approach to human problems spread so rapidly among the middle classes of the eighteenth century that the whole era became known as the "Age of Enlightenment," or the "Age of Reason." The dominant figure in this development was the philosopher John Locke (Figure 3.1), whose most important writings were published in the late seventeenth century. Strongly influenced by Descartes and Newton, Locke's work had a decisive impact on eighteenth-century thought.

3.1.1 The Enlightenment

Following Newtonian physics, Locke developed an atomistic view of society, describing it in terms of its basic building blocks, the individual human beings. As physicists reduced the properties of gases to the motion of their atoms, or molecules, so Locke attempted to reduce the phenomena observed in society to the behavior of its individuals. Thus he proceeded to study first the nature of the individual human being, and then tried to apply the principles of human nature to economic and political problems.

Locke's analysis of human nature was based on that of an earlier philosopher, Thomas Hobbes (1588–1679), who had declared that all knowledge was based on sensory perception. Locke adopted this theory of knowledge and, in a famous metaphor, compared the human mind at birth to a *tabula rasa*, a completely blank tablet on which knowledge is imprinted once it is acquired through sensory experience. This image was to have a strong influence on psychology as well as on political philosophy. According to Locke, all human

Figure 3.1 John Locke (1632–1727). iStockphoto.com/© pictore.

beings – "all men," as he would say – were equal at birth and depended in their development entirely on their environment. Their actions, Locke believed, were always motivated by what they assumed to be in their own interest.

When Locke applied his theory of human nature to social phenomena, he was guided by the belief that there were laws of nature governing human society similar to those governing the physical universe. As the atoms in a gas would establish a balanced state, so human individuals would settle down in society in a "state of nature." Thus the function of government was not to impose its own laws on the people, but rather to discover and enforce the natural laws that existed before any government was formed. According to Locke, these natural laws included the freedom and equality of all individuals as well as the right to property, which represented the fruits of one's labor.

Locke's ideas became the basis for the value system of the Enlightenment and had a strong influence on the development of modern economic and political thought. The ideals of individualism, property rights, free markets, and representative government, all of which can be traced back to Locke, contributed significantly to the thinking of Thomas Jefferson and are reflected in the Declaration of Independence and the American Constitution.

3.1.2 The positivist straitjacket

Social thought in the late nineteenth and early twentieth centuries was greatly influenced by positivism, a doctrine formulated by the social philosopher Auguste Comte (1798–1857). Its assertions include the insistence that the social sciences should search for general laws of human behavior, an emphasis on quantification, and the rejection of explanations in terms of subjective phenomena, such as intentions or purposes.

It is evident that the positivist framework is patterned after Newtonian physics. Indeed, it was Auguste Comte who called the scientific study of society at first "social physics" before introducing the term "sociology." The major schools of thought in early twentieth-century sociology can be seen as attempts at emancipation from the positivist straitjacket. In fact, as Baert (1998) shows in a concise review of twentieth-century social theory, most sociologists of that time positioned themselves explicitly in opposition to the positivist epistemology.

3.1.3 "Hard" and "soft" sciences

The triumph of Newtonian mechanics in the eighteenth and nineteenth centuries established physics as the prototype of a "hard" science against which all other sciences were measured. The closer scientists could come to emulating the methods of physics, and the more of its concepts they were able to use, the higher the standing of their discipline in the scientific community.

In the twentieth century, this tendency to model scientific concepts and theories after those of Newtonian physics became a severe handicap in many fields, but first and foremost in the social sciences. These were traditionally regarded as the "softest" among the sciences, and social scientists tried very hard to gain respectability by adopting the Cartesian paradigm and the methods of Newtonian physics. However, the Cartesian framework is often quite inappropriate for the phenomena they are describing, and consequently their models have become increasingly unrealistic. This is now especially apparent in economics. Our brief review of the history of economics in the following pages is based on an essay written by the economist and futurist Hazel Henderson (see Capra, 1982; see also Henderson, 1978, 1981).

3.1.4 The emergence of economics

Economic theory emerged with the Scientific Revolution and the Enlightenment, and found its classical formulation during the Industrial Revolution. Before the sixteenth century there was no isolation of purely economic phenomena from the fabric of life. Throughout most of human history food, clothing, shelter, and other basic resources were produced for use value and were distributed within tribes and groups on a reciprocal basis (see Polanyi, 1968). A national system of markets is a relatively recent phenomenon that arose in seventeenth-century England and spread from there over the entire world, resulting in today's interlinked "global market." Markets, of course, had existed since the Stone Age, but they were based on barter, not cash, and thus were bound to be local. Even early trading had little economic motivation but was more often a sacred and ceremonial activity related to kinship and family customs.

With the Scientific Revolution and the Enlightenment, critical reasoning, empiricism, and individualism became the dominant values, together with a secular and materialistic orientation that led to the production of worldly goods and luxuries, and to the manipulative mentality of the Industrial Age. The new customs and activities resulted in the creation of new social and political institutions and gave rise to a new academic pursuit: the theorizing about a set of specific *economic* activities – production, exchange, distribution, moneylending – which suddenly stood out in sharp relief and required not only description and explanation but also rationalization.

The first theorists of economic phenomena did not call themselves economists. They were politicians, administrators, and merchants, and they used the ancient notion of economy in the sense of managing a household, derived from the Greek *oikonomia* ("householding"). They applied this notion to the state as the household of the ruler, and thus their policies became known as "political economy." This term remained in use until the twentieth century, when it was replaced by the modern term "economics."

The rise of capitalism

One of the most important consequences of the shift of values at the end of the Middle Ages was the rise of capitalism in the sixteenth and seventeenth centuries. According to an ingenious thesis by Max Weber (1976/1905), the development of the capitalist mentality was closely related to the religious idea of a "calling," which emerged with Martin Luther and the Reformation, together with the notion of moral obligation to fulfill one's duty in worldly endeavors.

This idea of a worldly calling projected religious behavior into the secular world. It was emphasized even more forcefully by the Puritan sects, which saw worldly activity and the material rewards resulting from industrious behavior as a sign of divine predestination. Thus arose the famous Protestant "work ethic," in which hard, self-denying work and worldly success were equated with virtue. On the other hand, the Puritans abhorred all but the most frugal consumption, and consequently the accumulation of wealth was sanctioned, as long as it was combined with an industrious career. In Weber's theory these religious values and motives provided the essential emotional drive and energy for the rise and rapid development of capitalism.

Modern economics

Modern economics, strictly speaking, is a little over 350 years old. It was founded in the seventeenth century by Sir William Petty (1623–1687), professor of anatomy at Oxford University and of music at Gresham College in London, as well as physician to the army of Oliver Cromwell. Petty's *Political Arithmetick* seemed to owe much to Newton, Descartes, and Galileo, its method consisting of replacing words and arguments by numbers, weights and measures, and of using rational arguments to explain economic phenomena in terms of visible natural causes.

Along with Petty, John Locke laid the foundation of modern economics. One of Locke's most innovative theories had to do with prices. Whereas Petty held that prices and commodities should reflect justly the amount of labor they embodied, Locke came up with the idea that prices were also determined objectively, by demand and supply. This not only liberated the merchants of the day from the moral law of "just" prices; it also became another

Figure 3.2 Adam Smith (1723–1790). iStockphoto.com/© HultonArchive.

cornerstone of economics and was elevated to equal status with the laws of mechanics, where it stands even today in many economic analyses.

The law of supply and demand fit perfectly with the new mathematics of Newton and Leibniz – the differential calculus – since economics was perceived as dealing with continuous variations of very small quantities, which could be described most efficiently by this mathematical technique. This notion became the basis of subsequent efforts to turn economics into an exact mathematical science. However, the problem was – and is – that the variables used in these mathematical models cannot be rigorously quantified but are defined on the basis of assumptions that often make the models quite unrealistic.

A distinct school of eighteenth-century thought that had a significant influence on classical economic theory, and notably on Adam Smith, was that of the French Physiocrats. Physiocracy meant "the rule of nature," and the Physiocrats advocated the idea that natural law, if left unimpeded, would govern economic affairs for the greatest benefit of all. Thus the doctrine of "laissez faire" was introduced as another keystone of economics.

3.2 Classical political economy

3.2.1 Adam Smith: The Wealth of Nations

The period of "classical political economy" was inaugurated in 1776, when Adam Smith (Figure 3.2) published *An Inquiry into the Nature and Causes of the Wealth of Nations*. Smith, a Scottish moral philosopher, remains by far the most influential of all economists. His *Wealth of Nations* was the first full-scale treatise on economics. Its importance as the foundation of modern economic theory has been compared to that of Newton's *Principia* for physics and of Darwin's *Origin of Species* for biology.

Smith lived at a time when the Industrial Revolution had begun to change the face of Britain. When he wrote *The Wealth of Nations* the transition from an agrarian, handicraft economy to one dominated by steam power and by machines operated in large factories and mills was well under way. The spinning jenny had been invented and machine looms were used in cotton factories employing up to 300 workers. The new private enterprise, factories, and power-driven machinery shaped Smith's ideas to such an extent that he enthusiastically advocated the social transformation of his time and criticized the remnants of the land-based feudal system.

From the prevailing Newtonian idea of natural law Smith deduced that it was "human nature to barter and exchange," and he also thought it "natural" that workers would gradually facilitate their work and improve their productivity with the help of labor-saving machinery. At the same time the early manufacturers had a much darker view of the role of machines; they well understood that machines could replace workers and thus could be used to keep them afraid and docile.

3.2.2 The invisible hand

From the Physiocrats Smith adopted the theme of laissez faire, which he immortalized in the metaphor of the "invisible hand". According to Smith, the invisible hand of the market would guide the individual self-interest of all entrepreneurs, producers, and consumers for the harmonious betterment of all, "betterment" being equated with the production of material wealth. In this way a social result would be achieved that was independent of individual intentions, and thus an objective science of economic activity was made possible.

Smith believed in the labor theory of value, according to which the value of a product is derived only from the human labor required to produce it, but he also accepted the idea that prices would be determined in "free" markets by the balancing effects of supply and demand. He based his economic theory on the Newtonian notions of equilibrium, laws of motion, and scientific objectivity. One of the difficulties in applying these mechanistic concepts to social phenomena was the lack of appreciation for the problem of friction. Because the phenomenon of friction is generally neglected in Newtonian mechanics, Smith imagined that the balancing mechanisms of the market would be almost instantaneous. He described their adjustments as "prompt," "occurring soon," and "continual," while prices were "gravitating" in the proper direction. Small producers and small consumers would meet in the marketplace with equal power and information.

This idealistic picture underlies the "competitive model" widely used by economists today. Its basic assumptions include perfect and free information for all participants in a market transaction; the belief that each buyer and seller in a market is small and has no influence on price; and the complete and instant mobility of displaced workers, natural resources, and machinery. All these conditions are violated in the vast majority of today's markets, yet most economists continue to use them as the basis of their theories.

Smith thought that the self-balancing market system was one of slow and steady growth, with continually increasing demands for goods and labor. This idea of continual growth was adopted by succeeding generations of economists, who, paradoxically continued to use mechanistic equilibrium assumptions while at the same time postulating continuing economic growth. Smith himself predicted that economic progress would eventually come to an end when the wealth of nations had been pushed to the natural limits of soil and climate, but he thought this point was so far in the future that it was irrelevant to his theories. Today our global economy is fast approaching these natural limits, as we discuss in Chapter 17.

Smith alluded to social and economic structures like monopolies when he denounced people in the same trade who conspired to raise prices artificially, but he did not see the broad implications of such practices. The growth of these structures, and in particular of the class structure, was to become a central theme in Marx's economic analysis. Adam Smith justified capitalists' profits by arguing that they were needed to invest in more machines and factories for the common good. He noted the struggle between workers and employers and the efforts of both to "interfere with the market," but he never referred to the unequal power of workers and capitalists – a point that Marx would drive home with force.

3.2.3 Economic models

At the beginning of the nineteenth century, economists began to systematize their discipline in an attempt to cast it into the form of a science. The first and most influential among these systematic economic thinkers was David Ricardo (1772–1823), who introduced the concept of an "economic model," a logical system of postulates and laws, involving a limited number of variables that could be used to describe and predict economic phenomena.

The systematic efforts of Ricardo and other classical economists consolidated economics into a set of dogmas that supported the existing class structure and countered all attempts at social improvement with the "scientific" argument that "laws of nature" were operating and the poor were responsible for their own misfortune. At the same time, workers' uprisings were becoming frequent and the new body of economic thought engendered its own horrified critics long before Marx.

3.3 The critics of classical economics

3.3.1 John Stuart Mill

The greatest among the classical economic reformers was John Stuart Mill (1806–1873), a child prodigy who had absorbed most of the works of the philosophers and economists of his time by the time he was thirteen. At the age of forty-two he published his own *Principles of Political Economy*, a herculean reassessment that came to a radical conclusion. Economics,

Figure 3.3 Karl Marx (1818–1883). iStockphoto.com/© Rubén Hidalgo.

Mill wrote, had only one province: production and the scarcity of means. Distribution was not an economic but a political process. This narrowed the scope of political economy to "pure economics," later to be called "neoclassical," and allowed a more detailed focus on the "economic core process," while excluding social and environmental variables in analogy to the controlled experiments of the physical sciences.

After Mill, economics became split between the neoclassical, "scientific," and mathematical approach, on the one hand, and the "art" of broader social philosophy on the other. Eventually this split led to today's disastrous confusion between the two approaches, resulting in policy tools that are often derived from abstract, unrealistic mathematical models.

3.3.2 Karl Marx

The thought of Karl Marx (Figure 3.3), the most thorough and most eloquent critic of classical economics, has engendered worldwide intellectual fascination far beyond the field of economics. According to economic historian Robert Heilbroner (1978), this fascination is rooted in the fact that Marx was "the first to discover a whole mode of inquiring that would forever after belong to him." Marx's mode of inquiry was that of social critique. He referred to himself not as a philosopher, historian, or economist – though he was all of those – but as a social critic; and this is why his social philosophy and science continue to exert a strong influence on social thought.

As a philosopher, Marx taught a philosophy of action. "The philosophers," he wrote, "have only *interpreted* the world in various ways; the point, however, is to *change* it" (quoted in Tucker, 1972, p. 109). As an economist, Marx criticized classical economics more expertly and efficiently than any of its practitioners. His main influence, however, has been not intellectual but political. As Heilbroner (1980, p. 134) observed, if judged by the number of worshiping followers, "Marx must be considered a religious leader to rank with Christ or Mohammed."

While Marx the revolutionary was canonized by millions around the world, economists had to deal with – but more often ignored or misquoted – his embarrassingly accurate predictions, among them the occurrence of "boom" and "bust" business cycles and the tendency of market-oriented economies to develop "reserve armies" of unemployed.

Marx's main body of work, set forth in his three-volume *Das Kapital* ("Capital"), represents a thorough critique of capitalism. He viewed society and economics from the explicitly stated perspective of the struggle between workers and capitalists, but his broad ideas about social evolution allowed him to see economic processes in much larger contexts.

Marx recognized that capitalist forms of social organization would speed the process of technological innovation and increase material productivity, and he predicted that this, "dialectically" (i.e., by changing into its opposite), would change social relationships. Thus he was able to foresee phenomena like monopolies and depressions, and to predict that capitalism would foster socialism – as it did – and that it would, eventually, disappear – as it may.

In his "Critique of Political Economy," as he subtitled *Das Kapital*, Marx used the labor theory of value to raise issues of justice, and developed powerful new concepts to counter the reductionist logic of the neoclassical economists of his time. He knew that to a large extent wages and prices are politically determined. Starting from the premise that human labor creates all values, Marx observed that continuing labor must, at the very least, produce subsistence for the worker plus enough to replace the materials used up. But, in general, there will be a surplus over and above that minimum. The form this "surplus value" takes will be a key to the structure of society, its economy, and its technology.

In capitalist societies, Marx pointed out, the surplus value is appropriated by capitalists, who own the means of production and determine the conditions of labor. This transaction between people of unequal power allows the capitalists to make more money from the labor of the workers, and thus money is turned into capital. In this analysis Marx emphasized that the precondition for capital to arise was a specific social class relationship, itself the product of a long history.

Marx had a rich intellectual life with many insights that have decisively shaped our age. His social critique inspired millions of revolutionaries around the world, and the Marxian economic analysis, although now somewhat outdated (as we discuss in Chapter 17), is respected academically not only in former and current socialist countries but also in most other countries around the world. Even in the USA, Marxian thought is taught, with different emphases, in all leading universities, and some of the most prominent American social scientists – e.g., Michael Burawoy at Berkeley, David Harvey at City College of New

York, or Erik Olin Wright at Wisconsin – are known explicitly as Marxian scholars. In fact, it is interesting that from the 1970s on, Marxism has been growing among American academics while it has almost disappeared in France and other European countries, let alone Russia.

As the work of the scholars mentioned shows, Marxian thought is capable of a wide range of interpretations and thus continues to fascinate. Of particular interest for our review is the relation of the Marxian critique to the reductionist framework of the science of his time. Like most nineteenth-century thinkers, Marx was very concerned about being "scientific," using the term constantly in the description of his critical approach. Accordingly he often attempted to formulate his theories in Cartesian and Newtonian language. Still, his broad view of social phenomena allowed him to transcend the Cartesian framework in significant ways.

He did not adopt the classical stance of the objective observer, but fervently emphasized his role as participator by asserting that his social analysis was inseparable from social critique. In his critique he went beyond social issues and often revealed deeply humanistic insights. Finally, although Marx often argued for technological determinism (i.e., the belief that technological development determines social change), a fact which made his theory more acceptable as a science, he also had profound insights into the interrelatedness of all phenomena, seeing society as an organic whole in which ideology and technology are equally important.

3.4 Keynesian economics

3.4.1 Neoclassical models and the Great Depression

By the middle of the nineteenth century, classical political economy had branched into two broad streams. On the one side were the reformers: the Marxists and the minority of classical economists who followed John Stuart Mill. On the other side were the neoclassical economists, who concentrated on the economic core process and developed the school of mathematical economics. Some of them tried to establish objective formulas for the maximization of welfare, while others retreated into ever more abstruse mathematics to escape the devastating Marxist critique.

Much of mathematical economics was – and is – devoted to studying the "market mechanism" with the help of curves for demand and supply, always expressed as functions of prices and based on various assumptions about economic behavior, many of them highly unrealistic in today's world. For example, perfect competition in free markets, as postulated by Adam Smith, is assumed in many models.

As mathematical economists refined their models during the late nineteenth and early twentieth centuries, the world economy headed for the worst depression in its history, which shook the foundations of capitalism and seemed to verify all the Marxian predictions. However, after the Great Depression capitalism's fortunes were saved by a new set of social and economic interventions of governments. These policies were based on the theory of

John Maynard Keynes (1883–1946), who had a decisive influence on modern economic thought.

3.4.2 John Maynard Keynes: economics as policy

Keynes was keenly interested in the entire social and political scene and viewed economic theory as an instrument of policy. He bent the so-called value-free methods of neoclassical economics to serve instrumental purposes and goals, and in doing so made economics once again political, but this time in a new way. This, of course, involved giving up the ideal of the objective scientific observer, which neoclassical economists were very reluctant to do. But Keynes calmed their fears of interfering with the balancing operations of the market system by showing them that he could *derive* his policy interventions from the neoclassical model. To do this he demonstrated that economic equilibrium states were "special cases," exceptions rather than the rule in the real world.

To determine the nature of government interventions, Keynes shifted his focus from the microlevel to the macrolevel – to economic variables like the national income, total consumption and total investment, the total volume of employment, and so on. By establishing simplified relations between those variables, he was able to show that they were susceptible to short-term changes that could be influenced by appropriate policies. According to Keynes, these fluctuating business cycles were an intrinsic property of national economies.

In the twentieth century the Keynesian model became thoroughly assimilated into the mainstream of economic thought. Most economists have remained uninterested in the political problem of unemployment, and instead have continued their attempts to "fine-tune" the economy by applying the Keynesian remedies of printing money, raising or lowering interest rates, cutting or increasing taxes, and so on. However, these methods ignore the detailed structure of the economy and the qualitative nature of its problems, and hence their successes are very limited.

The flaws of Keynesian economics have now become evident. The Keynesian model is inadequate because it ignores so many factors that are crucial to understanding the economic situation in the twenty-first century. It concentrates on the domestic economy, dissociating it from the global economic networks and disregarding international economic agreements; it neglects the overwhelming political power of today's global corporations, pays no attention to political conditions, and ignores the social and environmental costs of economic activities. At best the Keynesian approach can provide a set of possible scenarios but cannot make specific predictions. Like most of Cartesian economic thought, it has outlived its usefulness.

3.5 The impasse of Cartesian economics

3.5.1 Narrow concepts and fragmented models

Contemporary economics is a mixed bag of concepts, theories, and models stemming from various epochs of economic history. They include various neoclassical schools using more sophisticated mathematical techniques but still based on classical notions, as well as neoclassical models with Keynesian tools grafted onto them to manipulate the so-called market forces while at the same time, schizophrenically, retaining old equilibrium concepts.

All these models and theories are still deeply rooted in the Cartesian paradigm (see Section 1.1.3). Their approaches are fragmentary and reductionist, and today the flaws of contemporary economic thought are ever more apparent. Economists generally fail to recognize that the economy is merely one aspect of the whole ecological and social fabric. They neglect this social and ecological interdependence, treating all goods equally without considering the many ways in which these goods are related to the rest of the world, and reducing all values to the single criterion of private profit making.

Most economists still measure a country's wealth in terms of its gross domestic product (GDP). This is a system in which all economic activities associated with monetary values are added up indiscriminately while all nonmonetary aspects of the economy are ignored. Social costs, like those of accidents, wars, litigation, and healthcare, are added as positive contributions to the GDP, as are "defensive expenditures" on mitigating pollution and similar externalities, and the undifferentiated growth of this crude quantitative index is considered to be the sign of a "healthy" economy.

The one-dimensional metric of the GDP has been adopted almost universally by governments, mass media, and academia to measure overall social progress, and is enshrined in the United Nations System of National Accounts (UNSNA). The media have played a huge role in perpetuating this outdated economic indicator, since most journalists and editors simply report GDP figures with little time or incentive to question their usefulness.

3.5.2 The illusion of unlimited economic growth

The outstanding characteristic of most of today's economic models – whether they are promoted by economists in government, in the corporate world, or in academia – is their assumption that perpetual economic growth is possible. The goal of most national economies is to achieve unlimited growth of their GDP through the continuing accumulation of material goods. Such undifferentiated und unlimited growth is seen as essential by virtually all economists and politicians, even though it should by now be abundantly clear that unlimited expansion on a finite planet can only lead to disaster. Since human needs are finite, but human greed is not, economic growth can usually be maintained through artificial creation of needs by means of advertising. The goods that are produced and sold in this way are often unneeded, and thus are essentially waste. The pollution and depletion of natural resources generated by this enormous waste of unnecessary goods is exacerbated by the waste of energy and materials in inefficient production processes.

Indeed, as we discuss in Chapter 17, the continuing illusion of unlimited growth on a finite planet is the fundamental dilemma at the roots of all the major problems of our time. It is the result of a clash between linear, reductionist thinking and the nonlinear patterns in our biosphere – the ecological networks and cycles that constitute the web of life.

3.5.3 Economics in crisis

The fragmentary approach of contemporary economists, their preference for abstract quantitative models, and their inability to see economic activities within their proper ecological context have resulted in a tremendous gap between theory and economic reality. As a consequence, economics today is in a profound conceptual crisis. This became strikingly apparent during the global financial crisis of 2008–9.

As the CBS journalist Steve Kroft (2008) showed in detail, the crisis was brought about by Wall Street bankers through a combination of greed, incompetence, and weaknesses inherent in the system. It began as a mortgage crisis, caused by the reckless marketing of risky "subprime" loans; then it slowly evolved into a credit crisis; and finally it became a global financial crisis. During the mortgage crisis, big Wall Street investment houses bought up millions of the least dependable mortgages, chopped them up into tiny bits and pieces, and repackaged them as exotic investment securities that hardly anyone could understand. For this repackaging they collected enormous fees.

These complex financial instruments which lay at the heart of the credit crisis were actually designed by mathematicians and physicists, who used computer models to reconstitute the unreliable loans in ways that were supposed to eliminate most of the risks. But their models turned out to be wrong, because physicists and mathematicians are not experts in human behavior, and human behavior cannot be modeled mathematically. In their misguided efforts, they followed a long tradition of economists modeling how consumers behave as rational actors and self-interested individuals, competing with each other to maximize their own gain. These narrow models, in which pure greed is the main ingredient, are mere caricatures of actual human behavior, and hence their failure is not surprising.

In the wake of the global financial crisis, two economics professors, Kamran Mofid and Steve Szeghi (2010), wrote a very sober, reflective essay, titled "Economics in Crisis: What Do We Tell the Students?" They argued that the standard economic theory being taught at our major universities may have been responsible not only for the striking failure to predict the timing and magnitude of the events that unfolded in 2008 but also even for the crisis itself. Their analysis led the authors to a stark conclusion:

> Now is the time to acknowledge the failures of standard theory and the narrowness of market fundamentalism. The times demand a revolution in economic thought, as well as new ways of teaching economics. In many respects this means a return to the soil in which economics was initially born, moral philosophy amid issues and questions of broad significance involving the fullness of human existence.

3.6 The machine metaphor in management

3.6.1 The mechanization of human organizations

In the centuries after Descartes and Newton, the view of the world as a mechanical system composed of elementary building blocks shaped people's perceptions not only of nature, the human organism, and society but also of the human organizations within society. As the metaphor of organizations as machines took hold, it generated corresponding mechanistic theories of management with the aim of increasing an organization's efficiency by designing it as an assemblage of precisely interlocking parts – functional departments such as production, marketing, finance, and personnel – linked together through clearly defined lines of command and communication.

As Morgan (1998) explains in his detailed review of mechanistic management theories, the machine metaphor became prominent during the Industrial Revolution when factory owners and their engineers realized that the efficient operation of the new machines required major changes in the organization of the workforce. With increasing specialization of manufacturing, the division of labor intensified, control of the machines was shifted from workers to their supervisors, and new procedures were introduced to discipline workers and force them to accept the rigorous routines of factory production.

As Morgan (1998) puts it, "Organizations that used machines became more and more like machines."

3.6.2 Classical management theories

During the nineteenth century, various attempts were made to represent and promote the new mechanistic view of human organizations in a systematic way, but it was only in the early twentieth century that coherent theories of organization and management were developed. One of the first organizational theorists was the influential social scientist Max Weber (1864–1920), whose theory about the origin of capitalism we discussed in Section 3.1.4. A keen observer of social and political phenomena, Weber emphasized the role of values and ideologies in shaping societies. Accordingly, he was very critical of the development of mechanistic forms of organization in parallel to that of actual machines.

Weber was not only one of the first observers of the parallels between the mechanization of industry and bureaucratic forms of organization, but also the first to offer a comprehensive definition of bureaucracy as a form of organization emphasizing precision, clarity, regularity, reliability, and efficiency. He was concerned about the psychological and social effects of the proliferation of bureaucracy – the mechanization of human life, the erosion of the human spirit, and the undermining of democracy.

Subsequent management theorists, by contrast, were firm advocates of bureaucratization. They identified and promoted detailed principles and methods through which organizations could be made to function with machine-like efficiency. These theories became known as "classical management theories" and "scientific management."

Frederick Taylor (1911), in particular, perfected the engineering approach to management in his *Principles of Scientific Management*. Taylor's principles, known today as Taylorism, provided the cornerstone of management theory during the first half of the twentieth century. As Morgan (1998, pp. 27–8) points out, Taylorism in its original form is still alive in numerous fast-food chains around the world. In these mechanized restaurants that serve hamburgers, pizzas, and other highly standardized products, "work is often organized in the minutest detail on the basis of designs that analyze the total process of production, find the most efficient procedures, and then allocate these as specialized duties to people trained to perform them in a very precise way. All the thinking is done by the managers and designers, leaving all the doing to the employees."

3.6.3 The machine metaphor today

In the second half of the twentieth century, the machine metaphor continued to have a profound impact on the theory and practice of management, and it was only during the last two decades that organizational theorists began to apply the systems view of life to the management of human organizations (as we discuss in Chapter 14). Even today, however, the mechanistic view of organizations is still widespread among managers.

A company, in this view, is created and owned by people outside the system. Its structure and goals are designed by management or by outside experts and are imposed on the organization. As a machine must be controlled by its operators to function according to their instructions, so the main thrust of management theory has been to achieve efficient operations through top-down control.

Seeing a company as a machine implies that it will eventually run down, unless it is periodically "serviced" and rebuilt by management. It cannot change by itself; all changes need to be designed by someone else. In the 1990s, a new mechanistic catch phrase – "re-engineering" – was invented to describe such redesign of human organizations. It sparked a whole movement dedicated to shift the focus from bureaucratic functions to key business processes.

The principles of classical management theory have become so deeply ingrained in the ways managers think about organizations that for most of them the design of formal structures, linked by clear lines of communication, coordination, and control, has become almost second nature. This largely unconscious embrace of the mechanistic approach to management has now become one of the main obstacles to organizational change.

3.7 Concluding remarks

As we move further into the twenty-first century, transcending the mechanistic view of organizations will be as critical for the survival of human civilization as transcending the mechanistic conceptions of health, the economy, or biotechnology. All these issues are linked, ultimately, to the profound scientific, social, and cultural transformation that is now under way with the emergence of the new systemic conception of life. In the following chapters, we shall discuss the rise of systems thinking in the twentieth century before turning to our detailed discussion of the biological, cognitive, social, and ecological dimensions of the systems view of life.

II

The rise of systems thinking

4

From the parts to the whole

As we mentioned in the Introduction, the tension between mechanism and holism has been a recurring theme throughout the history of Western science. In twentieth-century science, the holistic perspective became known as "systemic" and the way of thinking it implies as "systems thinking." In this chapter we shall review the origin and early development of systems thinking during the first three decades of the twentieth century.

The main characteristics of systemic thinking emerged in Europe during the 1920s in several disciplines. Systems thinking was pioneered by biologists, who emphasized the view of living organisms as integrated wholes. It was further enriched by Gestalt psychology and the new science of ecology, and it had perhaps the most dramatic effects in quantum physics.

4.1 The emergence of systems thinking

At the turn of the century, the triumphs of nineteenth-century biology – cell theory, embryology, and microbiology – had established the mechanistic conception of life as a firm dogma among biologists. And yet they carried within themselves the seeds of the next wave of opposition, the school known as organismic biology, or "organicism." While cell biology made enormous progress in understanding the structures and functions of many of the cell's subunits, it remained largely ignorant of the coordinating activities that integrate those operations into the functioning of the cell as a whole.

4.1.1 The debate between mechanism and vitalism

Before organismic biology was born, many outstanding biologists went through a phase of vitalism, and for many years the debate between mechanism and holism was framed as one between mechanism and vitalism. Vitalism and organicism were both opposed to the reduction of biology to physics and chemistry. Both schools maintained that, although the laws of physics and chemistry are applicable to living organisms, they are insufficient to fully understand the phenomenon of life. The behavior of a living organism as an integrated whole cannot be understood from the study of its parts alone. As the systems theorists would put it several decades later, the whole is more than the sum of its parts.

Vitalists and organismic biologists differed sharply in their answers to the question: In what sense exactly is the whole more than the sum of its parts? The vitalists asserted that some nonphysical entity, or force, must be added to the laws of physics and chemistry to understand life. Organismic biologists maintained that the additional ingredient is the understanding of "organization," or "organizing relations."

Since these organizing relations are patterns of relationships immanent in the physical structure of the organism, organismic biologists asserted that no separate, nonphysical entity is required for the understanding of life. Later on, the concept of organization was refined to that of "self-organization," which is still used in contemporary theories of living systems. Indeed, in these theories the understanding of the patterns of self-organization is the key to understanding the essential nature of life (as we discuss in Chapter 8).

4.1.2 Organismic biology

During the early twentieth century, organismic biologists, opposing both mechanism and vitalism, took up the problem of biological form with great enthusiasm. Some of the main characteristics of what we now call systems thinking emerged from their extensive reflections.

Ross Harrison (1870–1959), one of the early exponents of the organismic school, explored the concept of organization. He identified configuration and relationship as two important aspects of organization, which were subsequently unified in the concept of "pattern of organization" as a configuration of ordered relationships.

The biochemist Lawrence Henderson (1878–1942) was influential through his early use of the term "system" to denote both living organisms and social systems. From that time on, a system came to mean an integrated whole whose essential properties arise from the relationships between its parts, and "systems thinking" the understanding of a phenomenon within the context of a larger whole. This is, in fact, the root meaning of the word "system," which derives from the Greek *syn* + *histanai* ("to place together"). To understand things systemically means literally to put them into a context, to establish the nature of their relationships.

The biologist Joseph Woodger (1894–1981) asserted that organisms could be described completely in terms of their chemical elements, "plus organizing relations." This formulation had considerable influence on subsequent biological thought, and historians of science have stated that the publication of Woodger's *Biological Principles* in 1936 marked the end of the debate between mechanists and vitalists. Woodger and other organismic biologists also emphasized that one of the key characteristics of the organization of living organisms is its hierarchical nature.

Indeed, an outstanding property of all life is the tendency to form multileveled structures of systems within systems. Each of these forms a whole with respect to its parts while at the same time being a part of a larger whole. Thus, cells combine to form tissues, tissues to form organs, and organs to form organisms. These in turn exist within social systems and ecosystems. Throughout the living world, we find living systems nesting within other living systems.

The double role of living systems as parts and wholes requires the interplay of two opposite tendencies: an integrative tendency to function as part of a larger whole, and a self-assertive, or self-organizing tendency to preserve individual autonomy (see Chapter 7).

Since the early days of organismic biology, these multileveled structures of systems within systems have been called hierarchies. However, this term can be misleading, since it is derived from human hierarchies, which are fairly rigid structures of domination and control, quite unlike the multileveled order found in nature. We shall see in Section 4.1.5 that the important concept of living networks provides a new perspective on the so-called "hierarchies" of nature.

What the early systems thinkers recognized very clearly is the existence of different levels of complexity with different kinds of laws operating at each level. Thus the notion of "organized complexity" became another key concept. At each level of complexity, the observed phenomena exhibit properties that do not exist at the lower level. For example, the concept of temperature, which is central to thermodynamics, is meaningless at the level of individual atoms where the laws of quantum theory operate. In the early 1920s, the philosopher C.D. Broad (1887–1971) coined the term "emergent properties" for those properties that emerge at a certain level of complexity but do not exist at lower levels.

4.1.3 A new way of thinking

The ideas set forth by organismic biologists during the first half of the twentieth century helped to give birth to a new way of thinking – thinking in terms of connectedness, relationships, patterns, and context. According to the systems view, the essential properties of an organism, or living system, are properties of the whole, which none of the parts have. They arise from the interactions and relationships between the parts. These properties are destroyed when the system is dissected, either physically or theoretically, into isolated elements. Although we can discern individual parts in any system, these parts are not isolated, and the nature of the whole is always different from the mere sum of its parts. The systems view of life is illustrated beautifully and abundantly in the writings of Paul Weiss (1971, 1973), who brought systems concepts to the life sciences from his earlier studies of engineering and spent his whole life exploring and advocating a fully organismic conception of biology.

The emergence of systems thinking was a profound revolution in the history of Western scientific thought. The belief that in every complex system the behavior of the whole can be understood entirely from the properties of its parts is central to the Cartesian paradigm. This was Descartes' celebrated method of analytic thinking, which has been an essential characteristic of modern scientific thought. In the analytic, or reductionist, approach, the parts themselves cannot be analyzed any further, except by reducing them to still smaller parts. Indeed, Western science has been progressing in that way, and at each step there has been a level of fundamental constituents that could not be analyzed any further.

The great shock of twentieth-century science has been that living systems cannot be understood by analysis. The properties of the parts are not intrinsic properties, but can be understood only within the context of the larger whole. Thus the relationship between the parts and the whole has been reversed. In the systems approach, the properties of the parts can be understood only from the organization of the whole. Accordingly, systems thinking does not concentrate on basic building blocks but rather on basic principles of organization. Systems thinking is "contextual," which is the opposite of analytical thinking. Analysis means taking something apart in order to understand it; systems thinking means putting it into the context of a larger whole.

4.1.4 Gestalt psychology

When the first organismic biologists grappled with the problem of organic form and debated the relative merits of mechanism and vitalism, German psychologists contributed to that dialogue from the very beginning. The German word for organic form is *Gestalt* (as distinct from *Form,* which denotes inanimate form), and the much discussed problem of organic form was known as the *Gestaltproblem* in those days. At the turn of the century, the Austrian philosopher Christian von Ehrenfels (1859–1932) was the first to use *Gestalt* in the sense of an irreducible perceptual pattern, sparking the school of *Gestaltpsychologie*. In the subsequent decades, the Anglo-Saxon followers of this new discipline would continue to use the term "Gestalt" as an English technical term to denote an irreducible perceptual pattern. Ehrenfels (1960/1890) characterized a Gestalt by asserting that "the whole is more than the sum of its parts," a statement which would become the key formula of systems thinkers later on.

4.1.5 Ecology

While organismic biologists encountered irreducible wholeness in organisms and Gestalt psychologists in perception, ecologists encountered it in their studies of animal and plant communities. The new science of ecology emerged out of the organismic school of biology during the late nineteenth century, when biologists began to study communities of organisms.

Ecology – from the Greek *oikos* ("household") – is the study of the Earth Household. More precisely, it is the study of the relationships that interlink all members of the Earth Household. The term was coined by the German biologist Ernst Haeckel (1834–1919), who defined it as "the science of relations between the organism and the surrounding outer world" (Haeckel, 1866). The word *Umwelt* ("environment") was used for the first time by the Baltic German biologist and ecological pioneer Jakob von Uexküll (1909). In the 1920s, ecologists focused on functional relationships within animal and plant communities. In a pioneering book, *Animal Ecology*, Charles Elton (1927) introduced the concepts of food chains and food cycles, viewing the feeding relationships within biological communities as their central organizing principle.

Since the language of the early ecologists was very close to that of organismic biology, it is not surprising that they compared biological communities to organisms. For example, Frederic Clements (1874–1945), an American plant ecologist and pioneer in the study of succession, viewed plant communities as "superorganisms." This concept sparked a lively debate, which went on for more than a decade until the British plant ecologist A.G. Tansley (1871–1955) rejected the notion of superorganisms and coined the term "ecosystem" to characterize animal and plant communities. The ecosystem concept – defined today as "a community of organisms and their physical environment interacting as an ecological unit" – shaped all subsequent ecological thinking and, by its very name, fostered a systems approach to ecology.

The term "biosphere" was first used in the late nineteenth century by the Austrian geologist Eduard Suess (1831–1914) to describe the layer of life surrounding the Earth. A few decades later, the Russian geochemist Vladimir Vernadsky (1863–1945) developed the concept into a full-fledged theory in his pioneering book, *Biosphere*. Vernadsky (1986/1926) saw life as a "geological force" which partly creates and partly controls the planetary environment. Among all the early theories of the living Earth, Vernadsky's comes closest to our contemporary Gaia theory (see Section 8.3.3; see also Margulis and Sagan, 1995).

Ecological communities

The new science of ecology enriched the emerging systemic way of thinking by introducing two new concepts – community and network. By viewing an ecological community as an assemblage of organisms, bound into a functional whole by their mutual relationships, ecologists facilitated the change of focus from organisms to communities and back, applying the same kinds of concepts to different systems levels.

Today we know that most organisms are not only members of ecological communities but are also complex ecosystems themselves, containing a host of smaller organisms that have considerable autonomy and yet are integrated harmoniously into the functioning of the whole. So, there are three kinds of living systems – organisms, parts of organisms, and communities of organisms – all of which are integrated wholes whose essential properties arise from the interactions and interdependence of their parts.

The network concept

From the beginning of ecology, ecological communities have been seen as consisting of organisms linked together in network fashion through feeding relations. This idea is found repeatedly in the writings of nineteenth-century naturalists, and when food chains and food cycles began to be studied in the 1920s, these concepts were soon expanded to the contemporary concept of food webs.

As the network concept became more and more prominent in ecology, systemic thinkers began to use network models at all systems levels, viewing organisms as networks of cells, organs, and organ systems, just as ecosystems are understood as networks of individual organisms. Correspondingly, the flows of matter and energy through ecosystems were perceived as the continuation of the metabolic pathways through organisms.

The view of living systems as networks provides a novel perspective on the so-called "hierarchies" of nature. Since living systems at all levels are networks, we must visualize the web of life as living systems (networks) interacting in network fashion with other systems (networks). For example, we can picture an ecosystem schematically as a network with a few nodes. Each node represents an organism, which means that each node, when magnified, appears itself as a network. Each node in the new network may represent an organ, which in turn will appear as a network when magnified, and so on.

In other words, the web of life consists of networks within networks. At each scale, under closer scrutiny, the nodes of the network reveal themselves as smaller networks. We tend to arrange these systems, all nesting within larger systems, in a hierarchical scheme by placing the larger systems above the smaller ones in pyramid fashion. But this is a human projection. In nature, there is no "above" or "below," and there are no hierarchies. There are only networks nesting within other networks.

4.2 The new physics

The realization that systems are integrated wholes that cannot be understood by analysis was even more shocking in physics than in biology. Ever since Newton, physicists had believed that all physical phenomena could be reduced to the properties of hard and solid material particles. In the 1920s, however, quantum theory forced them to accept the fact that we cannot decompose the world into independently existing smallest units. As we shift our attention from macroscopic objects to atoms and subatomic particles, nature does not show us any isolated building blocks, but rather appears as a complex web of relationships between the various parts of a unified whole.

4.2.1 The strange reality of atomic phenomena

At the beginning of the "new physics" stands the extraordinary intellectual feat of one man – Albert Einstein (1879–1955). In two papers, both published in 1905, Einstein initiated two revolutionary trends in scientific thought. One was his special theory of relativity; the other was a new way of looking at electromagnetic radiation which was to become characteristic of quantum theory, the theory of atomic phenomena. The complete quantum theory was worked out twenty years later by a whole team of physicists under the leadership of Niels Bohr (1885–1962). Relativity theory, however, was constructed in its complete form almost entirely by Einstein himself. Einstein's scientific papers are intellectual monuments that mark the beginning of twentieth-century scientific thought.

Einstein strongly believed in nature's inherent harmony, and throughout his scientific life his deepest concern was to find a unified foundation of physics. He began to move toward this goal by constructing a common framework for electrodynamics and mechanics, the two separate theories of "classical physics." This framework is known as the special theory of relativity. It unified and completed the structure of classical physics, but at the same time it involved radical changes in the traditional concepts of space and time and thus undermined one of the foundations of the Newtonian worldview. Ten years later Einstein proposed his general theory of relativity, in which the framework of the special theory is extended to include gravity.

This is achieved by further drastic modifications of the concepts of space and time, as we discuss in Section 4.6.10 below.

The other major development in twentieth-century physics was a consequence of the experimental investigation of atoms. At the turn of the nineteenth century physicists discovered several phenomena connected with the structure of atoms, such as X-rays and radioactivity, which could not be explained in terms of classical physics. Besides being objects of intense study, these phenomena were used, in most ingenious ways, as new tools to probe deeper into matter than had ever been possible before. For example, the so-called alpha particles emanating from radioactive substances were perceived to be high-speed projectiles of subatomic size that could be used to explore the interior of the atom. They could be fired at atoms, and from the way they were deflected one could draw conclusions about the atom's structure.

This exploration of the atomic and subatomic world brought scientists in contact with a strange and unexpected reality that shattered the foundations of their worldview and forced them to think in entirely new ways. Nothing like that had ever happened before in science. Revolutions like those of Copernicus and Darwin had introduced profound changes in the general conception of the universe, changes that were shocking to many people, but the new concepts themselves were not difficult to grasp. In the twentieth century, however, physicists faced, for the first time, a serious challenge to their ability to understand the universe. Every time they asked nature a question in an atomic experiment, nature answered with a paradox, and the more they tried to clarify the situation, the sharper the paradoxes became.

In their struggle to grasp this new reality, scientists became painfully aware that their basic concepts, their language, and their whole way of thinking were inadequate to describe atomic phenomena. Their problem was not only intellectual but involved an intense emotional and even existential experience, as vividly described by Werner Heisenberg (1958, p. 42) in his classic account *Physics and Philosophy*: "I remember discussions with Bohr which went through many hours till very late at night and ended almost in despair; and when at the end of the discussion I went alone for a walk in the neighboring park I repeated to myself again and again the question: Can nature possibly be so absurd as it seemed to us in these atomic experiments?"

It took these physicists a long time to accept the fact that the paradoxes they encountered are an essential aspect of atomic physics, and to realize that they arise whenever one tries to describe atomic phenomena in terms of classical concepts. Once this was perceived, the physicists began to learn to ask the right questions and to avoid contradictions, and finally they found the precise and consistent mathematical formulation known as quantum theory, or quantum mechanics.

Quantum theory was formulated during the first three decades of he twentieth century by an international group of physicists including Niels Bohr from Denmark; Max Planck, Albert Einstein, and Werner Heisenberg from Germany; Louis de Broglie from France; Erwin Schrödinger and Wolfgang Pauli from Austria; and Paul Dirac from England. These men joined forces across national borders to shape one of the most exciting periods of modern science, one that saw not only brilliant intellectual exchanges but also dramatic human conflicts, as well as deep personal friendships, vividly portrayed in Heisenberg's (1958, 1969) narratives.

4.2.2 Physicists and mystics

The revolutionary changes in our concepts of reality that were brought about by the new physics were followed during the subsequent decades by conceptual revolutions in several other sciences out of which a coherent worldview is now emerging. It is a holistic and ecological view, which we are calling "the systems view of life" in this book. In the new systemic understanding of the living world, physics is no longer seen as the science providing the most fundamental description of reality, as we discussed in the Introduction. However, the new physics is an integral part of the systems view of life, being essential to the understanding of the behavior of molecules in living cells, of the propagation of nerve impulses in the brain, and of many other biological phenomena.

The systems view of life is an ecological view that is grounded, ultimately, in spiritual awareness. Connectedness, relationship, and community are fundamental concepts of ecology; and connectedness, relationship, and belonging are the essence of spiritual experience. Thus it is not surprising that the emerging systemic and ecological paradigm is in harmony with many ideas in spiritual traditions. In Chapter 13, we discuss the parallels between the basic concepts and ideas of physicists and Eastern mystics, which were explored in detail by Capra (1975) over thirty-five years ago.

In addition, we discuss in the same chapter more recent explorations of parallels between the systems view of life and Western mystical traditions, and between Buddhism and consciousness research, as well as explorations of the spiritual dimensions of psychology, economics, and politics. The gradual convergence of the worldviews underlying these scientific disciplines and various spiritual traditions makes it ever more apparent that mysticism, or the "perennial philosophy," as it is sometimes called, provides a consistent philosophical background to our contemporary scientific theories.

The basic concepts of the new physics have been discussed in considerable detail by Capra (1975), Davies (1983), Hawking (1988), Greene (1999), Levin (2002), and many other scientists. In the following pages we present merely a brief overview.

4.2.3 The uncertainty principle

The experimental investigation of atoms at the beginning of the twentieth century yielded sensational and totally unexpected results. Far from being the hard, solid particles of time-honored theory, atoms turned out to consist of vast regions of empty space in which extremely small particles – the electrons – moved around the nucleus, bound to it by electric forces.

When the physicists' attention turned from the atom to its constituents – the electrons, and the protons and neutrons in the nucleus – quantum theory made it clear that even those subatomic particles were nothing like the solid objects of classical physics. The subatomic units of matter are very abstract entities that have a dual aspect. Depending on how we look at them, they appear sometimes as particles, sometimes as waves; and this dual nature is also exhibited by light, which can take the form of electromagnetic waves or particles. The particles of light were first called "quanta" by Einstein – hence the term "quantum theory" – and are now known as photons.

This dual nature of matter and of light is very strange. It seems impossible to accept that something can be, at the same time, a particle – that is, an entity confined to a very small volume – and a wave, which is spread out over a large region of space. And yet this is exactly what physicists had to accept. The situation seemed hopelessly paradoxical until it was realized that the terms "particle" and "wave" refer to classical concepts which are not fully adequate to describe atomic phenomena.

An electron is neither a particle nor a wave, but it may show particle-like aspects in some situations and wave-like aspects in others. While it acts like a particle, it is capable of developing its wave nature at the expense of its particle nature, and vice versa, thus undergoing continual transformations from particle to wave and from wave to particle. This means that neither the electron nor any other atomic "object" has any intrinsic properties independent of its environment. The properties it shows – particle-like or wave-like – will depend on the experimental situation – that is, on the apparatus it is forced to interact with.

It was the great achievement of Werner Heisenberg (1901–1976) to express the limitations of classical concepts in a precise mathematical form, which is known as the uncertainty principle. It consists of a set of mathematical relations that determine to what extent classical concepts can be applied to atomic phenomena. These relations stake out the limits of human imagination in the atomic world. Whenever we use classical terms – particle, wave, position, velocity – to describe atomic phenomena, we find that there are pairs of concepts, or aspects, which are interrelated and cannot be defined simultaneously in a precise way. The more we emphasize one aspect in our description, the more the other aspect becomes uncertain, and the precise relation between the two is given by the uncertainty principle.

For a better understanding of this relation between pairs of classical concepts, Niels Bohr introduced the notion of complementarity. He considered the particle picture and the wave picture two complementary descriptions of the same reality, each of them only partly correct and having a limited range of application. Both pictures are needed to give a full account of the atomic reality, and both are to be applied within the limitations set by the uncertainty principle.

4.2.4 Patterns of probabilities

The resolution of the particle/wave paradox forced physicists to accept a situation that called into question the very foundation of the mechanistic worldview – the concept of the reality of matter. At the subatomic level, matter does not exist with certainty at definite places, but rather shows "tendencies to exist," and atomic events do not occur with certainty at definite times and in definite ways, but rather show "tendencies to occur." In the formalism of quantum theory, these tendencies are expressed as probabilities and are associated with quantities (i.e., mathematical functions) that take the form of waves; they are similar to the functions used to describe, say, a vibrating guitar string, or a sound wave. This is how particles can be waves at the same time. They are not "real," three-dimensional waves like water waves or sound waves. They are "probability waves" – abstract mathematical quantities with all the characteristic properties of waves – that are related to the probabilities of finding the particles at particular points in space and at particular times. All the laws of atomic physics are expressed in terms of these probabilities. We can never predict an atomic event with certainty; we can only predict the likelihood of its happening.

The discovery of the dual aspect of matter and of the fundamental role of probability has demolished the classical notion of solid objects. At the subatomic level, the solid material objects of classical physics dissolve into wave-like patterns of probabilities. These patterns, furthermore, do not represent probabilities of things, but rather of probabilities of interconnections. A careful analysis of the process of observation in atomic physics shows that the subatomic particles have no meaning as isolated entities but can be understood only as interconnections, or correlations, between various processes of observation and measurement. As Bohr (1934, p. 57) explained: "Isolated material particles are abstractions, their properties being definable and observable only through their interaction with other systems."

Subatomic particles, then, are not "things" but are interconnections among things, and these, in turn, are interconnections among other things, and so on. In quantum theory we never end up with any "things"; we always deal with interconnections. This is how the new physics reveals the oneness of the universe. It shows that we cannot decompose the world into independently existing smallest units. As we penetrate into matter, we do not perceive any isolated building blocks, but rather a complex web of relations between the various parts of a unified whole. In the words of Heisenberg (1958, p. 107), "The world thus appears as a complicated tissue of events, in which connections of different kinds alternate or overlap or combine and thereby determine the texture of the whole."

To some extent, the material universe can be divided into separate parts, into objects made of molecules and atoms, themselves made of particles. But here, at the level of particles, the notion of separate parts breaks down. The subatomic particles – and therefore, ultimately, all parts of the universe – cannot be understood as isolated entities but must be defined through their interrelations. According to the physicist Henry Stapp (quoted by Capra, 1982, p. 139), "An elementary particle is not an independently existing analyzable entity. It is, in essence, a set of relationships that reach outward to other things."

4.2.5 A new notion of causality

The fundamental role of probability in atomic physics implies a new notion of causality that has profound implications for all fields of science. Classical Newtonian science was constructed by the Cartesian method of analyzing the world into parts and arranging those parts according to causal laws. The resulting deterministic picture of the universe was expressed by the metaphor of nature as a clockwork. In atomic physics, such a mechanical and deterministic picture is no longer possible. Quantum theory has shown us that the world cannot be analyzed into independently existing, isolated elements. The notion of separate parts – atoms, molecules, or particles – is an idealization with only approximate validity; these parts are not connected by causal laws in the classical sense.

In quantum theory individual events do not always have a well-defined cause. For example, the jump of an electron from one atomic orbit to another, or the decay of a subatomic particle, may occur spontaneously without any single event causing it. We can never predict when and how such a phenomenon is going to happen; we can only predict its probability. This does not mean that atomic events occur in completely arbitrary fashion; it means only that they are not brought about by local causes. The behavior of any part is determined by its nonlocal connections to the whole, and since we do not know these connections precisely, we have to replace the narrow classical notion of cause and effect by the wider concept of statistical causality.

The laws of atomic physics are statistical laws, according to which the probabilities for atomic events are determined by the dynamics of the whole system. Whereas in classical mechanics the properties and behavior of the parts determine those of the whole, the situation is reversed in quantum mechanics: it is the whole that determines the behavior of the parts.

4.2.6 The observer as participator

Another insight of atomic physics with far-reaching consequences is the fact that the universal interwovenness revealed by quantum theory always includes the human observer and his or her consciousness. In atomic physics the observed phenomena can be understood only as connections, or correlations, between various processes of observation and measurement, and the end of this chain of processes always lies in the consciousness of the human observer.

The crucial feature of quantum theory is that the observer is not only necessary to observe the properties of an atomic phenomenon but is also necessary even to bring about these properties. My conscious decision of how to observe, say, an electron will determine the electron's properties to some extent. If I ask it a particle question, it will give me a particle answer; if I ask it a wave question, it will give me a wave answer. The electron does not *have* objective properties independent of my mind. This means that in atomic physics, the sharp Cartesian division between mind and matter, between the observer and the observed, can no longer be maintained. We can never speak about nature without, at the same time, speaking of ourselves. In the words of Werner Heisenberg (1958, p. 58), "What we observe is not nature itself, but nature exposed to our method of questioning."

Even though quantum mechanics has been extremely successful in describing a great variety of atomic and subatomic phenomena, the philosophical interpretation of its mathematical formalism still contains unresolved paradoxes. Many of them have to do with the analysis of the process of observation in atomic physics. The traditional starting point of such an analysis – in physics as well as in other sciences – is the division of the physical world into an observed system ("object") and an observing system ("observer"). In quantum theory, the observed system can be an atom, a subatomic particle, an atomic process, etc. The observing system consists of the experimental apparatus and will include one or several human observers.

In quantum theory, the core problem arises from the fact that these two systems are treated in different ways. The observing system is described in the terms of classical physics, but these terms cannot be used consistently for the description of the observed "object." We know that classical concepts are inadequate at the atomic level, yet we have to use them to describe our experiments and to state the results. There is no way we can escape this paradox. The technical language of classical physics is just a refinement of our everyday language, and it is the only language we have to communicate our experimental results.

The contrast between the two kinds of description – classical terms for the experimental arrangement and "probability waves" for the observed objects – leads to deep metaphysical problems that have not yet been resolved. In practice, these problems are circumvented by describing the observing system in operational terms – that is, in terms of instructions that permit scientists to set up and carry out their experiments. In this way, the measuring devices and the scientists are effectively joined into one complex system that has no distinct, well-defined parts, and a complete description of this observing system does not need to be part of the theory.

Most physicists today celebrate the tremendous success of quantum mechanics as a mathematical theory and are not very interested in discussing the persisting conceptual paradoxes. Others, however, believe that further progress in subatomic physics will not be possible until the foundational problems of quantum theory are resolved, as we discuss in Section 4.2.12 below.

4.2.7 The restlessness of matter

The conception of the universe as an interconnected web of relations is one of two major themes that recur throughout the new physics. The other theme is the realization that the cosmic web is intrinsically dynamic. The dynamic aspect of matter arises in quantum theory as a consequence of the wave nature of subatomic particles, and is even more relevant in relativity theory, which has shown us that the being of matter cannot be separated from its activity. The properties of its basic patterns, the subatomic particles, can be understood only in a dynamic context, in terms of movement, interaction, and transformation.

The fact that particles are not isolated entities but wave-like probability patterns implies that they behave in a very peculiar way. Whenever a subatomic particle is confined to a small region of space, it reacts to this confinement by moving around. The smaller the region of confinement, the faster the particle will "jiggle" around in it. This behavior is a typical "quantum effect," a feature of the atomic world that has no analogy in macroscopic physics: the more a particle is confined, the faster it will move around. (See Capra, 1975), for a more detailed discussion of this phenomenon and its relation to the uncertainty principle.) This tendency of particles to react to confinement with motion implies a fundamental "restlessness" of matter that is characteristic of the atomic world. In this world most of the material particles *are* confined; they are bound to the molecular, atomic, and nuclear structures, and therefore are not at rest but have an inherent tendency to move about.

According to quantum theory, matter is never quiescent. To the extent that things can be pictured to be made of smaller constituents – molecules, atoms, and particles – these constituents are in a state of continual motion. Macroscopically, the material objects around us may seem passive and inert, but when we magnify such a "dead" piece of stone or metal, we see that it is full of activity. The closer we look at it, the more restless it appears. All the material objects in our environment are made of atoms that link up with each other in various ways to form molecular structures which are not rigid and motionless but vibrate according to their temperature and in harmony with the thermal vibrations of their environment. Inside the vibrating atoms the electrons are bound to the atomic nuclei by electric forces that try to keep them as close as possible, and they respond to this confinement by whirling around extremely fast. In the nuclei, finally, protons and neutrons are pressed into a minute volume by the strong nuclear forces, and consequently race about at even higher velocities.

Modern physics thus pictures matter not at all as passive and inert but as being in a continuous dancing and vibrating motion whose rhythmic patterns are determined by the molecular, atomic, and nuclear configurations. There is stability, but this stability is one of dynamic balance, and the further we advance into matter, the more we need to understand its dynamic nature to understand its patterns.

4.2.8 Space, time, and energy

In this penetration into the world of submicroscopic dimensions, a decisive point is reached in the study of atomic nuclei, in which the velocities of protons and neutrons are often so high that they come close to the speed of light. This fact is crucial for the description of their interactions, because any description of natural phenomena involving such high velocities has to take the theory of relativity into account. To understand the properties and interactions of subatomic particles we need a framework that incorporates not only quantum theory but also relativity theory; and it is relativity theory that reveals the dynamic nature of matter to its fullest extent.

Einstein's theory of relativity has brought about a drastic change in our concepts of space and time. It has forced us to abandon the classical ideas of an absolute space as the stage of physical phenomena and of absolute time as a dimension separate from space. According to relativity theory, both space and time are relative concepts, reduced to the subjective role of elements of the language a particular observer uses to describe natural phenomena. To provide an accurate description of phenomena involving velocities close to the speed of light, a "relativistic" framework has to be used, one that incorporates time with the three space coordinates, making it a fourth coordinate to be specified relative to the observer. In such a framework space and time are intimately and inseparably connected and form a four-dimensional continuum called "space-time." In relativistic physics, we can never talk about space without talking about time, and vice versa.

The concepts of space and time are so basic for our description of natural phenomena that their radical modification in relativity theory entailed a modification of the whole framework we use in physics to describe nature. The most important consequence of the new relativistic framework has been the realization that mass is nothing but a form of energy. Even an object at rest has energy stored in its mass, and the relation between the two is given by Einstein's famous equation

$$E = mc^2 \tag{4.1}$$

in which c is the speed of light.

Once it has been seen to be a form of energy, mass is no longer required to be indestructible, but can be transformed into other forms of energy. This happens continually in the collision processes of high-energy physics, in which material particles are created and destroyed, their masses being transformed into energy of motion and vice versa. The collisions of subatomic particles are our main tool for studying their properties, and the relation between mass and energy is essential for their description. The equivalence of mass and energy has been verified innumerable times and physicists have become completely familiar with it – so familiar, in fact, that they measure the masses of particles in the corresponding energy units.

The discovery that mass is a form of energy has had a profound influence on our picture of matter and has forced us to modify our concept of a particle in an essential way. In modern physics, mass is no longer associated with a material substance, and hence particles are not seen as consisting of any basic "stuff," but are bundles of energy. Energy, however, is associated with activity, with processes, and this implies that the nature of subatomic particles is intrinsically dynamic.

To understand this better we must remember that these particles can be conceived only in relativistic terms – that is, in terms of a framework where space and time are fused into a four-dimensional continuum. In such a framework the particles can no longer be pictured as small billiard balls, or small grains of sand. These images are inappropriate not only because they represent particles as separate objects, but also because they are static, three-dimensional images. Subatomic particles must be conceived as four-dimensional entities in space-time. Their forms have to be understood dynamically, as forms in space and time. Particles are dynamic patterns, patterns of activity that have a space aspect and a time aspect. Their space aspect makes them appear as objects with a certain mass, their time aspect as processes involving the equivalent energy. Thus the being of matter and its activity cannot be separated; they are but different aspects of the same space-time reality.

The energy patterns of the subatomic world form the stable nuclear, atomic, and molecular structures that build up matter and give it its macroscopic solid aspect, thus making us believe that it is made of some material substance. At the macroscopic level this notion of substance is a useful approximation, but at the atomic level it no longer makes sense. Atoms consist of particles, and these particles are not made of any material stuff. When we observe them we never see any substance; what we observe are dynamic patterns continually changing into one another – a continuous dance of energy.

4.2.9 Gravity and curved space-time

The theory of relativity we have discussed so far is known as the "special theory of relativity." It provides a common framework for the description of the phenomena associated with moving bodies and with electricity and magnetism, the basic features of this framework being the relativity of space and time and their unification into four-dimensional space-time.

In the "general theory of relativity" the framework of the special theory is extended to include gravity. The effect of gravity, according to general relativity, is to make space-time curved. This, again, is extremely hard to imagine. We can easily imagine a two-dimensional curved surface, such as the surface of an egg, because we can see such curved surfaces lying in three-dimensional space. The meaning of curvature for two-dimensional curved surfaces is thus quite clear; but when it comes to three-dimensional space – let alone four-dimensional space-time – our imagination abandons us. Since we cannot look at three-dimensional space "from outside," we cannot imagine how it can be "bent in some direction."

To understand the meaning of curved space-time, we have to use curved two-dimensional surfaces as analogies. Imagine, for example, the surface of a sphere. The crucial fact which makes the analogy to space-time possible is that the curvature is an intrinsic property of that surface and can be measured without going into three-dimensional space. For example, on a plane the sum of the three angles in a triangle is always 180°, but on a sphere it is larger than 180°. A two-dimensional insect confined to the surface of the sphere and unable to experience three-dimensional space could nevertheless find out that the surface on which it is living is curved, provided that it can make such geometric measurements and compare the results to those predicted by Euclidean geometry. If there is a discrepancy, the surface is curved; and the larger the discrepancy is – for a given size of geometric figures – the stronger the curvature.

In the same way, we can define a curved three-dimensional space to be one in which Euclidean geometry is no longer valid. The laws of geometry in such a space will be of a different, "non-Euclidean" type. Such a non-Euclidean geometry was introduced as a purely abstract mathematical idea by the mathematician Georg Riemann (1826–1866), and it was not considered to be more than that until Einstein made the revolutionary suggestion that the three-dimensional space in which we live is actually curved.

According to Einstein's theory, the curvature of space is caused by the gravitational fields of massive bodies. Wherever there is a massive object, the space around it is curved, and the degree of curvature – that is, the degree to which the geometry deviates from that of Euclid – depends on the mass of the object.

Since space can never be separated from time in relativity theory, the curvature caused by gravity cannot be limited to three-dimensional space, but must extend to four-dimensional space-time. This is, indeed, what the general theory of relativity predicts. In a curved space-time, the distortions caused by the curvature affect not only the spatial relationships described by geometry but also the length of time intervals.

Time does not flow at the same rate as in "flat space-time," and as the curvature varies from place to place, according to the distribution of massive bodies, so does the flow of time. It is important to realize, however, that this variation of the flow of time can only be seen by an observer who remains at a different place from the clocks used to measure the variation. If the observer, for example, went to a place where time flows slower, all her clocks would slow down too and she would have no means of measuring the effect.

The equations relating the curvature of space-time to the distribution of matter in that space are called Einstein's field equations. They can be applied not only to determine the local variations of curvature in the neighborhood of stars and planets, but also to find out whether there is an overall curvature of space on a large scale. In other words, Einstein's equations can be used to determine the structure of the universe as a whole. Unfortunately, they do not give a unique answer. Several mathematical solutions of the equations are possible; they correspond to the various models of the universe studied by astrophysicists and cosmologists.

4.2.10 The unification of physics

The two basic theories of contemporary physics have transcended the principal aspects of the Cartesian worldview and of Newtonian physics. Quantum theory has shown that subatomic particles are not isolated grains of matter but are probability patterns, interconnections in an inseparable cosmic web that includes the human observer and his or her consciousness. Relativity theory has revealed the intrinsically dynamic character of this cosmic web by showing that its activity is the very essence of its being.

Current research in physics aims at unifying quantum theory and relativity theory into a complete theory of subatomic matter. Such a theory would need to give a full account of the four fundamental forces that operate at the subatomic level: electromagnetism (which binds electrons to the nucleus and controls all chemical processes), gravity, the strong nuclear force (which holds atomic nuclei together), and the weak nuclear force (which is responsible for radioactive decay). Physicists have not yet been able to formulate such a complete theory, but we now have several partial theories that describe some of the four fundamental forces and the phenomena associated with them very well (see Smolin, 2006).

4.2.11 Systems thinking and the new physics

After this brief review of the "new physics" we shall now return to our historical account of the rise of systems thinking during the 1920s and 1930s. As we have seen in the preceding sections, the quantum physicists in the 1920s struggled with the same conceptual shift from the parts to the whole that gave rise to the school of organismic biology. In fact, the biologists would probably have found it much harder to overcome Cartesian mechanism, had it not broken down in such a spectacular fashion in physics, which had been the great triumph of the Cartesian paradigm for three centuries. Heisenberg (1969) saw the shift from the parts to the whole as the central aspect of that conceptual revolution, and he was so impressed by it that he titled his scientific autobiography *Der Teil und das Ganze* ("The Part and the Whole").

4.3 Concluding remarks

By the 1930s most of the key characteristics of systems thinking had been formulated by organismic biologists, Gestalt psychologists, and ecologists. In all these fields the exploration of living systems – organisms, parts of organisms, and communities of organisms – had led scientists to the same new way of thinking in terms of connectedness, relationships, and context. This new thinking, moreover, was also supported by the revolutionary discoveries in quantum physics in the realm of atoms and subatomic particles, which led physicists to see the universe as an interconnected web of relationships whose parts can be defined only through their connections to the whole.

In our review of the key characteristics of systems thinking (see Box 4.1), we have emphasized several shifts of perspective. In fact, all these shifts of perspective are really just different ways of saying the same thing. Systems thinking means a shift of perception from material objects and structures to the nonmaterial processes and patterns of organization that represent the very essence of life. We should also add that the emphasis on relationships, qualities, and processes does not mean that objects, quantities, and structures are no longer important. When we talk of shifts of perspective, we do not imply that systems thinking completely eliminates one perspective in favor of the other, but rather that there is a

Box 4.1
Characteristics of systems thinking

Shift of perspective from the parts to the whole

The first, and most general, characteristic of systems thinking is the shift of perspective from the parts to the whole. Living systems are integrated wholes whose properties cannot be reduced to those of smaller parts. Their essential, or "systemic," properties are properties of the whole, which none of the parts have. They arise from patterns of organization that are characteristic of a particular class of systems. Systemic properties are destroyed when a system is dissected, either physically or conceptually, into isolated elements.

Inherent multidisciplinarity

Examples of living systems abound in nature. Every organism – animal, plant, microorganism, or human being – is an integrated whole, a living system. Parts of organisms – e.g., leaves, or cells – are again living systems; and living systems also include communities of organisms. These may be social systems – a family, a business organization, a village – or ecosystems. The systems view of life teaches us that all living systems share a set of common properties and principles of organization. This means that systems thinking is inherently multidisciplinary. It can be applied to integrate academic disciplines and to discover similarities between different phenomena within the broad range of living systems.

From objects to relationships

Throughout the living world we find systems nesting within larger systems. Cells are parts of tissues; tissues are parts of organs, organs parts of organisms; and living organisms are parts of ecosystems and social systems. At each level the living system is an integrated whole with smaller components, while at the same time being a part of a larger whole. Ultimately – as quantum physics showed so impressively – there are no parts at all. What we call a part is merely a pattern in an inseparable web of relationships. Therefore, the shift of perspective from the parts to the whole can also be seen as a shift from objects to relationships.

In a sense, this is a figure/ground shift, as illustrated in Figure 4.1. In the mechanistic view (a) the world is a collection of objects. The objects interact with one another, and hence there relationships between them. But the relationships (dotted lines) are secondary. In the systems view (b), we realize that the objects themselves are networks of relationships, embedded in larger networks. For the systems thinker, the relationships are primary. The boundaries (dotted lines) of the discernible patterns – the so-called "objects" – are secondary.

From measuring to mapping

The shift of perspective from objects to relationships does not come easily, because it is something that goes counter to the traditional scientific enterprise in Western culture. In science, we have been told, things need to be measured and weighed. But relationships cannot be measured and weighed; relationships need to be mapped. Thus, the perceptual shift from

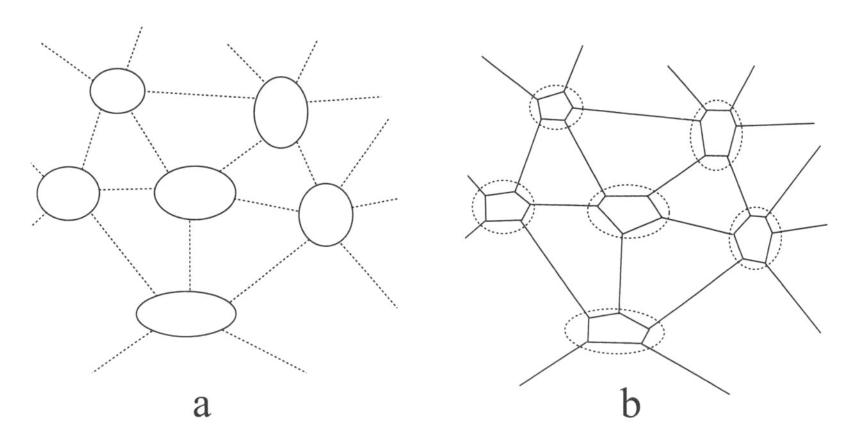

Figure 4.1 Figure/ground shift from objects to relationships (from Capra, 1996).

objects to relationships goes hand in hand with a change of methodology from measuring to mapping. When we map relationships, we find certain configurations that occur repeatedly. This is what we call a pattern. Networks, cycles, and boundaries are examples of patterns of organization that are characteristic of living systems and are at the center of attention in systems science.

From quantities to qualities

Mapping relationships and studying patterns is not a quantitative but a qualitative approach. Thus, systems thinking implies a shift from quantities to qualities. This has been particularly striking in the recent development of complexity theory (see Chapter 6). The new mathematics of complexity is a mathematics of visual patterns, and the analysis of these patterns is known as qualitative analysis.

From structures to processes

In the mechanistic framework of Cartesian science, there are fundamental structures, and then there are forces and mechanisms through which these interact, thus giving rise to processes. In systems science, every structure is seen as the manifestation of underlying processes. Systems thinking includes a shift of perspective from structures to processes. Since the earliest days of biology, scientists and philosophers have recognized that living form is more than shape, more than a static configuration of components in a whole. There is a continual flow of matter through a living system, while its form is maintained; there is growth and decay, regeneration and development. Hence, the understanding of living structure is inextricably linked to the understanding of metabolic and developmental processes.

From objective to epistemic science

The systemic conception of reality as an inseparable network of relationships has important implications not only for our view of nature, but also for our understanding of scientific knowledge. In Cartesian science, scientific descriptions were believed to be objective – that is, independent of the human observer and the process of knowing. Systems science, by contrast, implies that epistemology – the understanding of the process of knowing – has to be included explicitly in the description of natural phenomena.

This recognition entered into science with Werner Heisenberg (see Section 4.2.7) and is closely related to the view of physical reality as a web of relationships. If we imagine the network pictured in Figure 4.1 as much more intricate, perhaps somewhat similar to an ink blot in a Rohrschach test, we can easily understand that isolating a pattern in this complex network by drawing a boundary around it and calling it an "object" will be somewhat arbitrary. Different observers may do it in different ways.

To quote Heisenberg (1958, p. 58) once more: "What we observe is not nature itself, but nature exposed to our method of questioning." Thus systems thinking involves a shift from objective to "epistemic" science; to a framework in which epistemology – "the method of questioning" – becomes an integral part of scientific theories.

The realization that the subjective dimension is always implicit in the practice of science does not mean that we have to give up scientific rigor. When we speak of an "objective" description in science, we mean first and foremost a body of knowledge that is shaped, constrained, and regulated by the collective scientific enterprise, rather than being merely a collection of individual accounts. Such intersubjective validation is standard practice in science and need not be abandoned.

From Cartesian certainty to approximate knowledge

In the epistemic approach to science, nature is seen as an interconnected web of relationships, in which the identification of specific patterns as "objects" depends on the human observer and the process of knowing. This new approach immediately raises an important question. If everything is connected to everything else, how can we ever hope to understand anything? Since all natural phenomena are ultimately interconnected, in order to explain any one of them we would need to understand all the others, something which is obviously impossible.

What makes it possible to turn the systems approach into a proper science is the discovery that there is approximate knowledge. This insight is crucial to all of contemporary science. The mechanistic paradigm is based on the Cartesian belief in the certainty of scientific knowledge. In the systemic paradigm it is recognized that all scientific concepts and theories are limited and approximate. Science can never provide any complete and definitive understanding. In science, to put it bluntly, we never deal with truth, in the sense of a precise correspondence between our descriptions and the described phenomena. We always deal with limited and approximate knowledge.

This may sound discouraging, but over the last century scientists have become very used to the approximate nature of scientific knowledge. Indeed, the fact that we *can* formulate approximate but effective models and theories to describe an endless web of interconnected phenomena, and that we are able to improve our approximations over time, has been a source of confidence and strength in the scientific community.

complementary interplay between the two perspectives, a figure/ground shift, as illustrated in Figure 4.1.

In the 1940s, once the basic systemic concepts – system, systems level, organization, complexity, emergent properties, etc. – had been clarified, some systems thinkers began to formulate actual systems theories, meaning that they integrated systemic concepts into coherent theoretical frameworks describing some of the basic principles of organization of living systems. We shall discuss the development of these "classical" systems theories in the following chapter.

5
Classical systems theories

5.1 Tektology

The Austrian biologist Ludwig von Bertalanffy is commonly credited with the first formulation of a comprehensive theoretical framework describing the principles of organization of living systems. However, twenty to thirty years before Bertalanffy published the first papers on his "general systems theory," Alexander Bogdanov (1873–1928), a Russian medical researcher, philosopher, and economist, developed a systems theory of equal sophistication and scope, which unfortunately is still largely unknown outside of Russia.

Bogdanov called his theory "tektology," from the Greek *tekton* ("builder"), which can be translated as "the science of structures." Bogdanov's main goal was to clarify and generalize the principles of organization of all living and nonliving structures. In his own words (quoted in a detailed summary of tektology by Gorelik, 1975),

> Tektology must clarify the modes of organization that are perceived to exist in nature and human activity; then it must generalize and systematize these modes; further it must explain them, that is, propose abstract schemes of their tendencies and laws . . . Tektology deals with organizational experiences not of this or that specialized field, but of all these fields together. In other words, tektology embraces the subject matter of all the other sciences.

Tektology was the first attempt in the history of science to arrive at a systematic formulation of the principles of organization operating in living and nonliving systems. It anticipated the conceptual framework of Ludwig von Bertalanffy's general systems theory, and it also included several important ideas that were formulated four decades later, in a different language, as key principles of cybernetics by Norbert Wiener (see Section 5.3.2).

Bogdanov's goal was to formulate a "universal science of organization." He defined organizational form as "the totality of connections among systemic elements," which is virtually identical to our contemporary definition of pattern of organization. Using the terms "complex" and "system" interchangeably, Bogdanov distinguished three kinds of systems: organized complexes, where the whole is greater than the sum of its parts; disorganized complexes, where the whole is smaller than the sum of its parts; and neutral complexes, where the organizing and disorganizing activities cancel each other.

The stability and development of all systems can be understood, according to Bogdanov, in terms of two basic organizational mechanisms: formation and regulation. By studying both forms of organizational dynamics and illustrating them with numerous examples from natural and social systems, Bogdanov explored several key ideas pursued by organismic biologists *and* by cyberneticists.

The dynamics of formation consists of the joining of complexes through various kinds of linkages, which Bogdanov analyzed in great detail. He emphasized in particular that the tension between crisis and transformation is central to the formation of complex systems. Foreshadowing the work of Ilya Prigogine (see Chapter 8), Bogdanov showed how organizational crisis manifests itself as a breakdown of the existing systemic balance and at the same time represents a transition to a new state of balance. By defining categories of crises, Bogdanov even anticipated the concept of catastrophe developed in the 1960s by the French mathematician René Thom, which later on, under the name of "bifurcation," became a key concept of complexity theory (see Chapter 6).

Like Bertalanffy, Bogdanov recognized that living systems are open systems that operate far from equilibrium, and he carefully studied their regulation and self-regulation processes. A system for which there is no need of external regulation, because the system regulates itself, is called a "biregulator" in Bogdanov's language. Using the example of the centrifugal governor of a steam engine to illustrate self-regulation, as the cyberneticists would do several decades later, Bogdanov essentially described the mechanism defined as feedback by Norbert Wiener, which became a central concept of cybernetics.

Bogdanov did not attempt to formulate his ideas mathematically, but he did envisage the future development of an abstract "tektological symbolism," a new kind of mathematics to analyze the patterns of organization he had discovered. Half a century later such a new mathematics of complex systems would indeed emerge.

Bogdanov's pioneering book, *Tektology,* was published in Russian in three volumes between 1912 and 1917. A German edition was published and widely reviewed in 1928. However, very little is known in the West about this first version of a general systems theory and precursor of cybernetics. Even in Ludwig von Bertalanffy's *General System Theory*, published in 1968, which includes a section on the history of systems theory, there is no reference to Bogdanov whatsoever. It is difficult to understand how Bertalanffy, who was widely read and published all his original work in German, would not have come across Bogdanov's work.

5.2 General systems theory

Before the 1940s, the terms "system" and "systems thinking" had been used by several scientists, but it was Bertalanffy's concepts of an open system and a general systems theory that established systems thinking as a major scientific movement. With the subsequent strong support from cybernetics, the concepts of systems thinking and systems theory became integral parts of the established scientific language and led to numerous new methodologies and applications – systems engineering, systems analysis, systems dynamics, and so on.

Ludwig von Bertalanffy (1901–1972) began his career as a biologist in Vienna during the 1920s. He soon joined a group of scientists and philosophers, known internationally as the "Vienna circle", and his work included broader philosophical themes from the very beginning. Like other organismic biologists, he firmly believed that biological phenomena required new ways of thinking, transcending the traditional methods of the physical sciences. He set out to replace the mechanistic foundations of science with a holistic vision, which he discussed in a series of papers between 1940 and 1966, summarized in Bertalanffy (1968, p. 37):

> General system theory is a general science of "wholeness" which up till now was considered a vague, hazy, and semi-metaphysical concept. In elaborate form it would be a mathematical discipline, in itself purely formal but applicable to the various empirical sciences. For sciences concerned with "organized wholes," it would be of similar significance to that which probability theory has for sciences concerned with "chance events."

In spite of this vision of a future formal, mathematical theory, Bertalanffy sought to establish his general systems theory on a solid biological basis. He objected to the dominant position of physics within modern science and emphasized the crucial difference between physical and biological systems.

To make his point, Bertalanffy pinpointed a dilemma that had puzzled scientists since the nineteenth century, when the novel idea of evolution entered into scientific thinking. Whereas Newtonian mechanics was a science of forces and trajectories, evolutionary thinking – thinking in terms of change, growth, and development – required a new science of complexity. The first formulation of this new science was classical thermodynamics with its celebrated "second law," the law of the dissipation of energy. As we discussed in Section 1.2.3, the second law of thermodynamics presented scientists with the fundamental dilemma of two diametrically opposed views of evolutionary change – that of a living world unfolding toward increasing order and complexity, and that of an engine running down, a world of ever-increasing disorder. Ludwig von Bertalanffy could not resolve this dilemma, but he took the crucial first step by recognizing that living organisms are open systems that cannot be described by classical thermodynamics. He called such systems "open" because they need to feed on a continual flux of matter and energy from their environment to stay alive.

Unlike closed systems, which settle into a state of thermal equilibrium, open systems maintain themselves far from equilibrium in this "steady state" characterized by continual flow and change. Bertalanffy coined the German term *Fliessgleichgewicht* ("flowing balance") to describe such a state of dynamic balance. He recognized clearly that classical thermodynamics, which deals with closed systems at or near equilibrium, is inappropriate to describe open systems in steady states far from equilibrium.

In open systems, Bertalanffy speculated, entropy (or disorder) may decrease, and the second law of thermodynamics may not apply. He postulated that classical science would have to be complemented by a new thermodynamics of open systems. However, in the 1940s the mathematical techniques required for such an expansion of thermodynamics were not available to Bertalanffy. The formulation of the new thermodynamics of open systems had to wait until the 1970s. It was the great achievement of Ilya Prigogine (1917–2003), who used the new mathematics of complexity to re-evaluate the second law by radically rethinking traditional scientific views of order and disorder. This enabled him to resolve unambiguously the two contradictory nineteenth-century views of evolution, as we discuss in Chapter 8.

5.3 Cybernetics

While Ludwig von Bertalanffy worked on his general systems theory, attempts to develop self-guiding and self-regulating machines led to an entirely new field of investigation that had a major impact on the further development of the systems view of life. Drawing from several disciplines, the new science represented a unified approach to problems of communication and control, involving a whole complex of novel ideas, which inspired Norbert Wiener (1894–1964) to invent a special name for it – "cybernetics." The word is derived from the Greek *kybernetes* ("steersman"), and Wiener (1948) defined cybernetics as the science of "control and communication in the animal and the machine."

5.3.1 The cyberneticists

Cybernetics soon became a powerful intellectual movement, which developed independently of organismic biology and general systems theory. The cyberneticists were neither biologists nor ecologists; they were mathematicians, neuroscientists, social scientists, and engineers. They were concerned with a different level of description, concentrating on patterns of communication, especially in closed loops and networks. Their investigations led them to the concepts of feedback and self-regulation, and then, later on, to self-organization.

This attention to patterns of organization, which was implicit in organismic biology and Gestalt psychology, became the explicit focus of cybernetics. Wiener, especially, recognized that the new notions of message, control, and feedback referred to patterns of organization – that is, to nonmaterial entities – that are crucial to a fully scientific description of life. Later on, Wiener (1950, p. 96) expanded the concept of pattern, from the patterns of communication and control that are common to animals and machines to the general idea of pattern as a key characteristic of life. "We are but whirlpools in a river of ever-flowing water," he wrote. "We are not stuff that abides, but patterns that perpetuate themselves."

The cybernetics movement began during World War II, when a group of mathematicians, neuroscientists, and engineers – among them Norbert Wiener, John von Neumann, Claude Shannon, and Warren McCulloch – formed an informal network to pursue common scientific interests. Their work was closely linked to military research that dealt with the problems of tracking and shooting down aircraft, and was funded by the military, as was most subsequent research in cybernetics.

Around the same time, independently of the cybernetics group, the brilliant British mathematician and logician Alan Turing (1912–1954) developed an abstract logical system that formalized concepts like "algorithm" and "computation," which would become key concepts in the development of computer science. Turing's formulation involved a hypothetical computing device, now known as the Turing machine. Because of his groundbreaking concepts and ideas, Turing is widely considered to be the father of computer science and artificial intelligence (Turing, 1950; see also Teuscher, 2010). During World War II, he worked for Britain's secret code-breaking center at Bletchley Park where his computing devices and techniques were instrumental in breaking the code of the German Enigma machine.

The first cyberneticists (as they would call themselves several years later) set themselves the challenge of discovering the neural mechanisms underlying mental phenomena and expressing them in explicit mathematical language. Thus, while the organismic biologists were concerned with the material side of the Cartesian split, revolting against mechanism and exploring the nature of biological form, the cyberneticists turned to the mental side. Their intention from the beginning was to create an exact science of mind. Although their approach was quite mechanistic, concentrating on patterns common to animals and machines, it involved many novel ideas that exerted a tremendous influence on subsequent systemic conceptions of mental phenomena. Indeed, the contemporary science of cognition (discussed in Chapter 12), which offers a unified scientific conception of brain and mind, can be traced back directly to the pioneering years of cybernetics.

The conceptual framework of cybernetics was developed in a series of legendary meetings in New York City between 1946 and 1953, known as the Macy Conferences (see Heims, 1991). These meetings were extremely stimulating, bringing together a unique group of highly creative people who engaged in intense interdisciplinary dialogues to explore new ideas and ways of thinking. The participants fell into two core groups. The first formed around the original cyberneticists and consisted of mathematicians, engineers, and neuroscientists. The other group consisted of scientists from the humanities who clustered around Gregory Bateson and Margaret Mead. From the first meeting on, the cyberneticists made great efforts to bridge the academic gap between themselves and the humanities.

Norbert Wiener (1894–1964) was the dominant figure throughout the conference series, imbuing it with his enthusiasm for science and dazzling his fellow participants with the brilliance of his ideas and often irreverent approaches. Wiener was not only a brilliant mathematician but also an articulate philosopher. He was keenly interested in biology and appreciated the richness of natural, living systems. He looked beyond the mechanisms of communication and control to larger patterns of organization and tried to relate his ideas to a wide range of social and cultural issues.

John von Neumann (1903–1957) was the second center of attraction at the Macy Conferences. A mathematical genius, he had written a classic treatise on quantum theory, was the originator of the theory of games, and became world famous as the inventor of the digital computer (inspired by Turing's pioneering theoretical work).

Norbert Wiener had a strong influence on the anthropologist Gregory Bateson (1904–1980). Bateson's mind, like Wiener's, roamed freely across disciplines, challenging the

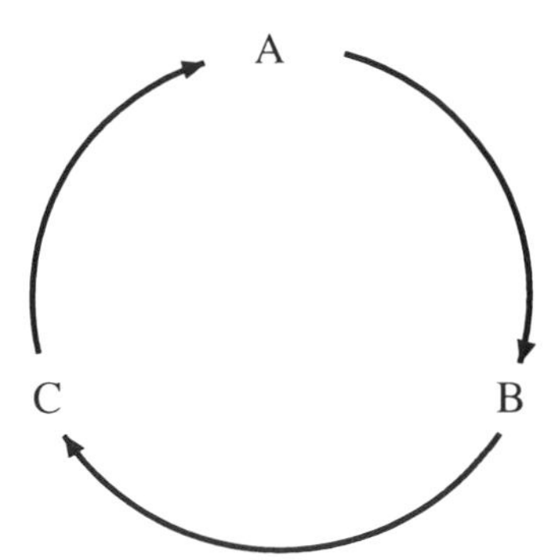

Figure 5.1 Circular causality of a feedback loop.

basic assumptions and methods of several sciences by searching for general patterns and powerful universal abstractions. Bateson thought of himself primarily as a biologist and considered the many fields he became involved in – anthropology, epistemology, psychiatry, and others – as branches of biology. The great passion he brought to science embraced the full diversity of phenomena associated with life, and his main aim was to discover common principles of organization in that diversity – "the pattern which connects," as he would put it many years later.

His dialogues with Wiener and the other cyberneticists had a lasting impact on Bateson's subsequent work. He pioneered the application of systems thinking to family therapy, developed a cybernetic model of alcoholism, and authored the double-bind theory of schizophrenia, which had a major impact on the work of R.D. Laing and many other psychiatrists. However, Bateson's most important contribution to science and philosophy may have been a concept of mind based on cybernetic principles, which he developed during the 1960s. This revolutionary work, which we discuss in Chapter 12, opened the door to understanding the nature of mind as a systems phenomenon and became the first successful attempt in science to overcome the Cartesian division between mind and body.

5.3.2 Feedback

The pioneering years of cybernetics resulted in an impressive series of concrete achievements, in addition to the lasting impact on systems thinking as a whole.

All the major achievements of cybernetics originated in comparisons between organisms and machines – in other words, in mechanistic models of living systems. However, the cybernetic machines are very different from Descartes' clockworks. The crucial difference is embodied in Norbert Wiener's concept of feedback and is expressed in the very meaning of "cybernetics." A feedback loop is a circular arrangement of causally connected elements, in which an initial cause propagates around the links of the loop, so that each element has an effect on the next, until the last "feeds back" the effect into the first element of the cycle (see Figure 5.1). The consequence of this arrangement is that the first link ("input") is affected by the last ("output"), resulting in self-regulation of the entire system, as the initial

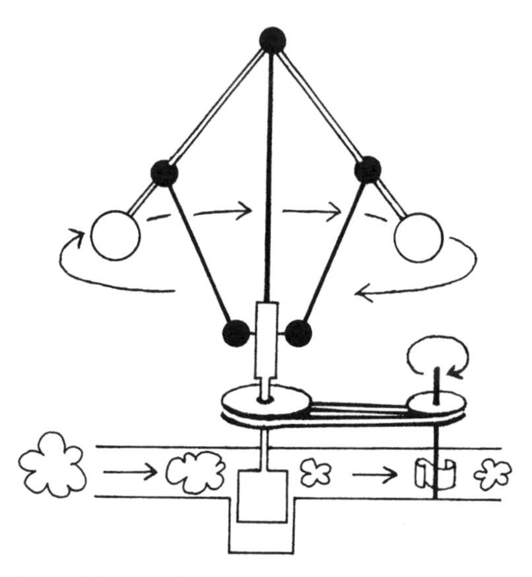

Figure 5.2 Centrifugal governor (from Capra, 1996).

effect is modified each time it travels around the cycle. In a broader sense, feedback has come to mean the conveying of information about the outcome of any process or activity to its source.

Wiener's original example of the steersman is one of the simplest examples of a feedback loop. When the boat deviates from the preset course – say, to the right – the steersman assesses the deviation and then counter-steers by moving the rudder to the left. This decreases the boat's deviation, perhaps even to the point of moving through the correct position and then deviating to the left. At some time during this movement the steersman makes a new assessment of the boat's deviation, counter-steers accordingly, assesses the deviation again, and so on. Thus he relies on continual feedback to keep the boat on course, its actual trajectory oscillating around the preset direction. The skill of steering a boat consists in keeping these oscillations as smooth as possible.

A similar feedback mechanism is in play when we ride a bicycle. At first, when we learn to do so, we find it difficult to monitor the feedback from the continual changes of balance and to steer the bicycle accordingly. Thus a beginner's front wheel tends to oscillate strongly. But as our expertise increases, our brain monitors, evaluates, and responds to the feedback automatically, and the oscillations of the front wheel smooth out into a straight line.

Self-regulating machines involving feedback loops existed long before cybernetics. The centrifugal governor of a steam engine, invented by James Watt in the late eighteenth century, is a classic example (see Figure 5.2). It consists of a rotating spindle with two weights ("flyballs") attached to it in such a way that they move apart, driven by the centrifugal force, when the speed of the rotation increases. The governor sits on top of the steam engine's cylinder, and the weights are connected with a piston, which cuts off the steam as they move apart. The pressure of the steam drives the engine, which drives a flywheel. The flywheel, in turn, drives the governor, and thus the loop of cause and effect is closed. An increase in the speed of the engine increases the rotation of the governor. This increases the distance between the weights, which cuts down the steam supply. As the steam supply decreases, the speed of the engine decreases as well; the rotation of the governor slows down; the weights move closer together; the steam supply increases; the engine speeds up again; and so on.

The engineers who designed these early feedback devices described their operations and pictured their mechanical components in design sketches, but they never recognized the pattern of circular causality embedded in them. In the nineteenth century, the famous physicist James Clerk Maxwell wrote a formal mathematical analysis of the steam governor without ever mentioning the underlying loop concept. Another century had to go by before the connection between feedback and circular causality was recognized. At that time, during the pioneering phase of cybernetics, machines involving feedback loops became a central focus of engineering and have been known as "cybernetic machines" ever since.

Wiener and his colleagues also recognized feedback as the essential mechanism of homeostasis, the self-regulation that allows living organisms to maintain themselves in a state of dynamic balance. When Walter Cannon (1932) introduced the concept of homeostasis a decade earlier in his influential book *The Wisdom of the Body*, he gave detailed descriptions of many self-regulatory metabolic processes but never explicitly identified the closed causal loops embodied in them. Thus the concept of the feedback loop introduced by the cyberneticists led to new perceptions of the many self-regulatory processes characteristic of life. Today we understand that feedback loops are ubiquitous in the living world, because they are a special feature of the nonlinear network patterns that are characteristic of living systems (see Chapter 8). These feedback loops not only have self-balancing effects but may also be self-amplifying. Cyberneticists, accordingly distinguished between "negative" and "positive" feedback, respectively.

Feedback in social systems

From the beginning of cybernetics, Norbert Wiener was aware that feedback is an important concept for modeling not only living organisms but also social systems. Thus he wrote in *Cybernetics* (Wiener, 1948, p. 24):

> It is certainly true that the social system is an organization like the individual, that is bound together by a system of communication, and that it has a dynamics in which circular processes of a feedback nature play an important role.

It was the discovery of feedback as a general pattern of life, applicable to organisms and social systems, which got Gregory Bateson and Margaret Mead so excited about cybernetics. As social scientists, they had observed many examples of circular causality implicit in social phenomena, and during the Macy Conferences the dynamics of these phenomena was made explicit in a coherent unifying pattern.

The importance of the feedback concept in social science has been analyzed in great detail by Richardson (1992), who points out that throughout the history of the social sciences, numerous metaphors have been used to describe self-regulatory processes in social life. The best known, perhaps, are the "invisible hand" regulating the market in the economic theory of Adam Smith (see Section 3.2.2), the "checks and balances" of the US Constitution, and the interplay of thesis and antithesis in the dialectic of Hegel and Marx (see Section 3.2.3). The phenomena described by these models and metaphors all imply circular patterns of causality that can be represented by feedback loops, but none of their authors made that fact explicit.

If the circular logical pattern of self-balancing feedback was not recognized before cybernetics, that of self-amplifying feedback had been known for hundreds of years in common parlance as a "vicious circle." The expressive metaphor describes a bad situation leading to its own worsening through a circular sequence of events. Perhaps the circular nature of such self-amplifying, "runaway" feedback loops was recognized explicitly much earlier, because their effect is much more dramatic than the self-balancing of the negative feedback loops that are so widespread in the living world.

There are other common metaphors to describe self- amplifying feedback phenomena. The "self-fulfilling prophecy," in which originally unfounded fears lead to actions that make the fears come true, and the "bandwagon effect" – the tendency of a cause to gain support simply because of its growing number of adherents – are two well-known examples.

In spite of the extensive knowledge of self-amplifying feedback in common folk wisdom, it played hardly any role during the first phase of cybernetics. The cyberneticists around Norbert Wiener acknowledged the existence of runaway feedback phenomena but did not study them any further. Instead, they concentrated on the self-regulatory, homeostatic processes in living organisms. Indeed, purely self-amplifying feedback phenomena are rare in nature, as they are usually balanced by negative feedback loops constraining their runaway tendencies.

In an ecosystem, for example, every species has the potential of undergoing an exponential population growth, but these tendencies are kept in check by various balancing interactions within the system. Exponential runaways will only appear when the ecosystem is severely disturbed. Then some plants will turn into "weeds," some animals into "pests," other species will be exterminated, and thus the balance of the whole system will be threatened.

5.3.3 Information theory

An important part of cybernetics was the theory of information developed by Norbert Wiener and Claude Shannon in the late 1940s. It originated in Shannon's attempts at the Bell Telephone Laboratories to define and measure amounts of information transmitted through telegraph and telephone lines in order to estimate efficiencies and establish a basis for charging for messages.

Shannon realized that, in order to develop an effective mathematical theory of information, communication signals must be treated independently of the meaning of the message. Thus, the term "information," as used in information theory, has nothing to do with meaning. It is a measure of the order, or nonrandomness, of a signal; and the main concern of information theory is the problem of how to get a message, coded as a signal, through a noisy channel.

To measure the order, and thus the information content, of a signal, Shannon borrowed the concept of entropy from thermodynamics, where it is defined as a measure of disorder (see Section 1.2.3). He used probability theory to express the accuracy of the transmission of a given amount of information under known conditions of noise and was able to derive a formula that shows how the capacity of a channel to carry signals depends on its bandwidth (i.e., its theoretical signal capacity) and its signal-to-noise ratio (the measure of interference).

Shannon made the surprising discovery that, even in the presence of noise, signals can be transmitted effectively, and that the capacity of the channel can be increased significantly by adopting various coding schemes. Thus information theory became an important theoretical framework for coding and data compression in communication theory and computer science.

5.3.4 Cybernetics of the brain

During the 1950s and 1960s, Ross Ashby (1903–1972) became the leading theorist of the cybernetics movement. Like McCulloch, Ashby was a neurologist by training, but he went much further than McCulloch in exploring the nervous system and constructing cybernetic models of neural processes. In his book *Design for a Brain*, Ashby (1952, p. 9) attempted to explain the brain's unique adaptive behavior, capacity for memory, and other patterns of brain functioning in purely mechanistic and deterministic terms. "It will be assumed," he wrote, "that a machine or an animal behaved in a certain way at a certain moment because its physical and chemical nature at that moment allowed no other action."

It is evident that Ashby was much more Cartesian in his approach to cybernetics than Norbert Wiener, who made a clear distinction between a mechanistic model and the non-mechanistic living system it represents. "When I compare the living organism with . . . a machine," wrote Wiener (1950, p. 32), "I do not for a moment mean that the specific physical, chemical, and spiritual processes of life as we ordinarily know it are the same as those of life-imitating machines."

In spite of his strictly mechanistic outlook, Ross Ashby advanced the fledgling discipline of cognitive science considerably with his detailed analyses of sophisticated cybernetic models of neural processes. In particular, he clearly recognized that living systems are energetically open while being – in today's terminology – organizationally closed: "Cybernetics might . . . be defined," wrote Ashby (1952, p. 4), "as the study of systems that are open to energy but closed to information and control – systems that are 'information-tight'."

When the cyberneticists explored patterns of communication and control, the challenge to understand "the logic of the mind" and express it in mathematical language was always at the very center of their discussions. Thus, for over a decade the key ideas of cybernetics were developed through a fascinating interplay between biology, mathematics, and engineering. Detailed studies of the human nervous system led to the model of the brain as a logical circuit with neurons as its basic elements. This view was crucial for the invention of digital computers, and that technological breakthrough in turn provided the conceptual basis for a new approach to the scientific study of mind. John von Neumann's invention of the computer and his analogy between computer and brain functioning are so closely intertwined that it is difficult to know which came first.

The computer model of mental activity became the prevalent view of cognitive science and dominated all brain research for the next thirty years. The basic idea was that human intelligence resembles that of a computer to such an extent that cognition – the process of knowing – can be defined as information processing – that is, as manipulation of symbols based on a set of rules.

After dominating brain research and cognitive science for thirty years, the information-processing dogma was finally questioned seriously. Critical arguments had been presented already during the pioneering phase of cybernetics. For example, it was argued that in actual brains there are no rules; there is no central logical processor, and information is not stored locally. Brains seem to operate on the basis of massive connectivity, storing information distributively and manifesting a self-organizing capacity that is not found in computers. However, these alternative ideas were eclipsed in favor of the dominant computational view, until they re-emerged in the 1970s when systems thinkers became fascinated by a new phenomenon with an evocative name – self-organization.

5.3.5 Self-organization

To understand the phenomenon of self-organization, we first need to understand the importance of pattern. The idea of a pattern of organization – a configuration of relationships characteristic of a particular system – became the explicit focus of systems thinking in cybernetics and has been a crucial concept ever since. From the systems point of view, the understanding of life begins with the understanding of pattern.

As we discussed in the Introduction, there has been a tension between two perspectives – the study of matter and the study of form – throughout the history of Western science and philosophy. The study of matter begins with the question, "What is it made of?"; the study of form asks, "What is its pattern?" Those are two very different approaches, which have been in competition with one another throughout our scientific and philosophical tradition.

The study of the constituents of matter began in Greek antiquity with the theory of the four classical elements – earth, air, fire, water. In modern times those were recast into the chemical elements – now numbering more than 100, but still a finite number of ultimate constituents out of which all matter was thought to be made. Then Dalton identified the elements with atoms, and with the rise of atomic and nuclear physics in the twentieth century the atoms were further reduced to subatomic particles.

Similarly, in biology the basic elements were first organisms, or species, and in the eighteenth century biologists developed elaborate classification schemes for plants and animals. Then, with the discovery of cells as the common elements in all organisms, the focus shifted from organisms to cells. Finally, the cell was broken down into its macromolecules – proteins, amino acids, etc. – and molecular biology became the new frontier of research. In all those endeavors the basic question had not changed since Greek antiquity: What is reality made of? What are its ultimate constituents?

At the same time, throughout the same history of philosophy and science, the study of form, or pattern, was always present. It began with the Pythagoreans in Greece and was continued by Leonardo da Vinci, Paracelsus, the Romantic poets, and various other intellectual movements. However, for most of the time the study of pattern was eclipsed by the study of matter until it re-emerged forcefully in our century, when it was recognized by systems thinkers as essential to the understanding of life.

The study of pattern is crucial to the understanding of living systems because systemic properties, as we discussed in Section 4.1.2, arise from a configuration of ordered relationships. Systemic properties are properties of a pattern. What is destroyed when a living organism is dissected is its pattern. The components are still there, but the configuration of relationships between them – the pattern – is destroyed, and thus the organism dies.

Once the importance of pattern for the understanding of life is appreciated, it becomes natural to ask: Is there a common pattern of organization that can be identified in all living systems? We shall see in Chapter 7 that this is indeed the case. As the early systems thinkers discovered, the most important property of this pattern of organization, common to all living systems, is that it is a network pattern. Whenever we encounter living systems – organisms, parts of organisms, or communities of organisms – we can observe that their components are arranged in network fashion. Whenever we look at life, we look at networks.

The appreciation of the importance of networks in living systems came into science in the 1920s, when ecologists began to study food webs. Soon after that, network models were extended to all systems levels. Cyberneticists, in particular, tried to understand the brain as a neural network and developed special mathematical techniques to analyze its patterns. The structure of the human brain is enormously complex. It contains about 10 billion nerve cells (neurons), which are interlinked in a vast network through 1,000 billion junctions (synapses). The whole brain can be divided into subsections, or subnetworks, which communicate with each other in network fashion. All this results in intricate patterns of intertwined webs, networks nesting within larger networks.

The first and most obvious property of any network is its nonlinearity – it goes in all directions. Thus the relationships in a network pattern are nonlinear relationships. In particular, an influence, or message, may travel along a cyclical path, which may become a feedback loop. In living networks, the concept of feedback is intimately connected with the network pattern.

Because networks of communication may generate feedback loops, they may acquire the ability to regulate themselves. For example, a community that maintains an active network of communication will learn from its mistakes, because the consequences of a mistake will spread through the network and return to the source along feedback loops. Thus the community can correct its mistakes, regulate itself, and organize itself. This is how the study of communication and feedback in living networks naturally leads to the notion of self-organization.

Emergence of the self-organization concept

The concept of self-organization originated in the early years of cybernetics, when scientists began to construct mathematical models representing the logic inherent in neural networks. In 1943, the neuroscientist Warren McCulloch and the mathematician Walter Pitts published a pioneering paper in which they introduced idealized neurons represented by binary switching elements – that is, elements that can switch "on" or "off" – and they modeled the nervous system as complex networks of those binary switching elements.

In such a McCulloch–Pitts network, the "on–off" nodes are coupled to one another in such a way that the activity of each node is governed by the prior activity of other nodes according to some "switching rule." For example, a node may switch on at the next moment only if a certain number of adjacent nodes are "on" at this moment. McCulloch and Pitts were able to show that although binary networks of this kind are simplified models, they are a good approximation of the networks embedded in the nervous system.

In the 1950s, scientists began to build actual models of such binary networks, including some with little lamps flickering on and off at the nodes. To their great amazement they discovered that after a short time of random flickering, some ordered patterns would emerge in most networks. They would see waves of flickering pass through the network, or they would observe repeated cycles. Even though the initial state of the network was chosen at random, after a while those ordered patterns would emerge spontaneously, and it was that spontaneous emergence of order that became known as "self-organization."

As soon as this evocative term appeared in the literature, systems thinkers began to use it widely in different contexts and with different meanings. Ross Ashby in his early work was probably the first to describe the nervous system as "self-organizing." The physicist and cyberneticist Heinz von Foerster (1911–2002) became a major catalyst for the self-organization idea in the late 1950s, organizing conferences around this topic, providing financial support for many of the participants, and publishing their contributions.

For two decades, Foerster maintained an interdisciplinary research group dedicated to the study of self-organizing systems at the University of Illinois. This group was a close circle of friends and colleagues who worked away from the reductionist mainstream and whose ideas, being ahead of their time, were not widely published. However, those ideas were the seeds of many of the successful models of self-organizing systems developed during the late 1970s and the 1980s.

5.4 Concluding remarks

During the 1950s and 1960s, systems thinking had a strong influence on engineering and management, where systemic concepts – including those of cybernetics – were applied to solve practical problems. These applications gave rise to the new disciplines of systems engineering, systems analysis, and systemic management (see Chapter 14). While the systems approach had a significant influence on these disciplines, its influence on biology, paradoxically, was almost negligible during that time. The 1950s was the decade of the spectacular triumphs of genetics (see Chapter 2), which eclipsed the systems view of life for almost three decades.

However, in the late 1970s two developments occurred that brought systems thinking to the fore again. One of these developments was the discovery of a new mathematics for the description and analysis of complex nonlinear systems; the other was the emergence of the concept of self-organization, which had been implicit in the early discussions of the cyberneticists but was not developed explicitly for another thirty years.

An important difference between the early concept of self-organization in cybernetics and the more elaborate later models is that the latter include the creation of new structures and new modes of behavior in the self-organizing process. For Ashby all possible structural changes took place within a given "variety pool" of structures, and the survival chances of the system depended on the richness, or "requisite variety," of that pool. There was no creativity, no development, no evolution. The later models, by contrast, include the creation of novel structures and modes of behavior.

This critical advance was possible in the 1980s when self-organizing systems were analyzed and modeled with the help of a much more sophisticated mathematics – the newly discovered mathematics of complexity, which we discuss in the following chapter. The application of complexity theory, technically known as nonlinear dynamics, raised systems thinking to an entirely new level and provided the conceptual basis for vastly more sophisticated formulations of the systems view of life.

6
Complexity theory

The view of living systems as self-organizing networks whose components are all interconnected and interdependent has been expressed repeatedly, in one way or another, throughout the history of philosophy and science. However, detailed models of self-organizing systems could be formulated only very recently when new mathematical tools became available that allowed scientists for the first time to describe and model the fundamental interconnectedness of living networks mathematically.

The intricacy of these networks defies the imagination. Even the simplest living system, a bacterial cell, is a highly complex network involving literally thousands of interdependent chemical reactions (see Figure 7.1). Before the 1970s, there was simply no way in which these networks could be modeled mathematically. But then powerful high-speed computers appeared on the scene, making it possible for scientists and mathematicians to develop a new set of concepts and techniques for dealing with that enormous complexity. During the subsequent two decades, these new conceptions coalesced into a coherent mathematical framework, popularly known as complexity theory. Its technical name is nonlinear dynamics, and it is sometimes also called "nonlinear systems theory," or "dynamical systems theory." Chaos theory and fractal geometry are important branches of this new mathematics of complexity, which has been discussed in several popular books (see Stewart, 2002, for an excellent nontechnical introduction), as well as in more technical textbooks (e.g., Hilborn, 2000; Strogatz, 1994). The discovery of nonlinear dynamics has led to major breakthroughs in our understanding of biological life and is widely regarded as the most exciting scientific development of the late twentieth century.

To avoid confusion, it is important to realize that scientists and mathematicians mean different things when they speak of a theory. A scientific theory, such as quantum theory or Darwin's theory of evolution, is an explanation of a well-defined range of natural phenomena, based on systematic observation and formulated in terms of a set of consistent but approximate concepts and principles (as we discussed in the Introduction). Complexity theory is not a *scientific* theory, but rather a *mathematical* theory, like calculus or the theory of functions. In the words of the mathematician Ian Stewart (2002, p. vii), a mathematical theory is "a coherent body of mathematical knowledge with a clear and consistent identity." This implies that complexity theory itself does not represent a scientific advance, but it can be – and has been – the basis for new scientific theories when used properly (and ingeniously) to explain nonlinear natural phenomena.

The new mathematics, as we shall see in detail, is one of relationships and patterns. When we solve a nonlinear equation with these new techniques, the result is not a formula but a visual shape, a pattern traced by the computer. The strange attractors of chaos theory and the fractals of fractal geometry are examples of such patterns. They are visual descriptions of the system's complex behavior. Nonlinear dynamics, then, represents a qualitative rather than a quantitative approach to complexity and thus embodies the shift of perspective that is characteristic of systems thinking – from objects to relationships, from measuring to mapping, from quantity to quality.

6.1 The mathematics of classical science

6.1.1 Geometry and algebra

To appreciate the novelty of the new mathematics of complexity it is instructive to contrast it with the mathematics of classical science. When Galileo (in the celebrated passage we quoted in Section 1.1.1) compared the universe to a great book written in mathematical language, he specified that the characters of this language were "triangles, circles, and other geometric figures." In other words, mathematics for Galileo meant geometry. He inherited this view from the philosophers of ancient Greece, who tended to geometrize all mathematical problems and to seek answers in terms of geometrical figures. Plato's Academy in Athens, the principal Greek school of science and philosophy for nine centuries, is said to have had a sign above its entrance, "Let no one enter here who is unacquainted with geometry."

Several centuries later, a very different approach to solving mathematical problems, known as algebra, was developed by Islamic philosophers in Persia, who in turn had learned it from Indian mathematicians. The word is derived from the Arabic *al-jabr* ("binding together") and refers to the process of reducing the number of unknown quantities by binding them together in equations. Elementary algebra involves equations in which letters – by convention taken from the beginning of the alphabet – stand for various constant numbers. A well-known example, which most readers will remember from their school years, is Equation 6.1,

$$(a+b)^2 = a^2 + 2ab + b^2. \tag{6.1}$$

Higher algebra involves relationships, called "functions," between unknown variable numbers, or "variables," which are denoted by letters taken by convention from the end of the alphabet. For example, in Equation 6.2,

$$y = x + 1, \tag{6.2}$$

the variable y is said to be "a function of x," which is written in mathematical shorthand as $y = f(x)$.

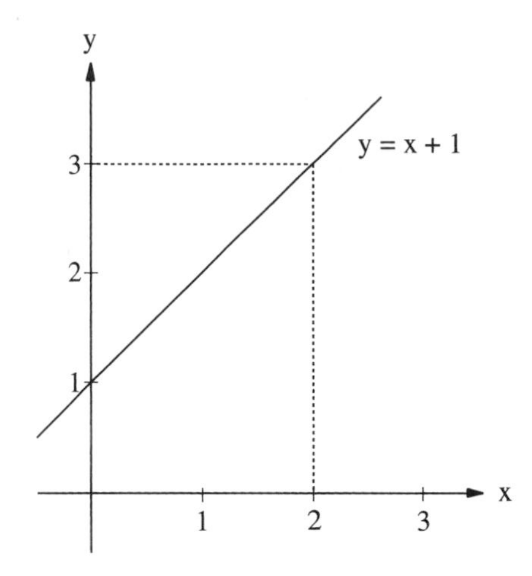

Figure 6.1 Graph corresponding to the equation $y = x + 1$. For any point on the straight line, the value of the y-coordinate is one unit more than that of the x-coordinate.

At the time of Galileo, then, there were two different approaches to solving mathematical problems, geometry and algebra, which came from different cultures. These two approaches were unified by René Descartes, who invented a method to make algebraic formulas and equations visible as geometric shapes. Descartes' invention, now known as analytic geometry, was the greatest among his many contributions to mathematics. It involves Cartesian coordinates, the coordinate system named after Descartes (Cartesius).

For example, when the relationship between the two variables x and y in our previous example, the equation $y = x + 1$, is pictured in a graph with Cartesian coordinates, we see that it corresponds to a straight line (Figure 6.1). This is why equations of this type are called "linear" equations.

Similarly, the equation $y = x^2$ is represented by a parabola (Figure 6.2). Equations of this type, corresponding to curves in the Cartesian grid, are called "nonlinear" equations. They have the distinguishing feature that one or several of their variables are squared or raised to higher powers.

6.1.2 Differential equations

With Descartes' new method, the laws of mechanics that Galileo had discovered could be expressed either in algebraic form as equations or in geometric form as visual shapes. However, there was a major mathematical problem, which neither Galileo nor Descartes nor any of their contemporaries could solve. They were unable to write down an equation describing the movement of a body at variable speed, accelerating or slowing down.

To understand the problem, let us consider two moving bodies, one traveling with constant speed, and the other accelerating. If we plot their distance against time, we obtain the two graphs shown in Figure 6.3. In the case of the accelerating body, the speed changes

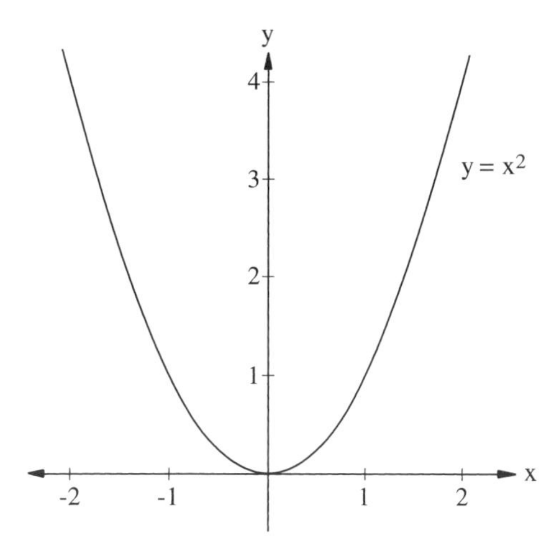

Figure 6.2 Graph corresponding to the equation $y = x^2$. For any point on the parabola, the y-coordinate is equal to the square of the x-coordinate.

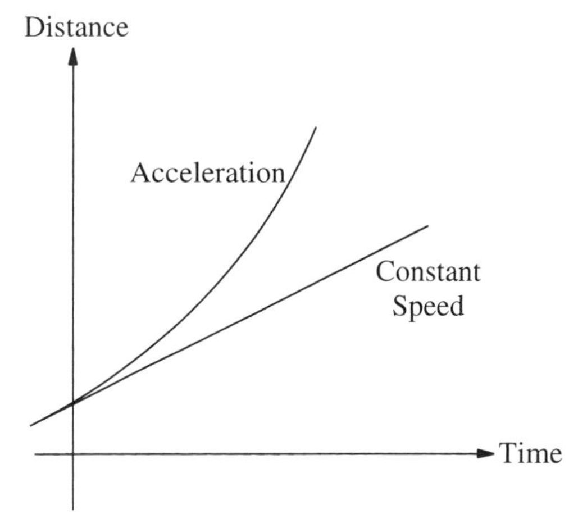

Figure 6.3 Graphs showing the motion of two bodies, one moving at constant speed, and the other accelerating.

at every instant, and this is something Galileo and his contemporaries could not express mathematically. In other words, they were unable to calculate the exact speed of the accelerating body at a given time.

This was achieved a century later by Isaac Newton, the giant of classical science, and around the same time by the German philosopher and mathematician Gottfried Wilhelm Leibniz. To solve the problem that had plagued mathematicians and natural philosophers for centuries, Newton and Leibniz independently invented a new mathematical method, which is now known as differential calculus, or "calculus" for short, and is considered the gateway to "higher mathematics."

For science, the invention of the differential calculus was a giant step. For the first time in human history the concept of the infinite, which had intrigued philosophers and poets from time immemorial, was given a precise mathematical definition, opening countless new possibilities for the analysis of natural phenomena. The power of this new analytical tool can be illustrated with the celebrated paradox of Zeno from the early Eleatic school of Greek philosophy (see Box 6.1). Greek philosophers and their successors argued about this paradox for centuries, but they could never resolve it because the exact definition of the infinitely small eluded them.

The precise definition of the limit of the infinitely small is the crux of calculus. The limits of infinitely small differences are called "differentials," and the calculus invented by Newton and Leibniz is therefore known as differential calculus. Equations involving differentials are called differential equations. In the seventeenth century, Isaac Newton used his calculus to describe all possible motions of solid bodies in terms of a set of differential equations, which have been known as "Newton's equations of motion" ever since.

6.1.3 Complexity in thermodynamics

During the eighteenth and nineteenth centuries, the Newtonian equations of motion were cast into more general, more abstract, and more elegant forms by some of the greatest minds in the history of mathematics. Successive reformulations by Pierre Laplace, Leonhard Euler, Joseph Lagrange, and William Hamilton did not change the content of Newton's equations, but their increasing sophistication allowed scientists to analyze an ever-broadening range of natural phenomena.

Applying his theory to the movement of the planets, Newton himself was able to reproduce the basic features of the solar system, though not its finer details. Laplace, however, refined and perfected Newton's calculations to such an extent that he succeeded in explaining the motion of the planets, moons, and comets down to the smallest details, as well as the flow of the tides and other phenomena related to gravity.

These impressive successes made scientists of the early nineteenth century believe that the universe was indeed a large mechanical system running according to the Newtonian laws of motion. Thus Newton's differential equations became the mathematical foundation of the mechanistic paradigm. The Newtonian world machine was seen as being completely causal and deterministic. All that happened had a definite cause and gave rise to a definite effect, and the future of any part of the system could – in principle – be predicted with absolute certainty if its state at any time was known in all details.

In practice, of course, the limitations of modeling nature through Newton's equations of motion soon became apparent. As Ian Stewart (2002, p. 38) points out, "to *set up* the equations is one thing, to *solve* them quite another." Exact solutions were restricted to a few simple and regular phenomena, while the complexity of vast areas of nature seemed to

Box 6.1
Zeno's paradox

According to Zeno, the great mythical warrior Achilles can never catch up with a tortoise in a race in which the tortoise is granted an initial lead (see Figure 6.4). For when Achilles has completed the distance corresponding to that lead, the tortoise will have covered a further distance; while Achilles covers that, the tortoise will have advanced again, and so on to infinity. Although Achilles' lag keeps decreasing, it will never disappear. At any given moment, the tortoise will always be ahead. Therefore, Zeno concluded, the swift Achilles can never catch up with the tortoise.

Figure 6.4 Zeno's paradox: the race between Achilles and a tortoise.

The flaw in Zeno's argument lies in the fact that, even though it will take Achilles an infinite number of *steps* to reach the tortoise, this does not take an infinite *time*. With the tools of Newton's calculus it is easy to show that a moving body will run through an infinite number of infinitely small intervals in a finite time. In mathematical terms, the limits of the distance at which the paths of Achilles and the tortoise converge is the finite sum of the infinite geometric series in Equation 6.3,

$$s = a + ar + ar^2 + ar^3 + \ldots = \sum_{K=0}^{\infty} ar^k = a/(1-r), \tag{6.3}$$

where a is the initial lead of the tortoise and r is the ratio between their speeds.

elude all mechanistic modeling. For example, the relative motion of two bodies under the force of gravity could be calculated precisely; that of three bodies was already too difficult for an exact solution; and when it came to gases with millions of particles, the situation seemed hopeless.

On the other hand, physicists and chemists had long observed regularities in the behavior of gases, which had been formulated in terms of so-called "gas laws" – simple mathematical relations between the temperature, volume, and pressure of a gas. How could this apparent simplicity be derived from the enormous complexity of the motion of the individual molecules?

In the nineteenth century, the great physicist James Clerk Maxwell found an answer. Even though the exact behavior of the molecules of a gas could not be determined, Maxwell argued that their *average* behavior might give rise to the observed regularities. Hence, Maxwell proposed to use statistical methods to formulate the laws of motion for gases. Maxwell's method was highly successful. It enabled physicists immediately to explain the basic properties of a gas in terms of the average behavior of its molecules.

For example, it became clear that the pressure of a gas is the force caused by the molecules' average push (the average force divided by the area the gas is pushing against, to be precise), while the temperature turned out to be proportional to the molecules' average energy of motion. Statistics and probability theory, its theoretical basis, had been developed since the seventeenth century and could readily be applied to the theory of gases. The combination of statistical methods with Newtonian mechanics resulted in a new branch of science, appropriately called "statistical mechanics," which became the theoretical foundation of thermodynamics, the theory of heat (see Section 1.2.3).

6.2 Facing nonlinearity

Thus, by the end of the nineteenth century, scientists had developed two different mathematical tools to model natural phenomena – exact, deterministic equations of motion for simple systems; and the equations of thermodynamics, based on statistical analysis of average quantities, for complex systems.

Although these two techniques were quite different, they had one thing in common. They both featured *linear* equations. The Newtonian equations of motion are very general, appropriate for both linear and nonlinear phenomena; indeed, every now and then nonlinear equations were formulated. But since these were usually too complex to be solved, and because of the seemingly chaotic nature of the associated physical phenomena – such as turbulent flows of water and air – scientists generally avoided the study of nonlinear systems.

A technical point should perhaps be made here. Mathematicians distinguish between dependent and independent variables. In the function $y = f(x)$, y is the dependent variable and x the independent variable. Differential equations are called "linear" when all *dependent* variables appear in the first power, while independent variables may appear in higher powers, and they are called "nonlinear" when *dependent* variables appear in higher powers.

Until recently, whenever nonlinear equations appeared in science, they were immediately "linearized" – that is, replaced by linear approximations. Thus, instead of describing the phenomena in their full complexity, the equations of classical science deal with *small* oscillations, *shallow* waves, *small* changes of temperature, and so forth. This habit became so ingrained that many equations were linearized *while they were being set up*, so that the science textbooks did not even include the full nonlinear versions. Consequently, most scientists and engineers came to believe that virtually all natural phenomena could be described by linear equations. "As the world was a clockwork for the eighteenth century," Ian Stewart (2002, p. 83) observes, "it was a linear world for the nineteenth and most of the twentieth century."

6.2.1 Explorations of nonlinear systems

The decisive change over the last three decades has been to recognize that nature, as Stewart puts it, is "relentlessly nonlinear." Nonlinear phenomena dominate much more of the inanimate world than we had thought, and they are an essential aspect of the network patterns of living systems. Nonlinear dynamics is the first mathematics that enables scientists to deal with the full complexity of these nonlinear phenomena.

The exploration of nonlinear systems over the past decades has had a profound impact on science as a whole, as it has forced us to re-evaluate some very basic notions about the relationships between a mathematical model and the phenomena it describes. One of those notions concerns our understanding of simplicity and complexity.

In the world of linear equations we thought we knew that systems described by simple equations behaved in simple ways, while those described by complicated equations behaved in complicated ways. In the nonlinear world – which includes most of the real world, as we have discovered – simple deterministic equations may produce an unsuspected richness and variety of behavior. On the other hand, complex and seemingly chaotic behavior can give rise to ordered structures, to subtle and beautiful patterns. In fact, in chaos theory the term "chaos" has acquired a new technical meaning. The behavior of chaotic systems only appears to be random but in reality shows a deeper level of patterned order. As we shall see in the following pages, the new mathematical techniques enable us to make these underlying patterns visible in distinct shapes.

Another important property of nonlinear equations, which has been very disturbing to scientists, is that exact prediction is often impossible, even though the equations may be strictly deterministic. We shall see that this striking feature of nonlinearity has brought about an important shift of emphasis from quantitative to qualitative analysis.

6.2.2 Feedback and iterations

The third important property of nonlinear systems is a surprising difference in cause-and-effect relationships. In linear systems, small changes produce small effects, and large effects are due either to large changes or to a sum of many small changes. In nonlinear systems, by contrast, small changes may have dramatic effects because they may be amplified repeatedly by self-reinforcing feedback. Such nonlinear feedback processes are the basis of the instabilities and the sudden emergence of new forms of order that are so characteristic of self-organization (see Chapter 8).

Mathematically, a feedback loop corresponds to a special kind of nonlinear process known as iteration (Latin for "repetition"), in which a function operates repeatedly on itself. For example, if the function consists of multiplying the variable x by 3 – that is, $f(x) = 3x$ – the iteration consists in repeated multiplications. In mathematical shorthand this is written as Equation 6.4:

$$\begin{aligned} x &\rightarrow 3x \\ 3x &\rightarrow 9x \\ 9x &\rightarrow 27x \\ &\text{etc} \end{aligned} \tag{6.4}$$

Each of these steps is called a "mapping." If we visualize the variable x as a line of numbers, the operation $x \rightarrow 3x$ maps each number to another number on the line. More generally, a mapping that consists of multiplying x by a constant number k is written as in Equation 6.5:

$$x \rightarrow kx. \tag{6.5}$$

An iteration found very often in nonlinear systems, which is very simple and yet produces a wealth of complexity, is the mapping Equation 6.6,

$$x \rightarrow kx(1 - x) \tag{6.6}$$

where the variable x is restricted to values between 0 and 1. This mapping, known to mathematicians as "logistic mapping," has many important applications. It is used by ecologists to describe the growth of a population under opposing tendencies and is therefore also known as the "growth equation."

As Stewart (2002) demonstrates in detail, exploring the iterations of various logistic mappings is a fascinating exercise, which can easily be carried out with a small pocket calculator. In Box 6.2, we demonstrate that an iteration of the logistic mapping with the value $k = 3$ corresponds to repeated stretching and folding of a particular segment of the number line, which is known as the "baker transformation." As the stretching and folding proceeds, neighboring points on the line segment will be moved further and further away from each other, and it is impossible to predict where a particular point will end up after many iterations.

The reason for this impossibility is that even the most powerful computers round off their calculations at a certain number of decimal points, and after a sufficient number of iterations even the most minute round-off errors will have added up to enough uncertainty to make predictions impossible. The baker transformation is a prototype of the nonlinear, highly complex, and unpredictable processes known technically as chaos.

Box 6.2
The baker transformation

To see the essential feature of the iterations of a logistic mapping, let us choose again the value $k = 3$, as in Equation 6.7:

$$x \to 3x(1 - x). \tag{6.7}$$

The variable x can be visualized as a line segment running from 0 to 1, and it is easy to calculate the mappings for a few points, as in Equation 6.8:

$$\begin{aligned} 0 &\to 0(1 - 0) = 0 \\ 0.2 &\to 0.6(1 - 0.2) = 0.48 \\ 0.4 &\to 1.2(1 - 0.4) = 0.72 \\ 0.6 &\to 1.8(1 - 0.6) = 0.72 \\ 0.8 &\to 2.4(1 - 0.8) = 0.48 \\ 1 &\to 3(1 - 1) = 0 \end{aligned} \tag{6.8}$$

When we mark these numbers on two line segments, we see that numbers between 0 and 0.5 are mapped to numbers between 0 and 0.75. Thus 0.2 becomes 0.48, and 0.4 becomes 0.72. Numbers between 0.5 and 1 are mapped to the same segment but in reverse order. Thus 0.6 becomes 0.72, and 0.8 becomes 0.48. The overall effect is shown in Figure 6.5. We see that the mapping stretches the segment so that it covers the distance from 0 to 1.5, and then folds it back over itself, resulting in a segment running from 0 to 0.75 and back.

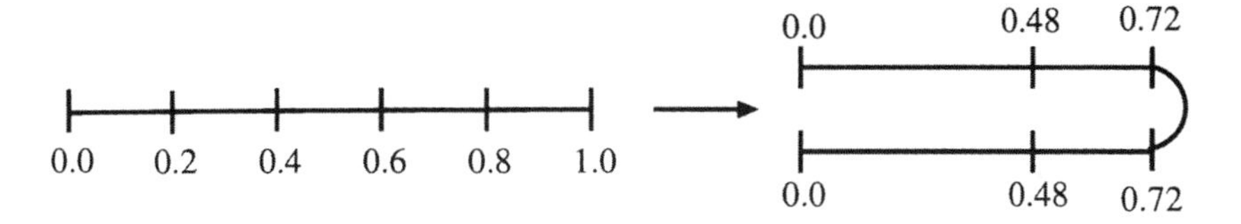

Figure 6.5 The logistic mapping $x \to 3x(1 - x)$.

An iteration of this mapping will result in repeated stretching and folding operations, much like a baker stretches and folds a dough over and over again. The iteration is therefore called, very aptly, the "baker transformation."

6.2.3 Poincaré and the footprints of chaos

Nonlinear dynamics, the mathematics that has made it possible to bring order into chaos, was developed very recently, but its foundations were laid at the turn of the nineteenth century by one of the greatest mathematicians of the modern era, Henri Poincaré (1854–1912). Among the mathematicians of the twentieth century, Poincaré was the last great generalist. He made innumerable contributions in virtually all branches of mathematics. His collected works run into several hundred volumes.

From the vantage point of the twenty-first century, we can see that Poincaré's greatest contribution was to bring visual imagery back into mathematics. From the seventeenth century on, the style of European mathematics had gradually shifted from geometry, the mathematics of visual shapes, to algebra, the mathematics of formulas. Laplace, especially, was one of the great formalizers. who boasted that his *Analytical Mechanics* contained no pictures. Poincaré reversed that trend, breaking the stranglehold of analysis and formulas that had become ever more opaque, and turning once again to visual patterns.

Poincaré's visual mathematics, however, is not the geometry of Euclid. It is a geometry of a new kind, a mathematics of patterns and relationships known as topology. Topology is a geometry in which all lengths, angles, and areas can be distorted at will. Thus a triangle can be continuously transformed into a rectangle, the rectangle into a square, the square into a circle, and so on. Similarly, a cube can be transformed into a cylinder, the cylinder into a cone, the cone into a sphere. Because of these continuous transformations, topology is known popularly as "rubber sheet geometry." All figures that can be transformed into each other by continuous bending, stretching, and twisting are called "topologically equivalent."

However, not everything is changeable by these topological transformations. In fact, topology is concerned precisely with those properties of geometric figures that do not change when the figures are transformed. Intersections of lines, for example, remain intersections, and the hole in a torus (doughnut) cannot be transformed away. Thus a doughnut may be transformed topologically into a coffee cup (the hole turning into a handle) but never into a pancake. Topology, then, is really a mathematics of relationships, of unchangeable, or "invariant," patterns.

Poincaré used topological concepts to analyze the qualitative features of complex dynamical problems and, in doing so, laid the foundations for the mathematical theory of complexity that would emerge a century later. Among the problems Poincaré analyzed in this way was the celebrated three-body problem in celestial mechanics – the relative motion of three bodies under their mutual gravitational attraction – which nobody had been able to solve. By applying his topological method to a slightly simplified three-body problem Poincaré was able to determine the general shape of its trajectories and found it to be of awesome complexity. In his own words (quoted by Stewart, 2002, p. 71):

> When one tries to depict the figure formed by these two curves and their infinity of intersections . . . [one finds that] these intersections form a kind of net, web, or infinitely tight mesh; neither of the two curves can ever cross itself, but must fold back on itself in a very complex way in order to cross the links of the web infinitely many times. One is struck with the complexity of this figure that I am not even attempting to draw.

What Poincaré pictured in his mind is now called a "strange attractor." In the words of Ian Stewart (2002, p. 72), "Poincaré was gazing at the footprints of chaos."

By showing that simple deterministic equations of motion can produce unbelievable complexity that defies all attempts at prediction, Poincaré challenged the very foundations of Newtonian mechanics. However, because of a quirk of history, scientists at the turn of the nineteenth century did not take up this challenge. A few years after Poincaré published his work on the three-body problem, Max Planck discovered energy quanta and Albert Einstein published his special theory of relativity. For the next half-century physicists and mathematicians were fascinated by the revolutionary developments in quantum physics and relativity theory, and Poincaré's groundbreaking discovery moved backstage. It was not until the 1960s that scientists stumbled again into the complexities of chaos.

6.3 Principles of nonlinear dynamics

6.3.1 Trajectories in abstract spaces

The mathematical techniques that have enabled researchers during the past four decades to discover ordered patterns in chaotic systems are based on Poincaré's topological approach and are closely linked to the development of computers. With the help of today's high-speed computers, scientists can solve nonlinear equations by techniques that were not available before. These powerful computers can easily trace out the complex trajectories that Poincaré did not even attempt to draw.

As most readers will remember from school, an equation is solved by manipulating it until you get a final formula as the solution. This is called solving the equation "analytically." The result is always a formula. Most nonlinear equations describing natural phenomena are too difficult to be solved analytically. But there is another way, which is called solving the equation "numerically." This involves trial and error. You try out various combinations of numbers for the variables until you find the ones that fit the equation. Special techniques and tricks have been developed for doing this efficiently, but for most equations the process is extremely cumbersome, takes a long time, and gives only very rough, approximate solutions.

All this changed when the new powerful computers arrived on the scene. Now we have programs for numerically solving an equation in extremely fast and accurate ways. With the new methods nonlinear equations can be solved to any degree of accuracy. However, the solutions are of a very different kind. The result is not a formula, but a large collection of values for the variables that satisfy the equation, and the computer can be programmed to trace out the solution as a curve, or set of curves, in a graph. This technique has enabled scientists to solve the complex nonlinear equations associated with chaotic phenomena and to discover order beneath the seeming chaos.

To reveal these ordered patterns, the variables of a complex system are displayed in an abstract mathematical space called "phase space." This is a well-known technique that was developed in thermodynamics at the turn of the nineteenth century. Every variable of the system is associated with a different coordinate in this abstract space, and every single point in phase space describes the entire system (see Box 6.3). As the system changes over time, the values of its variables change and thus the point traces out a trajectory, known as an attractor, which is a mathematical representation of the system's long-term behavior.

Over the past twenty years, the phase-space technique has been used to explore a wide variety of complex systems. In case after case scientists and mathematicians would set up

Box 6.3
Attractors in phase space

We shall illustrate the phase-space technique with a very simple example, a ball swinging back and forth on a pendulum. To describe the pendulum's motion completely, we need two variables: the angle, which can be positive or negative, and the velocity, which can again be positive or negative, depending on the direction of the swing. With these two variables, angle and velocity, we can describe the state of motion of the pendulum completely at any moment.

If we now draw a Cartesian coordinate system, in which one coordinate is the angle and the other the velocity (see Figure 6.6), this coordinate system will span a two-dimensional space in which certain points correspond to the possible states of motion of the pendulum. Let us see where these points are. At the extreme elongations the velocity is zero. This gives us two points on the horizontal axis. At the center, where the angle is zero, the velocity is at its maximum, either positive (swinging one way) or negative (swinging the other way). This gives us two points on the vertical axis. Those four points in phase space, which we have marked in Figure 6.8, represent the extreme states of the pendulum – maximum elongation and maximum velocity. The exact location of these points will depend on our units of measurement.

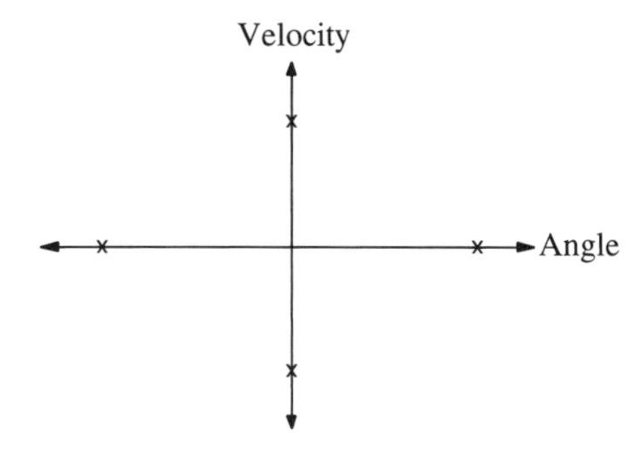

Figure 6.6 The two-dimensional phase space of a pendulum.

If we were to go on and mark the points corresponding to the states of motion between the four extremes, we would find that they lie on a closed loop. We could make it a circle by choosing our units of measurement appropriately, but in general it will be some kind of an ellipse (as shown in Figure 6.7). This loop is called the pendulum's trajectory in phase space. It completely describes the system's motion. All the variables of the system (two in our simple case) are represented by a single point, which will always be somewhere on this loop. As the pendulum swings back and forth, the point in phase space will go around the loop. At any moment, we can measure the two coordinates of the point in phase space, and we will know the exact state – angle and velocity – of the system. Note that this loop is not in any sense a trajectory of the ball on the pendulum. It is a curve in an abstract mathematical space, composed of the system's two variables.

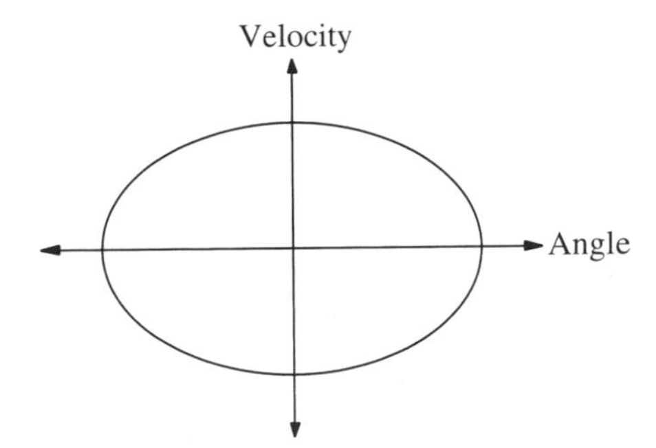

Figure 6.7 Trajectory of the pendulum in phase space.

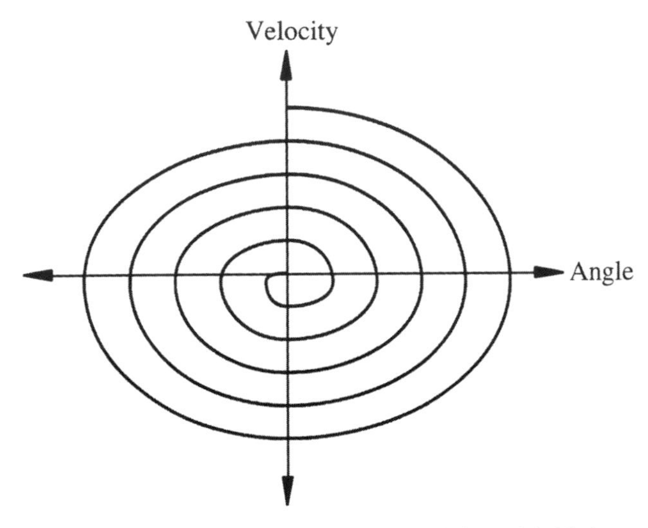

Figure 6.8 Phase space trajectory of a pendulum with friction.

So this is the phase-space technique. The variables of the system are pictured in an abstract space, in which a single point describes the entire system. As the system changes, the point traces out a trajectory in phase space – a closed loop in our example. When the system is not a simple pendulum but much more complicated, it will have many more variables, but the technique is still the same. Each variable is represented by a coordinate in a different dimension in phase space. If there are 16 variables, we will have a 16-dimensional space. A single point in that space will describe the state of the entire system completely, because this single point has 16 coordinates, each corresponding to one of the system's 16 variables.

Of course, we cannot visualize a phase space with 16 dimensions; this is why it is called an abstract mathematical space. Mathematicians do not seem to have any problems with such abstractions. They are just as comfortable in spaces that cannot be visualized. At any rate, as the system changes, the point representing its state in phase space will move around in that space, tracing out a trajectory. Different initial states of the system correspond to different starting points in phase space and will, in general, give rise to different trajectories.

Now let us return to our pendulum and notice that it was an idealized pendulum without friction, swinging back and forth in perpetual motion. This is a typical example of classical physics, where friction is generally neglected. A real pendulum will always have some friction that will slow it down so that, eventually, it will come to a halt. In the two-dimensional phase space, this motion is represented by a curve spiraling inward toward the center, as shown in Figure 6.8.

The shape of a system's trajectory in phase space is known as an "attractor." A closed-loop trajectory, such as the one representing the frictionless pendulum, is called a "periodic attractor," whereas a trajectory spiraling inward is called a "point attractor." The reason for this choice of metaphor is that the attractor represents the system's long-term dynamics. A complex system, typically, will move differently in the beginning, depending on how it starts off, but then will settle down to a characteristic long-term behavior, represented by its attractor. Metaphorically speaking, the trajectory is "attracted" to this pattern whatever its starting point may have been.

We emphasize the origin and correct definition of the term "attractor" because there seems to be a common misconception among nonscientists that an attractor is some entity distinct from the system, which attracts the system to a certain portion of phase space. This is incorrect. In complexity theory, an attractor is a mathematical representation of a dynamic (the system's long-term behavior) that is intrinsic to the system.

nonlinear equations, solve them numerically, and have computers trace out the solutions as trajectories in phase space. To their great surprise, these researchers discovered that there are a very limited number of different attractors. Their shapes can be classified topologically, and the general dynamic properties of a system can be deduced from the shape of its attractor.

There are three basic types of attractors: point attractors, corresponding to systems reaching a stable equilibrium; periodic attractors, corresponding to periodic oscillations; and so-called "strange attractors," corresponding to chaotic systems. A typical example of a system with a strange attractor is the "chaotic pendulum," studied first by the Japanese mathematician Yoshisuke Ueda in the late 1970s. It is a nonlinear electronic circuit with an external drive, which is relatively simple but produces extraordinarily complex behavior. Each swing of this chaotic oscillator is unique. The system never repeats itself, so that each cycle covers a new region of phase space.

However, in spite of the seemingly erratic motion, the points in phase space are not randomly distributed. Together they form a complex, highly organized pattern – a strange attractor, which now bears Ueda's name.

The Ueda attractor is a trajectory in a two-dimensional phase space that generates patterns that almost repeat themselves, but not quite. This is a typical feature of all chaotic systems. The picture shown in Figure 6.9 contains over 100,000 points. It may be visualized as a cut through a piece of dough that has been repeatedly stretched out and folded back on

Figure 6.9 The Ueda attractor (from Capra, 1996).

itself. Thus we see that the mathematics underlying the Ueda attractor is that of the "baker transformation."

One striking fact about strange attractors is that they tend to be of very low dimensionality, even in a high-dimensional phase space. For example, a system may have 50 variables, but its motion may be restricted to a strange attractor of three dimensions, a folded surface in that 50-dimensional space. This, of course, represents a high degree of order.

It is evident that chaotic behavior, in the new scientific sense of the term, is very different from random, erratic motion. With the help of strange attractors a distinction can be made between mere randomness, or "noise," and chaos. Chaotic behavior is deterministic and patterned, and strange attractors allow us to transform the seemingly random data into distinct visible shapes.

6.3.2 The "butterfly effect"

As we have seen in the case of the baker transformation, chaotic systems are characterized by extreme sensitivity to initial conditions. Minute changes in the system's initial state will lead over time to large-scale consequences. In chaos theory this is known as the "butterfly effect" because of the half-joking assertion that a butterfly stirring the air today in Beijing can cause a storm in New York next month.

The butterfly effect was discovered in the early 1960s by the meteorologist Edward Lorenz (1917–2008), who designed a very simple model of weather conditions consisting of three coupled nonlinear equations. He found that the solutions to his equations were extremely sensitive to the initial conditions. From virtually the same starting point, two trajectories would develop in completely different ways, making any long-range prediction impossible.

This discovery sent shock waves through the scientific community, which was used to relying on deterministic equations for predicting phenomena such as solar eclipses or the appearance of comets with great precision over long spans of time. It seemed inconceivable that strictly deterministic equations of motion should lead to unpredictable results. Yet, this was exactly what Lorenz discovered.

The Lorenz model is not a realistic representation of a particular weather phenomenon, but it is a striking example of how a simple set of nonlinear equations can generate enormously complex behavior. Its publication in 1963 marked the beginning of chaos theory, and the model's attractor, known as the Lorenz attractor ever since, became the most celebrated and most widely studied strange attractor. Whereas the Ueda attractor lies in two dimensions, the Lorenz attractor is three-dimensional (see Figure 6.10). To trace it out, the point in phase space moves in an apparently random manner with a few oscillations of increasing amplitude around one point, followed by a few oscillations around a second point, then suddenly moving back again to oscillate around the first point, and so on.

6.3.3 From quantity to quality

The impossibility of predicting which point in phase space the trajectory of the Lorenz attractor will pass through at a certain time, even though the system is governed by deterministic equations, is a common feature of all chaotic systems. However, this does not mean that chaos theory is not capable of any predictions. We can still make very accurate predictions, but they concern the qualitative features of the system's behavior rather than the precise values of its variables at a particular time. The new mathematics thus represents the shift from quantity to quality that is characteristic of systems thinking in general. Whereas conventional mathematics deals with quantities and formulas, nonlinear dynamics deals with qualities and patterns.

Indeed, the analysis of nonlinear systems in terms of the topological features of their attractors is known as "qualitative analysis." A nonlinear system can have several attractors, which may be of different types, both "chaotic," or "strange," and nonchaotic.

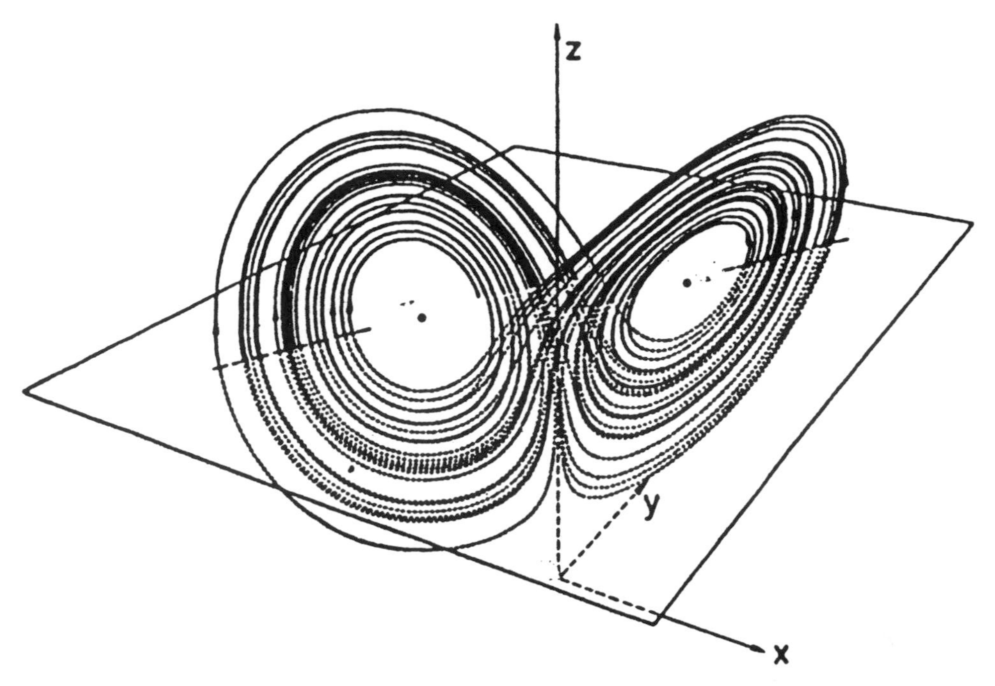

Figure 6.10 The Lorenz attractor. Image reproduced with permission from E. Mosekilde, J. Aracil, and P.M. Allen, "Instabilities and Chaos in Nonlinear Dynamics Systems," *System Dynamic Review*, 4, 14–15, 1988.

All trajectories starting within a certain region of phase space will lead sooner or later to the same attractor. This region is called the "basin of attraction" of that attractor. The phase space of a nonlinear system can often be partitioned into several basins of attraction, each embedding its separate attractor.

The qualitative analysis of a dynamic system, then, consists in identifying the system's attractors and basins of attraction, and classifying them in terms of their topological characteristics. The result is a dynamical picture of the entire system, called the "phase portrait." The mathematical methods for analyzing phase portraits are based on the pioneering work of Poincaré and were further developed and refined by the topologist Stephen Smale in the early 1960s.

Smale used his technique not only to analyze systems described by a given set of nonlinear equations but also to study how those systems behave under small alterations of their equations. As the parameters of the equations change slowly, the phase portrait – that is, the shapes of its attractors and basins of attraction – will usually go through corresponding smooth alterations without any changes in its basic characteristics. Smale used the term "structurally stable" to describe such systems, in which small changes in the equations leave the basic character of the phase portrait unchanged.

In many nonlinear systems, however, small changes of certain parameters may produce dramatic changes in the basic characteristics of the phase portrait. Attractors may disappear, or change into one another, or new attractors may suddenly appear. Such systems are said to be structurally unstable, and the critical points of instability are called "bifurcation points," because they are points in the system's evolution where a fork suddenly appears and the system branches off in a new direction. Mathematically, bifurcation points mark sudden changes in the system's phase portrait. Physically, they correspond to points of instability at which the system changes abruptly and new forms of order suddenly appear.

This spontaneous emergence of order at critical points of instability – often referred to simply as "emergence" – has been recognized as one of the hallmarks of life, as we discuss in Chapter 8. The elucidation of its underlying dynamics, pioneered by the physical chemist Ilya Prigogine, is perhaps the most important contribution of complexity theory to the systems view of life.

As there are only a small number of different types of attractors, so too there are only a small number of different types of bifurcation events, and like the attractors the bifurcations can be classified topologically. One of the first to do so was the French mathematician René Thom in the 1970s, who used the term "catastrophes" instead of "bifurcations" and identified seven elementary catastrophes. Today mathematicians know about three times as many bifurcation types. The chaos theorist Ralph Abraham (1982) and the graphic artist Christopher Shaw have created a series of visual mathematics books without any equations or formulas, which they see as the beginning of a complete encyclopedia of bifurcations.

6.4 Fractal geometry

6.4.1 "A language to speak of clouds"

While the first strange attractors were explored during the 1960s and 1970s, a new geometry, called "fractal geometry," was invented independently of chaos theory, which would provide a powerful mathematical language to describe the fine-scale structure of chaotic attractors. The author of this new language is the mathematician Benoît Mandelbrot (1924–2010). In the late 1950s, Mandelbrot began to study the geometry of a wide variety of irregular natural phenomena, and during the 1960s he realized that all these geometric forms had some very striking common features.

Over the next ten years, Mandelbrot invented a new type of mathematics to describe and analyze these features. He coined the term "fractal" to characterize his invention and published his results in a spectacular book, *The Fractal Geometry of Nature* (Mandelbrot, 1983), which had a tremendous influence on the new generation of mathematicians who were developing chaos theory and other branches of nonlinear dynamics.

One of the best introductions to fractal geometry is a video documentary by the mathematician Heinz-Otto Peitgen (Peitgen *et al.*, 1990), which contains stunning computer animation and a captivating interview with Benoît Mandelbrot. In this interview, Mandelbrot explains that fractal geometry deals with an aspect of nature that almost everybody had been aware of but that nobody was able to describe in formal mathematical terms. Some features of nature are geometric in the traditional sense. The trunk of a tree is more or less a cylinder; the full moon appears more or less as a circular disk; the planets go around the sun more or less in ellipses. But these are exceptions, Mandelbrot reminds us:

> Most of nature is very, very complicated. How could one describe a cloud? A cloud is not a sphere . . . It is like a ball but very irregular. A mountain? A mountain is not a cone . . . If you want to speak of clouds, of mountains, of rivers, of lightning, the geometric language of school is inadequate.

So Mandelbrot created fractal geometry – "a language to speak of clouds" – to describe and analyze the complexity of the irregular shapes in the natural world around us.

6.4.2 Self-similarity

The most striking property of these fractal shapes is that their characteristic patterns are found repeatedly at descending scales, so that their parts, at any scale, are similar in shape to the whole. Mandelbrot illustrates this property of "self-similarity" by breaking a piece out of a cauliflower and pointing out that, by itself, the piece looks just like a small cauliflower. He repeats this demonstration by dividing the part further, taking out another piece, which again looks like a very small cauliflower. Thus every part looks like the whole vegetable. The shape of the whole is similar to itself at all levels of scale.

There are many other examples of self-similarity in nature. Rocks on mountains look like small mountains; branches of lightning, or borders of clouds, repeat the same pattern again and again; coastlines divide into smaller and smaller portions, each showing similar arrangements of beaches and headlands. Photographs of a river delta, the ramifications of a tree, or the repeated branching of blood vessels may show patterns of such striking similarity that we are unable to tell which is which. This similarity of images from vastly different scales has been known for a long time, but before Mandelbrot nobody had a mathematical language to describe it.

When Mandelbrot published his pioneering book in the mid-1970s, he was not aware of the connections between fractal geometry and chaos theory, but it did not take long for his fellow mathematicians and him to discover that strange attractors are exquisite examples of fractals. If parts of their structure are magnified, they reveal a multilayered substructure in which the same patterns are repeated again and again. Thus it has become customary to define strange attractors as trajectories in phase space that exhibit fractal geometry.

6.4.3 Fractal dimensions

Another important link between chaos theory and fractal geometry is the shift from quantity to quality. As we have seen, it is impossible to predict the values of the variables of a chaotic system at a particular time, but we *can* predict the qualitative features of the system's behavior. Similarly, it is impossible to calculate the length or area of a fractal shape, but we can define the degree of "jaggedness" in a qualitative way.

Figure 6.11 Geometric operation for constructing a Koch curve.

Mandelbrot highlighted this dramatic feature of fractal shapes by asking a provocative question: How long is the coast of Britain? He showed that, since the measured length can be extended indefinitely by going to smaller and smaller scales, there is no clear-cut answer to the question. However, it is possible to define a number between 1 and 2 that characterizes the jaggedness of the coast. For the British coastline this number is approximately 1.58; for the much rougher Norwegian coast, it is approximately 1.70.

Since it can be shown that this number has certain properties of a dimension, Mandelbrot called it a fractal dimension. We can understand this idea intuitively by realizing that a jagged line on a plane fills up more space than a smooth line, which has dimension 1, but less than the plane, which has dimension 2. The more jagged the line, the closer its fractal dimension will be to 2. Similarly, a crumpled-up piece of paper fills up more space than a plane but less than a sphere. Thus, the more tightly the paper is crumpled, the closer its fractal dimension will be to 3.

This concept of a fractal dimension, which was at first a purely abstract mathematical idea, has become a very powerful tool for analyzing the complexity of fractal shapes, because it corresponds very well to our experience of nature. The more jagged the outlines of lightning or the borders of clouds, the rougher the shapes of coastlines or mountains, the higher are their fractal dimensions.

6.4.4 Models of fractal shapes

To model the fractal shapes that occur in nature, geometric figures can be constructed that exhibit precise self-similarity. The principal technique for constructing these mathematical fractals is iteration – that is, repeating a certain geometric operation again and again. The process of iteration, which led us to the baker transformation, the mathematical characteristic underlying strange attractors, thus reveals itself as the central mathematical feature linking chaos theory and fractal geometry.

One of the simplest fractal shapes generated by iteration is the so-called Koch curve, or "snowflake curve." The geometric operation consists of dividing a line into three equal parts and replacing the center section by two sides of an equilateral triangle, as shown in Figure 6.11. By repeating this operation again and again on smaller and smaller scales, a jagged snowflake is created (see Figure 6.12). Like a coastline, the Koch curve becomes infinitely long if the iteration is continued to infinity. Indeed, the Koch curve can be seen as a very rough model of a coastline (see Figure 6.13).

With these new mathematical techniques scientists have been able to construct accurate models of a wide variety of irregular natural shapes, and in so doing have discovered

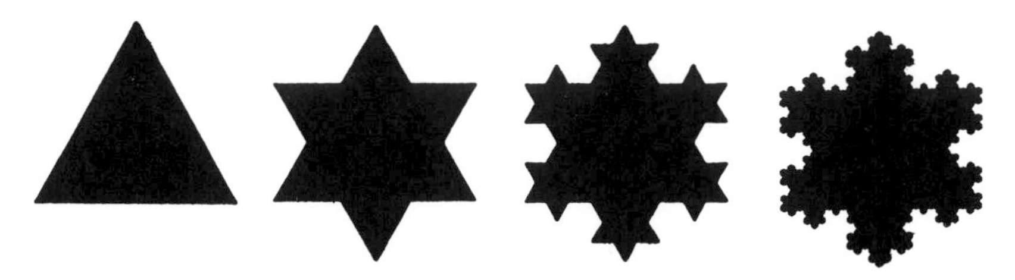

Figure 6.12 The Koch snowflake (from Capra, 1996).

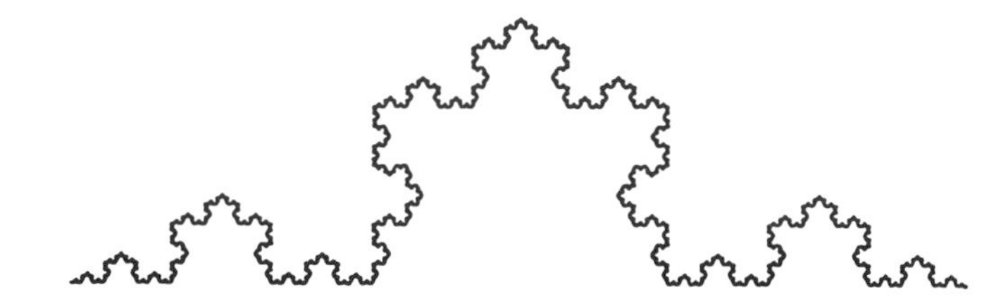

Figure 6.13 Modeling a coastline with the Koch curve (from Capra, 1996).

the pervasive appearance of fractals. Of all those, the fractal patterns of clouds, which originally inspired Mandelbrot to search for a new mathematical language, are perhaps the most stunning. Their self-similarity stretches over seven orders of magnitude, meaning that the border of a cloud magnified 10 million times still shows the same familiar shape.

6.4.5 *Patterns within patterns: complex numbers*

The culmination of fractal geometry has been Mandelbrot's discovery of a mathematical structure that is of awesome complexity and yet can be generated with a very simple iterative procedure. To understand this amazing fractal figure, known as the Mandelbrot set, we need to first familiarize ourselves with one of the most important mathematical concepts – complex numbers.

The discovery of complex numbers is a fascinating chapter in the history of mathematics (see, e.g., Dantzig, 2005). When algebra was developed in the Middle Ages and mathematicians explored all kinds of equations and classified their solutions, they soon came across problems that had no solution in terms of the set of numbers known to them. Equations like $x + 5 = 3$ led them to extend the number concept to negative numbers, so that the solution could be written as $x = -2$. Later on, all "real" numbers – positive and negative integers, fractions and irrational numbers (like square roots, or the famous number π) – were represented as points on a single, densely populated number line (shown in Figure 6.14).

With this expanded concept of numbers, all algebraic equations could be solved in principle except for those involving square roots of negative numbers. The equation

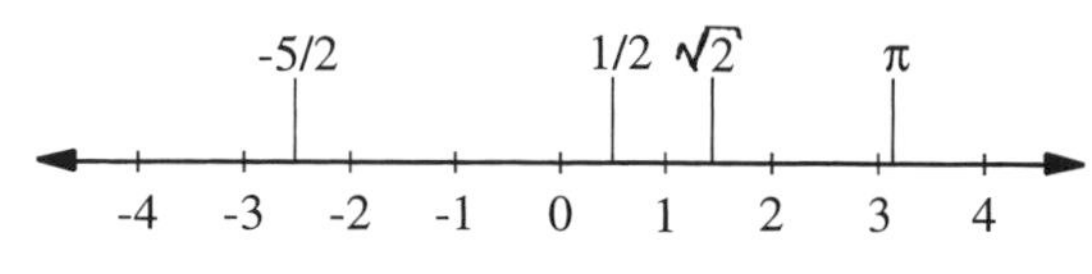

Figure 6.14 The number line.

$x^2 = 4$ has two solutions, $x = 2$ and $x = -2$; but for $x^2 = -4$ there seems to be no solution, because neither $+2$ nor -2 will give -4 when squared.

The early Indian and Arabic algebraists repeatedly encountered these equations, but they refused to write down expressions like $\sqrt{-4}$ because they thought them to be completely meaningless. It was not until the sixteenth century that square roots of negative numbers appeared in algebraic texts, and even then the authors were quick to point out that such expressions did not really mean anything.

Descartes called the square root of a negative number "imaginary" and believed that the occurrence of such "imaginary" numbers in a calculation meant that the problem had no solution. Other mathematicians used terms like "fictitious," "sophisticated," or "impossible," to label those quantities that today, following Descartes, we still call "imaginary numbers."

Since the square root of a negative number cannot be placed anywhere on the number line, mathematicians up to the nineteenth century could not ascribe any sense of reality to those quantities. The great Leibniz, inventor of the differential calculus, attributed a mystical quality to the square root of –1, seeing it as a manifestation of "the Divine Spirit" and calling it "that amphibian between being and not-being." A century later, Leonhard Euler (1707–1783), the most prolific mathematician of all time, expressed the same sentiment in his *Algebra* in words that, even though less poetic, still echo the same sense of wonder (quoted by Dantzig, 2005, p. 189):

> All such expressions as $\sqrt{-1}$, $\sqrt{-2}$, etc., are consequently impossible, or imaginary numbers, since they represent roots of negative quantities; and of such numbers we may truly assert that they are neither nothing, nor greater than nothing, nor less than nothing, which necessarily constitutes them imaginary or impossible.

In the nineteenth century another mathematical giant, Carl Friedrich Gauss (1777–1855), finally declared forcefully that "an objective existence can be assigned to these imaginary beings" (quoted by Dantzig, 2005, p. 190). Gauss realized, of course, that there was no room for imaginary numbers anywhere on the number line, and so he took the bold step of placing them on a perpendicular axis through the point zero, thus creating a Cartesian coordinate system. In this system, all real numbers are placed on the "real axis" and all imaginary numbers on the "imaginary axis" (as shown in Figure 6.15). The square root of –1 is called the "imaginary unit" and given the symbol i, and since any square root of a negative number can always be written as $\sqrt{-a} = \sqrt{-1}\,\sqrt{a} = i\sqrt{a}$, all imaginary numbers can be placed on the imaginary axis as multiples of i.

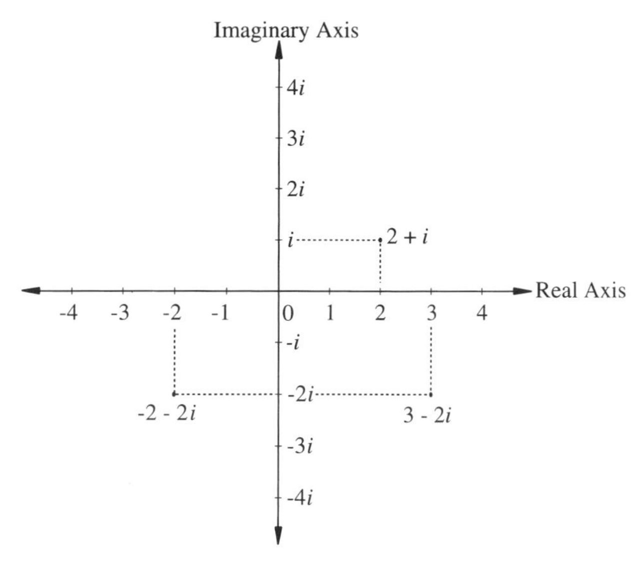

Figure 6.15 The complex plane.

With this ingenious device, Gauss created a home not only for imaginary numbers but also for all possible combinations of real and imaginary numbers, like $(2 + i)$, $(3 - i)$, etc. Such combinations are called "complex numbers" and are represented by points in the plane spanned by the real and imaginary axes, which is called the "complex plane." In general, any complex number can be written as Equation 6.9:

$$z = x + iy, \tag{6.9}$$

where x is called the "real part" and y the "imaginary part."

With the help of this definition Gauss created a special algebra of complex numbers and developed many fundamental ideas about functions of complex variables. Eventually this led to a whole new branch of mathematics, known as "complex analysis," which has an enormous range of applications in all fields of science.

Julia sets

The reason why we took this excursion into the history of complex numbers is that many fractal shapes can be generated mathematically by iterative procedures in the complex plane. In the late 1970s, after publishing his pioneering book, Mandelbrot turned his attention to a particular class of those mathematical fractals known as Julia sets (see, e.g., Peitgen and Richter, 1986). They had been discovered by the French mathematician Gaston Julia (1893–1978) during the early part of the century, but had soon faded into obscurity. In fact, Mandelbrot had come across Julia's work as a student, had looked at his primitive drawings (done at that time without the help of a computer) and had soon lost interest. Now, however, Mandelbrot realized that Julia's drawings were rough renderings of complex fractal shapes,

Box 6.4
How to generate Julia sets and the Mandelbrot set

A Julia set is generated with the mapping Equation 6.10:

$$z \to z^2 + c, \tag{6.10}$$

where z is a complex variable and c a complex constant. The iterative procedure consists of picking any number z in the complex plane, squaring it, adding the constant c, squaring the result again, adding the constant c once more, and so on. When this is done with different starting values for z, some of them will keep increasing and move to infinity as the iteration proceeds, while others will remain finite. The Julia set is the set of all those values of z, or points in the complex plane, that remain finite under the iteration.

To determine the shape of the Julia set for a particular constant c, the iteration has to be carried out for thousands of points, each time until it becomes clear whether their values will keep increasing or remain finite. If those points that remain finite are colored black, while those that keep increasing remain white, the Julia set will emerge as a black shape in the end. The entire procedure is very simple but very time-consuming. It is evident that the use of a high-speed computer is essential if we want to obtain a precise shape in a reasonable time. For each constant c we will obtain a different Julia set, so there is an infinite number of these sets.

The Mandelbrot set is the collection of all points of the constant c in the complex plane for which the corresponding Julia sets are single connected pieces. To construct the Mandelbrot set, therefore, we need to construct a separate Julia set for each point c in the complex plane and determine whether that particular Julia set is "connected" or "disconnected." For example, among the Julia sets shown in Figure 6.16, the three sets in the top row and the one in the center panel of the bottom row are connected (i.e., consist of a single piece), while the two sets in the side panels of the bottom row are disconnected (i.e., consist of several pieces).

and he proceeded to reproduce them in fine detail with the most powerful computers he could find. The results were stunning.

The basis of the Julia set is a simple mapping in the complex plane (see Box 6.4), which generates an astounding variety of fractal patterns. Some are single connected pieces; others are broken into several disconnected parts; and yet others look as if they have burst into dust (see Figure 6.16). All have the jagged look that is characteristic of fractals, and most of them are impossible to describe in the language of classical geometry.

This rich variety of forms, many of which are reminiscent of living things, is amazing enough. But the real magic begins when we magnify the contour of any portion of a Julia set. As in the case of a cloud or coastline, the same richness is displayed across all scales. With increasing resolution (that is, with more and more decimals of the number z entering into the calculation), more and more details of the fractal contour appear, revealing a fantastic sequence of patterns within patterns – all similar without ever being identical.

When Mandelbrot analyzed different mathematical representations of Julia sets in the late 1970s and tried to classify their immense variety, he discovered a very simple way of

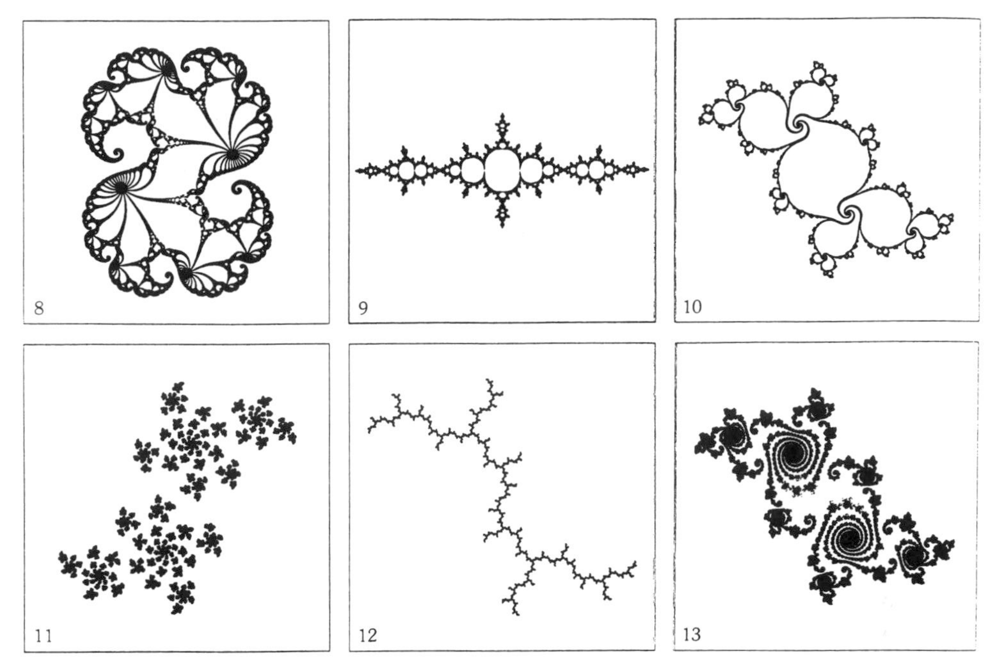

Figure 6.16 Varieties of Julia sets (from Peitgen and Richter, 1986).

creating a single image in the complex plane that would serve as a catalogue of all possible Julia sets (see Box 6.4). That image, which has since become the principal visual symbol of complexity theory, is the Mandelbrot set.

The Mandelbrot set

While there are an infinite number of Julia sets, the Mandelbrot set is unique. This strange figure is the most complex mathematical object ever invented. Although the rules for its construction are very simple, the variety and complexity it reveals upon close inspection is unbelievable. When the Mandelbrot set is generated on a rough grid, two disks appear on the computer screen; the smaller one approximately circular, the larger one vaguely heart-shaped. Each of the two disks shows several smaller disk-like attachments to its boundary, and further resolution reveals a profusion of smaller and smaller attachments looking not unlike prickly thorns (see Figure 6.17).

From this point on, the wealth of images revealed by increasing magnification of the set's boundary (i.e., by increasing resolution in the calculations) is almost impossible to describe. Such a journey into the Mandelbrot set, seen best on videotape (e.g., Peitgen *et al.*, 1990), is an unforgettable experience. As the camera zooms in and magnifies the boundary, sprouts and tendrils seem to grow out from it, which, upon further magnification, dissolve into a multitude of shapes – spirals within spirals, seahorses and whirlpools, repeating the same patterns over and over again (see Figure 6.18). At each scale of this fantastic journey – in

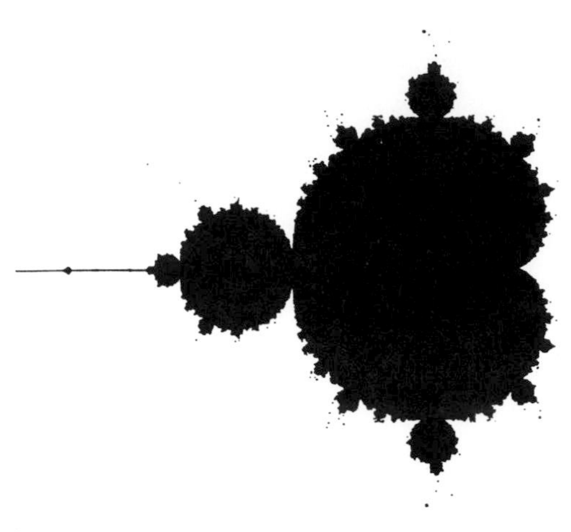

Figure 6.17 The Mandelbrot set (from Peitgen and Richter, 1986).

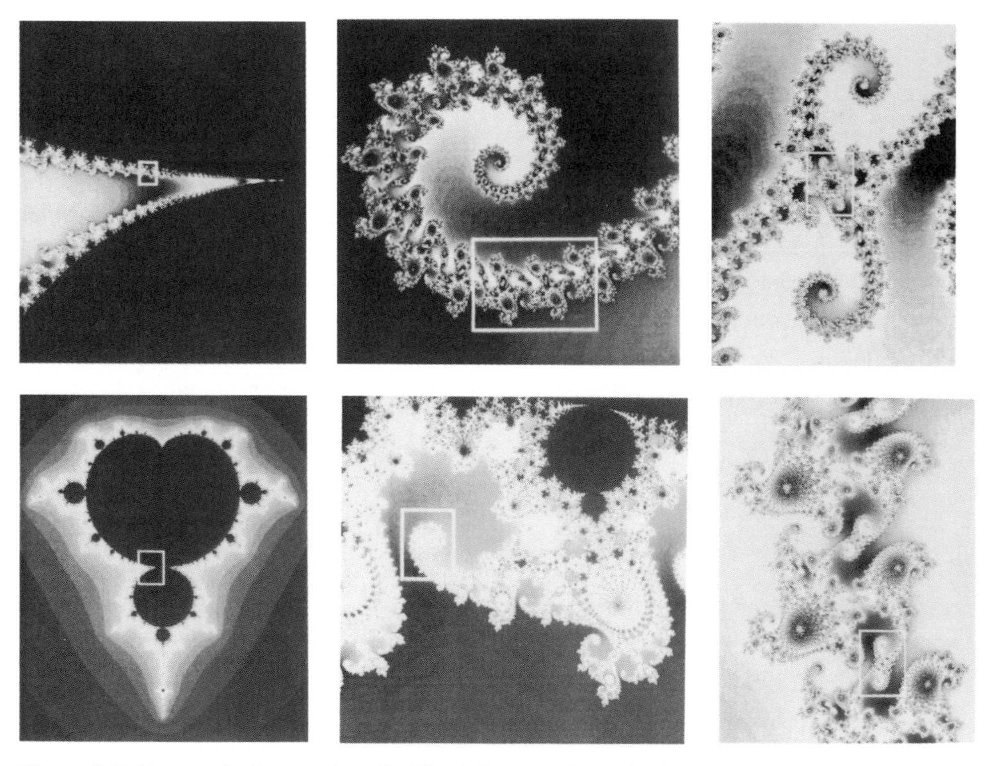

Figure 6.18 Stages of a journey into the Mandelbrot set. In each picture the area of the subsequent magnification is marked with a white rectangle (from Peitgen and Richter, 1986).

which present-day computer power can produce magnifications up to 100 million times! – the picture looks like a richly fragmented coast, but featuring forms that look organic in their never-ending complexity. And every now and then, we make an eerie discovery – a tiny replica of the whole Mandelbrot set buried deep inside its boundary structure.

The Mandelbrot set is a storehouse of patterns of infinite detail and variations. Strictly speaking, it is not self-similar because it not only repeats the same patterns over and over again, including small replicas of the entire set, but also contains elements from an infinite number of Julia sets. It is thus a "superfractal" of inconceivable complexity.

And yet this structure, whose richness defies the human imagination, is generated by a few very simple rules. Thus fractal geometry, like chaos theory, has forced scientists and mathematicians to re-examine the very concept of complexity. In classical mathematics, simple formulas correspond to simple shapes, complicated formulas to complicated shapes. In nonlinear dynamics the situation is dramatically different. Simple equations may generate enormously complex strange attractors, and simple rules of iteration give rise to structures more complicated than we can even imagine.

6.5 Concluding remarks

Since the Mandelbrot set appeared on the cover of *Scientific American* in August 1985, hundreds of computer enthusiasts have used the iterative program published in that issue to undertake their own journeys into the set on their home computers. Vivid colors have been added to the patterns discovered on those journeys, and the resulting pictures have been published in numerous books and shown in exhibitions of computer art around the world.

Mandelbrot saw the tremendous interest in fractal geometry outside the mathematics community as a very healthy development. He hoped that it would end the isolation of mathematics from other human activities and the consequent widespread ignorance of mathematical language even among otherwise highly educated people.

This isolation of mathematics is a striking sign of our intellectual fragmentation and as such is a relatively recent phenomenon. Throughout the centuries many of the great mathematicians have made outstanding contributions to other fields as well. In the eleventh century, the Persian poet Omar Khayyám, who is world-renowned as the author of the *Rubaiyat*, also wrote a pioneering book on algebra and served as the official astronomer at the caliph's court. Descartes, the founder of modern philosophy, was a brilliant mathematician and also practiced medicine. Both the inventors of the differential calculus, Newton and Leibniz, were active in many fields besides mathematics. Newton was a "natural philosopher" who made fundamental contributions to virtually all branches of science that were known at his time, in addition to studying alchemy, theology, and history. Leibniz is primarily known as a philosopher, but he was also the founder of symbolic logic and was active as a diplomat and historian during most of his life. The great mathematician Gauss was also a physicist and astronomer, and he invented several useful instruments, including the electric telegraph.

These examples, to which dozens more could be added, show that throughout our intellectual history mathematics was never separated from other areas of human knowledge and activity. In the twentieth century, however, increasing reductionism, fragmentation, and specialization led to an extreme isolation of mathematics, even within the scientific community. The great fascination exerted by chaos theory and fractal geometry on people in all disciplines – from scientists to managers to artists – may indeed be a hopeful sign that mathematics may be liberated from its recent isolation. Today, complexity theory is making more and more people realize that mathematics is much more than dry formulas, that the understanding of pattern is crucial to understand the living world around us, and that all questions of pattern, order, and complexity are essentially mathematical.

III

A new conception of life

7
What is life?

7.1 How to characterize the living

It is a common understanding that it is impossible to provide a scientific definition of life which is universally accepted. This stems from the fact that the background of scientists dealing with the question – biologists, chemists, computer scientists, philosophers, astrobiologists, engineers, theologians, social scientists, ecologists (just to cite a few) – differs considerably from one another, depending on one's conceptual framework. In this book, we will not dwell so much on the question of a unique definition of life – a single sentence catching all the various aspects of life – but rather, we will consider the more general question: what are the essential characteristics of a living system? This task is more amenable to a scientific inquiry, and we will show that the systems view of life represents a step forward within the horizon of the life sciences. In doing that, we will rely in good part on the conceptual scheme of the autopoiesis theory, as developed by Humberto Maturana and Francisco Varela (1980/1972, 1980). These are two Chilean biologists whose school is often referred to as the "Santiago school." Maturana is the senior scientist, and Varela was his student and later his colleague at the University of Santiago de Chile. Francisco, to whom we have dedicated this book, died prematurely in 2001; Humberto is presently still very active in Santiago.

Maturana had become famous already in the early 1960s for his work on the frog retina, which was the seed for his later work on visual perception and cognition. Both scientists are famous mostly for the theory of autopoiesis, which responds to the general and ambitious question "What is life?" by specifying how to characterize the living organism from a merely biological and phenomenological point of view, starting from the uniqueness of the biological cell. "Autopoiesis" is a term coined by Maturana and Varela in the 1970s. *Auto*, of course, means "self" and refers to the autonomy of self-organizing systems; and *poiesis* (which shares the same Greek root as the word "poetry") means "making." So, *autopoiesis* means "self-making".

According to Maturana and Varela, the main characteristic of life is self-maintenance due to the internal networking of a chemical system that continuously reproduces itself within a boundary of its own making. For the two authors of the Santiago school, together with the question "What is life?", there is another important concomitant question: "What is cognition [the processs of knowing]?" As used by Maturana and Varela, cognition is inseparable from autopoiesis, as we shall explain later on in this chapter.

7.2 The systems view of life

What does the term "systems view" mean when it is applied to life? It implies looking at a living organism in the totality of its mutual interactions. To make things clearer under this respect, let us look at the simplest possible living system, a unicellular organism.

We will do so by using a phenomenological approach – namely, one based on observations made at the level of our experience. Purposely, we do not want to start with a theoretical framework about information theory, or negative entropy, or any other theoretical a priori construct. Let us just observe the life of a simple microorganism as it is. The term "simple" becomes dubious, however, as soon as we observe the biochemical complexity of the metabolic network of one simple bacterium.

This is illustrated in Figure 7.1. Each point in the figure represents a chemical compound, each line represents a chemical reaction, and each chemical reaction is catalyzed by a specific enzyme, so that we are dealing with a three-dimensional network of extreme complexity. Every perturbation in any one point can affect, in principle, the entire system. What is not apparent from the graph of this metabolic network is the cellular compartmentalization and its implication. Let us then transform this complexity into a more manageable cellular concept. This can be represented by the simple sketch of Figure 7.2.

The sketch shows the spherical semipermeable membrane, which discriminates the inside world from the outside world, and in this way identifies the "self." Inside the compartment, many reactions and corresponding transformations take place. We can learn a lot from this drawing. In particular, there are four main phenomenological observations that we can make, and whose complementarity gives an answer to our question – "What is life?" – at the biological level.

7.2.1 Self-maintenance

The first observation stemming from the simple sketch of Figure 7.2 is the apparent contradiction between the internal changes and the constancy of the overall structure. In other words, there are very many transformations continuously taking place; however, there is cellular self-maintenance – the fact that the cell maintains its individuality. A yeast cell remains a yeast cell, a liver cell remains a liver cell, in the sense that the average concentration of the cellular components, as well as the overall structure, remain the same during the whole homeostatic period, the state of dynamic balance characteristic of the normal life of an individual cell. Actually, we could say, with the proponents of the autopoiesis theory (Maturana and Varela, 1980/1972, 1980) that the cell's main function is to maintain its own individuality despite the myriad of chemical transformations taking place in it.

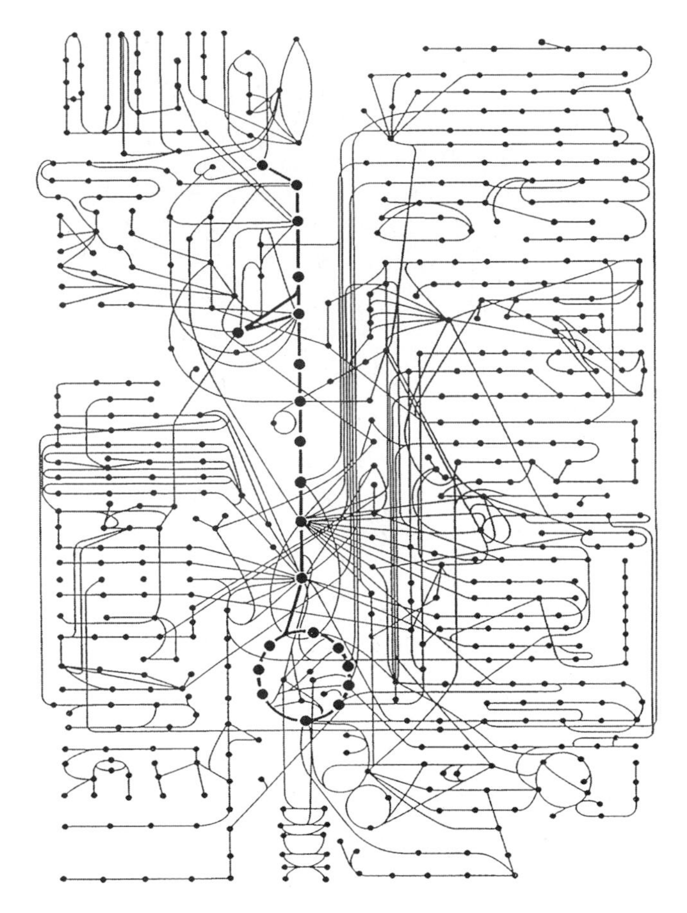

Figure 7.1 A section of the metabolic network of a "simple" bacterium. Note that each point (each chemical compound) is linked to any other point via the complexity of the network.

This apparent contradiction between change and constancy is explained by the fact that the cell regenerates from within the components that are consumed away – be they ATP (adenosine triphosphate) or glycogen, glucose or transfer-RNA. This, of course, takes place at the expense of nutrients and energy flowing inside the cell – a point that we will consider more in detail later.

What we have said for the microorganism is also valid for an elephant. Here, too, the phenomenological observation is the self-maintenance, despite the myriads of transformations taking place at all levels within the elephant. This can be illustrated with the "game of the two lists" (Luisi, 2006). Look at the following list of living things (presented by examples on the left) *vis-à-vis* a list of nonliving things (on the right):

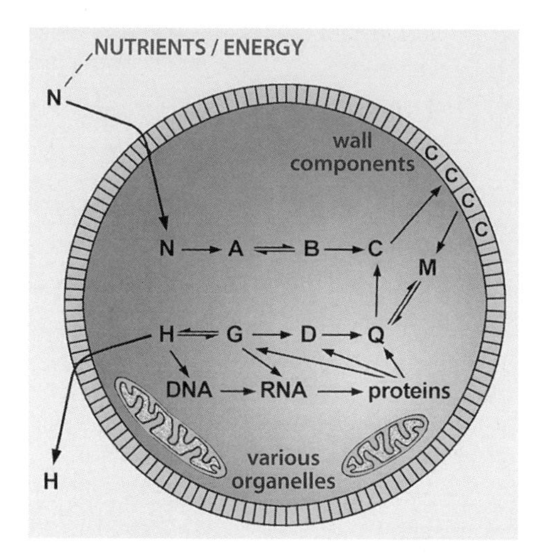

Figure 7.2 Schematization of the working of a cell as an open system. One important feature is the boundary, which is created by the internal network of reactions (a boundary of its own making). The network of reactions brings about a large series of transformations; however, under homeostatic conditions all material that disappears is generated again by the internal machinery. Thus, the cell (and by inference, life) can be seen as a factory concerned with self-maintenance.

List of living things	List of nonliving things
a fly	a radio
a tree	an automobile
a mute	a robot
a baby	a crystal
a mushroom	the Moon
an amoeba	a computer

Now ask the question: what, beyond the great biodiversity of the organisms in the left list, is their common denominator? Something which cannot be present in any of the elements in the nonliving list? The answer is: self-maintenance via a mechanism of self-regeneration from within. Life is a factory that makes itself from within.

7.2.2 Nonlocalization

Still looking at the graphs in Figures 7.1 and 7.2, we consider now the question: where is cellular life localized? Is there a particular reaction, a particular magical spot, where we can put a tag to say: here is life? There is an obvious and very important answer to this question: life is not localized; life is a global property, arising from the collective interactions of the molecular species in the cell.

This is true not only for the simple cell, but for any other macroscopic form of life. Where is the life of an elephant, or of a given person, localized? Again, there is no localization; the life of any large mammal is the organized, integrated interaction of heart, kidneys, lungs, brain, arteries and veins. And each of these organs, which are connected in a network, can be seen in turn as a network of several different tissues and specialized organelles; and each tissue and each organelle is the networking of many different kinds of cells – each cell being then the basic network illustrated in Figures 7.1 and 7.2.

7.2.3 Emergent properties

None of the single molecular species involved in the networking shown in Figures 7.1 and 7.2 are *per se* living. Life, then, is an emergent property – a property that is not present in the parts and originates only when the parts are assembled together. Emergence, in the most classic interpretation, means in fact the arising of novel properties in an ensemble, novel in the sense that they are not present in the constituent parts (see Section 4.1.2). This is a notion which had arisen already in the mid nineteenth century with the school of "British emergentism" (Bain, 1870; Mill, 1872), in an inquiry which continued throughout the twentieth century (see Alexander, 1920; Broad, 1925; Kim, 1984; McLaughlin, 1992; Schröder, 1998; Sperry, 1986; Wimsatt, 1972, and many others). We will come back to the notion of emergence, as part of the larger phenomenon of self-organization, in the next chapter of this book. In order to indicate here something of philosophical flavor, notice the difference between the "emergentist" view and the simple reductionist view.

As we explained in Chapter 2, reductionism in its most classic interpretation means that the whole can be reduced to its components, a view that we can accept if this is restricted to the physical structure. Yes, a cell is composed of a large ensemble of molecules. However, the fact that the parts *compose* the *structure* of a living cell, does not imply that the properties of life can be reduced to those of the single components. The properties of life are emergent properties which cannot be reduced to the properties of the components. The difference between structure and properties is fundamental at this level: reductionism, then, is fine when it limits itself to structure and composition. Emergence assumes its real value at the level of properties, and its very notion is based upon the proposition that the emergent *properties* cannot be reduced to the *properties of the parts*. This is a somewhat subtle point: on the one hand, we are saying that biological life is chemistry only; on the other hand, we also state that the arising of life as a *property* cannot be reduced to the properties of the single chemical components.

7.2.4 Interaction with the environment

The cell, like any living organism, does not need any information from the environment to be itself: all information needed for a fly to be a fly is contained inside the fly, and the

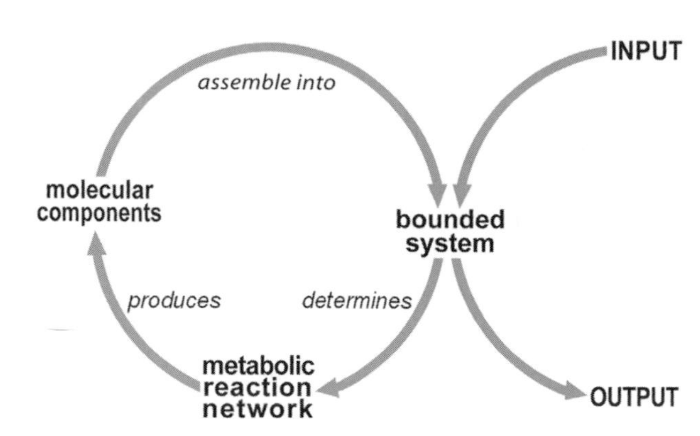

Figure 7.3 The cyclical logic of cellular life. The cell, an autopoietic unit, is an organized, bounded system that determines a network of reactions that produces molecular components that are assembled into the organized system that determines the reaction network that . . . and so on. The terms "input" and "output" – in observance of the fact that the cell is an open system – represent respectively the incoming of nutrients and energy from the outside, and the outgoing of waste products. The circularity illustrated in the figure corresponds to the notion of operational closure, giving rise to the broader notion of biological autonomy.

same is true for the elephant. In the language of epistemology, we say that the cell, and by inference every living organism, is an *operationally closed* system.

This illustrates another apparent contradiction of the living thing: it does not need any information from outside to be what it is, but it is strictly dependent on outside materials in order to survive. This means that we are not in a situation of static equilibrium. In somewhat more precise language, we can say that the cell – the living being – is a *thermodynamically open system:* the living being needs nutrients and energy, and these acquisitions are parts of its own life. According to Maturana and Varela (1980, 1998), the organism interacts with the environment in a "cognitive" way whereby the organism "creates" its own environment and the environment permits the actualization of the organism.

We need to dwell a little more on this concept of interaction between the living being and the environment, and in order to do it properly, we have to start by briefly reviewing the basic notion of autopoiesis.

7.3 The fundamentals of autopoiesis

An autopoietic unit is the most elementary organization of the organism. It can be defined as a system capable of sustaining itself due to a network of reactions which continuously regenerate the components – and this from within a boundary "of its own making." We can say, in other words, that the product of an autopoietic system is its own self-organization. We can also say that this scheme corresponds to a cyclical logic, the cyclical logic of the self (Luisi, 1997, 2006; Maturana and Varela, 1980, 1998; Varela, 2000; Varela *et al.*, 1974). This can be schematized as shown in Figure 7.3.

Again, what is valid for cellular life can be considered valid for any form of life. The primary literature distinguishes between first-order and second-order (multicellular) autopoietic systems. Thus, an organ like the heart can be seen as an autopoietic system, as it is capable of self-sustainment through a series of processes which regenerate all components within its own boundary. On the other hand, this complex autopoietic system is composed of smaller autopoietic units, down to the single cells of various kinds; and the entire human being can also be seen as an autopoietic system (Luisi *et al.*, 1996). For us, here it is important to see the relation to the systems view of life: we can now say that life, more precisely, may be seen as a system of interlocked autopoietic systems.

The relation between autopoiesis, operational closure, circular logic, and biological autonomy is also important. Autopoiesis is the particular self-organization of life that specifies the processes which, within a circular logic, permit the regeneration of the components. The notion of biological autonomy then also means that the living organism is an operationally closed system with a circular logic. Figure 7.3 combines all these notions.

7.4 The interaction with the environment

We have mentioned that "What is life?" and "What is cognition?" are the two main questions in the agenda of the Santiago school. As we have already anticipated with Figure 7.3, the living organism must necessarily be considered in relation to its environment, and it is now proper to consider the aspects of this interaction in more detail. This will lead us to the three important notions of *structural coupling*, *cognition*, and *structural determinism.*

Let us start by citing Maturana and Varela (1998, p. 95):

> Consider first that we have distinguished the living system as a unity from its background as a definitive organization . . . On the other hand, the environment appears to have a structural dynamics of its own, operationally distinct from the living being . . . Between these two systems there is a necessary structural congruence, but the perturbations of the environment do not determine what happens in the living – rather, it is the structure of the living being that determines what occurs in it. In other words, the disturbing agent brings about a change simply as a trigger, but the change is determined by the structure of the disturbed system. The same holds true for the environment: the living being is the source of perturbation and not of instructions . . . We can deal only with unities which are structurally determined.

7.4.1 Structural coupling

At this level the authors also introduce the notion of "structural coupling": some of these interactions will take place on a recurrent or more stable basis. According to the theory of autopoiesis, a living system relates to its environment *structurally* – that is, through recurrent interactions, each of which triggers structural changes in the system. For example, a cell membrane continually incorporates substances from its environment; an organism's nervous system changes its connectivity with every sensory perception.

7.4.2 Structural determinism

Structural coupling, as defined by Maturana and Varela, establishes a clear difference between the ways living and nonliving systems interact with their environments. For example, if you kick a stone, it will *react* to the kick according to a linear chain of cause and effect, and its behavior can be calculated by applying the basic laws of Newtonian mechanics. If you kick a dog, the dog will *respond* with structural changes according to its own nature and nonlinear pattern of organization – the resulting behavior is generally unpredictable.

Note that to say "structurally determined" does not mean to say "predictable," as the above example of the dog makes clear. This is another important point: the nonpredictability means that the ontogenetic (i.e., developmental) structural changes of a living being in an environment always occur as a structural "drift" congruent with the structural drift of the environment. In this way, by linking the notion of structural coupling with the notion of structural drift, we arrived at the basic mechanisms of evolution. For example, adaptation – the compatibility of the organism with its environment – is a term that is correlated to the structural coupling.

As it keeps interacting with its environment, a living organism will undergo a sequence of structural changes, and over time it will form its own, individual pathway of structural coupling. At any point on this pathway, the structure of the organism is a record of previous structural changes and thus of previous interactions. In other words, all living beings have a history. Living structure is always a record of prior development.

Now, since an organism's structure at any point in its development is a record of previous structural changes, and since each structural change influences the organism's future behavior, this implies that the behavior of the living organism is dictated by its structure. In Maturana's terminology, the behavior of living systems is structurally determined.

This notion of structural determinism sheds new light on the age-old philosophical debate about freedom and determinism. According to Maturana, the behavior of a living organism is determined. However, rather than being determined by outside forces, it is determined by the organism's own structure – a structure formed by a succession of autonomous structural changes. Hence, the behavior of the living organism is both determined and free.

Maturana and Varela emphasize this concept also at the level of the nervous system, suggesting that the operation of the nervous system as part of an organism is structurally determined. Again, the structure of the environment cannot specify changes; it can only trigger them. It follows that the nervous system should also be seen an operationally closed system, a concept very dear to Maturana (Maturana and Poerkson, 2004).

7.5 Social autopoiesis

In addition to representing the blueprint of life at many biological levels, the concept of autopoiesis has enjoyed considerable success also in the social sciences. Its extension to the social domain is not straightforward, however, since human social systems exist not only in the physical domain but also in a symbolic social domain. While behavior in the physical domain is governed by the "laws of nature," behavior in the social domain is governed by rules generated by the social system itself.

Thus, two questions arise: is it meaningful to apply the concept of autopoiesis to these domains at all, and, if so, to which domain should it be applied? There have been lively discussions of these issues among biologists and social scientists (Luhmann, 1984; Luisi, 2006; Mingers, 1992, 1997). We shall summarize them in Chapter 14, and we shall discuss in particular the concept of social autopoiesis developed by the sociologist Niklas Luhmann, who defines autopoietic networks in the social domain as networks of communications.

From these generalizations emerges the important insight that social networks exhibit the same general principles as biological networks. There is an organized ensemble with internal rules that generates both the network itself and its boundary (a physical boundary in biological networks, and a cultural boundary in social networks). Each social system – a political party, a business organization, a city, or a school – is characterized by the need to sustain itself in a stable but dynamic mode, permitting new members, materials, or ideas to enter the structure and become part of the system. These newly entered elements will generally be transformed by the internal organization (i.e., the rules) of the system.

The observation that the "bio-logic," or pattern of organization, of a simple cell is the same as that of an entire social structure is highly nontrivial. It suggests a fundamental unity of life, and hence also the need to study and understand all living structures from such a unifying perspective.

7.6 Criteria of autopoiesis, criteria of life

The circularity expressed in Figure 7.4 is valid in the limited interval of homeostasis, life *hic et nunc,* as we said before. In a longer range, there are two obvious disturbances. One, at the level of individual life, is aging; the other, at the level of the progression of generations in time, is evolution. These are the two irreversible arrows of time, and each one has its own characteristic features. Evolution will be discussed in great detail in Chapters 8–10, so let us now turn to aging. The aging of an autopoietic organism is interesting in one respect: the overall organization of the living being does not change, but some structural features do. This goes well with the general predicament of the theory of Maturana and Varela, who emphasize that in the autopoietic mechanism there is an *invariant property*, and this is the autopoietic self-organization of cyclic production of components and systems processes that make such components (see Figure 7.4); and then there is a *variable property*, which is the actual structure that can vary from cell to cell, and from one individual to another, depending upon the actual structure of the cell organisms and other circumstances, of which aging is one. The end of aging is death, a process by which all molecular components are yielded back to the environment, and used for other purposes.

One important question is whether autopoiesis is equivalent to life – namely, whether it is the necessary and sufficient condition for biological life. The primary literature answers this question affirmatively, while more recent studies (Bitbol and Luisi, 2004; Bourgine

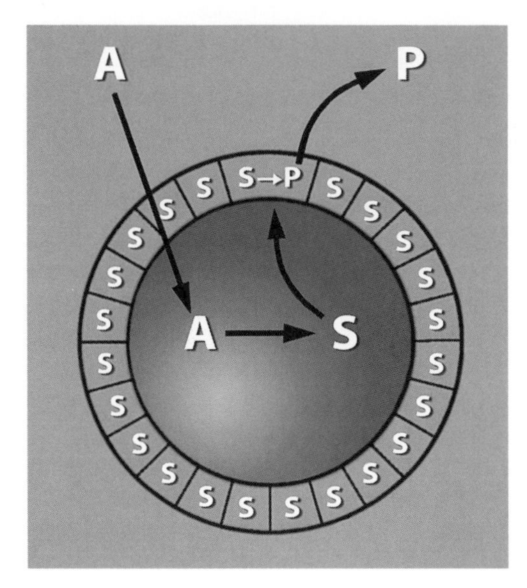

Figure 7.4 Schematization of the three dynamic modes of an autopoietic unit, representing the homeostasis, reproduction, and death of a cell; see text for details.

and Stewart, 2004) have shown that it is more proper to say that autopoiesis is only the necessary condition and not a sufficient one, as a few artificial systems have been found that are autopoietic but not living. In these recent papers, particularly in Bitbol and Luisi (2004) the relation between autopoiesis and metabolism is also discussed. However, for the entire realm of the biological world, the equivalence between autopoiesis and life holds true, and we may safely adopt this generalization. This means that, in order to determine whether a given system is living or not, it will be sufficient to see whether it is autopoietic.

What, then, are the criteria for autopoiesis? Generally, the simplest criterion is to see whether the system is capable of sustaining itself due to self-generating processes taking place within its boundary, the boundary being of its own making. Take a cell: it satisfies these criteria, but a virus does not. A virus alone, in a test tube, is not capable of making its own coat, or the DNA or RNA inside itself; it is not autopoietic, and hence not living. The same holds for a crystal. The case of Gaia, the planet Earth, may be a little more complex. Is the Earth a system capable of sustaining itself by regenerating all components from within? One may tend to answer affirmatively. Does this happen within the boundary, and is the boundary self-made? In fact, what is the boundary of Gaia? These are important questions, to which we shall return in much more detail in Chapter 16.

It is important to emphasize that autopoiesis does not indicate reproduction as a criterion for life. According to the main philosophy of the Santiago school, reproduction is a property of life that can be present, or not, depending upon conditions. Of course, nobody denies that reproduction is the main process for biodiversity and life's unfolding on Earth, but the point should be made that, before talking about reproduction, one must have the "container" and the pattern of self-organization to make reproduction possible, and autopoiesis is this preliminary setup. Within that, one can have reproduction as one of the kinetic modes of the autopoietic being, or not (for example, when the organism is in homeostasis, or sterile, or does not need to reproduce). Take a colony of bacteria which have lost the capability of reproduction, but are self-sustaining and are provided with a normal metabolism. Would one not consider this colony living? Or, take babies or old people who cannot reproduce – are they not living?

7.7 What is death?

We have mentioned in the Introduction that a universal definition of life is very difficult to achieve; and so it may not be surprising that the same is true for the notion of death. There is in fact a rich literature on the issue (for references, see, e.g., the paper by Ramellini, in a recent book edited by Father Alfonso Aguilar, 2009; this book includes some of the most important reflections on death at the scientific, philosophical, and theological levels).

In molecular and structural terms, we can describe death as the disintegration of the autopoietic organization that characterizes life. The essence of life is integration; namely, the linking of the various organs – heart and kidneys, brain and lungs, etc. – with one another. When this mutual linkage disappears, the system is no longer an integrated unity, and death occurs (papers by Damiano and Luisi, in Aguillar, 2009).

However, at the point where the main link disappears, some autopoietic parts may be still operating, being supported by the residual nutrients still present in the body; and when they also disintegrate, individual organelles or, finally, single cells may continue to live for a while, depending on the relative internal supply of nutrients and energy.

Thus, death, seen in this way, is a progressive process, and corresponds to the destruction of the emergent properties of the various levels characterizing the complexity of the entire organism. In fact, in a recent study, the death of a living organism has been described in terms of the concept of "neg-emergence" (paper by Damiano and Luisi in Aguillar, 2009).

The question posed by this kind of reasoning is the following: when should one consider an individual organism as dead? Should one wait until the last cell has disintegrated? Or may the disintegration of the networking of the major organs be sufficient? This brings us to the much debated issue of the criteria for death in humans. In previous times, the arrest of the heartbeat was taken as the most direct and simplest criterion (papers by Gini and Rossi in Aguillar, 2009). More recently, in order to provide a criterion that is more objectively visible, the notion of "cerebral death" has been used: from the inspection of an electroencephalogram (EEG) chart, everybody can see that the person is dead when the EEG is "flat." All this goes back to a report by a committee at Harvard Medical School, published in the summer of 1968 a few months after the first heart transplant operation performed by Christiaan Barnard. This report resulted in the famous Harvard Protocol, which equated human death to a flat brain EEG. The most fundamentalist wing of the Vatican and of Catholic writers appeared to be against, or have serious doubts about, the Harvard definition of death.

It may sound surprising that some voices have spoken openly and rather vehemently against the EEG criterion (see papers by Rossi and Gini in Aguilar, 2009). One main reason for this opposition has to do with the question of organ transplants. The claim is that immediately after the EEG becomes flat, an organ can be removed and taken away when the body is still warm and, for some, has not yet reached the stage of real death.

From the systemic perspective, we can accept the idea that all the main organs have a connection with the brain, and when the brain does not receive any more signals from the organs, and does not send any signal back, their networking is no longer in operation.

It is important to mention that the point of definition of death in humans has become in the last years an acute political issue, being linked to the ethical problems of defining the point at which, for example, artificial nourishment should be stopped in a comatose patient, and to the legislation to regulate if and when. Here again, the problem, from the social point of view, is made difficult by the interference of the established religious authorities, the Vatican in particular, that are defending the position that "the plug" should never be pulled. Those who consider the dignity of life superior to a vegetative, comatose life do not agree, obviously.

Let us now return to autopoiesis, integrating cellular death with the normal kinetic pathway in a very simple way, as indicated in Figure 7.4. In this simplified sketch, cellular life, as well as biological life in general, is represented as a closed autopoietic system characterized by only two chemical reactions: the transformation of an entering nutrient A into an element S, which is part of the autopoietic structure, and the transformation of S into an element P, which is being expelled. (both transformations, $A \rightarrow S$ and $S \rightarrow P$, may represent a series of reactions). The system is in dynamic equilibrium when the rates of these two transformations are equal. If the rate of $A \rightarrow S$ is larger than that of $S \rightarrow P$, then there can be growth of the unit and eventually self-reproduction. And if the rate of decay, $S \rightarrow P$, is greater than the rate of growth, then the unit is destined to die. Thus, the simple sketch of Figure 7.4 represents three different operational moods of a cell, or of any form of life: homeostasis, growth (and eventually reproduction), and death, according to the interplay of only two types of chemical reactions.

The drawing of Figure 7.4 has practical importance: it can give the chemist a hint on how to build an autopoietic unit in the laboratory. In fact, some simple autopoietic systems, based on the self-reproduction of micelles and vesicles, have been realized (Bachmann *et al.*, 1992; Luisi, 2006), thus initiating the field of *chemical autopoiesis*. These considerations are relevant to the general field of cellular models of life, and the attempts to implement the modes represented in Figure 7.4 in the laboratory with chemical structures is leading to possible forms of artificial life and models of cellular life. In fact, this has been partly attempted already (Luisi, 2006; Walde *et al.*, 1994), as we shall discuss in more detail in Chapter 10.

7.8 Autopoiesis and cognition

We should now come back to the interaction of the living organism with the environment. We have said before that the living organism is characterized by biological autonomy but at the same time is strictly dependent upon the external medium for its survival. The interaction with the environment is structurally determined; namely, it is determined by the internal organization of the living organism. In turn, as already mentioned, such a structural determination for each particular organism is due to biological evolution, and in fact we can see the environment and the living organism as co-evolving.

At this point we can introduce the additional qualification that the environment is "created" by the living organism through a series of recursive interactions, which in turn have been produced during mutual co-evolution. The term "create" may sound forced, but it is not. It may be proper in this respect to cite a well-known biologist, Lewontin, who has been working quite outside the realm of autopoiesis. Mentioning that the atmosphere that we all breathe was not on Earth before living organisms, he adds (Lewontin, 1991, p. 109):

> [T]here is no "environment" in some independent and abstract sense. Just as there is no organism without an environment, there is no environment without an organism. Organisms do not experience environments. They create them. They construct their own environments out of the bits and pieces of the physical and biological world, and they do so by their own activities.

The emphasis and overall concern here is not to define cognition in terms of an input from the external world acting on the perceiver, but rather to explain cognition and perception in terms of the internal structure of the organism. In this view, which we shall discuss in more detail in Chapter 12, autopoiesis and cognition are closely linked. The important feature of both is that they represent a general pattern applicable to all levels of life.

Cognition, then, operates at various levels, and as the sophistication of the organism grows, so does its sensorium for the environment, and so does the extent of co-emergence between organism and environment. Thus we go from unicellular to multicellular organisms, where we can have flagella and light- or sugar-sensitive receptors, to the development of sensitive tentacles in the first aquatic organisms, and up to the higher cognitive functions in fish. In all these cases, the organisms contribute to the "creation" of their environments. For example, the onset of photosynthetic organisms may have indeed created a novel, oxygen-rich environment. Similarly, the spider's web, the woody constructions of the beaver, and the cities constructed by mankind modify the structure of their environments. In all these cases, the environment is created by the organism, and this creation permits the existence of the living organism.

At a certain point in the evolution of the sensorium's sophistication, there is the development of a nervous system and with it, eventually, the emergence of consciousness (see Chapter 12). But from the flagella up to the brain, the same basic mechanism is operative: acts of cognition and mutual co-emergence with the familiar environment. In this process of "enaction," a term proposed by Varela (Varela, 2000), or co-emergence, as we can say more generally, the organic living structure and the mechanism of cognition are two facets of the same phenomenon of life (Varela, 2000).

At this point, having recognized that life has no meaning without cognition, we need a way to represent the whole of this complex situation. In a first approximation, this can be done with the illustration shown in Figure 7.5. Here we see life represented not only in terms of the autopoietic unit but also in terms of a trilogy, where the living organic structure

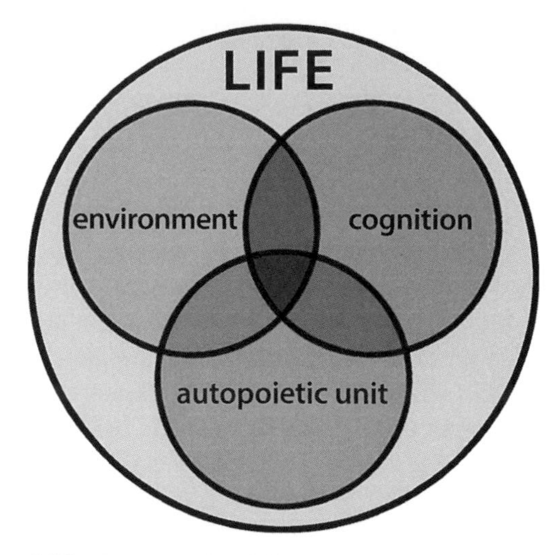

Figure 7.5 The trilogy of life. The organic, living structure interacts with the environment via a cognition sensorium, which is a specific product of its development and evolution. It does not make sense to consider each of these three domains as independent of one another. Life is the synergy of the three domains, as the notion of the "embodied mind" implies.

(autopoietic unit) interacts cognitively with the environment, its process of cognition being a product of evolution. As we said above, each specific organism has its own cognitive sensorium. When one substitutes the common term "mind" for cognition, one arrives at the notion of the "embodied mind."

This notion, proposed by Varela already in the 1990s (Varela *et al.*, 1991), is now widely accepted in cognitive science (see Chapter 12). This means that it makes no sense to talk about mind in an abstract way. Mind is always present in a bodily structure; and, vice versa, a truly living organism must be capable of cognition (the process of knowing). The same holds for human consciousness. Consciousness is not a transcendent entity, but it is always manifest within an organic living structure, as we discuss in more detail in Chapter 12.

In this way, we have described a design that goes from the cell up to the realm of consciousness, while remaining in the realm of biology and without using any transcendent or metaphysical aspects. This entire, wide spectrum appears as a product of immanence – namely, a construction from within. In closing the present section, we should mention that, at the human level, the interactions between organisms and their environment include both interactions among humans (the domain of the social sciences) and interactions with nature (the domain of ecology). Those will be discussed in separate chapters later on in this book.

7.9 Concluding remarks

We have started discussing the question "What is life?" at the level of the simplest microorganisms, but we could show that the same arguments and concepts are valid also for the higher forms of life: the principles underlying life are the same at all levels. This is an important point. To reiterate this, let us look again at Figure 7.1, which shows a complex network representing the life of a microorganism, each point indicating one molecular compound. However, a similar graph may describe the macroscopic life of a mammal if the points indicate the organs that are linked to each other in a three-dimensional array of interactions. And if we consider each point to be a human being, the network may describe a society with people interacting with each other. In all these cases, it is apparent from this kind of picture that the perturbation in one single point can, in principle, be felt in the entire web, as each interaction depends on all the others.

The maze of Figure 7.1 does not express the relations with the environment, relations that we have seen as fundamental for understanding the living organism. The term "environment" can represent quite different things, depending on the levels of life we consider: it can be the milieu in which cells swim, or the habitat where animals live, or the urban environment of humans. In all cases, as in the case of the bio-logic of life, there is a conceptual similarity: the interaction between the living organism and the environment is a dynamic one based on co-emergence, where the living organism and the environment become one through cognitive interactions. Within cognition we have recognized the notions of recursive structural coupling and structural determination, which relate to biological evolution. In this way the "trilogy of life," expressed in Figure 7.5, acquires a dynamic and historical perspective.

8

Order and complexity in the living world

8.1 Self-organization

8.1.1 Introducing the field

Having discussed the systemic conception of life in terms of self-generating, autopoietic networks, we now want to consider the question of life's origin and evolution from the same systemic perspective. Today, there is a broad consensus among biologists that biological evolution was preceded by a molecular, or "prebiotic," type of evolution, in which the transition from nonliving to living matter was brought about by a gradual and spontaneous increase of molecular complexity until the first living cells emerged about 3.5 billion years ago (as we discuss in Chapter 10).

This idea, first stated boldly by the Russian chemist Alexander Oparin in 1924 (see Section 10.1), appears to be at odds with the second law of thermodynamics (discussed in Section 1.2.3) and with the common belief that natural processes preferentially are accompanied by an increase of entropy, or disorder.

However, there are in fact quite a few processes that bring about an increase of molecular complexity in perfect agreement with thermodynamics. The general term for such processes is "self-organization," sometimes also called "self-assembly" (Whitesides and Boncheva, 2002; Whitesides and Grzybowski, 2002).

Some of the chemical processes of self-organization are "under thermodynamic control," which means that they take place "spontaneously" by themselves without the necessity of imposing external forces (the term "spontaneous" is not uncontroversial from a strictly orthodox thermodynamic point of view, but here we will use this term in the common meaning of the word, indicating those reactions that, once given the initial conditions, proceed by themselves).

Another important clarification we need to make right away concerns the difference between static and dynamic aspects of self-organization. Reactions or processes under thermodynamic control usually lead to a final equilibrium situation where all is still, in the sense that the relative concentrations do not change anymore. Self-organization is, however, important also in dynamic systems that operate far from equilibrium, such as the Bénard convection and the Belousov–Zhabotinsky reaction (see Section 8.3.2).

Indeed, the most interesting self-organizing systems, including many centrally important to the life of the cell, are dynamic; that is, they are nonequilibrium systems that form their characteristic order while dissipating entropy. In this chapter we shall deal first with static self-organization, and then pass on to its dynamic aspects.

Self-organization is an extremely broad field of investigation, and in this chapter we will be able to review only a few examples. For a wider review of the literature, see Birdi (1999), Riste and Sherrington (1996), and Westhof and Hardy (2004), as well as the early review by Jantsch (1980).

The phenomenon of emergence must be considered together with self-organization. The term "emergence" refers to the arising of novel properties of the organized structure, novel in the sense that they are not present in the parts or components. Although self-organization and emergence go hand in hand, for heuristic reasons emergence will be discussed separately in this chapter. A particular combination of self-organization and emergence gives rise to self-reproduction, which we shall also discuss separately.

The last point we wish to make in this introduction concerns the qualification of the term "self." In the context of this chapter, and in the field of life science in general, "self" denotes a process that is endogenous – that is, dictated by the "internal rules" of the system. Conversely, when the structure is organized by external, imposing forces, it is not self-organizing.

To give just a few simple and clear examples: the assembly of a TV set, or the pagination of a manuscript, are certainly not processes of self-organization, as they are imposed from outside. Protein folding, the formation of the DNA double helix, the formation of soap bubbles, and crystallization, by contrast, are all examples of self-organization under thermodynamic control.

A subtle point in this field is when we are dealing with complex biological systems, like the assembly of a virus or of a beehive, as well as all other animal constructions. Those are cases in which the organization is the result of genomic and enzymatic activity, and the term "self-organization" is still correct, since we consider the determining genes, or the enzymes, as part of the system itself.

8.1.2 Some basic examples of molecular self-organization

The simplest examples of molecular self-organization are based on molecules, such as those of soaps and the fatty and oily substances known as lipids, that tend to associate with each other. Soaps and lipids are typical "amphiphilic" substances – that is, their molecules contain a hydrophilic (water-loving) head group, and one or more long hydrophobic (water-hating) chains (see Figure 8.1a). Because of this "schizophrenic" character, amphiphilic molecules, when put in water, do not make normal solutions, but instead tend to associate with each other, spontaneously forming ordered structures which, depending upon conditions, can take different forms and names, such as bilayers, micelles, reverse micelles, vesicles, and so on (see Figure 8.1b). The forces which hold them together are usually the hydrophobic forces.

These structures are well known in the literature and are important both in basic science and in applied technology. Generally, they change the properties of water surfaces and are therefore known as "surfactants" (see Box 8.1).

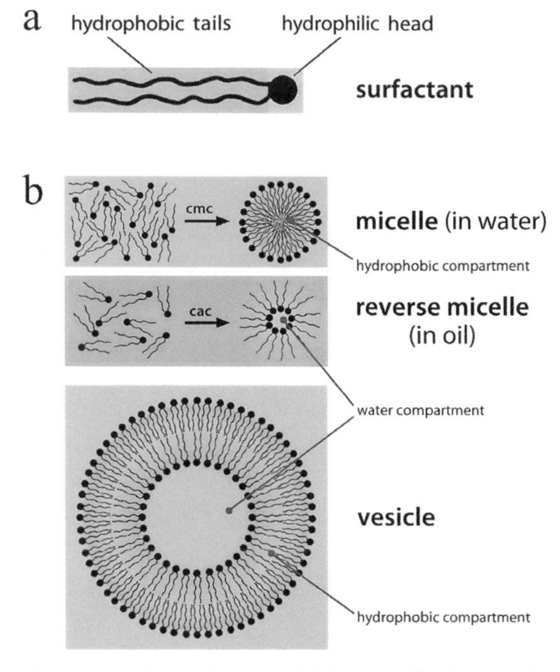

Figure 8.1 (a) Schematic picture of a surfactant with hydrophilic head and two hydrophobic tails. (b) At a critical concentration, called critical micelle concentration (cmc) or generally critical aggregate concentration (cac), soap molecules form spherical micelles or the much larger vesicles (in cross section, and not drawn in scale), giving rise also to a series of emergent properties – e.g., the formation of distinct compartments and collective dynamic movements.

Why do these substances aggregate? The simple sketch of Figure 8.2 explains why. In the upper panel, we see two oil droplets floating on a surface of water. After a while, spontaneously, the droplets come together and fuse to make a single, larger droplet of oil. The driving force is the decrease of the total surface exposed to water. This decrease of surface "liberates" water molecules from an energetically unfavored contact with oil. As a consequence, there is an overall increase of entropy (or disorder) due to the "liberation" of water molecules, which makes the process thermodynamically favorable. There are two remarkable aspects of this process: first, the spontaneous formation of local order, attended by an overall increase of entropy (or disorder), and second, the formation of spherical compartments (see Figure 8.1), which, as we shall see later on, are very important as a model for biological cells.

Now, the same happens with lipids. Lipids, in particular, phospholipids (see Box 8.1), are the main components of all our biological membranes. The oily, hydrophobic parts tend to assemble and form an oily microenvironment, known as a "microphase," by expelling water from their surroundings. Therefore, the formation of membranes, just like the formation of micelles and vesicles (see Figure 8.3), is a thermodynamically favored process. Again,

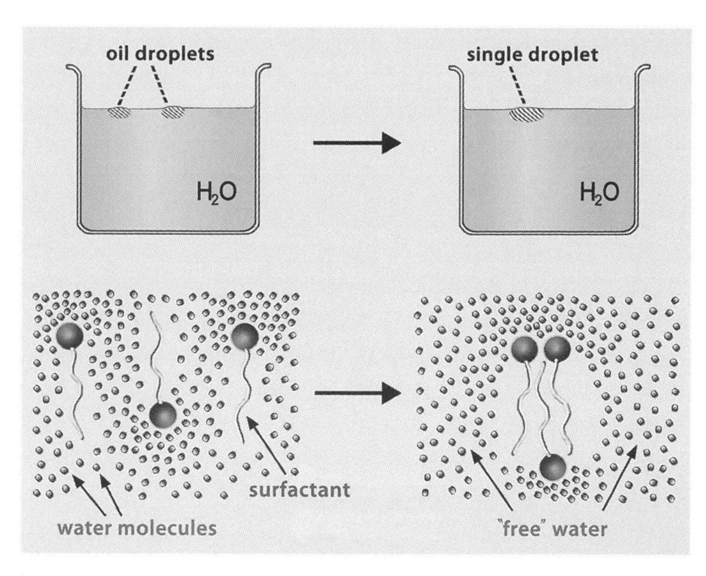

Figure 8.2 In the upper panel, two oil droplets spontaneously come together to form a single droplet having a smaller total surface than the sum of the two, a process which is thermodynamically favored. In the lower panel, the same situation is shown in the case of three (or more) surfactant molecules. Their assembly proceeds with an increase of entropy caused by the expulsion of water molecules from the formed hydrophobic assembly.

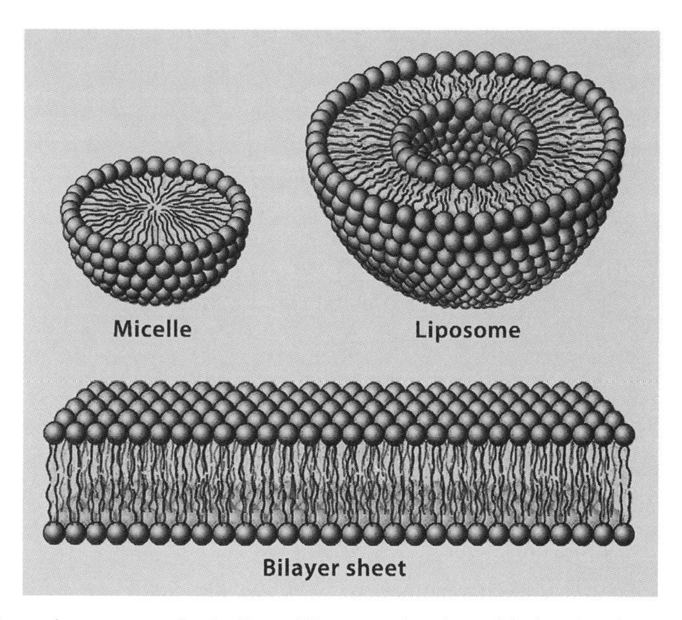

Figure 8.3 Schematic structure of micelle and liposome (not in scale) showing the structural analogy between liposomes and the membrane bilayers. In all cases, water molecules are excluded from the internal hydrophobic microphase.

Box 8.1
Surfactants, lipids, and liposomes

Surfactants are organic compounds that are amphiphilic, meaning they contain both hydrophobic groups (their "tails") and hydrophilic groups (their "heads"). The name is a contraction of "surface active agents." Because of their property of reducing the surface tension of water, surfactants play an important role as soaps and as wetting, dispersing, emulsifying, foaming, and antifoaming agents in many practical applications (annual global production 13 million metric tons in 2008).

Surfactants can be classified as cationic (positive charge in the hydrophilic head), anionic (negative charge), zwitterionic (both a positive and a negative charge), and nonionic (no charges in head). When dispersed in water, surfactants associate to form ordered assemblies, such as bilayers, micelles, or vesicles).

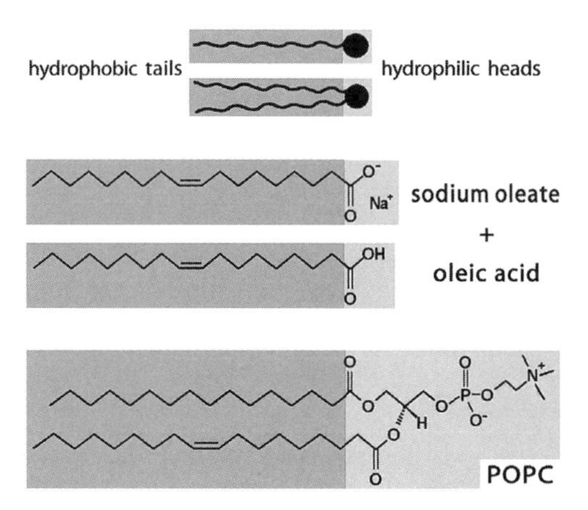

Figure 8.4 A typical surfactant with the indication of the hydrophilic and hydrophobic parts; POPC is the abbreviation of the best-known phospholipid, palmytoyl-oleoyl-phosphatydil choline. The middle panel shows the structure of a simpler surfactant, oleic acid/oleate, of the class of fatty acids, very diffuse in the natural world, and important in prebiotic chemistry (see later on the section on autocatalysis, where fatty acid plays a most important role).

Lipids (from the Greek *lipos* – "fat" or "grease") constitute a broad group of naturally occurring molecules, some of which are surfactants. They include fats, waxes, sterols, fat-soluble vitamins (such as vitamins A, D, E, and K), and glycerolipids (see Figure 8.5). The latter are composed of fatty acids and glycerol, joined by ester bonds. The best known are the so-called triglycerides shown in the figure. The main biological functions of lipids include energy storage as structural components of cell membranes.

Although the term "lipid" is sometimes used as a synonym for fats, fats are a subgroup of lipids that should be called triglycerides. An important class of glycerolipids are glycerophospholipids, also referred to simply as phospholipids (see Figure 8.5), which are ubiquitous in nature and are key components of the lipid bilayers of cells, as well as being involved in cellular metabolism and cell signaling.

Liposomes are vesicles constituted by lipid surfactants, and generally by phospholipids.

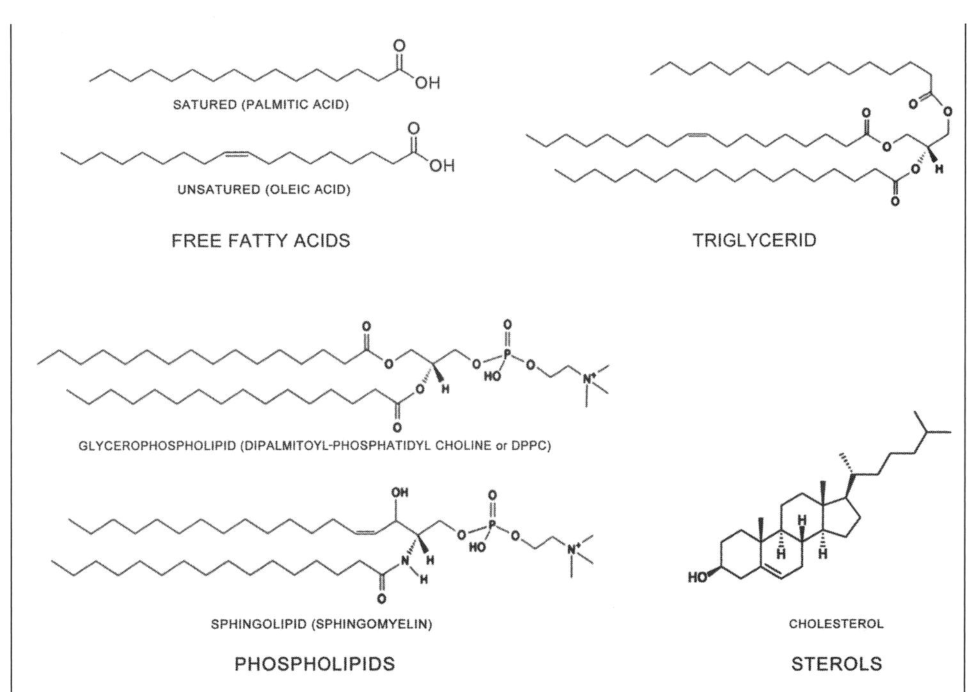

Figure 8.5 Some typical lipids, including the triesters of glycerol with fatty acids (carboxylic acid with a long chain) and a typical phospholipid.

there is an increase of order in spite of an overall increase of entropy. As we shall see in our discussion of the dynamic aspects of self-organization in Section 8.3, such "islands of order in a sea of disorder" are characteristic of the "dissipative structures" of living systems. In our examples of lipids, the whole structure remains compatible with water, because its surface is made by hydrophilic compounds, as shown schematically in Figure 8.3.

Notice in this regard the apparent contradiction between our water-based life, and the existence of so many hydrophobic (water-hating) compounds and substances in our biological structures, like the lipids, steroids, alkaloids, fats, and waxes. Nature solves this problem by the self-organization of these substances, with the formation of microphases that exclude water from their interior, but are made compatible with water by an external layer of hydrophilic head groups, as shown in Figure 8.3.

8.1.3 Self-organization in biological systems

The same principles of organization found in lipids can also be observed in nucleic acids. The two complementary strands of DNA come together to form a duplex in which the hydrophobic complementary bases escape from the contact with water, building a kind of lipid interior and leaving the hydrophilic "fingers" of the phosphate groups to take care of

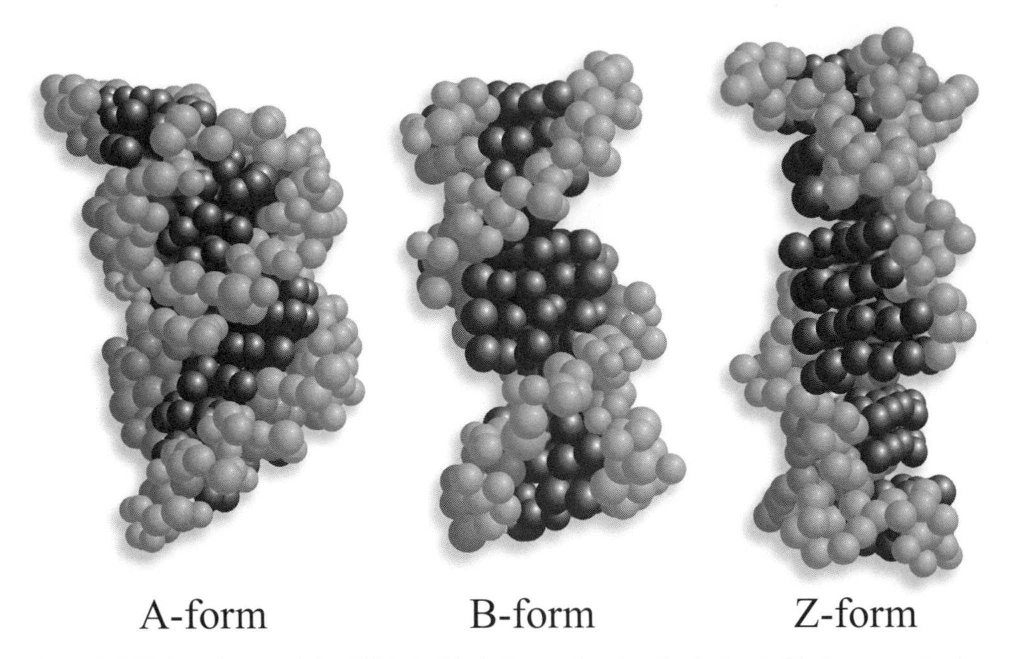

Figure 8.6 Various forms of the DNA double helix; notice that the hydrophobic bases are in the interior of the structure, avoiding as much as possible contact with water, while water solubility is guaranteed by the hydrophilic phosphate groups at the exterior.

the water solubility (see Figure 8.6). Here the internal rule is the complementarity of the A–T and G–C bases.

The folding of proteins can also be considered a self-organization process. The specific folding procedure of the polypeptide chain into the native structure is dictated by the primary sequence of amino acids – the "internal rule" – and the native folding gives rise to a set of specific, novel properties, such as the binding site of an enzyme, the microenvironment that permits catalysis.

It is also important to remark that the native structure of the protein, which is the one displaying the biological activity, is generally the most stable from the thermodynamic point of view. This has been demonstrated by Chris Anfinsen (Anfinsen *et al.*, 1954) from the observation of "reversible denaturation": the native folding can be disrupted with mild reagents – like urea, for example – with complete loss of biological activity. Such a protein is called "denatured." But when urea is removed from the reaction glass, the protein acquires again its native folding *in vitro*, thus demonstrating that the process of folding is under thermodynamic control.

8.1.4 Self-organization and autocatalysis

When surfactant molecules aggregate in water, the process is often slow at the beginning and gets faster with time: the more surface bilayer is formed, the more the process speeds

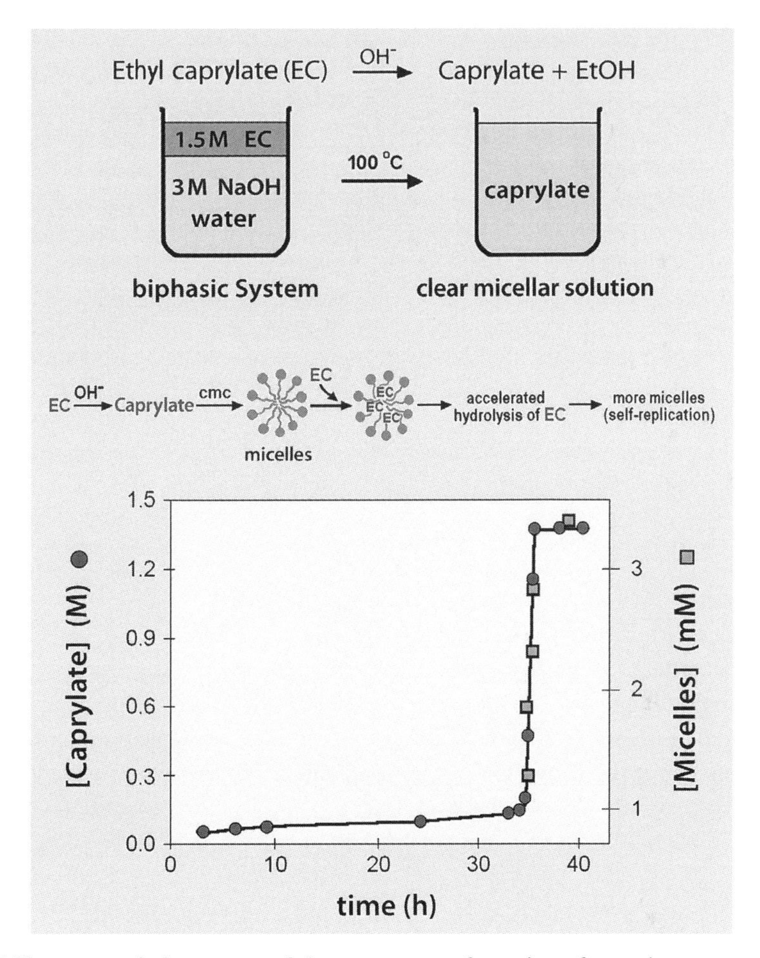

Figure 8.7 The autocatalytic process of the spontaneous formation of caprylate aqueous micelles starting from the insoluble precursor ethyl caprylate (EC). As soon as the first micelles are formed, more EC is solubilized in their core and hydrolyzed, and then more micelles are formed – and the more micelles that are formed, the more will be formed – a typical autocatalytic process (Bachmann *et al.*, 1992).

up, because there is more and more active surface where the next steps of aggregation can take place. In cybernetic terms, positive feedback loops come into play (see Section 5.3.2), and the whole process is known as autocatalysis.

A beautiful example of self-organization and autocatalysis is shown in Figure 8.7. Here, a water-insoluble surfactant precursor, ethyl caprylate (EC)), is overlaid on an aqueous solution with an alkaline pH value that is capable of hydrolyzing EC. The reaction is very slow at the beginning, but then, all of a sudden, it accelerates exponentially.

What has happened? Aqueous micelles can be seen as small, oily droplets, which readily uptake and solubilize water-insoluble substances, like the grease of our hands when we

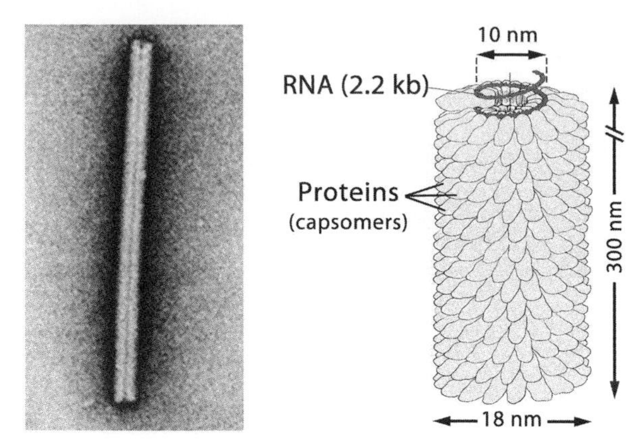

Figure 8.8 Tobacco mosaic virus (TMV), formed by a core of helical RNA and a mantle of identical protein subunits in a helical array.

wash our hands with soap. In the same way, EC is rapidly solubilized in the oily core of the newly formed micelles, and then hydrolyzed inside, and thus the surfactant caprylate is formed *in situ*, and this forms more micelles, which in turn solubilize more and more EC – a typical autocatalytic process. This process has been then extended to vesicles (Walde *et al.*, 1994). Notice that in this way there is the spontaneous self-reproduction of compartments, leading the authors to suggest that this process, which is in principle prebiotic, might have been at the origin of the formation of the first cellular compartments (Bachmann *et al.*, 1992).

8.1.5 Self-organization in complex biological systems

Many other examples of self-organization can be found in our complex biological world. For example, the many cases of protein–protein interaction in the formation of active "oligomeric" complexes are generally processes of self-organization. A well-known example (to be discussed in more detail in Section 8.2.2) is that of hemoglobin, the protein that carries oxygen in mammals.

Another interesting example of self-organization in protein–protein interactions is the structure of the tobacco mosaic virus (Figure 8.8), formed by a helical RNA macromolecule surrounded by thousands of identical globular protein subunits: its assembly is a self-organization process under thermodynamic control. In fact, Fraenkel-Conrat and Williams (1955) demonstrated that the tobacco mosaic virus can be reconstituted reversibly *in vitro* starting from the two main components, the purified RNA and the proteins.

Assemblies of different proteins are also important in the formation of quite particular tissue structures. The example of the muscle proteins is one of the most interesting in this respect. The point here is that this organization permits the functioning of a molecular motor, that of our muscle movement.

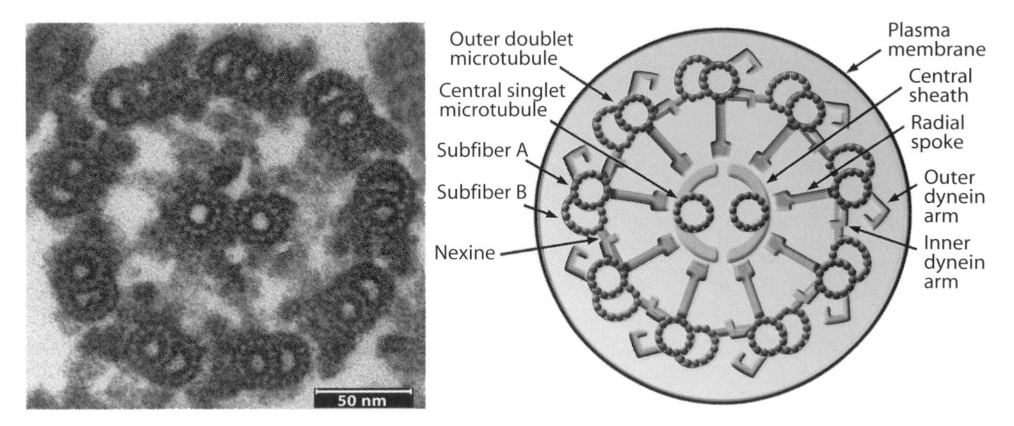

Figure 8.9 Schematic diagram of the cross section of an axoneme, the microtubule-based cytoskeleton of cilia and flagella. Each geometric form represents a different kind of protein with its particular form of assemblage (from Stryer, 1975).

Speaking of molecular motors, let us look at the beautiful object, depicted in Figure 8.9: an axoneme, the inner skeletal core of a bacterial flagellum. This is really a wonderful construction, reminiscent of an Oriental mandala. It involves the coordination of dozens of different, specialized parts, each part being in turn the organized structure of a bundle of specific proteins. The perfect juxtaposition of all these components is the result of a calibrated, sequential ordering, a self-organizing process in which thermodynamic and kinetic factors produce a concerted mechanism of utmost complexity.

The folding of actin and of myosin, as well as several other examples given until now, are cases of processes under *thermodynamic control,* where the product forms because it is more stable than the starting components. Not all self-organization processes in nature are of this type. There are also cases of *kinetic control*, in which the products form because the reaction velocity to arrive at them is much larger than the velocity to arrive at the more stable products. In the language of chemistry, the activation energy barrier (the energy barrier that must be overcome for a chemical reaction to occur) is much lower than the energy barrier to arrive at the thermodynamically more stable forms. This is why the products under kinetic control form preferentially.

In the biological world, such self-organization under kinetic control is determined by the action of specific enzymes. At the origin of life (see Chapter 10), when enzymes did not yet exist, simple peptides or metal catalysts most probably permitted the catalytic formation of compounds that were not particularly stable from the thermodynamic point of view. This is quite an interesting field in the study of prebiotic chemistry, on which we cannot dwell here. But we would like to mention that it has been recently discovered that certain simple dipeptides can catalyze the formation of the peptide bond – that is, they permit the coupling of amino acids with each other (Gorlero *et al.*, 2009).

The study of the self-organization of complex biological systems raises quite a few intriguing questions, such as whether and to what extent ribosomes (the cellular structures where proteins are synthesized), or even the biological cell itself, can be reconstituted from their components. A positive answer would mean that we are dealing here with self-organization processes under thermodynamic control.

It now appears that the reconstitution of ribosomes from their components (proteins and RNA) is indeed possible (Alberts *et al.*, 1989). But can the cell be reconstituted?

One finds in the literature several attempts to reconstitute cells. Examples are the reassembling of *Amoeba proteus* (Jeong *et al.*, 2000), and of *Acetabularia mediterranea,* a green alga that was reassembled from the cell wall, nucleus, and cytoplasm (Pressman *et al.*, 1973). Kobayashi and Kanaizuka (1977) reassembled *Bryopsis maxima* from its two dissociated components. In addition, there were mammal experiments with mouse cells (Veomett *et al.*, 1974). This experiment was also reported as successful.

However, looking at the protocol of these and similar cellular reconstitution experiments, one finds that the reassembly is not taking place spontaneously, but, at least in one or two critical steps, is "guided" by the help of micromanipulations; for example, specific enzymes and reagents are added manually during the course of the reconstitution process. Thus, the notion of "self" is not really respected completely.

These experiments, however, clearly indicate that cellular life can be reconstituted from nonliving components, even though the cellular reconstitutions are based on parts which are already very large and complex structures. Each of these parts is the product of complex enzymatic reactions, all under kinetic control. Thus, it is safe to state that the reconstitution of a cell from its single molecular components is not possible. It is not a process under thermodynamic control; furthermore, the construction of a cell in nature has to follow a sequential pathway where the parts are synthesized and assembled one after the other in a precise order, all under kinetic control.

8.2 Emergence and emergent properties

8.2.1 Introducing the main issues

In the previous pages we have seen the importance of self-organization for the assembly of large polymer structures (see Box 9.4) by living organisms. Obviously, a simple increase of size or complexity is not enough for life to evolve from nonliving matter. It must be accompanied by the onset of new functions and novel properties. In fact, as we have already noted in Section 8.1.1, self-organization should always be considered in conjunction with the corresponding novel properties that arise as a consequence of the assembly of the smaller components. And here we encounter the notion of emergence, or emergent properties, one of the most important concepts in the modern theory of complexity and, more generally, in the systemic conception of life.

Emergent properties are the novel properties that arise when a higher level of complexity is reached by putting together components of lower complexity. The properties are novel in the sense that they are not present in the parts: they *emerge* from the specific relationships and interactions among the parts in the organized ensemble. The early systems thinkers expressed this fact in the celebrated phrase, "The whole is more than the sum of its parts" (see Section 4.1).

The study of emergence has been an active field of inquiry in the philosophy of science for a long time. In fact, "British emergentism" can be dated back to Mill (1872) and Bain (1870); and it has continued until the present (see, e.g., Bedau and Humphreys, 2007; Primas, 1998; Schröder, 1998; Sperry, 1986; Wimsatt, 1976).

In our time, emergence is being considered not only in chemistry and biology but also in quite a variety of other research fields, such as cybernetics, artificial intelligence, nonlinear dynamics, information theory, social science, and the theory of music (the harmony arising from a musical phrase is obviously not present in the single notes). Applications to language, painting, memory, biological evolution, and the nervous system are mentioned and discussed by Farre and Oksala (1998). A useful insight into physics, including superconductivity and other collective phenomena, is provided by Coleman (2007). Emergence at the human level, including ethics and religious naturalism, is discussed by Goodenough and Deacon (2006). In view of this great variety of studies, it is not surprising that the term "emergence" often has confusing connotations.

8.2.2 A few examples

Let us begin with emergence in simple systems. Consider again, for example, the formation of micelles and vesicles from surfactants pictured in Figure 8.1. The compartmentation, as well as the collective motions of the surfaces, are emergent properties, and so are novel physical properties, such as the lower acidity of the fatty acids constituting the assembly. Similarly, the biological membranes shown in Figure 8.3 display a series of collective properties that are not present in the single lipid molecules. Finally, Figure 8.7 illustrates a very remarkable emergent property of surfactant aggregates, the capability of self-reproduction.

Emergent properties can already be observed in geometry, where a line can be seen as emerging from the motion of a point, a two-dimensional surface from the motion of a line, and so on. At each level of complexity, novel, emergent properties arise – angles, surface, volume, etc. – which are not present at the lower level.

Turning to a more complex chemical structure, let us consider the proteins myoglobin (Mb) and hemoglobin (Hb), pictured in Figure 8.10. As is well known, those are the proteins responsible for the transport and storage of oxygen in mammals and even in certain invertebrates. In mammals, Mb is localized in muscle tissues and does not circulate, whereas Hb circulates, transporting oxygen from the lungs to the muscles and returning charged with CO_2.

To begin with, the specific ability to bind oxygen is an emergent property of the entire molecular ensembles in both proteins. This binding property is usually measured in terms

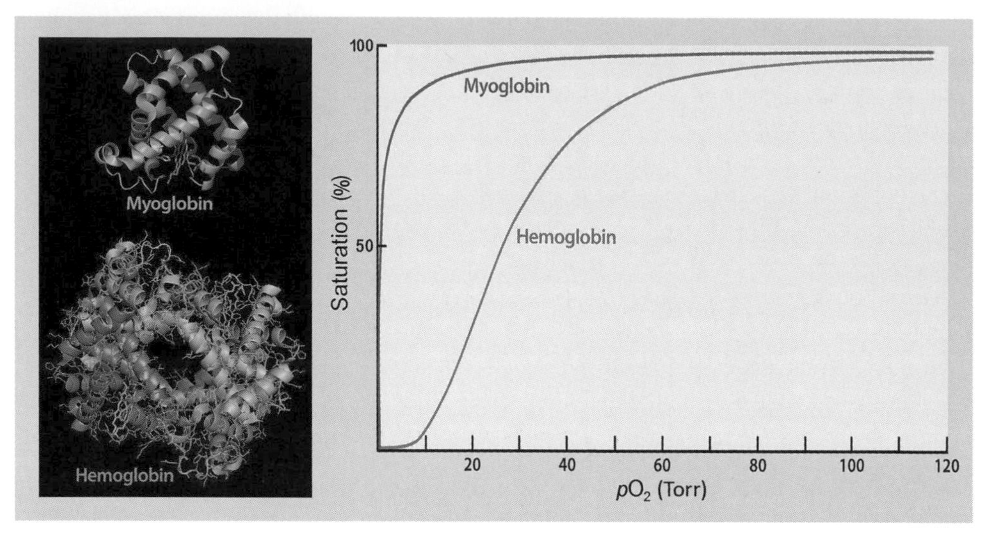

Figure 8.10 The single chain of myoglobin and the four chains (two α and two β) of hemoglobin. The cooperative behavior of hemoglobin (*S*-shaped curve) can be seen as an emergent property, resulting from the assembly of the four chains (from Stryer, 1975).

of the percentage of the chain being saturated by oxygen, as shown in the lower panel of Figure 8.13. The comparison of this property between Mb and Hb permits us to illustrate an even more striking example of emergence.

Mb is a single polypeptide chain (see Box 9.4), while Hb is formed by four chains, each of them very similar to that of myoglobin. Whereas the binding of oxygen to myoglobin (single chain) gives the usual saturation curve, in the case of mammal hemoglobin the saturation curve is *S*-shaped. This means that affinity for oxygen changes during the very process of binding – it is low when oxygen binds initially, but the binding in the first chain induces structural changes in the neighboring chains so that the affinity becomes better and better. The resulting sigmoid behavior is the very basis of mammal respiration: at the low concentration of oxygen that we have in our muscle tissues, Hb loses its affinity for oxygen, which can thus be up taken by Mb and distributed to the tissue cells. The *S*-shape of the saturation curve for Hb can be viewed as an emergent property arising from the interaction of the four chains.

We can now turn to all examples of self-organization illustrated in Section 8.1.3, and discover in each case the arising of emergent properties. Thus, Figure 8.6 shows the case of the DNA double helix, which acquires the property of self-replication; Figure 8.7 illustrates the folding of proteins, which brings forth binding and catalytic properties.

Figure 8.8, showing the tobacco mosaic virus, is a reminder that the emergent infectious properties of the virus are only present when the two basic components, RNA and the protein mantle, come together. In the case of the muscle structure, as we have already noted, a very particular property emerges: the macroscopic movement of the muscle. And the axoneme of Figure 8.9 is another complex case, specific mechanical properties being acquired by assembling a certain number of protein subunits in a complex arrangement.

At the level of the cell, finally, we make the important discovery, as we have mentioned already, that life itself is an emergent property. In fact, none of the basic constituents of a living cell – nucleic acids, proteins, lipids, polysaccharides, etc. – are alive by themselves. But when they all come together in a particular space/time situation, life – the ultimate emergent property – arises.

However, the notion of emergence goes beyond life at the individual level, extending to living colonies and, in general, to social life. Indeed, the structures created by social insects – beehives, anthills, and so on – emerge at the social level. In a beehive, for example, each bee appears to behave as an independent element, acting apparently on its own account, but the whole population of bees produces a highly sophisticated structure emerging from their collective activities.

The relation between emergent properties and the properties of the basic components has been, and still is, much debated in the literature. One school of thought claims that the properties of the higher hierarchic level are *in principle* not deducible from the components of the lower level. This is the "strong emergence" or "radical emergence" (see, e.g., Schröder, 1998). Opposed to this "strong emergence" is the school of "weak emergence," which asserts more pragmatically that the relationship between the whole and the parts cannot be determined, simply because of technical difficulties, such as the lack of computational power or insufficient progress of our skills, and not as a matter of principle. Atmanspacher and Bishop (2002) discuss this point at length.

The view that the emergent properties of molecules are not explicable as a matter of principle on the basis of the components is opposed by several scientists, who argue that this is tantamount to assuming a mysterious force of some undefined nature – a kind of vitalistic principle. The systems view of life takes a third position, asserting that there is no need to assume any mysterious force to explain emergent properties, but that the focus on relationships, patterns, and underlying processes is essential. Once this is accepted, practical difficulties will still be relevant, and the distinction between strong and weak emergence may not always make sense.

8.2.3 Downward causation

It is generally accepted that the development of emergent properties, which is an upward (or bottom-up) causality, is attended by a downward – or top-down – causality stream. This means that the higher hierarchic level affects the properties of the lower components (see, e.g., Schröder, 1998). Discussion continues in philosophical literature on the relation between emergence and downward causation – also called macrodeterminism – see, e.g., Bedau (1997), Schröder (1998), Thompson and Varela (2001), and Thompson (2007).

Generally, the point can be made that molecular sciences and chemistry in particular offer very clear examples of downward causation, as defined above. In chemistry, any form of chemical reaction modifies the original structural properties of the atomic components. Of course, examples from molecular science are not the only ones to show the effect of downward causation. Consider the progression of social hierarchic levels that go from the individuals to the family, to the tribe, to the nation. It is clear that once the individuals are in a family, the rules of the family affect and change the behavior of the individuals; likewise, belonging to a tribe affects the behavior of the family, and so on.

We shall see in the following chapter (Section 9.6.4) that downward causation provides an important argument (employed eloquently by Noble, 2006) against "genetic determinism," the reductionist view of genetics. As shown in Figure 9.6, the upward stream of emergence goes from the genes to the proteins, and on to the tissues, organs, and the organism – the genes being the primary cause of this upward stream. However, it is the entire organism that determines which proteins should be built and when; it is the downward causation that is the primary source of biological functions and behavior.

8.3 Self-organization and emergence in dynamic systems

8.3.1 Theoretical and historical basis

The chemical and biological examples of static self-organization given above are familiar to all chemists and biologists. As already mentioned in Section 8.1.1, for a growing community of scientists working in the area of complex systems (see Section 6.2.1), self-organization and emergence acquire their full potential in dynamical systems – that is, systems that change over time. The key characteristic of such dynamical systems is that they generally operate far from equilibrium, and yet are capable of producing stable, self-organizing structures. The analysis and mathematical description of these, at first highly counterintuitive, situations is associated, first and foremost, with the physical chemist Ilya Prigogine and his colleagues at the Free University of Brussels.

The crucial breakthrough occurred for Prigogine during the early 1960s, when he realized that systems far from equilibrium must be described by nonlinear equations. The clear recognition of this link between "far from equilibrium" and "nonlinearity" opened an avenue of research for Prigogine that would culminate a decade later in his theory of "dissipative structures" (see Nicolis and Prigogine, 1977; Prigogine, 1980; Prigogine and Glansdorff, 1971; for a nontechnical review, see Prigogine and Stengers (1984); see also Capra, 1996).

In order to solve the puzzle of stability far from equilibrium, Prigogine did not study living systems but turned to much simpler physical and chemical nonequilibrium systems, such as special forms of heat convection, known as "Bénard cells," and special chemical oscillations, now known as "Brusselators," which we review in Section 8.3.2 below. By applying the newly developed mathematics of complexity, or nonlinear dynamics (see

Chapter 6), to these systems, Prigogine and his colleagues were able to develop a new nonlinear thermodynamics to describe the self-organization phenomenon in open systems far from equilibrium. "Classical thermodynamics," Prigogine and Stengers (1984, p. 143) explain, "leads to the concept of 'equilibrium structures' such as crystals. Bénard cells are structures too, but of a quite different nature. That is why we have introduced the notion of 'dissipative structures,' to emphasize the close association, at first paradoxical, in such situations between structure and order on the one side, and dissipation . . . on the other." In classical thermodynamics, the dissipation of energy in heat transfer, friction, etc., is always associated with waste. Prigogine's concept of a dissipative structure introduced a radical change in this view by showing that in open systems dissipation becomes a source of order.

According to Prigogine's theory, dissipative structures not only maintain themselves in a stable state far from equilibrium but may also even evolve. When the flow of energy and matter through them increases, they may go through new instabilities and transform themselves into new emergent structures of increased complexity. In the language of nonlinear dynamics, the system encounters bifurcation points at which it may branch off into entirely new states, each characterized by a specific attractor, where new structures and new forms of order emerge.

This is illustrated in Figure 8.11 with two so-called bifurcation diagrams representing the behavior of the chemical oscillation known as the Brussellator (see Section 8.3.2). The first diagram (a) shows the basic (primary) bifurcation. At concentration C_{eq}, the system is at or near equilibrium and is described by linear equations. As the distance from equilibrium increases, the system becomes increasingly nonlinear, reaching the first bifurcation point at λ_c. The two solid lines represent two possible stable states of the system for $\lambda > \lambda_c$. Diagram (b) shows the full series of bifurcations. Note that at the bifurcation point the system must "choose" between two possible pathways. The continuation of the initial pathway, indicated by a broken line, represents the region of instability.

Prigogine's detailed analysis of the dynamic process of emergence shows that, while dissipative structures receive their energy from outside, the instabilities and jumps to new forms of organization are the result of fluctuations amplified by positive feedback loops. Thus, amplifying "runaway" feedback, which had always been regarded as destructive in cybernetics (see Section 5.3.2), appears as a source of new order and complexity in the theory of dissipative structures. In fact, both types of feedback play important roles in the self-organization of dynamic systems. Self-balancing (negative) feedback loops maintain the system in a stable but continually fluctuating state, whereas self-amplifying (positive) feedback loops may lead to new emergent structures.

One of Prigogine's greatest achievements has been to resolve the paradox of the two contradictory views of evolution in physics and biology – one of an engine running down, and the other of a living world unfolding toward increasing order and complexity. In Prigogine's theory, the second law of thermodynamics (the law of ever-increasing entropy, or disorder) is still valid, but the relationship between entropy and disorder is seen in a new light. At bifurcation points, states of greater order may emerge spontaneously without

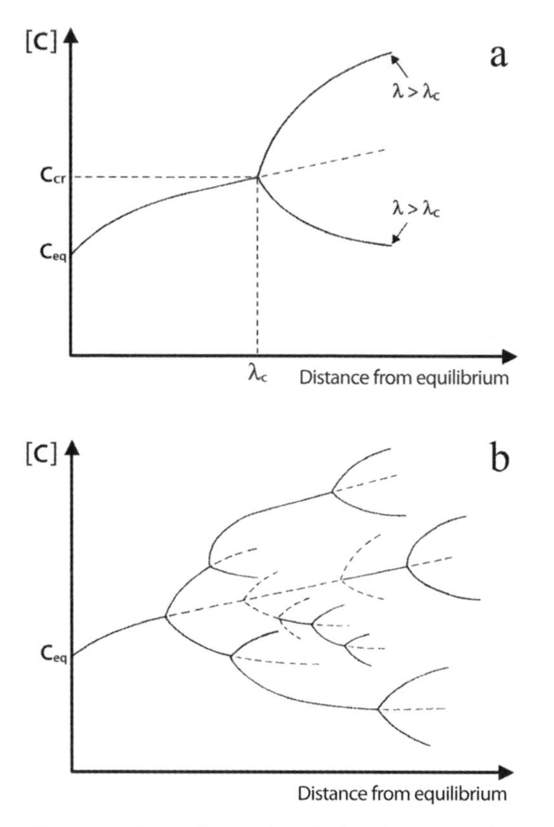

Figure 8.11 Bifurcation diagrams of a nonlinear chemical system according to the Prigogine school, in which the concentration C of a chemical system is plotted against its distance from equilibrium. Panel (a) shows the primary bifurcation; C_{eq} is the concentration at or near equilibrium; λ_c is the distance of the bifurcation point at the critical concentration C_{cr}. Panel (b) shows the whole series of bifurcations.

contradicting the second law of thermodynamics. The total entropy of the system keeps increasing, but this increase in entropy is not a uniform increase in disorder. In the living world, order and disorder are always created simultaneously.

According to Prigogine, dissipative structures are islands of order in a sea of disorder, maintaining and even increasing their order at the expense of greater disorder in their environment. For example, living organisms take in ordered structures (food) from their environment, use them as resources for their metabolism, and dissipate structures of lower order (waste). In this way, order "floats in disorder," as Prigogine (1989) puts it, while the overall entropy keeps increasing in accordance with the second law.

During the decades after Prigogine's pioneering work, the study of emergence in dynamic systems became a rich and lively field, as we have mentioned. Today, the spontaneous emergence of order at critical points of instability is one of the most important concepts of the new understanding of life. Emergence is one of the hallmarks of life. It has been

Figure 8.12 General view and detail of the hexagonal convection patterns known as Bénard cells in a thin layer of silicone oil (1 mm thickness). Image adapted, with modifications, from Coveney and Highfield, 1990.

recognized as the dynamic origin of development, learning, and evolution. In other words, creativity – the generation of new forms – is a key property of all living systems. And since emergence is an integral part of the dynamics of open systems, open systems develop and evolve. Life constantly reaches out into novelty.

Moreover, as we shall discuss in Chapter 14, emergence is a key characteristic not only of biological life but also of social systems, with important implications for economics, management, and other social sciences.

8.3.2 Bénard cells and Brusselators

The classical case of self-organization in nonequilibrium systems, and the first studied by Prigogine, is the striking phenomenon of heat convection known as "Bénard instability." At the beginning of the twentieth century, the French physicist Henri Bénard discovered that the heating of a thin layer of certain liquids (like silicone oil) may result in strangely ordered structures (Figure 8.12). Prigogine's detailed analysis of these "Bénard cells" showed that, as the system moves further away from equilibrium (i.e., from a state with uniform temperature throughout the liquid), it reaches a critical point of instability, at which the ordered hexagonal pattern emerges (see Coveney and Highfield, 1990; Prigogine and Stengers, 1984).

Another amazing self-organization phenomenon studied extensively by Prigogine and his colleagues is the "chemical clocks." These are reactions far from chemical equilibrium, which produce very striking periodic oscillations (see Prigogine and Stengers, 1984). The first chemical oscillation of this kind had been discovered in the 1950s by the chemist Boris Belousov, and was studied in more detail later by Anatoly Zhabotinsky. Accordingly, the whole family of oscillating chemical reactions is now known as the Belousov–Zhabotinsky reaction. The complex reaction mixture contains potassium bromate, sulfuric acid, and cerium ions, and (as shown by later studies) may involve about thirty chemical intermediates and subreactions, some of them of autocatalytic nature (Winfree, 1984).

In the meantime, quite a number of chemical systems have been studied, which show this regular oscillating behavior, in addition to similar systems in biology and engineering (see Strogatz, 1994, 2001; see also Jimenez-Prieto *et al.*, 1998; Noyes, 1989). The theoretical model describing these reactions (a set of nonlinear differential equations) was proposed by Ilya Prigogine and his colleagues at the Free University of Brussels and has become known as the Brusselator in their honor.

To these examples of emergence in dynamical systems we would like to add one particular phenomenon observed in the bacterial world, known as "quorum sensing." It is a signaling mechanism, used by many species of bacteria (Funqua *et al.*, 2001). Quorum sensing controls several important functions in different bacterial species, including biofilm formation, the production of so-called virulence factors by pathogens, and the capability of bacteria to colonize higher organisms.

Quorum sensing is a dynamic intercellular signaling, based on the production of special molecules, which activate genes that in turn produce special proteins. There are even studies referring to the "semantics" and "syntax" of these signaling processes (Ben Jacob *et al.*, 2004), as well as to interkingdom signaling (Shiner *et al.*, 2005). Particularly fascinating is the onset of bioluminescence in *Vibrio fischeri* (Miller and Bassler, 2001; Smith and Iglewski, 2003).

More recently, a totally different type of microbial communication networks has been discovered by microbiologists at the Craig Venter Institute in San Diego and the University of Southern California in Los Angeles. These newly discovered networks are electrochemical networks, and their main function seems to be to coordinate the microbes' cellular respiration.

Like quorum sensing, the appearance of these electromicrobial networks is an emergent phenomenon. Certain bacteria respond to a scarcity of oxygen (or other elements in the case of anaerobic bacteria) by growing tiny "hairs," known as nanowires, that conduct electricity. Through these nanowires the bacteria interconnect their electron flows to optimize the vital access to oxygen, effectively forming an integrated community that engages in coordinated respiration (El-Naggar *et al.*, 2010; Ntarlagiannis *et al.*, 2007).

Observations have now shown that microbial networks of nanowires exist in a variety of environments. In the soil, they may extend underground for hundreds of meters. Some of these networks are extraordinarily dense and look strikingly like neural networks. Indeed, research has shown that the biochemical processes in these microbial communities are completely analogous to brain chemistry. This research has opened a whole new field, which the microbiologist Yuri Gorby has called "electromicrobiology" and which has many fascinating scientific and technological applications.

Another interesting case of dynamic self-organization and emergence in the biological realm is the regular patterns of flying birds that can easily be observed in certain seasons of the year, known as "swarm intelligence." For a more detailed view of this fascinating phenomenon, see Bonabeau *et al.* (1999) and Eberhart *et al.* (2001).

From the examples of dynamic self-organization on a small scale – in chemical systems and in the bacterial world – we shall now turn to the largest scale of self-organization known to date, that of our planet as a whole.

8.3.3 Gaia – the self-organizing Earth

During the early 1960s, while Ilya Prigogine realized the crucial link between nonequilibrium systems and nonlinearity (see Section 8.3.1), and Humberto Maturana puzzled over the organization of living systems (see Section 12.1.2), the atmospheric chemist James Lovelock had an illuminating insight that led him to formulate a model that is perhaps the most surprising and most beautiful expression of self-organization – the idea that the planet Earth as a whole is a living, self-organizing system.

The origins of Lovelock's daring hypothesis lie in the early days of the NASA space program. While the idea of the Earth being alive is very ancient and speculative theories about the planet as a living system had been formulated several times, as we mentioned in the Introduction, the space flights during the early 1960s enabled human beings for the first time to look at our planet from outer space and perceive it as an integrated whole. This perception of the Earth in all its beauty – a blue and white globe floating in the deep darkness of space – moved the astronauts deeply and, as several have since declared, was a profound spiritual experience that forever changed their relationship to the Earth (see Kelley, 1988). The magnificent photographs of the whole Earth that they brought back provided the most powerful symbol for the global ecology movement.

At that time, NASA invited James Lovelock to the Jet Propulsion Laboratories in Pasadena, California, to help design instruments for the detection of life on Mars (see Lovelock, 1979). In contemplating this problem, Lovelock found that the fact that all living organisms take in energy and matter and discard waste products was the most general characteristic of life he could identify. Lovelock assumed that life on any planet would use the atmosphere and oceans as fluid media for raw materials and waste products. Therefore, he speculated, one might be able, somehow, to detect the existence of life by analyzing the chemical composition of a planet's atmosphere.

The terrestrial atmosphere contains gases like oxygen and methane, which are very likely to react with each other but coexist in high proportions, resulting in a mixture of gases far from chemical equilibrium. Lovelock realized that this special state must be due to the presence of life on Earth. Plants produce oxygen constantly and other organisms produce other gases, so that the atmospheric gases are being replenished continually while they undergo chemical reactions. In other words, Lovelock recognized the Earth's atmosphere as an open system, far from equilibrium, characterized by a constant flow of energy and matter – the telltale sign of life identified by Prigogine around the same time.

The process of self-regulation is the key to Lovelock's idea. He knew from astrophysics that the heat of the Sun has increased by 25% since life began on Earth and that, in spite of this increase, the Earth's surface temperature has remained constant, at a level comfortable for life, during those 4 billion years. What if the Earth were able to regulate its temperature, he asked, as well as other planetary conditions – the composition of its atmosphere, the salinity of its oceans, and so on – just as living organisms are able to self-regulate and keep their body temperature and other variables constant? Lovelock (1991) realized that this hypothesis amounted to a radical break with conventional science. Rather than seeing the Earth as a dead planet, composed of inanimate rocks, oceans, and atmosphere, he proposed to consider it as a complex system, "comprising all of life and all of its environment tightly coupled so as to form a self-regulating entity" (Lovelock 1991, p. 12).

In 1969 Lovelock presented his hypothesis of the Earth as a self-regulating system for the first time at a scientific meeting in Princeton. Shortly after that, a novelist friend, recognizing that Lovelock's idea represents the renaissance of a powerful ancient myth, suggested the name "Gaia hypothesis" in honor of the Greek goddess of the Earth. Lovelock gladly accepted the suggestion and in 1972 published the first extensive version of his idea in a paper titled "Gaia as seen through the atmosphere" (Lovelock, 1972).

At that time, the microbiologist Lynn Margulis was studying the very processes Lovelock needed to understand – the production and removal of gases by various organisms, including especially the myriad bacteria in the Earth's soil.

The scientific backgrounds and areas of expertise of James Lovelock and Lynn Margulis turned out to be a perfect match. Margulis had no problems answering Lovelock's many questions about the biological origins of atmospheric gases, while Lovelock contributed concepts from chemistry, thermodynamics, and cybernetics to the emerging Gaia theory. Thus the two scientists were able to gradually identify a complex network of feedback loops which – so they hypothesized – bring about the self-regulation of the planetary system (Lovelock and Margulis, 1974).

The outstanding feature of these feedback loops is that they link together living and nonliving systems. We can no longer think of rocks, animals, and plants as being separate. Gaia theory shows that there is a tight interlocking between the planet's living parts – plants, microorganisms, and animals – and its nonliving parts – rocks, oceans, and the atmosphere. The feedback cycles interlinking these living and nonliving systems regulate the Earth's climate, the salinity of its oceans, and other important planetary conditions. In view of the threats of climate change and other global environmental predicaments, the understanding of the Gaia system is now not only a subject of great intellectual fascination, but has also become a matter of great urgency (as we shall discuss in Chapters 16 and 17).

Gaia theory looks at life in a systemic way, bringing together geology, microbiology, atmospheric chemistry, and other disciplines whose practitioners are not used to communicating with each other. Lovelock and Margulis challenged the conventional view that those are separate disciplines, that the forces of geology set the conditions for life on Earth, and that the plants and animals were mere passengers who by chance found just the right conditions for their evolution. According to Gaia theory, life creates the conditions for its own existence.

At first the opposition of the scientific community to this new view of life was fierce. It is intriguing that of all the theories and models of self-organization, Gaia theory encountered by far the strongest resistance. One is tempted to wonder whether this highly irrational reaction by the scientific establishment was triggered by the evocation of Gaia, the powerful archetypal myth.

Scientists claimed that Gaia theory was unscientific because it was teleological – that is, implying the idea of natural processes being shaped by a purpose – although Lovelock and Margulis never made such a claim. The scientific establishment attacked the theory as teleological, because they could not imagine how life on Earth could create and regulate the conditions for its own existence without being conscious and purposeful. "Are there committee meetings of species to negotiate next year's temperature?", those critics asked with malicious humor (quoted in Lovelock, 1991).

Lovelock responded with an ingenious mathematical model, called "Daisyworld" and published in collaboration with the marine and atmospheric scientist Andrew Watson. The model is a computer simulation of a vastly simplified Gaian system, in which it is absolutely clear that the temperature regulation is an emergent property of the system that arises automatically, without any purposeful action, as a consequence of feedback loops between the planet's organisms and their environment (Watson and Lovelock, 1983).

During subsequent years, Lovelock and his colleagues designed several more sophisticated versions of Daisyworld, which generated lively discussions among biologists, geophysicists, and geochemists (see Harding, 2006; Schneider *et al.*, 2004). In fact, Daisyworld has become the mathematical basis for many other simulations of Gaia in the multidisciplinary fields of Earth system science and biogeochemistry. In particular, such models have been applied increasingly to studies of climate change at the prestigious Hadley Centre for Climate Prediction and Research in the UK and at other similar institutions.

Because of the central role of the Daisyworld simulation in Gaia theory, we have asked one of Lovelock's collaborators, the ecologist Stephan Harding, to discuss the model in some detail in a guest essay (see p. 166), and to show how it has contributed to the transformation of the Gaia idea from a controversial hypothesis into a respected theory.

To be considered as truly alive, the Gaia system must be shown to satisfy the various criteria of life that we discuss in this book. In this regard, it is useful to recall that there are three different levels of organization we have to consider in the complex systems of life. The first is self-organization, the capability of assuming an organized structure thanks to the inner rules of the system. The second level is autopoiesis, when the self-organization is such that it can regenerate from within all its own components (this is the necessary conditions for life itself). Finally, there is the level of the living organism, when autopoiesis becomes associated with cognition, and we have therefore both the necessary and sufficient conditions for life. In the literature, including the literature on Gaia theory, these three levels are not always clearly distinguished, and occasionally we find some confusion between them, as, for example, when self-organization is assumed to be equivalent to life.

These are fascinating issues, for the details of which we have to wait until Chapter 16, since we first need to gain a more comprehensive understanding of cognition (Chapter 12) and also of ecosystems (Chapter 16), which are the living systems most similar to the system of the planet as a whole, known to ecologists as the Earth system.

Guest essay

Daisyworld

Stephan Harding
Schumacher College, Dartington, Devon, UK

Daisyworld is a computer model of a planet, warmed by a sun with steadily increasing heat radiation, and with only two species growing on it – black daisies and white daisies. Seeds of these daisies are scattered throughout the planet, which is moist and fertile everywhere, but daisies will grow only within a certain temperature range (between 5 °C and 40 °C, with optimal growth at temperatures near 22 °C).

Lovelock programmed his computer with the mathematical equations, well known from thermodynamics, that correspond to all these conditions; chose a planetary temperature at the freezing point for the starting condition, and then let the model run on the computer. "Will the evolution of the Daisyworld ecosystem lead to the self-regulation of climate?" was the crucial question he asked himself.

The results were spectacular. As the model planet warms up (in later two-dimensional versions of the model), at some point the equator becomes warm enough for plant life. The black daisies appear first because they warm themselves by absorbing solar energy better than the white daisies and are therefore more fit for survival and reproduction. Thus in its first phase of evolution Daisyworld shows a ring of black daisies scattered around the equator (Figure 8.13).

As the planet warms up further, the equator becomes too hot for the black daisies and they begin to colonize the subtropical zones. At the same time, white daisies appear around the equator. Because they are white, they reflect solar energy and hence cool themselves, allowing them to survive better in hot zones than the black daisies. In the second phase, then, there is a ring of white daisies around the equator, and the subtropical and temperate zones are filled with black daisies, while it is still too cold around the poles for any daisies to grow.

Then the sun gets brighter still and plant life becomes extinct at the equator, where it is now too hot even for the white daisies. In the meantime, white daisies have replaced the black daisies in the temperate zones, and black daisies are beginning to appear around the poles. Thus the third phase shows the equator bare, the temperate zones populated with white daisies, and the zones around the poles filled with black daisies with just the pole caps themselves without any plant life.

In the last and final phase, vast regions around the equator and the subtropical zones are too hot for any daisies to survive, while there are white daisies in the temperate zones and black daisies at the poles. After that, it becomes too hot on the model planet for any daisies to grow and all life becomes extinct.

This is the basic dynamics of the two-dimensional Daisyworld system, which also applies in the most basic initial model with zero dimensions. The crucial property of the model that brings about the emergent self-regulation is that the black daisies, by absorbing solar energy, warm not only themselves but also the planet. Similarly, while the white daisies reflect solar energy and cool themselves, they also cool the planet. Thus solar energy is absorbed and reflected throughout the evolution of Daisyworld, depending on which species of daisies are present.

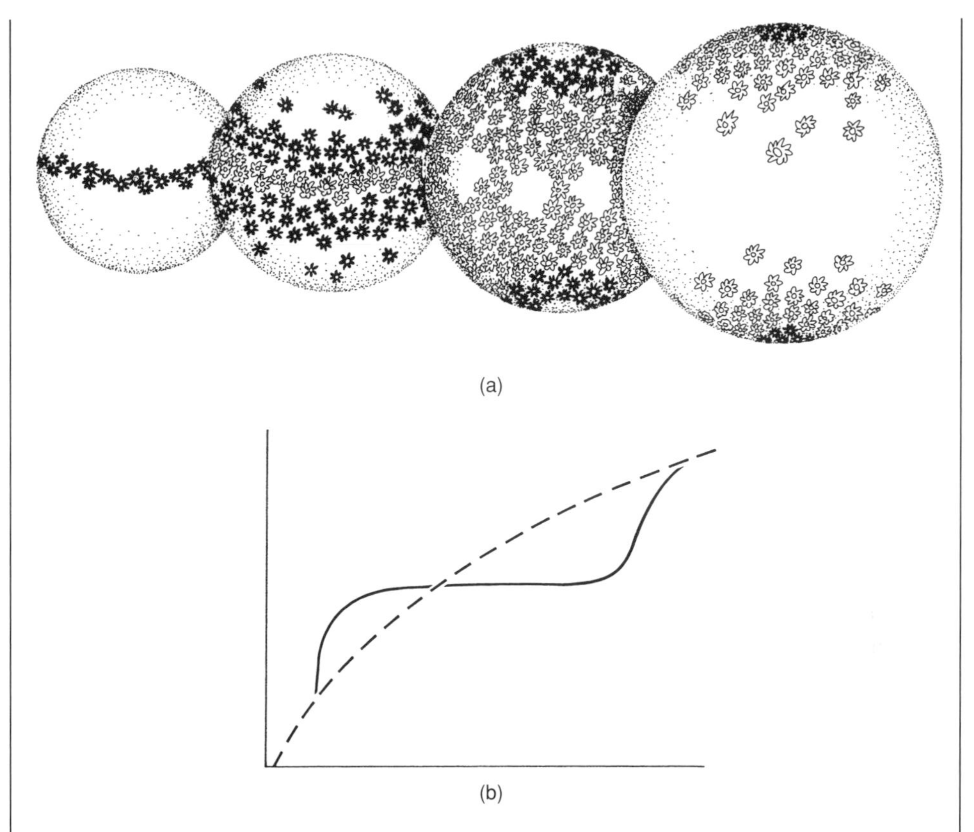

Figure 8.13 The four evolutionary phases of Daisyworld. Panel (a) shows the evolution of temperature on Daisyworld. (b) The dashed curve shows the rise of temperature with no life present; the solid curve shows how life maintains a constant temperature (Lovelock, 1991).

When Lovelock plotted the changes of temperature on the model planet throughout its evolution, he got the striking result that the planetary temperature is kept constant for a vast span of time (Figure 8.13). When the sun is relatively cold, Daisyworld increases its own temperature through solar energy absorption by the black daisies; as the sun gets hotter, the temperature is gradually lowered because of the progressive predominance of energy-reflecting white daisies. Thus Daisyworld, without any foresight or planning, regulates its own temperature over a vast time range by the dance of the daisies.

What amazed and delighted Lovelock most was that his system of nonlinear equations, modeling the tight coupling between the planet's nonliving environment and the growth of the two daisy species, produced two startling emergent properties. First, the overall temperature of the model planet remained remarkably constant over a vast period of time in spite of the shifting daisy populations and the ever-brightening sun; second, the temperature settled on a value just below the optimum for daisy growth.

My own work on Daisyworld, conducted with James Lovelock as a guide and mentor, involved designing more complex ecological communities on the model planet to explore how the increase in complexity would affect the stability of the planet's temperature. We introduced many species of daisies with varying pigments, instead of just two; in some models the daisies evolve and change color; in others rabbits eat the daisies and foxes eat the rabbits, and so on (see Harding, 2004).

The net result of these highly complex models was that the small temperature fluctuations that were present in the original Daisyworld simulation flattened out, and self-regulation became more and more stable as the model's complexity increased. In addition, we put catastrophes into our models that wipe out 30% of the daisies at regular intervals. We found that Daisyworld's self-regulation is remarkably resilient under these severe disturbances.

These extensive explorations confirmed that feedback loops linking environmental influences to the growth of daisies, which in turn affect the environment, are an essential feature of the Daisyworld model. When we broke some of these cycles, so that there was less influence of the daisies on the environment, the daisy populations began to fluctuate wildly and the whole system went chaotic. But as soon as the feedback loops were restored, linking the daisies back to the environment, the model stabilized and its self-regulation emerged again. These simulations showed me in an impressive way that more complex ecological communities are, in general, more stable, as ecologists had long suspected (see Elton, 1958; MacArthur, 1955; Odum, 1953).

Another interesting feature of the model that may have disturbing implications for the real Gaia is what happens at the moment when Daisyworld dies of overheating. Just before life disappears, the light daisies cope with small increments of solar energy by increasing their cover on what little bare soil remains. But under a very bright sun, with no more bare soil available, a small increase in solar energy extinguishes life with sudden rapidity. A similar event takes place in more complex versions of the model. Could it be that just an extra increment of pollution or habitat destruction might trigger an equally dramatic shift toward a new and potentially inhospitable climate regime on our real Earth?

8.4 Mathematical patterns in the living world

8.4.1 Chirality – an asymmetry of nature

In the preceding sections, we have seen how the twin phenomena of self-organization and emergence create a wide variety of intricate and beautiful patterns in the molecular realms studied by physicists and biochemists – from microscopic micelles, folded proteins, and the DNA double helix to the easily visible tessellations and spirals of Bénard cells and chemical clocks.

We shall now turn our attention to another striking and widespread principle of organization and order in nature, known as chirality, or "handedness" (from the Greek *chiros* – "hand"). We are, of course, quite familiar with handedness from our bodily experience. Our two hands (and feet) are not identical but are mirror images of each other. In the language of mathematics (see Section 8.4.3 below), such mirror images are said to be asymmetric under reflection, or "chiral."

To see how chirality is related to self-organization and order in nature, we need to reconsider the chemistry of life, and in particular the chemistry of the carbon atom, which

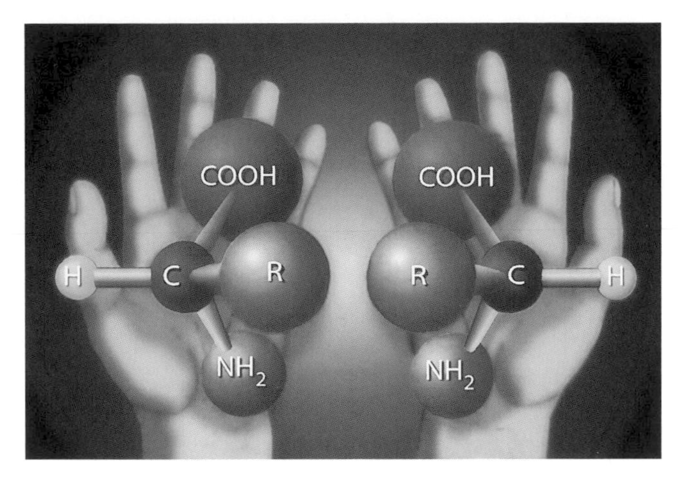

Figure 8.14 Two different α-amino acids that are chiral objects; the two mirror images are not superimposable.

is the chemical backbone of all important biomolecules, from the amino acids to the sugars, and from the lipids to the nucleic bases. The normal carbon atom, in the language of chemistry, is *tetravalent*, which means that it can be bound to four chemical groups. For example, in methane (CH_4), carbon is bound to four hydrogen atoms; in chloroform ($CHCl_3$) to one hydrogen and three chlorine atoms.

Something interesting happens when carbon is bound to four different groups, as is the case of the natural α-amino acids (see Figure 8.14). When this happens, the resulting molecule can exist in two different forms – that is, as two different molecules that are mirror images of each other: the two forms are chiral; they are not superimposable on each other. Of the two forms of α-amino acid, intriguingly, only the left-handed version (the so-called L-form) is present in nature, with negligibly few exceptions.

In the language of chemistry, compounds with the same molecular formula but different structures are known as isomers, and isomers differing in the spatial order are generally called stereoisomers, while the two chiral mirror images are called "enantiomers," or "optical isomers." The last term refers to their distinctive optical property of rotating plane polarized light in different directions – clockwise or counterclockwise. Two distinct chiral isomers are indistinguishable in terms of all other physical properties.

When chemists synthesize α-amino acids (or any other molecules containing asymmetric carbon atoms) by normal laboratory procedures, they produce automatically a 50:50 mixture of the two enantiomers. (The synthesis of one pure enantiomer is possible but extremely difficult – e.g., by using asymmetric catalysts – and the same can be said about the separation of one enantiomer from the other.) However, in nature (with very few exceptions) only the L-form of the α-amino acid is present, and the biochemical reactions in living organisms only produce the L-form of the α-amino acid. Biochemists speak of "homochirality" to indicate situations in which all compounds exhibit the same type of chirality. All our proteins are homochiral, being constituted solely by L-amino acids, and this intriguing asymmetry in nature is not restricted to this class of compounds. All natural sugars and their polymers are chiral, asymmetric molecules existing in nature only as one type of enantiomer (see Box 9.4 for definitions of polymers, peptides, and proteins).

It seems, then, that nature is intrinsically asymmetric, and the question is: why? What is the evolutionary advantage of this asymmetry? To answer this question, let us consider the example of a polypeptide hormone made of a linear sequence of, say, ten, amino acids. Biologically, this hormone works because it has a very specific interaction with a biological "receptor," generally a membrane protein that recognizes the hormone's structure and spatial form. Now, suppose that both L- and D-forms (D for *dextrorotatory*, the mirror image of the L-form) of amino acids were present in nature, being synthesized and incorporated into a growing chain with the same probability. The polypeptide would then exist in 2^{10}, or about 1,000, possible forms. It is obvious that the specificity of interaction with the receptor would be extremely difficult, if not impossible. Extrapolating this calculation to a protein with 50 amino-acid residues, we would obtain the astronomical number of 2^{50} protein isomers, approximately 10^{15}, or 1,000 trillion.

However, these astronomical numbers of hypothetical proteins with combinations of L- and D-amino acids are reduced to *a single protein* simply by having only one optical isomer – that is, by using only the L-amino acids. What a trick! It is evident that this amounts to a huge evolutionary advantage. There is no doubt, then, that homochirality is an extremely powerful principle to bring order and simplicity into the structures of life. Without homochirality, life as we know it would be impossible; and this consideration also suggests that, most probably, this molecular asymmetry was present in the very first steps of the origin of life.

This brings us to the important question: what is the origin of chirality in nature? In other words, what induced the symmetry breaking that favored one kind of chiral molecule over the other? We will discuss this question later on (in Section 8.4.6) within the broader context of symmetry and symmetry breaking. Here we just want to add the interesting observation that in nature the basic chiral asymmetry at the molecular level is generally attended by a high degree of symmetry at the macroscopic level.

The splendid symmetrical patterns exhibited by flowers, insects, and higher organisms, including the bilateral symmetry of mammals, are well known. The relation between symmetry and order is quite apparent, and it is also evident that symmetry in our living world corresponds to an economical strategy of nature: to make a flower with several identical petals, or a butterfly with identical wings, the organism needs just one set of genes, repeated several times. Indeed, the relation between molecular asymmetry and macroscopic symmetry is a fascinating aspect of order in nature.

Of course, symmetry in nature also has an evolutionary value. In many animals it is directly related to beauty as a mating attractor – think, for example, of the spectacular display of the peacock's feathers – and, more generally, it can serve as a recognition pattern, also among different species. In human civilization, symmetry is highly valued in all forms of art, from the architecture of the most primitive temples to modern painting and computer design. We shall return to the intriguing role of symmetry and beauty in evolution in a subsequent chapter when we discuss the basic characteristics of human nature (see Section 11.3.3).

8.4.2 "Biomathematics" – a new mathematical frontier

Asymmetry in nature, however, is not restricted to the molecular level but is conspicuous also in the macroscopic world. A spiral, or a helix, for example, can be right-handed or left-handed. Spirals, in particular, seem to be ubiquitous in nature, appearing in the growth patterns of many plants and animals, as well as in the vortices of turbulent flows of water and air, and the accumulations of stars in giant spiral galaxies.

Indeed, spiral patterns in the growth of leaves and flower petals, as well as in the pigments of seashells and other animals, have long been known by botanists and zoologists; and it is not surprising that mathematicians, too, became fascinated with these extraordinary markings on the skins and exoskeletons of animals, and tried to find mathematical explanations.

One of the first to do so was the Scottish mathematician and biologist D'Arcy Thompson in the nineteenth century. In his pioneering book *On Growth and Form*, Thompson (1917) took his inspiration from the successful use of mathematics to understand nature's patterns in the physical sciences, and advocated a similar approach in biology. He identified numerous mathematical patterns in the living world – the spiral shapes of shells, stripes of zebras, and numerical regularities of plant growth – and he tried to explain them in terms of underlying abstract principles. He failed to do so, however, because (as we know today) the mathematics of life is much more subtle and hidden than that of the nonliving world, and thus Thompson's book, although widely regarded as a classic today, had no significant influence on mainstream biology.

With the advent of complexity theory (see Chapter 6), which is essentially a mathematics of patterns, the situation changed dramatically. The techniques of nonlinear dynamics opened up exciting possibilities of modeling and explaining many details in the emergence of biological forms and revealed a variety of new connections between mathematics and biology. Indeed, in the 1990s, the mathematician Ian Stewart (1998, p. xii) argued forcefully that "biomathematics" would be the new mathematical frontier in the twenty-first century:

> I predict – and I am by no means alone – that one of the most exciting growth areas of twenty-first-century science will be biomathematics. The next century will witness an explosion of new mathematical concepts, of new *kinds* of mathematics, brought into being by the need to understand the patterns of the living world.

The methods used in this new discipline include those of nonlinear dynamics, group theory, and topology – even knot theory. What they all have in common is that they are qualitative approaches, dealing with patterns, order, and complexity. In this section we shall discuss only one mathematical concept, which is of central importance in contemporary physics and is now being used increasingly also in biology: the concept of symmetry.

8.4.3 Symmetry in physics and biology

We have already mentioned the pervasive occurrence of symmetry in nature. As Stewart (2011) explains, the symmetry of an object, to a mathematician, is not a thing but a transformation, whose application leaves the object looking exactly the same. For example, we can rotate a square about its center through one or more right angles, and we will always end up with an identical square. We can also reflect it along one of its diagonals (or along the lines joining the midpoints of opposite sides) with the same result: an identical square. Mathematicians say that the square has eight symmetries. Moreover, these transformations exhibit an important property of "closure": any two operations performed successively are equivalent to a single transformation belonging to the same eight symmetries. They are said to form a group, and hence the mathematical theory dealing with symmetries is known as group theory.

In modern physics, symmetry has been recognized as a fundamental principle that provides structure and coherence to the laws of nature. The requirement that the equations of physics should look the same to different observers (moving with different velocities relative to the observed events) was the foundation on which Einstein built his theory of relativity; and symmetry principles have played a major role in particle physics for the last 50 years, from quarks to string theory.

The key question asked by physicists is how a material universe exhibiting perfect symmetries – its laws being the same everywhere in space and time – can give rise to a great variety of structures and behaviors; for example, different particles governed by different fundamental forces. The answer turns out to be another general principle known as symmetry breaking. When a symmetric system encounters small disturbances, the resulting instability may break the symmetry and give rise to a diversity of patterns that are less symmetric than the system was originally. By trying to identify the detailed dynamics of this process, physicists hope to discover how the great diversity of material particles and the forces between them arose spontaneously from the highly symmetric state of the primordial big bang.

A similar approach is pursued today by biologists who try to understand the emergence of biological patterns and forms – not in the distant past but right now in the growth of the seeds and embryos of plants and animals. The basic idea is the same: a symmetric situation is disturbed, becomes unstable, and consequently gives rise to striking and often complex patterns.

8.4.4 The numerology of plant growth

The geometry and numerology of plant growth, known to botanists as phyllotaxis, is perhaps the oldest example of mathematical patterns recognized in biology. Indeed, D'Arcy Thompson devoted an entire chapter of his book *On Growth and Form* to these striking patterns – the arrangement of leaves on a stem, the numbers of petals on different flowers, the interpenetrating spirals formed by the seed heads of sunflowers, the packing of hexagons on the surface of pineapples, the scales of pine cones, and so on.

What is most remarkable in these diverse patterns is that they often feature spirals and, moreover, that many of them involve a curious sequence of numbers known as the Fibonacci sequence, in which each term is the sum of the previous two (see Sequence 8.1).

$$1, 1, 2, 3, 5, 8, 13, 21, 34, \ldots \tag{8.1}$$

This sequence was discovered in the thirteenth century by Leonardo di Pisa, who was also known as Fibonacci (short for *filius Bonacci*, the son of Bonaccio), in an attempt to model the growth of rabbit populations. Fibonacci, perhaps the greatest mathematician of the European Middle Ages, was most influential in introducing the Hindu-Arabic number system in Europe and demonstrating its superiority over the Roman numerals for arithmetic.

To see how the Fibonacci numbers appear in the growth of plants, let us look at a very common pattern of phyllotaxis: the arrangement of leaves around a stem in a helix in which successive leaves are spaced by the same angle. We might think that any angle should be possible, but in fact nature has chosen a very limited number of angles. When we express them as fractions of the full circle, the actual angles observed in plants form the Sequence 8.2:

$$1/2, 1/3, 2/5, 3/8, 5/13 \ldots \tag{8.2}$$

We can see immediately that both the numerators and denominators of these fractions follow the Fibonacci sequence in such a way that in each fraction they are spaced by two steps. Similar Fibonacci patterns can easily be identified in the packing of sunflower seeds and many other examples of phyllotaxis (see Huntley, 1970; Runion, 1990).

Over the last two centuries, many mathematicians have tried to explain the frequent occurrence of the Fibonacci numbers in phyllotaxis in terms of the underlying dynamics of plant growth. Victorian mathematicians discovered several critical features, but a full explanation was found only in the late twentieth century (see Stewart, 2011).

The starting point is a well-known and very intriguing property of the Fibonacci sequence. The fractions formed by successive numbers (Sequence 8.3)

$$1/1, 2/1, 3/2, 5/3, 8/5, 13/8 \ldots \tag{8.3}$$

get closer and closer to a particular irrational number, $1.618\ldots$ The exact value is $(1 + \sqrt{5})/2$. This is the famous golden section, usually denoted by the Greek letter Φ (see Box 8.2). Moreover, the fractions of angles observed in the helical patterns of phyllotaxis (Sequence 8.4) also approach a specific value related to the golden section:

$$1/2, 1/3, 2/5, 3/8, 5/13 \ldots \sim 1/\Phi^2 \tag{8.4}$$

This value, $1/\Phi^2$, is known as the "golden angle." It is obtained by dividing the full circle into two arcs that are in the golden section (also called golden ratio). The smaller arc (expressed as a fraction of the full circle) then defines the golden angle $1/\Phi^2$. Numerically, it is very close to $137.5°$. This means that the fractions of the angles observed in phyllotaxis

Box 8.2
The golden section

The golden section, also known as "golden ratio," was first defined by Euclid as a proportion derived from the division of a line into two unequal segments (see Livio, 2002). In Euclid's words: "As the whole line is to the greater segment, so is the greater to the lesser" (see Figure 8.15).

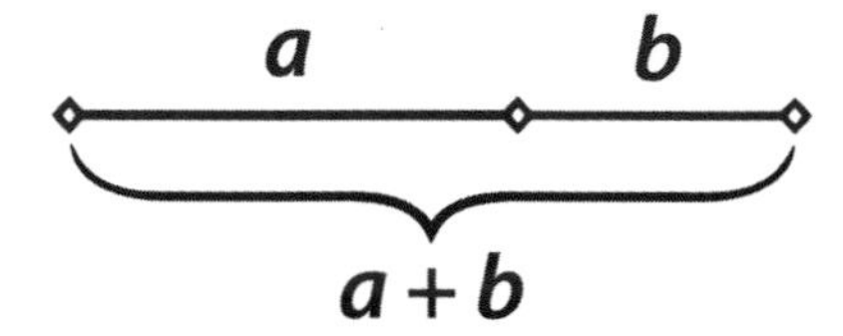

Figure 8.15 The golden ratio.

Expressed algebraically, the proportion reads as in Equation 8.1,

$$\Phi = (a + b)/a = a/b. \quad (8.1)$$

The golden ratio is generally denoted by the Greek letter Φ (phi) in honor of the sculptor Phidias. Φ is an irrational number (Equation 8.2) whose value can easily be calculated.

$$\Phi = (1 + \sqrt{5})/2 = 1.618\ldots \quad (8.2)$$

From its algebraic definition, we can also derive two special properties of Φ (Equations 8.3 and 8.4):

$$\Phi^2 = \Phi + 1, \quad (8.3)$$

$$1\Phi = \Phi - 1. \quad (8.4)$$

One of the classical constructions of the golden section is to inscribe a square into a semicircle (see Figure 8.16). The radius of the semicircle cuts the extended baseline of the square in the proportion of the golden ratio on both sides.

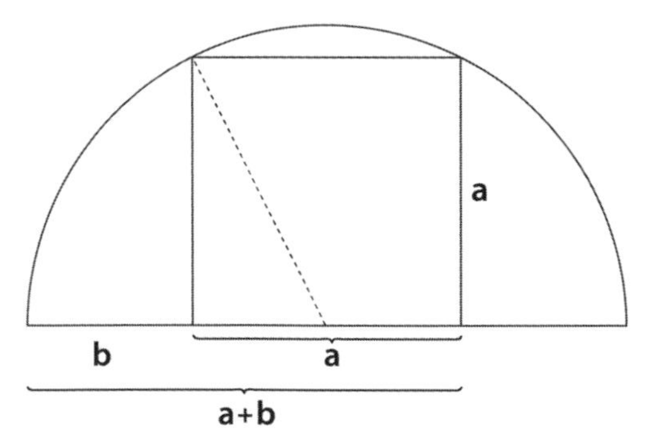

Figure 8.16 Classical construction of the golden section.

The golden ratio plays a crucial role in the symmetry properties of two regular solids, the dodecahedron (with twelve pentagonal faces) and the icosahedron (with twenty triangular faces). In both cases, these properties are based on a remarkable symmetry of the pentagon: each of its five diagonals cuts two diagonals with a golden section. In other words, the golden section is displayed in the well-known regular pentagram, in which each intersection divides both lines in the golden ratio. In addition, the proportion between the pentagon's diagonal and its edge is again equal to Φ (see Figure 8.17).

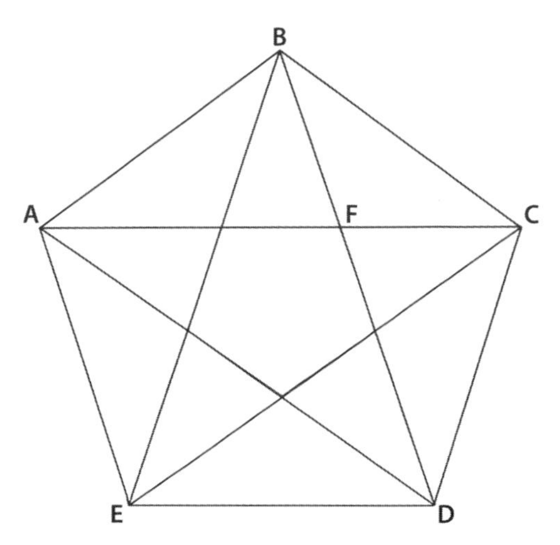

Figure 8.17 Golden ratios in the pentagon; $\Phi = AF/FC = BE/ED$.

The rectangle with sides a and b in the golden proportion is called the golden rectangle. It has the unique property that the smaller rectangle, generated by cutting off a square from the original rectangle, is again a golden rectangle. Moreover, when this procedure is continued, the points dividing the sides of the rectangles in golden ratios are connected by a logarithmic spiral, known appropriately as the "golden spiral." It can easily be constructed by inscribing quarter circles into the "whirling squares" (Figure 8.18).

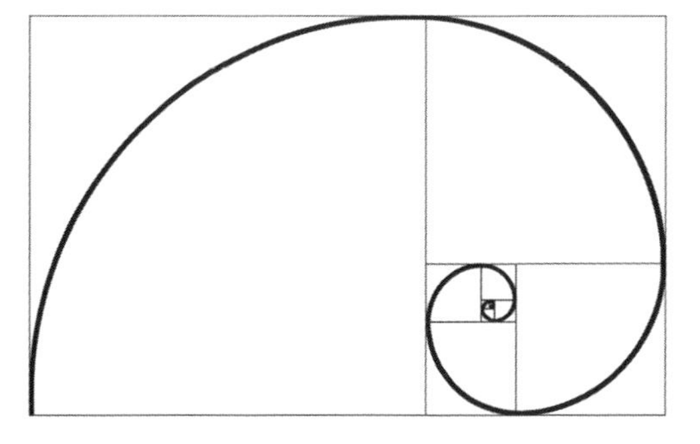

Figure 8.18 Golden rectangles and golden spiral.

In view of all these remarkable properties, it is no wonder that the golden section has fascinated mathematicians, philosophers, and artists throughout the ages. In the Renaissance, it was known as the "divine proportion," and was exalted, together with the square and the circle, as one of the three classical symbols of perfection. Renaissance artists and architects considered it as the proportion most pleasing to the eye, and many of them incorporated approximate golden ratios into their works (see Livio, 2002).

can be interpreted as the best approximations to the golden angle for a given size of the denominator. The problem, then, is to explain why the golden angle is so special in plant growth.

In 1868, the botanist Wilhelm Hofmeister closely observed how the first leaves appear on the tiny green shoot of a plant, and he noted that the basic pattern of leaf development is determined by what happens at the growing tip of the shoot. At the center of the tip, there is a circular region in which small lumps of new cells, known as primordia, are formed by cell divisions and then migrate outward. Each clump will eventually become a leaf, and thus the position of the leaves is determined by the interplay of forces between successive primordia.

In the late nineteenth century, several mathematicians tried to construct mathematical models of the growth dynamics discovered by Hofmeister, but detailed modeling of the actual physical forces between the primordia had to wait for another hundred years, until computer simulations using the formalism of complexity theory could be developed. These techniques finally allowed mathematicians to demonstrate precisely that the golden angle and the corresponding approximations of the Fibonacci fractions are indeed the organizing principle underlying the helical growth patterns of phyllotaxis (see Stewart, 2011).

8.4.5 Nature's spirals

The Fibonacci sequence and the golden section are closely associated with logarithmic spirals (see Box 8.2), which, as we have noted, are ubiquitous in the living world. The logarithmic spiral has several unique properties that help us understand why it appears so frequently in nature. It is defined mathematically as a curve that is magnified by the same factor (known as its growth rate) in successive turns through a constant angle around its origin (or "pole"). In other words, the spiral's radius (a straight line between the pole and a point on the curve) increases in geometric progression with each turn. Different growth rates will produce different geometric progressions, and hence different logarithmic spirals. The golden spiral is a particular logarithmic spiral that grows by a factor of Φ (the golden ratio) for every quarter turn.

As a consequence of this special geometry, the logarithmic spiral has the unique property known as self-similarity: it does not alter its shape as its size increases. As the astrophysicist Mario Livio (2002) points out, this is precisely the property required for many growth

Figure 8.19 The spiral form of *Nautilus*, a snail mollusk. iStockphoto.com/© FlamingPumpkin.

phenomena in nature. For example, the mollusk inside the *Nautilus* shell (Figure 8.19) grows in fixed proportions, and so does his "home" in successive chambers of the shell. By the way, the growth rate of the *Nautilus* shell is different from that of the golden spiral, meaning that there is no significant relationship between the *Nautilus* and the golden ratio, as is sometimes stated erroneously (see Stewart, 2011).

We have already mentioned the striking growth pattern of sunflower seeds, which features two sets of interpenetrating spirals, one running clockwise and the other counterclockwise (Figure 8.20). Typically, the number of spirals in each set turns out to be two consecutive Fibonacci numbers. This means that the golden angle is the generative principle of this pattern, just as it is in the helical phyllotaxis (see Section 8.4.4). In 1979, the biophysicist Helmut Vogel created a mathematical model representing the growth patterns of the corresponding primordia and was able to show that only the golden angle produces a tight packing of the seed heads. Even a slight change of the angle causes the pattern to break up into a single family of spirals with gaps between the seeds (see Stewart, 2011).

The artists of the Renaissance were fascinated not only by the golden section but also by the logarithmic spiral. For Leonardo da Vinci, the spiral form was the archetypal code for the ever-changing and yet stable nature of living forms (see Capra, 2013). He saw it in the growth patterns of plants and animals, in curling locks, and above all in the swirling vortices of water and air. Leonardo accurately depicted these spiral patterns in countless drawings, and his fascination with spiral movements can also be seen in many of his paintings, especially in the portraits. With his frequent use of spiral body configurations, Leonardo created the form of the serpentine figure that became one of the fundamental forms of classical elegance. In his *Lady with an Ermine* (Figure 8.21), for example, the model turns her face by 90° to look over her shoulder in the direction of the light that is illuminating her. As the art historian Daniel Arasse (1998, p. 397) observed, "The pose is particularly ingenious in that, in spite of being twisted, the figure remains supple. This

Figure 8.20 Seed heads in a sunflower, packed tightly in two sets of interlocking logarithmic spirals. iStockphoto.com/© Nicholas Belton.

impression of suppleness is further emphasized by the curving movement of the necklace and, most of all, by the movement of the animal. As arranged by Leonardo, the two figures participate in the same spiraling curve that is finally divided by the direction of their gazes."

8.4.6 Chirality and symmetry breaking

The emergence of the Fibonacci sequence in phyllotaxis, and of the properties associated with it, can be traced back to specific dynamics of symmetry breaking in the growth pattern of the primordia (the first clumps of cells) at the tip of the plant's tiny shoot (see Section 8.4.4). However, a corresponding explanation of the asymmetry embodied in the logarithmic spirals that are so widespread in nature has not yet been found. While the *Nautilus* shell is symmetric (the section shown in Figure 8.19 is one of two halves coiling in opposite directions), the shells of most snails in nature are coiled either to the right or to the left, depending on the species, with the great majority of species bearing right-handed shells. The search for an explanation of this apparently intrinsic asymmetry of nature is a subject of intense investigation (Schilthuizen and Davison, 2005)

As mentioned earlier, the key question is whether homochirality in nature is due to "chance," or whether there is a basic physical principle that demands the preference for one type of enantiomer over the other. There have been several attempts to develop theoretical

Figure 8.21 Leonardo da Vinci, *Lady with an Ermine*, *c.* 1490. With permission from Czartoryski Museum, Kraków, Poland.

models of chiral symmetry breaking at the molecular level (Mason and Tranter, 1983; Quack, 2002; Quack and Stohner, 2003; Tranter, 1985). However, the required calculations are not easy because of the very slight, almost negligible, energy differences between two enantiomers. Consequently, most chemists and biologists in the field remain skeptical about these models.

However, there is an interesting chemical observation that may bring support to the idea that the origin of chirality may be due to some fundamental principle. In certain meteorites, some derivatives of α-amino acids (called α-methyl amino acids) have been found to have a higher proportion of the L-form over the D-form (Cronin and Pizzarello, 1997). This may be consistent with the idea that particular conditions in outer space may favor one form over the other, thereby inducing a breaking of the symmetry. It is, of course, not possible to assess whether these chiral compounds in meteorites were the seeds for the homochirality of life on Earth (Bada, 1997).

One reason why many chemists and biologists are skeptical about these and other subtle physical effects is that the breaking of symmetry can be realized rather simply in the chemistry laboratory. This has been shown by Meir Lahav, one of the leading researchers in the field, working with crystals as agents of symmetry breaking (Weissbuch *et al.*, 2003). Similar experiments were performed by Kondepudi and collaborators (Kondepudi *et al.*, 1990; McBride and Carter, 1991), who were able to show that, starting from a mixture of a compound that can crystallize in two chiral forms, chance effects may induce the selective crystallization of only one of the two forms.

The argument is that these kinds of effects might have happened on the early Earth, and would have produced an asymmetric template on which the first chemical reactions started, thus giving the molecular imprint of chirality. In other words, the L- and D-forms had the same probability of occurring; it just so happened that, because of some accidental conditions, life started with the L-form.

8.5 Concluding remarks

The synergy between self-organization and emergence shapes and determines the structures and functions of life's molecular complexes, and, as we shall see in Chapter 14, it is also of crucial importance in social life. In static systems, self-organization and the resulting emergent properties are relatively simple concepts, well explained by chemistry and physics, but in dynamic systems the processes of self-organization and emergence are subtle and complex; and their outcomes are often unforeseeable, both in biological and in social life. In a way, this carries a positive message. New structures, technologies, and new forms of social organization may arise quite unexpectedly in situations of instability, chaos, or crisis.

The systems view of life is essential for understanding these phenomena. Instead of being a machine, nature at large turns out to be more like human nature – unpredictable, sensitive to the surrounding world, and influenced by small fluctuations. Accordingly, the appropriate way of approaching nature to learn about her complexity and beauty is not through domination and control but through respect, cooperation, and dialogue. Indeed, Ilya Prigogine and Isabelle Stengers (1984) gave their popular book, *Order Out of Chaos,* the subtitle "Man's New Dialogue with Nature."

In the deterministic world of Newton, there is no history and no creativity. In the living world of self-organizing and emergent structures, history plays an important role, the future is uncertain, and this uncertainty is at the heart of creativity. Thus Prigogine (1989), one of the architects of this new scientific perspective, reflected in a beautiful essay titled "The Philosophy of Instability":

> Today, the world we see outside and the world we see within are converging. This convergence of two worlds is perhaps one of the important cultural events of our age.

Another important point about emergence is that life itself can be seen as an emergent property – a consideration that gives the notion of emergence a particularly poignant significance. No vitalistic principle, no transcendent force, is invoked to arrive at life. As we have mentioned already, this has two consequences: (1) cellular life, at least in principle, can be explained in terms of molecular components and their complex nonlinear interactions; (2) it becomes conceivable to make some simple forms of life in the laboratory.

Life, as we have seen again and again, is one of those phenomena that cannot be explained in reductionistic terms. One could never grasp the essence of a rose by saying that it is composed of atoms and molecules. An "emergentist" approach to understanding the essence of the rose would be to consider its ontogeny (development), pausing at each level of growing complexity, in order to study the corresponding emergent properties – from the formation of the various flower cells to the interactions between all these cells, and up to the characteristics of the complex organs, such as petal and stem, including odor and color. We would then consider the rose as the final "flowering" of all these emergent properties.

The notion that one arrives at in the end is that the rose is an ensemble of various emergent properties – the colors, the perfume, the symmetry – without any central localization where the essence of the rose would be condensed. We have already encountered this concept of an ensemble with nonlocalized global properties when asking the question "What is life?" And we shall encounter it again when we discuss the nature of mind and consciousness in Chapter 12. Indeed, many cognitive scientists today would agree that the very notion of "I" is an emergent property arising from the simultaneous occurrence and resonance of feelings, memories, and thoughts, so that the "I" is not localized anywhere, but rather is an organized pattern without a center. In the words of one of the pioneers in this field, Francisco Varela (1999):

> This is one of the key ideas, and a stroke of genius in today's cognitive science. There are the different functions and components that combine and together produce a transient, nonlocalizable, relationally formed self, which nevertheless manifests itself as a perceivable entity . . . we will never discover a neuron, a soul, or some core essence that constitutes the emergent self of Francisco Varela or some other person.

9

Darwin and biological evolution

Looking at life around us we notice two important general qualities. One is the constancy of form from one generation to another: roses from roses, elephants from elephants. The other is that life is characterized by an amazing variety of different species – microbes, insects, mammals, fish, birds, and flowers – and within each species, there are hundreds or thousands of different forms. All this yields the rich biodiversity of our planet. These two apparently contradictory aspects of life – constancy of form and the existence of so many different forms – make up life on Earth.

The emphasis of this chapter is on biological evolution. For a proper historical start we have to go back to the early decades of the nineteenth century, in England, to introduce the time and the work of one of the greatest scientists, Charles Darwin (Figure 9.1).

9.1 Darwin's vision of species interlinked by a network of parenthood

At the time of Darwin, it was commonly believed that the different life forms were given, once and for all, by God's creation. And reproduction was the way by which these forms, created by God, would perpetuate their species on Earth. To doubt the credo that the biological forms were fixed once and forever, was close to blasphemy – and this is what Darwin did.

In fact, the foundation of Darwinism is the idea that we all come from a common ancestor with modifications, and that is tantamount to saying that all living forms, from trees to fish, and from mammals to birds – since they all come from the same primordial ancestor – are linked to each other by a network of parenthood. There is nothing more holistic and systemic than this notion of Darwinian biological evolution: all living creatures are intrinsically linked to each other and form one single family. But let us proceed in order.

Let us go back to the very young geologist Charles and his voyage on the *Beagle* in 1831–3. He was sharing the cabin with the captain of the ship, Robert Fitzroy, a learned man himself, and an extremely conservative believer – apparently still holding on to the idea of Noah's Ark and the immutability of species, created by God once and for all time. By contrast, the first observations and statements of Darwin after the first long months of travel and observations – something that tormented Captain Fitzroy very much – was that *species are not fixed, but change with time.*

Figure 9.1 Charles Darwin (1809–1882). iStockphoto.com/© Carina Lochner.

How could that be? Since the environment changes naturally, the likely result of the changes must have been that, in order to survive, the living species with time had to adapt and therefore change. Therefore the organisms must display *adaptation due to environmental changes.* However, the notion of adaptation must be attended by the consideration that not all members would be able to cope equally well with such environmental changes. The group of individuals that adapted more efficiently would be able to reproduce and survive better; so that the next generation would be extremely enriched by individuals characterized by these positive traits. This, basically, is evolution via *natural selection.*

A new species is one that cannot interbreed with the original one. This is the canonical, textbook definition, but the question "What is a species?" is still debated among biologists (see, e.g., Margulis and Sagan, 2002, pp. 4ff.). By the mechanism explained above, several new species could originate and even live in parallel to the older one. They all would be relatives sharing *a common ancestor* from which they all would have originated with modifications. Important for the generation of a new species by the mechanism of adaptation would be geographical isolation, as caused by some geological accident, or by weather-induced migration.

Box 9.1
Darwin's "tree of life"

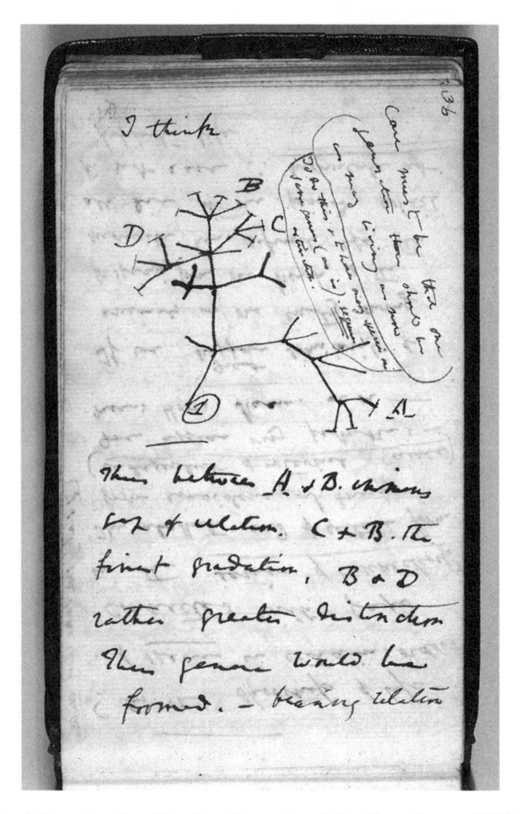

Figure 9.2 The original drawing by Charles Darwin of the first "tree of life," from his 1837 notebook. Classmark P382.c.367.2. Reproduced by kind permission of the Syndics of Cambridge University Library.

Darwin maintained that one should not use the metaphor of a tree of life, for which the roots must be still living – but rather that of coral. In fact, a tree is a hierarchical structure, where the roots, or the crown, is more important than the rest. In contrast, coral grows on parts that are dead, and no one branch is more important than any other. It should be stressed that Darwin arrived at all these conclusions – including the tree of life – without any knowledge of genetics. As the story goes, Gregor Mendel had developed a theory based on the study of genetic traits in plants, and after the publication of Darwin's *The Origin of Species* in 1859, when Darwin was already very famous, Mendel wrote to Darwin to expound his genetic theory. But Darwin never read these letters.

9.2 Darwin, Mendel, Lamarck, and Wallace: a multifaceted interconnection

The notion of evolution was already in the scientific world before Darwin, and in this respect one should pay tribute to Lamarck. Jean-Baptiste Pierre Antoine de Monet, Chevalier de Lamarck (1744–1829), usually known as *Lamarck*, was a French biologist. A towering intellectual figure of his time, he is credited with the first use of the word *biology* (1802).

Lamarck introduced the notion of biological evolution before Darwin by suggesting that the evolution of a species, which he called *transformation*, takes place as the result of "*a new need that continues to make itself felt*," and that characteristics acquired during an organism's life can be inherited by the organism's offspring. The most familiar example of this theory is giraffes: Lamarck suggested that giraffes who, through stretching to reach tall trees, make their necks longer would then pass on longer necks to their offspring. The theory of the inheritance of acquired characteristics is called *soft inheritance* or Lamarckism. The famous expression for Lamarck's theory of evolution is that *functions create organs* and heredity determines the change in offspring.

This idea of soft inheritance was a reflection of the folk wisdom of the time, accepted by many natural historians. Although the details of the Lamarckian theory had to be abandoned later on, he was the first to propose a coherent theory of evolution. His reversal of the traditional taxonomy – "turning the ladder of explanation upside down," as Gregory Bateson put it – was a tremendous feat (see Section 1.2.3).

To the narrative about Darwin and the theory of evolution belongs also the well-known fact that Darwin did not publish his observations for twenty years, as he was well aware, and afraid, of the effects that his ideas would have on the Christian society of his time – on his friends, and his wife as well.When asked to publish his book, he actually answered once that such a publication would equate to "murder." According to several scholars of Darwinism, the murder was not only the notion that species are not fixed but also the notion that nature proceeds rather randomly in its development – that is, not obeying any predetermined plan. To the narrative belongs also the fact that a younger colleague, Alfred Russel Wallace, was about to publish a paper on evolution based on very similar principles to Darwin's, and this hastened Darwin's own publication. The relation between the two scientists remains a beautiful example of gentlemanliness in the history of science. Not that they were always in agreement with each other; on the contrary, there were profound philosophical differences.

For the purpose of the present book, since we will comment later on the notion of "intelligent design," it is proper to mention that Wallace, at least in later years, adopted a point of view that today would be defined as belonging to the ideology of intelligent design. In fact he wrote (in a letter to Darwin on March 17, 1869): "Natural selection would have endowed the savage with a brain a little superior to that of an ape . . . and we must therefore admit the possibility that in the development of human race a Higher Intelligence has guided the same laws . . . for nobler ends." To which Darwin responded by scribbling a series of exclamations of "No! No!" in his personal notebook, and then writing to Wallace something that, for the cautious gentleman he was, sounds very strong: "I differ grievously

Box 9.2

Gregor Mendel's genetic experiments

Gregor Johann Mendel was born to peasant parents in a small rural town in Austrian Silesia, now in the Czech Republic. During his childhood he worked as a gardener, and in 1843 he entered an Augustinian monastery in Brünn, now Brno. His famous genetic experiments were performed with peas. He crossed peas of different varieties and saw that the traits were inherited in certain numerical ratios. In particular, Mendel selected twenty-two different varieties of peas and interbred them, keeping track of seven different traits, such as pea texture – smooth or wrinkled.

Mendel found that when he hybridized smooth and wrinkled peas, he produced peas that were all smooth. But if he then produced a new generation of peas from the hybrids, a quarter of the peas were wrinkled. He then came up with the idea of the dominance and segregation of genes and set out to test it meticulously. From his studies, Mendel derived certain basic laws of heredity, which are not easy to render in a simple form. Basically, hereditary factors do not combine, but are passed intact; each member of the parental generation transmits only half of its hereditary factors to each offspring (with certain factors "dominant" over others); and different offspring of the same parents receive different sets of hereditary factors.

from you. I can see no necessity for calling in an additional and approximate cause in regard to man. I hope that you have not murdered your own and my child" (this and the above statement by Wallace cited in Pievani, 2009). All these statements bring Darwin very close to our modern thinking – he was really a pioneer in all respects.

We have mentioned Lamarck as a forerunner of Darwin, at least as far as the general concept of evolution was concerned. Another important scientist in this field, who was actually his contemporary, was the Austrian monk Gregor Mendel (1822–1884). As we discussed in Section 2.2.3, Darwin had struggled with the question of how organisms pass traits on to their offspring. Why did some traits seem to be passed on and others not? How did the traits of the parents work together in the offspring – did they compete or combine? Mendel's work helped answer these questions (see Box 9.2). It was only some fifteen years after his death in 1884 that scientists realized that Mendel had discovered the answer to one of heredity's greatest mysteries. Mendel's work became the foundation of modern genetics.

In the scientific paradigm of the time, Darwin's theory of evolution was an earthquake with profound consequences in society, as well as in daily life. And it was also the beginning of a series of other evolutions. It reinforced the geologists' views on the drift of continents and oceans, an idea that had been previously expressed by Charles Lyell, the foremost geologist of the time, who actually had a strong influence on the young Darwin; so that the very geography of our Earth became the scene of evolution. Darwin's book was also instrumental in bringing about Oparin's scenario of prebiotic molecular evolution (as we shall see in our next chapter), according to which inorganic and organic matter had evolved to produce living cells. Later on, astronomers would describe the evolution of stars and galaxies – the view that our entire cosmos was also the place of evolution. After Darwin, these early examples of evolutionary thinking were taken much more seriously. Nothing was static anymore; all was evolving.

Within the framework of Darwinism, we can ask a more basic question: is natural selection all there is in evolution? This is an important question, which can be dealt with by considering the next step of classic Darwinism, the so-called modern synthesis.

9.3 The modern evolutionary synthesis

Darwin's *The Origin of Species* introduced two main concepts: the first, that all organisms have descended with modifications from a common ancestor; the second, that natural selection is the mechanism of evolution. Whereas the first point was accepted by most biologists of the time, the mechanism of evolution was not, and the publication of Darwin's book was followed by a period of uncertainty and confusion. Gregor Mendel's work was rediscovered only at the beginning of the twentieth century, and initially his genetics, based on the idea of distinct hereditary units (now called genes), was seen as in opposition to Darwin's views – until the evolutionary biologist R.A. Fisher (1930) was able to prove the contrary.

There was also an intense period of the development of population genetics, represented by the works of T.H. Morgan, R.A. Fisher, J.B.S. Haldane, and S. Wright. Actually, the development of population genetics was instrumental in the creation of what is known as the "modern synthesis" (also known as "modern evolutionary synthesis," or "new synthesis"; this synthesis is also referred to as "neo-Darwinism" in common language), in which the books by Julian Huxley (1942), the grandson of Darwin's contemporary Thomas Huxley, and Ernst Mayr (1942) were milestones. The incorporation of population genetics permitted us to recognize the importance of mutation and variation within a population, so that the alteration of the frequency of genes within a population defines evolution. It was accepted that characteristics are inherited as discrete entities, called genes, and that speciation is (usually) due to the gradual accumulation of small genetic changes – macroevolution is simply a lot of microevolution.

Thus the modern synthesis is a theory about how evolution works at the level of genes, phenotypes (i.e., the actual appearance of the living and its behavior), and populations. The major controversy was, and partly still is, about the relation between micro- and macroevolution, a controversy arising, for example, from the objection that the fossil record at any one site does not show gradual change but instead long periods of stasis followed by rapid speciation. The model that accounts for this phenomenon is called punctuated equilibrium and is now generally accepted. The importance of random mutations is also generally accepted; for details, see Futuyma, 1998). For a more recent discussion of the modern synthesis, see the book by Pigliucci and Müller (2010).

Another important aspect is the field of sociobiology, introduced by E.O. Wilson (1975) in the mid 1970s with the idea that behaviors like aggression, altruism, or love are determined by genes and are a product of evolution. This is now widely accepted, but the extension to human behavior stirred a lot of controversy at the time.

9.3.1 The genetic code

The next big step in the understanding of the evolutionary process came at the molecular level: the discovery of the structure and function of DNA. There are two basic functions of DNA that can be understood on the basis of its structure. The first is the capability of self-replication – that is, to make identical copies of itself. Not that DNA alone is capable of replicating itself; it needs for that a large number of enzymes and a precise biological context. But DNA is the only macromolecular structure that contains information on how to make copies of itself. It is not the place here to repeat basic notions of molecular biology, but, as a reminder, we offer in Box 9.3 and the corresponding figure a schematic representation of the DNA double-helix complementarity and replication mechanism.

The other important function of DNA is the capability of a DNA sequence to "code for" a polypeptide sequence. This means, that a linear sequence of DNA contains the information to produce a linear sequence of amino acids linked to each other, a polypeptide sequence. However, the transformation of a DNA sequence into a protein proceeds through an intermediate RNA macromolecule, called messenger RNA (generally abbreviated as m-RNA). The translation of the linear information of the DNA sequence into a linear polypeptide sequence (or protein; see Box 9.4) is based on a "genetic code." It is a triplet code, according to which one triplet of DNA codes for one given amino-acid residue in a sequential order that has a precise start signal in the sequence to be "read" – that is, recognized and processed – by the ribosome machinery (see Box 9.3 and Figure 9.3 for more details).

The ordered ensemble of the triplets coding for the various amino acids is the famous genetic code. The double-stranded sequence of DNA that codes for a given protein is called a gene. The ensemble of genes in one given organism is the genome.

When a fragmented gene is copied into an RNA strand, the copy must be processed before the assembly of the protein can begin. Special enzymes come into play that remove the noncoding segments and then splice the remaining coding segments together to form a mature transcript. In other words, the messenger RNA is edited on its way to protein synthesis.

9.3.2 Neutral drift in evolution

The modern theory of genetic evolution still maintains that natural selection is one of the main driving forces. A couple of qualifications are relevant in this respect. When a gardener chooses some particular seeds of roses in order to obtain a species with a more brilliant color, he makes a selection with a precise plan in mind; and so does a horse breeder. In natural selection, by contrast, there is no one who makes a selection, and there is no a priori plan.

If mutations are not directed towards an aim, then, conceivably, certain steps in evolution are probably random – that is, they do not obey any criteria of better fitness. They just happen, and can be accepted if they are not harmful.

Box 9.3
The DNA double helix and the genetic code

DNA is the nucleic acid that carries the cell's hereditary information. The DNA molecule is a long, two-stranded chain of four different monomers, called nucleotides, each containing a sugar, a phosphate group, and one of the four "genetic bases." The four bases are adenine, guanine, thymine, and cytosine, denoted A, G, T, and C, respectively. The self-replication of DNA is based on the double helix (duplex) structure and the complementarity between the bases which constitute DNA, the famous adenine–thymine (A–T) and guanine–cytosine (G–C) pairing. The two DNA strands are coiled in a double helix in such a way that each nucleotide on one strand is bound to a nucleotide on the other strand by (weak) hydrogen bonds between their bases (see Figure 9.3). There are only two types of base pairings, A–T and G–C, meaning that the two DNA strands are complementary.

From Figure 9.3, it is easy to understand the replication mechanism: from each single strand, the complementary strand can be constructed on the basis of the pairing complementarity.

The other thing that DNA is capable of doing is "coding" for proteins – that is, actualizing a correspondence between the linear sequence of DNA and the linear structure of proteins. This correspondence is based on the genetic code, according to which each triplet of DNA codes for one given amino-acid residue. See the text for more details.

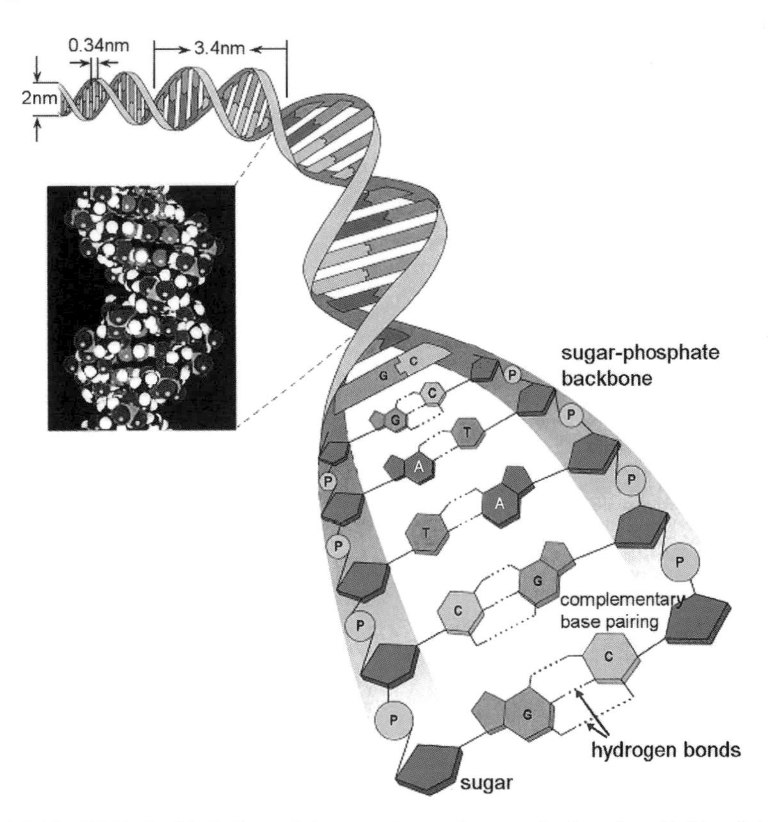

Figure 9.3 The DNA double helix and the strand complementarity based on C–G and A–T recognition.

Box 9.4
Polymers, peptides, and proteins

A polymer is a linear (occasionally also branched) sequence of the same structural unit, called a monomer unit or repeat unit. The chemical compound yielding the monomer unit is called a monomer. For example, in poly(propylene), represented as $H–[CH_2–CH(CH_3)–]n–H$, the monomer is propylene, $CH_2 = CH(CH_3)$, and the monomer unit is $–CH_2–CH(CH_3)–$, whereas n represents the polymerization degree – namely, the average number of repeat units in the chain. In the case of proteins, the monomer is the amino acid $NH_2–CH(R)–COOH$ and the repeat unit is the amino-acid residue –NH–CH(R)–CO–.

Proteins are linear sequences of α-amino acids, which are linked together as shown below (formally losing a molecule of water) to form the so-called peptide bond, the unit –CO–NH–. There are 20 common amino acids in nature, which differ in the chemical structure of the group R. If R = H, we have glycine, with $R = –CH_3$ we have alanine; when $R = CH_2–OH$ we have serine, etc. A polypeptide chain is thus a sequence of amino-acid residues, –CO CH(R)–NH–, and if all R groups are the same, we have a polymer – in particular, a poly-α-amino acid; e.g., a poly(alanine). In nature, we have generally sequences with a combination of different residues.

a

$$H_3\overset{+}{N}-\underset{\alpha}{\overset{R^1}{CH}}-\underset{O}{\overset{}{C}}-OH + H-\overset{H}{N}-\underset{\alpha}{\overset{R^2}{CH}}-COO^-$$

$$\xrightleftharpoons[H_2O]{H_2O}$$

$$H_3\overset{+}{N}-\underset{\alpha}{\overset{R^1}{CH}}-\underset{O}{C}-\overset{H}{N}-\underset{\alpha}{\overset{R^2}{CH}}-COO^-$$

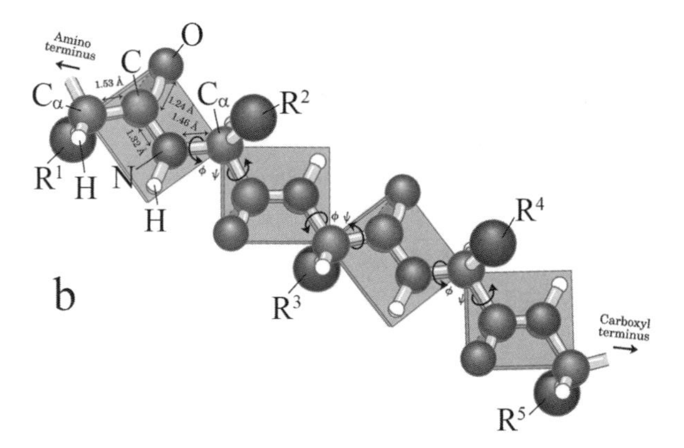

Figure 9.4 Chemical condensation of two amino-acid residues with side chains R^1 and R^2 to yield a dipeptide via elimination of one water molecule (panel a); section of a poly(alanine) chain, with $R = –CH_3$ (panel b); here also the torsion angles are shown; see text.

For relatively short sequences, the term "peptide" is used; for longer ones, the term *polypeptide* (snake or bee venom has, for example, polypeptides of 15–18 residues). Much longer ones give rise to proteins – the smallest proteins have 45–50 residues.

Figure 9.4 also illustrates the "torsion angles" around which the chain can assume its flexibility. There are two torsion angles (called Φ and Ψ) for each residue, so that, in principle, a chain with 100 residues can assume an extremely large number of forms (chain conformations). Figure 9.4 gives an example of the well-known proteins responsible for the storage and transport of oxygen in our body, myoglobin and one of the chains of hemoglobin (this is constituted by four chains, two α- and two β-chains).

It is this consideration that forms the basis of the "neutral theory of molecular evolution," as introduced and developed by Motoo Kimura (1968, 1983), starting in the 1960s. The theory regards most genomic features as neither subject to, nor explicable by, natural selection. In a way, then, this view can be seen as an offshoot of the modern evolutionary theory. The theory has also been applied to niche ecology (Hubbell, 2006), and was considered by Gould and Eldredge (1977) in their theory of punctuated equilibria.

This notion is also taken up by Maturana and Varela (1980), who prefer, however, the term "natural drift," which they see as the process of conservation of autopoiesis and adaptation. They add, in accordance with what we have emphasized, that no guiding force is needed to explain the directionality of the changes, that evolution is "like a sculptor with a *wanderlust*." At this point it is proper to recall a famous poem by the Spanish poet Antonio Machado, *Caminante no hay camino*, which our friend Francisco Varela loved and often used to illustrate the random walk of life:

Caminante, son tus huellas	Wanderer, your footsteps are
el camino y nada más;	the road, and nothing more;
Caminante, no hay camino,	wanderer, there is no road,
se hace camino al andar.	the road is made by walking.

9.3.3 The "central dogma"

In its most classic form, the direct relation one gene–one protein is referred to as the "central dogma of molecular biology" (as we discussed in Section 2.3.5):

$$\text{DNA} \rightarrow \text{RNA} \rightarrow \text{protein}$$

However, it soon turned out that the linear chain described by the central dogma is far too simplistic to accurately represent the biological reality, as we shall discuss in Section 9.6. For example, as Keller (2000) explains, genes that code for proteins in higher organisms tend to be fragmented rather than forming continuous sequences. They consist of coding segments interspersed with long repetitive, noncoding sequences whose function is still unclear. The proportion of coding DNA (called "extrons") varies a great deal and in some

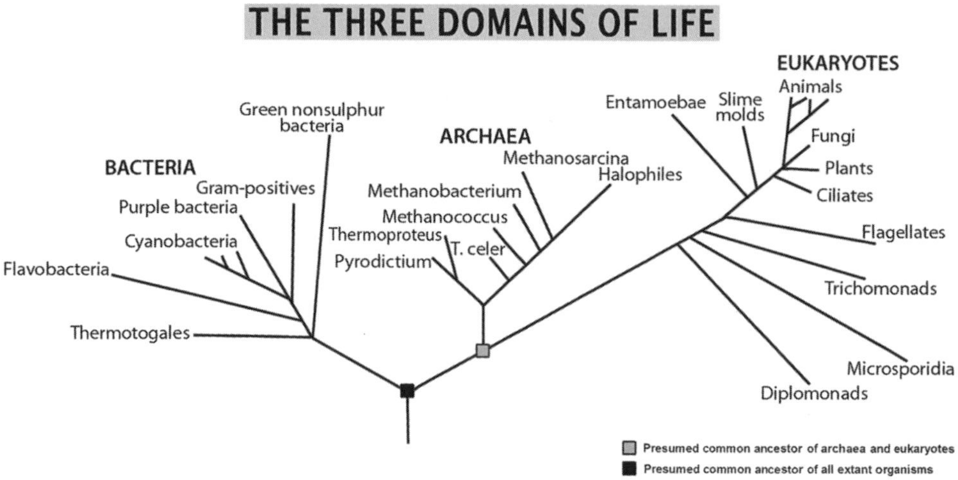

Figure 9.5 The three domains of the tree of life, branching off from the last common universal ancestor (LUCA).

organisms can be as low as 1–2%. The rest (the "introns") in previous years was often referred to as "junk DNA." However, since natural selection has preserved these noncoding segments throughout the history of evolution, it is reasonable to assume that they play an important, though still largely unknown, role. In fact, molecular biologists are discovering more and more functions of these introns, often related to epigenetics and to gene–gene interplay.

9.3.4 The three domains of life

Studies based on nucleic-acid sequences have permitted us to clarify that, starting from the last universal common ancestor (LUCA), three different branches of life have originated (as illustrated in Figure 9.5): the *archea,* the *bacteria*, and the *eukaryotes* (to which we belong). In this way, as already mentioned, all species on Earth are linked to each other – a universal systemic network that spans all existing species and goes back over 3.5 billion years.

During the first 2 billion years of biological evolution, bacteria and other microorganisms were the only life forms on the planet. During those 2 billion years, bacteria continually transformed the Earth's surface and atmosphere, and established the global feedback loops for the self-regulation of the Gaia system (see Section 8.3.3). In so doing, they invented all of life's essential biotechnologies, including fermentation, photosynthesis, nitrogen fixation, respiration, and various devices for rapid motion. Recent research in microbiology has made it clear that, as far as the processes of life are concerned, the planetary network of bacteria has been the main source of evolutionary creativity.

9.3.5 Avenues of evolution

These and other observations have led systems biologists to an understanding of evolution that is considerably richer and more diverse than the modern synthesis. According to this new systemic understanding, the unfolding of life on Earth proceeded through three major avenues of evolution. The first, but perhaps least important, is the random mutation of genes, the centerpiece of the neo-Darwinian theory. These gene mutations, caused by chance errors in the self-replication of DNA, do not seem to occur frequently enough to explain the evolution of the great diversity of life forms, given the well-known fact that most mutations are harmful and very few result in useful variations. In the case of bacteria the situation is different, because bacteria divide so rapidly that billions of them can be generated from a single cell within days. Because of this enormous rate of reproduction, a single successful bacterial mutation can spread rapidly through its environment, and thus mutation is indeed an important evolutionary avenue for bacteria.

Bacteria have also developed a second avenue of evolutionary creativity that is vastly more effective than random mutation. They freely pass hereditary traits from one to another in a global exchange network of incredible power and efficiency. The discovery of this global trading of genes, technically known as DNA recombination, must rank as one of the most astonishing discoveries of modern biology.

This gene transfer takes place continually, with many bacteria changing up to 15% of their genetic material on a daily basis. Since all bacterial strains can potentially share hereditary traits in this way, some microbiologists argue that bacteria, strictly speaking, should not be classified into species (see Sonea and Panisset, 1993). In other words, all bacteria are part of a single microscopic web of life.

A third major avenue of evolution, which has profound implications for all branches of biology, is evolution through symbiosis, also known as symbiogenesis, to be discussed in detail in Section 9.6.3.

Thus, through the evolutionary process a rich biodiversity appeared on our planet. The study of the relationships between genome structures across different biological species is a fascinating new discipline, known as comparative genomics. For example, limiting the analysis to our human species, we have discovered that more than 99% of our genes have a related copy in the mouse – despite more than 500 million years of evolutionary separation. Moreover, the differences between human races worldwide are thought to be coded by only 0.1% of the human genome. This means that molecular genetics has demonstrated that there are no significant differences among the various human races.

9.4 Applied genetics

The molecular biology of nucleic acids is important not only for basic science – the understanding of the main mechanisms of genetics – but also for practical, biotechnological aspects, known as "bioengineering," "genetic engineering," or, more recently, "synthetic biology." The basic concept is very simple: once we insert the gene of, say, insulin in the genome of a bacterium, this bacterium will construct – in addition to all its other proteins – also insulin. And this can, in principle, be turned into an industrial production.

The desire to manipulate genes in living organisms in such a manner arose soon after the discovery of the physical structure of DNA. But it took molecular biologists another twenty years to develop two crucial techniques that would allow them to realize their dream of genetic engineering. The first, known as "DNA sequencing," is the ability to determine the exact sequence of genetic bases along any stretch of the DNA double helix. The second crucial technique, "gene-splicing," is the cutting and joining together of pieces of DNA with the help of special enzymes isolated from microorganisms.

Because of the evocative term "genetic engineering," the public usually assumes that the manipulation of genes is an exact, well-understood mechanical procedure. The reality of bioengineering, however, is much more messy, and the process of inserting genes into living organisms is inherently hazardous, as we shall discuss in more detail in Chapter 18. We shall see that most of the problems surrounding bioengineering today, ultimately, are a consequence of the genetic determinism (see Section 2.3.5) that still pervades the field of biotechnology. Similar considerations apply to the medical applications of genetic engineering, as we discuss in Section 15.1.2.

Another set of genetic applications culminated in the field of synthetic biology, an offspring of genetic engineering, which has the ambition of creating in the laboratory forms of new life – that is, alternatives to the natural life forms – mostly by gene manipulation of extant bacteria. Later on in this book (in Section 10.5), we shall dwell in more detail on this recent and fascinating branch of the life sciences.

9.5 The Human Genome Project

A detailed discussion of all the successes and hazards of molecular biology would take us too far away from our track. Among the applications that are more closely related to this book, however, we should mention the Human Genome Project, the ambitious endeavor of identifying and mapping the complete genetic sequence of the human species.

The Human Genome Project began in 1990 as a collaborative program among several teams of leading geneticists that was coordinated by James Watson and funded by the US government to the tune of $3 billion. During subsequent years, the efforts of these research teams turned into a fierce race between the government-funded project that made its discoveries available to the public and a competing private group of geneticists, funded by venture capitalists, who kept their data secret in order to patent them and sell them to biotechnology companies. In its final dramatic phase, the race was decided by an unlikely hero, a young graduate student, James Kent, who single-handedly wrote the decisive computer program that helped the public project win the race by three days, and thus prevented private control of the scientific understanding of human genes (see *New York Times*, February 13, 2001).

The successful mapping of the human genome revealed a complex genetic landscape with many surprises, some of which contained intriguing clues about human evolution. To their amazement, scientists discovered a kind of genetic fossil record consisting of "jumping genes" that broke away from their chromosomes in our distant evolutionary past, replicated themselves independently, and then reinserted their copies into various sections of the main genome. Their distribution indicates that some of the genome's noncoding sequences may contribute to the overall regulation of genetic activity. In other words, they are not "junk" at all.

The ensemble of genes in the human organism consists of a sequence of 3 billion pairs of bases, and each of them has been identified. The human genome has captured the mass media, and is even seen as "the book of life." However, as we shall show in the next section, this notion is very problematic. There are in the human genome around 25,000 genes, and since human life is based on a considerably larger number of proteins, the classic notion of the "central dogma of molecular biology" – one gene/one protein – does not hold true anymore. Indeed, the Human Genome Project has been a major impetus for a conceptual revolution in genetics, to which we shall turn in the following pages.

9.6 Conceptual revolution in genetics

9.6.1 Problems with the central dogma

As we have seen, the central dogma of molecular biology describes a linear causal chain from DNA to RNA, to proteins (enzymes), and to biological traits. In the colloquial paraphrase that has become popular among molecular biologists, "DNA makes RNA, RNA makes proteins, and proteins make us." The central dogma includes the assertion that its linear causal chain defines a one-way flow of information from the genes to the proteins, without the possibility of any feedback in the opposite direction. During the last decades, this framework has led to a kind of genetic determinism, generating a host of powerful metaphors – DNA being referred to as the genetic "program" or "blueprint of life," the genetic code as the "language of life," and the human genome as "the book of life."

Certainly, the notion that the gene is the central aspect of life seems to be well embedded in our culture. One reads in the popular literature – and not only there – about the gene for obesity, the gene for aggressiveness, and, of course, the gene for longevity. Such a gene-centered view of life represents an exaggerated form of genetic determinism, a kind of novel reductionism. Partly responsible for that is the notion of "the selfish gene," expounded by Richard Dawkins (1976), which is based on the idea that a gene that confers an evolutionary advantage tends to ensure its own survival and transmission. We are opposed to this genetic determinism, and in Section 9.6.4 below we will make clear why.

One main problem with the central dogma became apparent during the late 1970s, when biologists extended their genetic research beyond bacteria. They soon found out that in higher organisms the simple correspondence between DNA sequences and sequences of amino acids in proteins no longer exists, and that the elegant principle of "one gene–one protein" had to be abandoned. Indeed, it seems – perhaps not unreasonably – that the processes of protein synthesis become increasingly complex as we move to more complex organisms.

As already mentioned, when a fragmented gene is copied into an RNA strand, the copy must be processed before the assembly of the protein can begin: in other words, the messenger RNA is edited on its way to protein synthesis. It turns out that this editing process is not unique. The coding sequences can be spliced together in more than one way, and each alternative splicing will result in a different protein. Thus, many different proteins can be produced from the same primary genetic sequence, sometimes as many as several hundred according to recent estimates.

This means that geneticists had to give up the principle that each gene leads to the production of a specific enzyme (or other protein). A clear example is given by the human genome, which, according to current estimates, contains only about 25,000 genes, while our body works with at least three times more different proteins. In other words, which protein is produced can no longer be deduced from the genetic sequence in the DNA. According to Keller (2000, p. 63), this portends a major shift of perspective in genetic research.

Another recent surprise has been the discovery that the regulatory dynamics of the cell determines not only which protein will be produced from a given fragmented gene but also how this protein will function. In short, cellular dynamics may lead to the emergence of many proteins from a single gene and of many functions from a single protein – a far cry indeed from the linear causal chain of the central dogma. And yet another blow to the classic view of the central dogma comes from the field of epigenetics – genetic inheritance without modification of the DNA primary sequences – to which we shall now turn.

9.6.2 Epigenetics

Epigenetics is the study of heritable changes in phenotype, or gene expresssion, caused by mechanisms other than changes in the underlying DNA sequence; hence the name *epi-* (Greek: *επί-* "over," "above") *genetics*. These changes may remain through cell divisions for the remainder of the cell's life and may also last for multiple generations. However, there is no change in the underlying DNA sequence of the organism; instead, nongenetic factors cause the organism's genes to behave (or "express themselves") differently.

There are actually at least two different meanings of this term, which correspond to two different processes. One refers broadly to cell differentiation and development; the other to chemical modification of DNA structure which does not affect the primary sequence.

The first concept was due mostly to the work of the geneticist Conrad Waddington (1905–1975), who actually coined the term "epigenetics" (Waddington, 1953) to refer to the study of the "causal mechanisms" by which "the genes of the genotype bring about phenotypic effects." Thus, epigenetics for Waddington was a subject similar to what we would now call developmental biology. For Waddington, the course of development was determined by the interaction of many genes with each other and with the environment. (Waddington, 1953).

Box 9.5
Genotype and phenotype

The genotype of an organism is its full genetic makeup, or genome, which contains all the hereditary information. The phenotype is the organism's outward physical appearance – that is, the totality of its physical and behavioral characteristics. The differences between genotype and phenotype are due to the fact that two organisms may have identical genomes, but their patterns of gene activity, or gene expression, will generally be different. These patterns of gene expression depend on many epigenetic mechanisms – mechanisms that go beyond the genome and involve the organism's entire metabolic network (see text for more details).

One example of epigenetic changes is the process of cellular differentiation. When cells divide in the development of an embryo, each new cell receives exactly the same set of genes, and yet the cells specialize in very different ways – stem cells becoming fully developed muscle cells, blood cells, nerve cells, and so on. These cell types differ from one another not because they contain different genes, but because different genes are active in them. In other words, the structure of the genome is the same in all these cells, but the patterns of gene expression are different. As Keller (2000) puts it, "Genes do not simply *act*: they must be *activated*."

A similar situation arises when we compare the genomes of different species. Recent genetic research has revealed surprising similarities between the genomes of humans and chimpanzees, and even between those of humans and mice. In fact, geneticists now believe that the basic body plan of an animal is built from very similar sets of genes across the entire animal kingdom. And yet the result is a great variety of radically different creatures. Another example is identical twins. The DNA sequences of their genes are exactly the same; they have the same genome. Yet, physically, identical twins become increasingly different over time.

As we said, the term "epigenetics" (see Haig, 2004) has another meaning. Molecular biologists are in fact more familiar with the definition of epigenetics as "the study of heritable changes in gene function that cannot be explained by changes in DNA sequence" (Riggs *et al.*, 1996). For them, epigenetic mechanisms would include chemical modifications of DNA (in particular, "DNA methylation") and modification of the proteins bound to DNA (the so-called histones). DNA and histones make up what is called chromatin.

Epigenetics is a rapidly expanding area of research with important implications for our understanding of development, evolution, and human health. In his guest essay on p. 198, the developmental biologist Patrick Bateson discusses some of the recent advances in this fascinating field.

9.6.3 Evolution is also symbiosis, symbiogenesis, cooperation, and altruism

As we have seen, there are several reasons to doubt the central dogma and the corresponding genetic determinism. The notion of the "selfish gene" – as we have already mentioned – is

Guest essay

The rise and rise of epigenetics

Patrick Bateson

University of Cambridge

Epigenetics is a term that has had multiple meanings since it was first coined by Waddington (1957). He used the term, in the absence of molecular understanding, to describe processes by which the inherited genotype could be influenced during development to produce a range of phenotypes. More recently, the term *epigenetics* has been used for the molecular processes by which traits, specified by a given profile of gene expression, can persist across the division of each cell without involving changes in the nucleotide sequence of the DNA. In this more restricted sense, epigenetic processes are those that result in the silencing or activation of gene expression through such modification of the roles of DNA or its associated RNA and protein. The term has, therefore, come to describe those molecular mechanisms through which both dynamic and stable changes in gene expression are achieved, and ultimately how variations in extracellular input and experience by the whole organism of its environment can modify regulation of DNA expression (Jablonka and Lamb, 2005).

The growth of interest in the molecular aspects of epigenetics has been extraordinary. In 1960, four papers included the word "epigenetics," according to the Web of Science. By the year 2000, 415 papers were published in that year alone with Waddington's word in their titles. In 2010, only a decade later, an astonishing 3,577 papers used "epigenetics" in their titles. It should be noted, however, that some authors, myself among them, continue to use Waddington's broader definition of epigenetics to describe all the developmental processes that bear on the character of the organism (Bateson, 2012; Jablonka and Lamb, 2010). In all these usages, epigenetics usually refers to what happens within an individual developing organism. Whether a broad or restricted view of epigenetics is taken, a revolution in thinking about the importance of developmental processes has occurred (Carey, 2011).

The molecular processes involved in the development of an organism's characteristics were initially worked out for the regulation of cellular differentiation and proliferation (Gilbert and Epel, 2009). All cells within the body contain the same genetic sequence information, yet each lineage has undergone specializations to become a skin cell, hair cell, heart cell, and so forth. These phenotypic differences are inherited from mother cells to daughter cells. The process of differentiation involves the expression of particular genes for each cell type in response to cues from neighboring cells and the extracellular environment, and the suppression of others. Genes that have been silenced at an earlier stage remain silent after each cell division. Such gene silencing provides each cell lineage with its characteristic pattern of gene expression. Since these epigenetic marks are faithfully duplicated across each cell division, stable cell differentiation results. These processes are likely to play many other roles in development, including the mediation of many aspects of developmental plasticity.

Mechanisms

Variation in the context-specific expression of genes, rather than in the sequence of genes, is critical in shaping individual differences in phenotype. This is not to say that differences in the sequences of particular genes between individuals do not contribute to phenotypic differences, but rather that individuals carrying identical genotypes can diverge in phenotype if they experience separate environmental experiences that differentially and permanently alter gene expression.

A variety of mechanisms are involved in the activation or silencing of genes. One of the silencing mechanisms involves a process known as methylation. Chromosomes consist of strands of chromatin. DNA is organized along chromatin in packets known as nucleosomes. These have a molecule with a hydrogen atom on one of its arms. If this is replaced by a methyl molecule, the nucleosomes close up and the DNA is less able to be expressed as messenger RNA, which in turn forms the template for synthesizing protein. Conversely, if the methyl molecule is replaced by a hydrogen atom, the DNA on the affected nucleosomes can be expressed.

Another important mechanism involves small molecules of noncoding RNA (Mattick, 2011). These are synthesized from DNA found in the part of the genome previously thought to have no function and misleadingly and incorrectly described as "junk." When the small molecules termed micro-RNA are expressed they bind onto messenger RNA with the result that the gene that expressed the messenger RNA loses its capacity to code for protein and is effectively silenced. These molecules that regulate the expression of the genome are extremely numerous and play an important role in the rapidly expanding field of epigenetics. The regulators have themselves to be regulated, and unraveling the networks will take a great deal of research, but the general principles involved in producing differences in cell lines are already apparent.

In their important survey of the growth of epigenetics, Gilbert and Epel (2009) noted the impact on medicine. The susceptibility to a wide variety of diseases is affected not just by genes but also by whether or not environmental influences impact on those genes to silence or activate them. A new field of the developmental origins of health and disease grew out of the observation that children who were born small and were well-adapted to a lean environment were much more likely to develop heart disease in later life if they grew up in an affluent environment (Barker, 1995; Bateson, 2001). As Gluckman *et al.* (2009) argued, the growing body of evidence has important implications for public health measures. The way to treat a small baby may not be to provide him or her with a rich diet but with one that is better tuned to its metabolic adaptations.

Epigenetics and evolution

Many biologists still believe that understanding evolutionary processs does not require any knowledge of development. The argument runs as follows. Genes influence the characteristics of the individual; if individuals differ because of differences in their genes, some may be better able to survive and reproduce than others, and, as a consequence, their genes are perpetuated. The extreme alternative to this view is a caricature of Lamarck's views about biological evolution and inheritance. If a blacksmith develops strong arms as a result of his work, it was argued, his children will have stronger arms than would have been the case if their father had been an office worker. This view has been ridiculed by essentially all contemporary biologists. Nevertheless, as so often happens in polarized debates, the excluded middle ground concerning the evolutionary significance of development and plasticity has turned out to be much more interesting and potentially productive than either of the extreme alternatives. This view was developed at length by West-Eberhard (2003), who argued that developmental plasticity was crucial in biological evolution. These same ideas are expressed superbly in Gilbert and Epel's (2009) book and are developed further in the book edited by Pigliucci and Müller (2010). Moreover, some of Lamarck's thinking, discredited by its association with the spurious blacksmith argument, has been rehabilitated (Gissis and Jablonka, 2011).

A growing body of evidence suggests that phenotypic traits established in one generation by epigenetic mechanisms may be passed directly via micro-RNA or indirectly through to the next (Gissis and Jablonka, 2011). One example of indirect transmission across generations comes from laboratory studies of rats. A mother rat that licks her pups a lot has offspring which, when adult, lick their offspring a lot. Conversely, mothers who are low groomers have offspring who grow up to be low groomers. In this way, a characteristic style of maternal behavior is transmitted from one generation to the next. Cross-fostering a pup born to a low-grooming mother to a naturally high-grooming mother switched the adult pattern of the pup to that of the foster mother (Champagne *et al.*, 2003), showing that this is not a genetically transmitted trait but an acquired one. The differences produced by maternal behavior arise due to variations in brain development induced by epigenetic modification of gene expression in the brain (Champagne, 2010). Receiving high levels of licking neonatally is associated with reduced levels of DNA methylation of the promoter region of a particular gene, which is established by 6 days after birth and persists throughout adulthood.

None of the evidence for transmission across generations relates in itself to the thinking about biological evolution because the transgenerational epigenetic effects could wash out if the conditions that triggered them in the first place did not persist. The crucial question is to ask how epigenetic changes that are not stable could lead to genetic changes.

The Galápagos finches are a clear example of how, in a relatively short space of time, birds arriving from the mainland were able to radiate out into many different habitats (Grant, 1986). Tebbich *et al.* (2010) discuss how the finches' capacity to respond to environmental challenges, for which they provide some evidence, could have played an important role in this process. None of this challenges the evolutionary mechanism postulated by Charles Darwin and Alfred Russel Wallace. The evolutionary process requires variation, differential survival and reproductive success, and inheritance. Three questions for the modern study of epigenetics arise from this formulation. First, what generates variation in the first place? Second, what leads to differential survival and reproductive success? Third, what factors enable an individual's characteristics to be replicated in subsequent generations? In answering all of these questions, an understanding of development is crucial.

The decoupling of development from evolutionary biology could not hold sway forever. Whole organisms survive and reproduce differentially, and the winners drag their genotypes with them (West-Eberhard, 2003). The way they respond phenotypically during development may influence how their descendants' genotypes will evolve and become fixed (Bateson and Gluckman, 2011). This is one of the important engines of evolution, and it is the reason why it is so important to understand how whole organisms behave and develop.

The characteristics of an organism may be such that they constrain the course of subsequent evolution, or they may facilitate a particular form of evolutionary change. The theories of biological evolution have been reinvigorated by the convergence of different disciplines. The combination of developmental and behavioral biology, ecology, and evolutionary biology has shown how important the active roles of the organism are in the evolution of its descendants.

The combination of molecular biology, palaeontology, and evolutionary biology has shown how important an understanding of developmental biology is in explaining the constraints on variability and the direction of evolutionary change.

Conclusions

The revolutionary changes in biology, through the impact of epigenetics, have enhanced greatly the understanding of what happens as an individual develops. The linear causal view of how genes are involved in the underlying processes has been replaced by much more holistic approaches to the dynamics of development. This change is leading to the bringing together of research from different levels of analysis. Advances in molecular biology have been stunning, but these have been accompanied by a growing respect for the whole organism. So epigenetics will continue to rise and will impact increasingly on medicine and the study of evolution.

References

Barker, D.J. (1995). The fetal origins of adult disease. *Proceedings of the Royal Society of London. Series B*, **262**: 37–43.

Bateson, P. (2001). Fetal experience and good adult design. *International Journal of Epidemiology*, **30**: 928–34.

(2012). The impact of the organism on its descendants. Article ID 640612. *Genetics Research International*, doi: 10.1155/2012/640612.

Bateson, P. and P. Gluckman (2011). *Plasticity, Robustness, Development and Evolution.* Cambridge University Press.

Carey, N. (2011). *The Epigenetics Revolution: How Modern Biology Is Rewriting Our Understanding of Genetics, Disease and Inheritance*. London: Icon Books.

Champagne, F.A. (2010). Epigenetic influence of social experiences across the lifespan. *Developmental Psychobiology*, **55**: 33–41.

Champagne, F.A., D.D. Francis, A. Mar, and M.J. Meaney (2003). Variations in maternal care in the rat as a mediating influence for the effects of environment on development. *Physiology & Behavior*, **79**: 359–71.

Gilbert, S.F. and D. Epel (2009). *Ecological Developmental Biology: Integrating Epigenetics, Medicine and Evolution*. Sunderland, MA: Sinauer.

Gissis, S.B. and E. Jablonka (2011). *Transformations of Lamarckism: From Subtle Fluids to Molecular Biology*. Cambridge, MA: MIT Press.

Gluckman, P.D., M.A. Hanson, P. Bateson, A.S. Beedle, C.M. Law, Z.A. Bhutta, *et al.* (2009). Towards a new developmental synthesis: adaptive developmental plasticity and human disease. *Lancet*, **373**: 1654–7.

Grant, P.R. (1986). *Ecology and Evolution of Darwin's Finches*. Princeton University Press.

Jablonka, E. and M.J. Lamb (2005). *Evolution in Four Dimensions*. Cambridge, MA: MIT Press.

(2010). Transgenerational epigenetic inheritance, in Pigliucci and Müller, *Evolution – the Extended Synthesis*, pp. 137–74.

Mattick, J.S. (2011). The central role of RNA in human development and cognition. *FEBS Letters*, **585**: 1600–16.

Pigliucci, M. and G.B. Müller, eds. (2010). *Evolution – the Extended Synthesis*. Cambridge, MA: MIT Press.

Tebbich, S., K. Sterelny, and I. Teschke (2010). The tale of the finch: adaptive radiation and behavioural flexibility. *Philosophical Transactions of the Royal Society of London. Series B*, **365**: 1099–1109.

Waddington, C.H. (1957). *The Strategy of the Genes*. London: Allen & Unwin.

West-Eberhard, M.J. (2003). *Developmental Plasticity and Evolution*. New York: Oxford University Press.

fallacious in many ways. For example, it conveys the idea that one gene works in isolation, going about its own selfish business. In fact, it does not make sense to consider a gene in isolation as being responsible for a complex function. For each biological function, there is always a series of genes working together. Cooperation of genes with each other is the main operational basis of genetics, and therefore of evolution. In addition, one should remember that each gene is read by proteins and is synthesized by proteins, so that a complex genetic function must be seen in terms of a network of genes connected to a network of proteins. This is the systems view of genetics and evolution. We are back to a systemic conception of life.

Cooperation is clearly visible also at many levels of living organisms. At the level of any multicellular organism, we see cooperation among the different cells and tissues. An insect may consist of dozens of different cells, and obviously the life of such an organism is based on the harmonious cooperation of all its parts. One question often asked here is this: "How do these different parts 'know' that they belong to the same unit?" In a literal sense, this is a fallacious question, as the parts cannot "know," but it is an interesting question from the heuristic point of view, as it obliges one to think in terms of systems biology, as well as in terms of cooperation in evolution, since these parts clearly have "learned" to positively interact with each other due to natural selection and adaptive pressure. This goes back to the notion of "cognition," as understood by Maturana and Varela, to which we shall return in Chapter 12.

Another very important aspect of cooperation is denoted by the term "symbiosis." Symbiosis, the tendency of different organisms to live in close association with one another, and often inside one another (like the bacteria in our intestines), is a widespread and well-known phenomenon. It is enough to think of our own body, which hosts such a large number of microorganisms that about 95% of the cells of our body are not human (of course this is not true in terms of weight). We have in our guts several hundred grams of *E. coli*, which lives happily inside us, performing at the same time important functions for our body. Throughout the animal kingdom we see countless examples of symbiotic life: birds with pachyderms, and small fish with larger fish, and the examples are even more numerous in plants. Again, all these cooperative arrangements are the result of millions of years of evolutionary pathways.

When certain bacteria merged symbiotically with larger cells and continued to live inside them as organelles, the result was a giant evolutionary step, which prepared the way

Box 9.6
A short microbial glossary

prokaryotes, from Greek *pro* ("before") + *karyon* ("kernel"): a group of microorganisms that lack a membrane-bound nucleus and also lack membrane-bound organelles; thus, their DNA is openly accessible within the cell; prokaryotes comprise two domains: bacteria and archaea.
eukaryotes, from Greek *eu* ("good") + *karyon* ("kernel"): organisms that have a nucleus, organelles, and genetic material organized in chromosomes; eukaryotes may be unicellular (protists, fungi) or multicellular (plants, animals).
archaea: a group of prokaryotes that were previously classified with bacteria but have recently been identified as a distinct domain of life; archaea inhabit some of the most extreme environments on the planet (hot springs, extremely saline waters, etc.).
cyanobacteria, from Greek *kyanos* ("blue"): photosynthesizing bacteria, also known as "blue-green bacteria," which in the distant past evolved into chloroplasts in plants and into eukaryotic algae through symbiogenesis (see text for details).
stromatolites, from Greek *stroma* ("stratum") + *lithos* ("rock"): rock-like accretions of microbial mats formed in hypersaline. shallow water by the trapping, binding, and cementation of sedimentary grains by microorganisms, especially cyanobacteria.
organelles: specialized, membrane-bound, subunits of eukaryotic cells, which are analogous to body organs; examples are the mitochondria (the sites of energy production through cellular respiration) and chloroplasts (the sites of photosynthesis in plants).

for the evolution of the complex, sexually reproducing, higher organisms we now see in our environment. The idea of bacteria and other microorganisms living inside larger cells, with this symbiosis leading to new forms of life, has been emphasized in particular by the microbiologist Lynn Margulis. Margulis published her revolutionary hypothesis first in the mid 1960s, and over the years developed it into a full-fledged theory, now known as "symbiogenesis," which sees the creation of new forms of life through permanent symbiotic arrangements as the principal avenue of evolution for all higher organisms (see Margulis and Sagan, 2002).

Her idea is, for example, that the eukaryotic cell came about through the fusion of a prokaryote with another microorganism in a kind of physical cooperation initially, which eventually gave rise to a new form of life (as an emergent property) that could self-reproduce more efficiently. Likewise, the chloroplasts in modern plants are the descendants of ancient symbiotic cyanobacteria, which installed themselves in plant cells about a billion years ago. The recent mapping of the human genome has provided more support for the theory of symbiogenesis, as geneticists discovered that the genome of higher animals contains numerous microbial gene sequences – very likely a signature of ancient symbiogenesis.

At the level of the animal kingdom, the notion of cooperation may acquire the noble aspect of altruism. This is indeed a very active field of research for evolutionary biologists and cognitive psychologists, as it is at first sight not apparent why a single individual may sacrifice its own life for the benefit of the group. However, it has now become accepted that altruism is also a way to defend and preserve the genetic patrimony of the entire group or species.

Altruism and cooperation are, of course, widely displayed at the social level in the formation of groups of animals – families, packs, herds, flocks – including human communities with families and all kinds of social institutions where mutual cooperation is essential. These social groupings, too, are the result of evolution, since the coming together in groups, as opposed to free individuals, is a more secure way of protection and survival. Physical cooperation probably preceded mental cooperation; indeed, there is an interesting interplay between these two domains (Tomasello, 1999).

Cooperation also has some political connotations. For example, the nineteenth-century Russian anarchist and aristocrat Piotr Kropoktin wrote a book with the title *Mutual aid* (1902), in which he argued that evolution results more in cooperation than harsh competition. And modern game theories show that in some cases cooperation is the winning strategy (Axelrod, 1984).

Speaking of the sociopolitical level, we should mention two important names, often linked to Darwinism and not always in a positive light: Malthus and Spencer. Let us consider first the common expression that many associate with Darwinism: the notion of "survival of the fittest." This expression was not used by Darwin, who, however, used the expression "struggle for life." The term "survival of the fittest" goes back to the sociologist Herbert Spencer and his social Darwinism (Spencer, 1891/1854), according to which those best fit in human society are the richer and the better educated people. These would and should go ahead and survive, whereas the weaker and the poor should be left to their fate. The economist Thomas Malthus, another important figure of that Victorian era, tended to defend that idea, which he considered to be the natural scenario of evolution (Malthus, 1798).

Darwinism actually had nothing to do with these ideas of "social Darwinism," which are an extrapolation based on the false assumption that the evolution of human society proceeds according to the biological evolution of simple organisms, and on the additional false assumption that all that is "natural" – seen in the scenario of life in nature – should be right from the moral point of view. However, even if we discard social Darwinism and the conservative aspects of sociobiology, the fact remains that natural selection, generally, can be seen as some sort of competition among different living groups. We will come back to this point in the next chapter in connection with the determinants of being human.

9.6.4 We are not our genes!

Let us now return to the issue of genetic determinism. The main criticism of the gene-centered view comes from the simple consideration that the commonly propagated relation "one gene – one function" does not exist. It is very difficult, if not impossible, to identify single genes with a defined function; or rather, this is possible to some extent with low-level functions, like the production of proteins, but is more difficult to do so with high-level functions, as at the level of behavior.

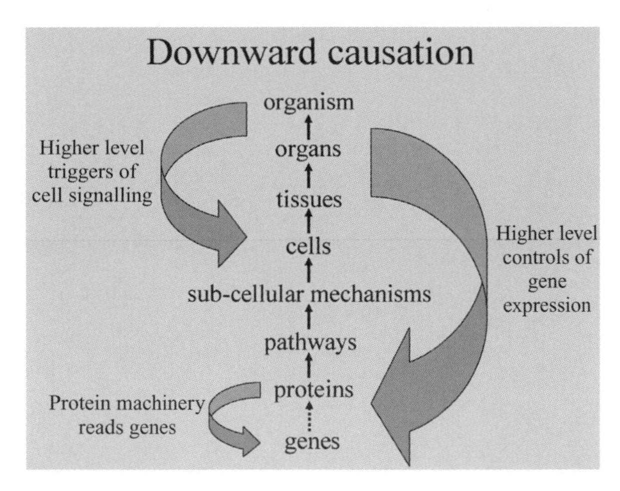

Figure 9.6 Downward forms of causation (large arrows). The higher levels trigger cell signaling and gene expression; the smaller arrow, in particular, indicates that it is the protein machinery that reads and interprets gene coding (from Noble, 2006, with permission).

A single gene can "express" (i.e., induce the synthesis of) a protein. However, any complex function is determined by a large number of proteins, so that we should at least consider, as we have said before, a network of genes and proteins always connected with each other. Even so, the idea that genes are the linear cause of biological functions, and of life itself, is a very rough approximation of the biological reality. In our criticism of the gene-centered view, we shall follow the ideas of the eminent systems biologist Denis Noble (2006), who advanced convincing arguments to suggest that the gene-centered deterministic view is simply wrong.

To begin with, one should consider that DNA alone does not do anything; it must be read by proteins in order to produce other proteins. Noble, very aptly, makes the analogy with a CD: the DNA is like a CD, a pattern of instructions, which does not make any sense without the CD reader. Specific proteins are the CD readers. There are several dozen proteins to do this reading, and many others to express a gene. And where do all these proteins come from? From the metabolism of the cell – namely, from life itself. Thus, the relevance of DNA needs to be understood in terms of systems biology, according to which it is the organism itself that determines which genes will be read and expressed. It is a top-down more than a bottom-up view. The conventional bottom-up view of a chain of causality – DNA causes proteins; proteins cause cells; and, finally, DNA "causes life" – can be seen as a form of reductionism; at the very least, it must be understood in terms of the arising of emergent properties as complexity increases.

A more systemic view involves the concept of downward causation. We reproduce here (Figure 9.6) a chart from Noble's book showing the linear chains of causality going from the genes to the function (the reductionist/deterministic view) and how this should be modified by feedback controls coming from the higher levels of organization. Based on these considerations, Noble (2006, p. 51) goes on to criticize the notion of the human genome as "the book of life":

> This is my primary reason for opposing the otherwise colorful metaphor describing the genome as "the book of life." A book may describe, explain, illustrate, and many other things, but if we open it up to find just strings of numbers, like the machine code of a computer program . . . we would say that we had only been given a database. My central point will be that the book of life is life itself; it cannot be reduced to just one of its databases.

Another important concept emphasized by Noble (2006, p. 18) relates to his opposition to the notion of the selfish gene: "Genes are captured entities, no longer having a life of their own independent from the organism. They are forced to cooperate with many other genes to stand any chance of survival." This is related to our notion of operational networks of genes and proteins and reinforces the idea that genes must cooperate with each other in order to ensure a functional organism. No gene is selected in isolation. It is interesting to recall here that there is also a book by Dawkins (2003) in which he says, "Genes aren't us."

There is another line of argument against the gene-centered view: this is the now accepted observation that the three-dimensional arrangement of the DNA threads is important to how the DNA sequences are read. We have already mentioned this notion in the previous section about epigenetics. Research on this subject is presently very actively carried out in several laboratories around the world, and although it is too early to draw a general conclusion, it is apparent that this will be an important, new chapter of molecular biology and genetics.

All this makes the relation between the genome and the organism an extremely complex matter. It is understandable that the mass media tends to simplify this picture, but this generates a series of misconceptions. One of them is that the knowledge of the genome sequence can tell us a lot about the nature of the organism, its state of health, and possibly even its behavior. It is important instead to emphasize that the genome is just a mute sequence of letters. Does the genome tell you why a dog is a dog, and why a man is a man? The answer is simply: no.

Another, more subtle, aspect of genetic determinism is contained in the metaphor of the genome as the software for the program of constructing life. In a way, this is a bottom-up view, with the software forming the basis of the hardware for the construction of proteins, and then all the rest, up to the cell. In accordance with the systemic view, Noble refuses the notion of a program and agrees with the assertion by Enrico Coen (1999) that organisms are not simply manufactured according to a set of instructions, as in fact there is no easy way to separate instructions from the way they are carried out, to distinguish plan from execution. Following strictly this point of view, one could say that in life there is no software, there is only the hardware: molecules that interact with each other in a complex networking system.

Note that this systemic view is in complete agreement with our exposition of cellular life as an integrated collective system, as discussed in the previous chapter. There we advanced the notion of a collective ensemble without a center of localization. The refusal to localize the center of life in the genome, or in the DNA in general, corresponds to this view. Life itself in its totality is the expression of an operationally closed system.

We would like to conclude this section with Noble's (2006, p. 19) forceful final assessment:

> Any intelligence the system has, is at the level of the organism, not at the level of genes. Also to say that this intelligence is encoded in the program of the genes, is not correct, because . . . there is no such a thing as a program. We are the system that allows the code to be read.

9.7 Darwinism and creationism

We have mentioned the reluctance of Charles Darwin to publish his theories, as he was aware that church authorities and the whole Christian cultural environment would be shaken and would take up arms against him and his notion of species changing due to natural selection. This was indeed what happened, and the controversy that followed was no surprise. The surprise is rather the fact that this controversy is still present after more than 150 years from the publication of *The Origin of Species*. Many people still hold on to the idea that species are fixed entities, created once and for all by God, and all according to the description of Genesis. This in itself would not be so much of a problem; the real problem lies in the fact that something has been created that took the name of creation science, or scientific creationism, which attempts to provide *scientific* support to the Genesis mythology that the world was created in six days *ex nihilo* a few thousand years ago. This attempt challenges the geological and astrophysical evidence for the age of the Earth and of the universe, as well as the relevance of the fossil record, and, of course, all forms of Darwinian evolution.

As the physicist and philosopher Carl Friedrich von Weizsäcker put it succinctly: "The Bible must be taken either seriously, or literally." Generally, creationists do not recognize that the biblical account of creation, like all religious creation myths, is just that: mythology. Such a fundamentalist blindness could even be tolerated, were it not for the fact that this kind of creationism has turned into a political movement that challenges science curricula in the school systems. Many school boards and lawmakers, mostly in the USA, have been persuaded to include the teaching of "creation science" alongside Darwinian evolution in the science curriculum, starting in Louisiana and Arkansas (Wilder-Smith, 1968, 1987).

This was actually the beginning of an exasperating series of trials. The 1982 ruling in *McLean* v. *Arkansas* found that creation science fails to meet the essential characteristics of science and that its chief intent is to advance the Christian religion. This was not sufficient to quieten the creationists, however, who kept pushing and lobbying. In 1987, there was finally a US Supreme Court decision (after the Louisiana trial), stating that the teaching of creation science alongside evolution is unconstitutional because its sole true purpose is to advance a particular religious belief.

But even this was not the end of the story. In fact, from this point on, creationism changed its stripes, becoming more subtle and covert by morphing into the ideology of "intelligent design" (ID). ID does not explicitly mention the biblical account. It does not even negate certain forms of evolution, but it maintains that even the simplest living systems are far too complex to have developed by natural, unguided processes. This ideology appeared also in a controversial and famous book, *Of Pandas and People: The Central Question of Biological Origins*, conceived as a textbook (Davis and Kenyon, 1989, 1993) Here, there was no reference to a "creator" nor to "creation," and therefore it was supposed to be in agreement with the Supreme Court ruling. The terms "creator" and "creation" were replaced by *designer* and *design*. Clever enough. However, again parents filed suit (in Dover, Pennsylvania) to halt the teaching of ID. There were other court trials (e.g., *Kitzmiller* v. *Dover Area School District*), but by this point ID had become a powerful political movement, and had extended to Australia and even to Europe. And this despite the fact that the US National Academy of Sciences (1999) stated that "creation science is in fact not science and should not be presented as such," and that "the claims of creation science lack empirical support and cannot be meaningfully tested." Moreover, mainstream Christian clergy criticized creation science on theological grounds, asserting either that religious faith alone should be a sufficient basis for belief in the truth of creation, or that efforts to prove the Genesis account of creation on scientific grounds are inherently futile because reason is subordinate to faith and cannot thus be used to prove it. For more literature on this controversy, see Behe (1996), Dembski (1999), and Petto and Godfrey (2007) .

There is nothing new in the arguments of ID. Actually the notion has a classic precursor in the writings of William Paley, the Anglican vicar who introduced one of the most famous metaphors in the philosophy of science, the image of the watchmaker (Paley, 1802, pp. 1–2):

> When we come to inspect the watch, we perceive . . . that its several parts are framed and put together for a purpose, e.g. that they are so formed and adjusted as to produce motion, and that motion so regulated as to point out the hour of the day; that if the different parts had been differently shaped from what they are, or placed after any other manner or in any other order than that in which they are placed, either no motion at all would have been carried on in the machine, or none which would have answered the use that is now served by it . . . [T]he inference we think is inevitable, that the watch must have had a maker – that there must have existed, at some time and at some place or other, an artificer or artificers who formed it for the purpose which we find it actually to answer, who comprehended its construction and designed its use.

Living organisms, Paley argued, are even more complicated than watches, and thus only an intelligent designer could have created them, just as only an intelligent watchmaker can make a watch. According to Paley, "*That designer must have been a person. That person is God.*" It is clear that ID is already implicit in this single passage. The modern ID ideologists have added nothing conceptually, only the ambition – and the illusion – of transforming this ideology into science.

Paley's metaphor was already opposed in his time by David Hume and other contemporary philosophers, and much more emphatically by molecular and evolutionary biologists in the twentieth century. In the 1970s Jacques Monod, one of the founders of molecular biology and famous (along with his colleague François Jacob) for his work on the control of gene expression, created a big stir with his book *Chance and Necessity* (1971), in which he stated unequivocally that the origin and evolution of life are a product of pure chance.

We shall return to Monod's radical statement and qualify it within a systemic perspective in Section 9.8 below.

More recently, Richard Dawkins, perhaps the most prominent representative of an uncompromising mechanistic view of life in our time, revived Paley's famous metaphor in the title of his book *The Blind Watchmaker* (1986). Dawkins reminds us that nature proceeds without an a priori purpose, according to the "tinkering" mechanisms advocated by Monod's colleague, François Jacob (1982), a few years earlier.

This is a very important, critical point: it is not the function that determines the structure but, rather, the opposite, the structure that determines the function. This is the very basis of natural selection, and at the same time it is counter-intuitive. For example, the flagellum of bacteria was not "created" and projected in order to permit the bacteria to swim – rather, out of the original bacteria without flagella, a fiber protein originated from a mutation; the organisms so provided could move better, gather more food, and reproduce faster, so that succeeding generations had a higher frequency of this new species. Then other mutations transformed this fiber protein into a bundle of proteins, and eventually into a more complex mosaic provided with motor-like activity, until the final structure emerged – all in the course of millions of years of ceaseless trials and errors. And so it was for the eye, which originated most likely from a single light-sensitive protein in some bacteria, and developed eventually, through chance mutations, into more and more complex structural associations.

Creationists would argue that humans have eyes because they have the need of eyesight; refusing, of course, to recognize the counterintuitive evolutionary mechanism. It is interesting, at this point, to mention that even Aristotle, in his classic text *De partibus animalium* ("On the Parts of Animals"), taught that "nature adapts the organ to the function, and not the function to the organ"; and Lucretius wrote in his celebrated epic poem *De rerum natura* ("On the Nature of Things") that "nothing in the body is made in order that we may use it. What happens to exist is the cause of its use."

The foresight of these philosophical geniuses is indeed astonishing. Lucretius' last sentence – "What happens to exist is the cause of its use" – could have been said in our time by Stephen Jay Gould or any modern evolutionary biologist. It is certainly remarkable that certain minds over two thousand years ago could see things that our modern ID advocates fail to see.

The fact that the arguments in favor of Darwinian evolution are counterintuitive and occasionally difficult to explain to a broad public, whereas those of the creationists are easy ("the eye has been fabricated in order to see"), explains in part the spreading of the ID movement. The other argument used by ID proponents pivots on the idea that, if Darwinism is right, life would have no purpose and would be deprived of moral codes. (We shall return to the issue of morals in our discussion of science and spirituality in Chapter 13.)

The problem, as already mentioned, is that ID is spreading, carrying with itself ignorance and misconceptions about science. Its success is due to a combination of factors; but, basically, it rests on the fact that preaching ignorance and fear is rather efficient in convincing uncritical, simple-minded listeners; and this is combined with a philosophy well rooted in fundamentalism, and turned into a political movement, with political lobbies, usually associated with wealthy, right-wing politicians. To fight against this cloud of falsehood should be the duty of any scientist who cares for accuracy and the respect of reason. It is the old battle between Galileo and the clergy, the eternal fight against blindness and superstition.

9.8 Chance, contingency, and evolution

As mentioned above, one of the arguments of the proponents of ID is that acceptance of Darwinian evolution would make all life, including humanity, the product of chance. Let us dwell on this critical point, and to do so let us start from two milestones of modern evolutionary thinking.

The first is the view that evolution proceeds without programmed plans; we have mentioned earlier Jacob's concept of "tinkering." If there are no programmed plans, then the arising of humanity cannot be considered programmed either. This means that humans might not have evolved – we might not be here.

The second point concerns the notion of chance. Is it true that all there is in nature, including human beings, is the product of chance? For this second question, let us consider again Figure 9.5, according to which the entire biodiversity on the planet originated from a single protocellular species, the LUCA (last universal common ancestor). The three domains of Figure 9.5, with all their branchings and subfamilies, originated from random mutations and from the other complex genetic exchange mechanisms briefly outlined earlier. Again, these three main directions of life were not programmed. All chance, then? Indeed, this is the view so forcefully expressed by Jacques Monod in his famous book, *Chance and Necessity* (1971, p. 122): "Chance alone is at the source of every innovation, of all creation in the biosphere."

Monod has a point, but it needs to be qualified. If we take it in an absolute sense, we get a very restricted and partly distorted view of nature and evolution. The first partial corrective concerns the notion of structural determinism, which we have already discussed (see Section 7.4.2); the second has to do with the replacement of the word "chance" by the word "contingency," in the sense of modern evolution thinkers, such as Stephen Jay Gould or Ernst Mayr, who consider contingency to be the main driving force of evolution (Gould, 1989, 2002; Mayr, 2000; see also Oyama *et al.*, 2003).

What is contingency? We will have more to say about this important concept and its relation with determinism in the next chapter of this book in connection with the origin of life on Earth. Here, we wish to state in anticipation that contingency, defined in the dictionary as "an unforeseen occurrence," is the simultaneous occurrence of factors which are independent of one another but together determine a specific event in a precise temporal and spatial situation. Take the example of a tile that falls from an old roof, hits a flying pigeon, and kills it. Chance? Yes, but this "chance event" is due to a series of independent factors, such as the velocity at which the bird was flying, the age of the roof, the density of the tile, the presence or absence of wind, and so on. Each of these factors is causally determined: the tile falls obeying gravity; the roof ages as a consequence of erosion; the flying velocity of the bird depends on its dimensions and age. But all these deterministic factors intermingle with each other in an unforeseeable way.

There is one additional important consideration in this simple example: change only one of these factors, (the speed of the pigeon, the age of the roof, the density of the tile, or the amount of wind) and the accident would have not happened. This is indeed the main statement that characterizes contingency – *It might have not happened!* – and this is, of course, in opposition to the idea of absolute determinism, of some kind of predetermined pathway.

What is the relevance of this example to biological evolution? This point has been discussed by several leading evolutionary theorists, including first and foremost Stephen Jay Gould, one of the most eloquent authors on contemporary evolutionary thought (see, e.g., Gould, 1980, 1991). In the memorable words of Gould (1989, p. 48):

> Since human intelligence arose just a geological second ago, we face the stunning fact that the evolution of self-consciousness required about half of the earth's potential time. Given the errors and uncertainties, the variation of rates and pathways and other runs of the tape, what possible confidence can we have in the eventual origin of our distinctive mental abilities? Run the tape again, and even if the same general pathways emerge, it might take twenty billion years to reach self-consciousness this time – except that the earth would be incinerated billions of years before. Run the tape again, and the first step from prokaryotic to eukaryotic cell might take twelve billion instead of two billion years; and stromatolites, never awarded the time needed to move on, might be the highest mute witnesses to Armageddon.

The implication of this passage is clear: the onset of multicellular organisms, and therefore the subsequent steps, all the way to the evolution of humans, may have arisen much later – or may never have arisen. This is contingency in the clearest form. Understandably, this crucial role of contingency in evolution is what worries the proponents of ID, as it clearly implies that humanity might never have arisen. We shall come back to this concept in the next chapter in connection with the origin of life, a field in which the relation between contingency and absolute determinism (the view that life is an absolute imperative on Earth) is particularly critical.

Here, going back to our notion of "chance," we want to emphasize that, once the confluence of contingent factors that triggered the emergence of eukaryotes has taken place, the products cannot be considered due to chance. They are due to the particular structure of the starting unicellular organisms. In other words, the inner organization of the starting prokaryotes determines, by the structural determinism we have discussed previously, and with the particular environmental conditions, the resulting product: the product cannot be an arbitrary, totally random structure.

Let us illustrate this point with some other examples. About 2 billion years ago, the "blue-green" cyanobacteria invented the type of photosynthesis that we see today, which produces oxygen as a byproduct. This was due to a mutation, and in principle there was no reason why this mutation should have happened. By accident then, oxygen production was "invented," and this mutation determined the course of all subsequent evolution. If this mutation had not occurred, there would be no oxygen, no chlorophyll, and no photosynthesis. In this case, our world would still be populated peacefully by prokaryotes alone.

Was the event that triggered oxygen production a totally random mutation? True, it might not have occurred. On the other hand, this particular mutation, which involved the interplay of several different genes, was due to the particular inner organization of the original prokaryotes, and also due to their particular interaction with the environment. So, again, it was a random event, but it took place against the background of structural determinism and consonant environmental interaction. It took place because it was, in a way, consonant with the inner structure of the original organisms.

On the basis of the contingency arguments discussed in this section, we cannot say that the human species is a deterministic outcome of the first algae and fish. Evolution could have gone into quite different avenues. The statement that humanity might not be here may be a bitter pill for many, but one that must be swallowed if we want to accept the Darwinian scenario.

For most advocates of ID, this means that, since nobody has programmed us, life is not worth living. We believe that this is indeed a strange, almost sickening conclusion. On the contrary, moral conduct, love for our fellow human beings, and the desire for peace and happiness assume the greatest value precisely because they are not the result of some program, but values discovered by our own humanity.

9.9 Darwinism today

Darwinian evolution is a theory capable of explaining the biodiversity of our planet, from microorganisms to mammals, and from fish to plants. It is a grand canvas that unifies the living in all its forms and behavior. As Charles Darwin himself observed in his *Origin of Species*,

> It is interesting to contemplate a tangled bank, clothed with many plants of many kinds, with birds singing on the bushes, with various insects flitting about, and with worms crawling through the damp earth, and to reflect that these elaborately constructed forms, so different from each other, and dependent upon each other in so complex a manner, have all been produced by laws acting around us . . . There is grandeur in this view of life, with its several powers, having been originally breathed into a few forms or into one; and that, whilst this planet has gone cycling on according to the fixed law of gravity, from so simple a beginning endless forms most beautiful and most wonderful have been, and are being, evolved.
>
> *(Darwin, 1859, p. 490)*

Darwin's view evidently is a perfect representation of the systems view of life. In his view, all living beings are linked to each other by a historical thread from the very beginning of life, in a gigantic network of relations that make all of us part of a unique family all over the globe, from immemorial time past, to all foreseeable future. There really is a "grandeur in this view of life," and in this sense the assertion of the creationists – that life has no dignity if one follows Darwinism – shows great ignorance.

Of course, Darwinism, as we have noted, also has its shadow sides. We mentioned, for example, the drift of neo-Darwinism into sociobiology and genetic determinism. According to the latter, behavior, including human behavior, is the result of one particular gene. Here too one needs to be careful. Of course, the lion's behavior differs from that of the gazelle because of their different genomes. This is undeniable, but to ascribe a behavioral trait to a single gene, as is often done in the mass media or, more maliciously, by certain pharmaceutical companies, is a different thing altogether and is generally fallacious. This is why in we emphasized the systemic downward approach in this chapter, highlighting that it is not the single gene that is the cause of the organism's function, but rather the entire organism that activates the gene.

Some of the problems with Darwinism, even in modern times, derive from the fact that some of the basic concepts are not as easy as they seem at first sight. Take, for example, the notion of chance mutation and its relation to natural selection. To clarify this notion, the concept of structural determinism, discussed in Chapter 7, is very helpful – but again, anything but trivial. Also, the arguments for punctuated equilibrium and, at another level, for the origin and development of human cognition and consciousness (see Chapter 12) are not easy to accept. On the other hand, one should not forget that Darwinism, like all other branches of science, is a theory in progress, far from having reached in all aspects a complete and satisfactory answer.

There is no doubt that today we are living in the age of the gene; the gigantic project on the human genome is the best and most celebrated evidence for it. In February 2001, *Nature* published a 62-page article with the initial sequencing and analysis of the human genome, and former US President Bill Clinton called it "the most important, most wondrous map ever produced by humankind," with his science adviser stating that we now had the possibility of achieving all that we ever hoped for from medicine (*Nature* editorial, 2001).

An editorial of the same journal ten years later is much more sobering, being aptly subtitled: "Ten years after the human genome was sequenced, its promise is still to be fulfilled" (*Nature* editorial, 2011). And in what is called an "updated vision" of the prospects for genomic medicine, the geneticists Eric Green and Mark Guyer of the US National Human Genome Research Institute (2011) write that we cannot expect profound improvements for many years. The biologist Eric Lander, the first author of the cited 2001 article with the sequencing of the human genome, echoes similar concepts (Lander, 2011). Indeed, it is a well-accepted notion today that the importance of the sequencing of the human genome has been, until now at least, more on the side of basic science – inducing a new concept of the gene in particular – than on the side of application. And it is also clear that the necessary step for successful applications is a better understanding of the DNA structure in the genome, including, in particular, the three-dimensional structure/function relationship of the noncoding regions.

We have mentioned that this "age of the gene" has brought about the simplistic philosophy of genetic determinism, and although from the academic side there is nowadays considerable opposition to it, the mass media is still filled with news about, say, the gene for obesity, the gene for longevity, or the gene for happiness. We will probably have to cope with that for long time, as the gap between good science and the mass media is generally quite big.

At any rate, genetic determinism is not so "politically" dangerous as creationism and its offshoot ID which claim to be scientific movements and want to change the education system in schools. Against this wave of ideology and ignorance, scientists are really called to battle.

9.10 Concluding remarks

We have mentioned that one of the reasons why the proponents of ID are opposed to Darwinism lies in their assumption that evolution is equivalent to nature being completely in the hands of chance. We have cited the frightening sentence by Monod, and we can add here another of this author's famous statements (Monod, 1971, p. 114): "Man at last knows he is alone in the unfeeling immensity of the universe, out of which he has emerged only by chance."

Yes, mutations are random, and so is gene mixing, but evolution as a whole does not proceed randomly at all. Nature is very choosy in selecting a viable mutation. In order to be "accepted," the mutation has to respect several conditions and constraints. First of all, there is the principle of structural determinism (see Section 7.4.2), which implies that only those changes can be accepted that are consistent with the existing inner structure and organization of the living organism. Moreover, the mutation must produce a minimal perturbation so as to respect the main function of the living cell, which is self-maintenance. The cell's proper individuality must be preserved: a liver cell tends to remain a liver cell; a nerve cell remains a nerve cell; and so on. In addition, the mutation must permit adaptations to environmental changes; and, finally, it has to comply with the laws of physics and chemistry that govern the cell's metabolism.

Mutations, as we have seen, are only one type of event leading to variation and evolution. The other two types are lateral exchanges of genes among bacteria and the ingestion of microbial genomes by larger organisms (see Section 9.3.5). However, the arguments do not change when we consider these other mechanisms. In each case, the new structures need to be integrated into their genetic and epigenetic environments. This involves the complex, nonlinear dynamics of a network of chemical reactions in which only a limited number of new forms and functions are possible. All this is a far cry from Monod's dictum of "chance alone."

Moreover, the view of evolution we described is not fully represented by determinism either, since it always allows for multiple choices. This is where the concept of contingency becomes relevant – a constellation of deterministic factors giving rise to events that are unpredictable and yet can be fully explained after they have occurred (see Section 9.8).

In the view of evolution we have presented here, there is always a subtle interplay between contingency and determinism. Although chance events trigger evolutionary changes, the emerging new forms of life are not the result of these chance events alone, but of a complex, nonlinear dynamics – a web of factors involving not only genetics but also the constraints of the physical structure of the organism and its context, as well as the ever-changing environment. As we have mentioned, in these complex processes every invention of contingency has to comply with the laws of physics and chemistry, and this is where determinism is critical. Evolutionary theorists have often used the image of water flowing down the irregular surface of a hill to illustrate the interplay between determinism and contingency. The water's downward movement is determined by the law of gravity, but the irregular terrain with its rocks and crevices determines the actual pathway.

Evolutionary changes are triggered by randomness and contingency, but the integration of the emergent genetic structures into their environment is far from random; it is a complex and highly ordered process. According to the systems view, the expression of life's creativity in the process of evolution must be seen as an aspect of the much broader process of life, which is closely associated with cognition (as we discuss in detail in Chapter 12). Evolution, then, is a process that is complex, highly ordered, and ultimately cognitive. It is an integral part of life's self-organization.

10

The quest for the origin of life on Earth

The evolutionary unfolding of life on Earth is a breathtaking story. Driven by the creativity inherent in all living systems, expressed through three distinct avenues – mutations, the trading of genes, and symbiosis – and honed by natural selection, the planet's living patina expanded over billions of years and intensified in forms of ever-increasing diversity. The story has been told many times. For a beautiful, detailed, and sophisticated account we recommend the book *Microcosm* by Lynn Margulis and Dorion Sagan (1986). Rather than retelling the whole story, we shall concentrate on its beginning, the origin of life, in this chapter, and on its most recent stage, the evolution of the human species in Chapter 11.

10.1 Oparin's molecular evolution

What can science say about the origin of life on Earth? It is fair to say from the start that we do not have an answer to the question on how life originated on Earth. It remains one of the great mysteries on scientists' agenda. It is generally accepted that life on Earth originated from inanimate matter via a very long series of chemical steps which brought about a spontaneous and continuous increase of molecular complexity and functionality, until the emergence of the first compartmentalized structures (protocells) capable of making copies of themselves at the expense of their surroundings (see Figure 10.1). It was a Russian chemist, Alexander I. Oparin (1894–1980), who put these ideas in writing, in a short, seminal book, *The Origin of Life*, published in Moscow in 1924.

Oparin was much influenced by dialectic materialism, but also by Darwin's *The Origin of Species*. In fact, Oparin's idea is, in a way, one of an evolution. Indeed, the sequence pictured in Figure 10.1 is known today as prebiotic molecular evolution.

The scientific view of the origin of life excludes, of course, divine creation and other beliefs in miraculous events. But the relation between science and religion is an important issue, to which we will come back later on in this book.

10.2 Contingency versus determinism in the origin of life

Now let us rather ask the question: why, or by which kind of forces, should inanimate matter organize itself in the glorious way that led to living cells? Here we are faced immediately

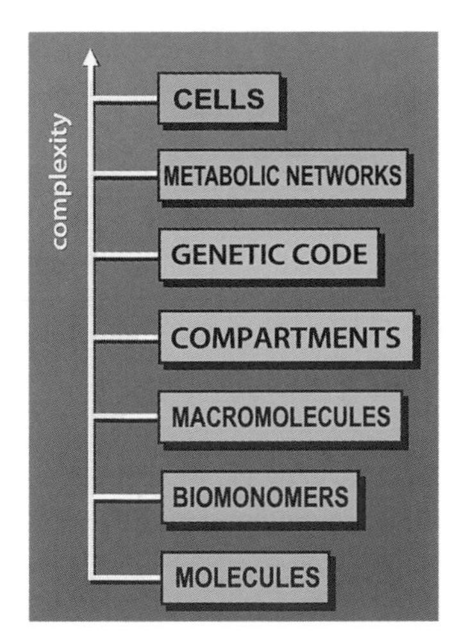

Figure 10.1 A simplified chart of the increase of biocomplexity leading from inanimate matter to the first form of cellular life.

with a very important question: is the pathway shown in Figure 10.1 obligatory, or is it rather due to a series of chance events? We have discussed this question already in the previous chapter when we introduced the dichotomy between contingency and the deterministic view.

The deterministic school of thought – perhaps better called the school of "absolute determinism" – holds that the origin of life on Earth was an inescapable outcome of the original initial conditions on the planet. The school of absolute determinism has the support of several important authors. For example, the biochemist and Nobel Laureate Christian de Duve (2002, p. 298) writes:

> I favor the view that life was bound to arise under the physical-chemical conditions that surrounded its birth.

The idea of the high probability of the occurrence of life on Earth, although phrased differently and generally with less emphasis, is put forth also by Harold Morowitz. In his well-known book, *Beginnings of Cellular Life* (1992), he states:

> We have no reason to believe that biogenesis was not a series of chemical events subject to all of the laws governing atoms and their interactions.
>
> *(p. 12)*

And he concludes with a clear plea against the early idea of contingency, or chance, advanced by Monod:

> We also reject the suggestions of Monod that the origin requires a series of highly improbable events.
>
> *(p. 13)*

This seems to lead to the idea that life on Earth was inescapable, and indeed Christian de Duve (2002), in opposition to Monod, restates this concept (p. 298):

It is self-evident that the universe was pregnant with life and the biosphere with man. Otherwise, we would not be here. Or else, our presence can be explained only by a miracle.

The view of the universe being "pregnant with life" is shared by other authors. For example, the physicist Freeman Dyson (1985, p. 7) cautiously, perhaps even reluctantly, offers the following reflection:

As we look out in the universe and identify the many accidents of physics and astronomy that have worked together to our benefit, it almost seems as if the Universe must in some sense have known that we were coming.

Opposed to this school of thought is the view based on contingency, according to which life might have never started, or might have taken another of many pathways, without ever arriving at humans. We discussed the concept of contingency and its implications in some detail in our previous chapter, where we quoted Stephen Jay Gould's memorable passage ("run the tape again"), and hence we do not need to dwell on it here.

However, the fact that the emergence of life on Earth would not have been possible if the physical and chemical properties of the universe had been only slightly different is extraordinary and deeply mysterious. This mystery has led some scientists and philosophers to posit something they call "the anthropic principle," to which we shall now briefly turn.

10.2.1 The anthropic principle

It has been known for some time that life on Earth is possible only if the mass of the Sun (which determines its luminosity) falls within the very small range between 1.6 and 2.4×10^{30} kg. Otherwise it would be either too cold or too hot. The fact that the mass of the sun is 2×10^{30} kg can be taken as a fortunate coincidence. But there are many of these "coincidences." If their protons were 0.2% heavier, our atoms would not be stable enough to exist; if the electromagnetic force were 4% weaker, there would be no hydrogen and no normal stars (Tegmark, 2003) – and the list goes on with the value of the gravitational force, the speed of light, the mass of the Moon (a large moon stabilizes the planet's wobbling, which otherwise would result in extreme seasonal climate variations), and down to the plate tectonics that produce the carbon cycle. All this is shown in greater detail in the book *Rare Earth*, by the palaeontologist Peter Ward and the astronomer Donald Brownlee (2000).

Of course, the values of all these constants are related to the chemistry of life we discussed earlier. In fact, our current understanding of prebiotic evolution makes it evident that the membrane-bounded vesicles that evolved into the protocells could form only because of the existence of amphiphilic molecules, which in turn required that water molecules exhibit an electric polarity (their electrons staying closer to the oxygen atom than to the hydrogen atoms, so that they would leave an effective positive charge on the latter and a negative charge on the former). Without this subtle electric property of water, the formation of vesicles, and thus the emergence of life as we know it, would not have been possible. In some sense, then, the possibility of life in the universe was implicit from the formation of water and, indeed, from the very formation of the elements hydrogen and oxygen shortly after the big bang.

The fact that the universe seems to have been hospitable to life from its beginning has been formulated as the anthropic principle. It has been expressed in many different ways, but the basic idea is that the universal constants of nature and the basic physical and chemical parameters of the universe have the observed values in order for life, and in particular for human life, to develop (Barrow, 2001; Barrow and Tipler, 1986; Carter, 1974; Davies, 2006; Weinberg, 1987). We wish to make only two comments here. First, the term "anthropic" is a misnomer, because the fact that the universe is hospitable to life does not imply the emergence of human life (except to an absolute determinist). Second, the argument that life as we know it could not have evolved with different values of various parameters in the universe is valid in itself. To formalize it into an "anthropic principle" provides no additional explanation (see Smolin, 2004). In other words, the anthropic principle is a philosophical, not a scientific, argument.

10.2.2 Parallel universes

As we have mentioned, the "fine-tuning" of the cosmological constants, seemingly so as to make life possible on our planet, is extraordinary and deeply mysterious. However, there is an alternative hypothesis, which may sound mind-boggling but is seriously discussed among astrophysicists and cosmologists. Rather than assuming one unique universe with a series of extraordinary coincidences, these theorists postulate that our universe might be just one in a vast ocean of different universes, each possibly having quite different cosmological and physical constants. This is the notion of a "multiverse," or of parallel universes.

The various universes within the multiverse are generally called parallel universes. Is this a totally abstract way of thinking? The cosmologist Tegmark (2003) makes an interesting observation: if you check into a hotel and are assigned a room number corresponding to a very particular year, you may say, "Oh, what a strange coincidence!" But at the same time you realize that the hotel must have many rooms, so that it is actually not impossible that you are given one room with a particular set of characteristics. Similarly, we humans might be in one particular universe out of the multiverse ensemble, an universe characterized by "our" particular set of constants.

According to contemporary cosmologists, this is no longer a metaphor, but a piece of hard science, debated in terms of different scientific models – scientific in the sense that each one can make predictions and can be falsified (Barrow, 2001; Carr, 2007; Ellis *et al.*, 2004). It is also interesting to observe that, according to these researchers, the idea of parallel universes respects the principle of Occam's razor (that "assumptions should not be multiplied beyond necessity"), since an entire ensemble is often "simpler" than one singularity.

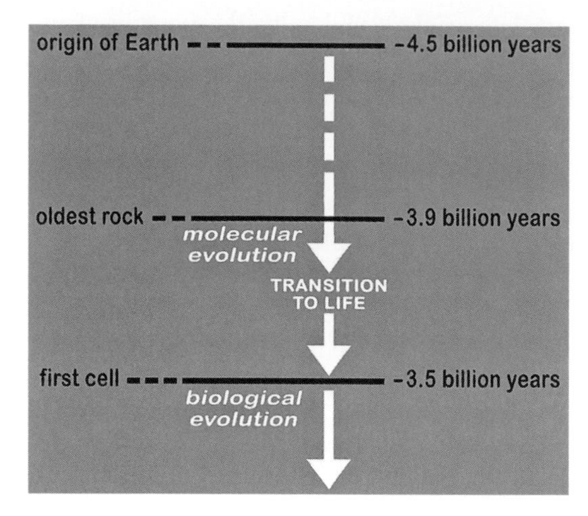

Figure 10.2 A simplified time flow of the origin of life. The solar system is supposed to have formed 4.5 billion years ago, around 9 million years after the big bang. Apparently the Earth was cold enough to host the first mild organic reactions around 3.9 billion years ago, and the first fossils are 3.45 billion years old (adapted from Schopf, 1993).

10.3 Prebiotic chemistry

After this panorama of various philosophical views on the origin of life, including some highly abstract ones, let us now return to basic science and reconsider the origin of life on Earth from the chemical point of view. A first question is this: when did it all start?

The transition to life may have started on our planet in a time window between 3.5 and 3.9 billion years ago, as schematized in the illustration of Figure 10.2. The oldest microfossils, from western Australia, were dated by J.W. Schopf (Schopf, 1992, 1993, 2002) at 3.465 billion years ago. Other microfossils from South Africa and North America have a similar age. And despite some disagreement with Schopf (Brasier *et al.*, 2002), most researchers accept that microorganisms already existed on our planet 3.4–3.5 billion years ago.

But do we know *how* this transition from nonlife to life took place? We do not have as yet a clear answer to this question either. Nobody has been able to reproduce life in the laboratory starting from simple molecules (the so-called bottom-up approach). Moreover, from the conceptual point of view, there is no specific theory on the succession of chemical events that made life possible. A number of people subscribing to the hypothesis of the "RNA world" might object, claiming that we can describe how life started from a self-replicating and mutating RNA molecular family, but the present authors, and many others, do not consider this hypothesis as scientifically grounded. The question, "and where does the self-replicating RNA come from?", is still up in the air (see Section 10.3.2 below on the nonplausibility of the "prebiotic-RNA" world).

There are two good reasons for why we are still so ignorant about the origin of life on Earth. One is, that we have no fossils or other signs of chemical intermediates left by the evolving protolife to rely upon. The second reason is that the progress and direction of the series of steps leading to life is a historical, zigzagging pathway which has been determined by contingency, the concept illustrated in the previous chapter. It is impossible to reconstruct the precise conditions operating at each step at those prebiotic times – temperature, pressure, pH, concentration of salinity, etc. Is, then, the study of the origin of life an empty enterprise?

A possible answer to this question is pointed out by the two Swiss chemists Eschenmoser and Kisakürek (1996, p. 1258):

> The aim of an experimental aetiological chemistry is not primarily to delineate the pathway along which our (natural) life on earth could have originated, but to provide decisive experimental evidence, through the realization of model systems ('artificial chemical life'), that life can arise as a result of the organization of the organic matter.

In fact, there are today very many investigators actively working along these lines. There are many schools of thought, and several quite different approaches, but generally all investigators adhere to the following set of four principles:

(1) Life originated from inanimate matter by a spontaneous and continuous increase of molecular complexity.
(2) The chemical processes in the transition to life can be reproduced in the laboratory with the presently available chemicals and chemical techniques.
(3) This can be implemented in a reasonable experimental time span (hours, or days) once we know the right conditions and starting materials.
(4) Since there is no documentation on how things really happened in nature, there is no obligatory research pathway.

The last item of our list is quite interesting. Since we do not know the pathway used by nature, each researcher in the field is free to choose his or her own in order to actualize the problem of Kisakürek and Eschenmoser. Do as you wish, but the important thing is to show that it is possible to create life from inanimate matter under prebiotic conditions.

The chemistry dealing with the origin of life is usually called prebiotic chemistry, a science that had an electrifying start with the famous 1953 experiments of Stanley Miller. He filled a flask with the four gaseous components assumed by Oparin to be the constituents of the prebiotic atmosphere: hydrogen, ammonia, methane, and water vapor – a so-called reductive atmosphere (no oxygen). Stanley Miller was able to witness the formation of several α-amino acids (such as alanine, glycine, glutamic acid, and aspartic acid) and other relatively complex substances of biological importance, such as urea, acetic acid, and lactic acid (see Figure 10.3).

The experiment was published (Miller, 1953) the same year of the discovery of the double helix by Watson and Crick, and at about the same time Frederick Sanger showed that proteins are linear sequences of amino acids: a memorable year indeed for the life sciences.

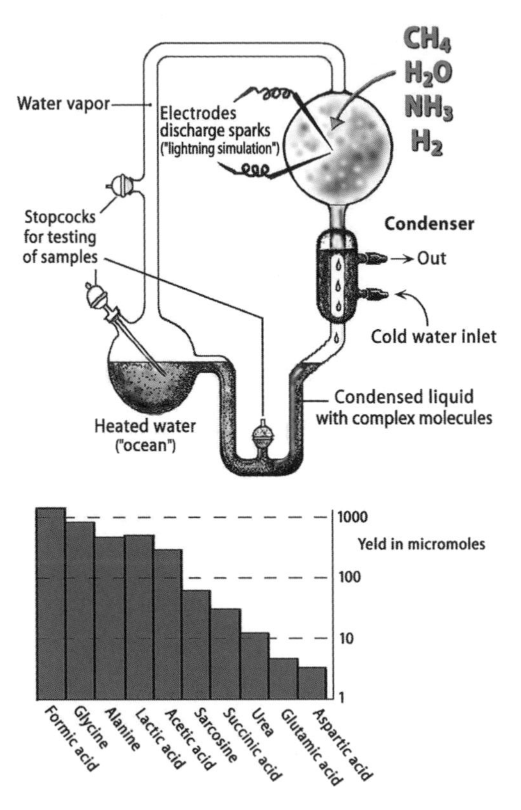

Figure 10.3 The famous 1953 Stanley Miller experiment, which has shown that starting from a mixture of primitive gaseous components, and primordial forms of energy, relatively complex biomolecules can be synthesized, as shown in the right hand side of the panel.

There is only limited agreement on whether the primitive atmospheric conditions were really the ones proposed by Oparin. However, this is not the key point. The key point is that relatively complex biochemicals can be formed from a mixture of very simple gaseous components in a chemical pathway that can indeed be regarded as prebiotic.

Following Miller's experiment, many chemists successfully attempted the synthesis of other compounds of biochemical relevance under prebiotic conditions – including sugars and the basis of nucleic acids, and quite recently even the mononucleotides themselves – namely, the entire monomer units of nucleic acids (Matthew *et al.*, 2009). All this demonstrated convincingly that several molecular bricks of life might have been synthesized on the prebiotic Earth. A considerable enrichment of prebiotic molecules may have come from cosmic bombardment (many organic molecules are synthesized in space – e.g., methane, ammonia, cyanidric acid, fatty acids, amino acids), and from submarine vents and other hydrothermal sources.

One question is important here: why do certain compounds form, and not others? The answer lies in the thermodynamics of chemical reactions: α-amino acids form because they are the most stable compounds that can be formed under those given conditions; they are the result of spontaneous reactions. As chemists would say, these reactions are *under thermodynamic control.*

These discoveries showed that on the primeval Earth there were enough organic materials to start life, as well as the conditions to activate spontaneous reactions leading to complex biomonomers. This is a necessary condition. But is it sufficient to start life?

10.3.1 The macromolecular conundrum

The answer to the question that ended the previous section is "no." Life as we know it today is based on a large number of compounds that are not formed under thermodynamic control. It is enough to think of the most important functional biopolymers, the proteins and nucleic acids. Those are chains with a specific sequence. For example, the digestive enzyme α-chymotrypsin we have in our gut is a long chain with 241 amino-acid residues, and the enzyme lysozyme has 129 amino-acid residues. In order to be active, these enzymes must possess their specific sequence (usually called "primary sequence") – that is,the very precise order of amino-acid residues along the chain – and so it is with each of the very many proteins of our cells.

In order to understand how remarkable the existence of a specific linear sequence is, take the example of lysozyme with its 129 amino-acid residues, and ask the question: in how many ways can a chain with 129 amino-acid residues exist, considering that we have 20 different amino acids for each position of the chain? The answer is that such a chain can exist in 20^{129} different forms, or 10^{168}, which corresponds to 10 followed by 168 zeroes. Note that all these compounds have the same composition – they simply differ in the order of the amino-acid residues along the chain: in the language of chemistry they are *isomers*. Lysozyme is just *one* of those!

Lysozyme is not with us because it is the most stable of all those isomers. And the same can be said about all the other biologically functional macromolecules: the enzymes, all RNAs, and all DNAs in life's metabolism. Clearly, each of these active biopolymer forms was not formed spontaneously as the most stable form, as in the case of Stanley Miller's flask amino acids. Rather, they are the products of a long series of contingency events, each possibly corresponding to a segmental growth of the chain, whereby each step is determined by the contingent set of parameters affecting the reaction at that particular moment. We have learned in the previous chapter how important contingency is in determining life's structures. And it is contingency that shapes a specific sequence of each of our biopolymers (proteins and nucleic acids).

The relevance of contingency to the formation of life biopolymers explains another important point in the study of the origin of life: why we are not able to reconstruct in the laboratory the prebiotic synthesis of α-chymotrypsin, or of any biologically relevant nucleic acid. The reason is that we do not know, and will never know, the contingent conditions (pressure, temperature, concentrations, salinity, pH, etc.) which have attended the origin and the growth in many steps of each of these macromolecules.

This also sets a conceptual limit to the idea of constructing life in the laboratory, starting from the amino acids and other monomer forms. If contingency shaped the specific sequences and forms of biopolymers, then contingency prevents us from reconstructing its precise pathway.

10.3.2 The fallacy of the prebiotic RNA world

There is an apparent (but fallacious) objection: that a random polymerization of amino acids may give, just as mere chance, the "good" structure of, say, a self-replicating RNA molecule. And if a single self-replicating RNA molecule should be formed by chance, so the argument goes, then, as a consequence, all the rest of the origin of life could unfold spontaneously by continuous cycles of replication and mutations. Novel RNA molecules with enzymatic activity would be formed ("ribo-enzymes," abbreviated as "ribozymes"), including those which catalyze the synthesis of DNA and proteins. The primordial conundrum – which came first, informational polynucleotides or functional polypeptides – would be avoided by the simple but elegant compaction of both genetic information and catalytic function into the same molecule.

The term "RNA world" needs a clarification at this point. There is a real RNA world, described by modern molecular biology, in which extant RNA acts in conjunction with the genetic code, and often with the help of natural DNA and enzymes. This kind of RNA world has given rise to some of the most beautiful work in modern synthetic biology; see the work by Joyce and collaborators (Hirao and Ellington, 1995; Lehman and Joyce, 1993; Lincoln and Joyce, 2009), or that of the group of Jack Szostak (Ekland *et al.*, 1995; Green and Szostak, 1992).

When we talk about the fallacy of the RNA world, we certainly do not refer to this elegant part of modern molecular biology, but to the "prebiotic RNA world," as it has been dubbed (Luisi, 2006). This refers to a prebiotic scenario, prior to the genetic code, prior to DNA and enzymes, a scenario in which RNA molecules – and, in particular, self-replicating RNA ribozymes – must have popped out by themselves and started – according to what Orgel and Joyce (1993) call the "molecular biologist's dream" – a scenario which has actually not been taken too seriously by the people working in the field.

A random polymerization from a mixture of all monomers gives in principle an astronomical number of different chains, as calculated above, and, yes, it is in principle possible that one of those corresponds to one single molecule of self-replicating RNA. However, this argument, based on the action of a single original molecule, is fallacious for a number of reasons.

The main reason lies in the strict laws of chemistry, which demand that in order to have reactions it is necessary that reagents are provided with a finite concentration. It does not make any sense, then, to talk about one single molecule A: we need at least two of them, so as to have a "dimeric" complex A_2 (a complex consisting of two components), so that one recognizes the other and makes copies of it. Actually, we need a relatively large number of these molecules so as to maintain the dimer A_2 as a stable entity in solution.

Box 10.1

About concentration in chemistry

In chemistry, concentrations in solution are measured in terms of *molarity*. One mole of a substance is the amount in grams corresponding to the molecular weight of the substance. One molar solution is a solution having 1 mole of that substance in 1 liter. For example, glucose $C_6H_{12}O_6$ has a molecular weight (the sum of the atomic weight of all atoms) of 180, and this means that one molar solution contains 180 g of glucose. One *micromole* (indicated as 10^{-6} mol) contains 0.000 180 g per liter.

Now, let us go back to the question of the prebiotic RNA world, and to the claim that we can start all machinery of macromolecular replication once we have produced by chance one single molecule of a self-replicating RNA molecule.

Suppose, then, we have a molecule A endowed, in principle, with the capability of self-replication. This means, that A can make copies of itself. In order to do that in a normal solution, as already mentioned in the text, A must bind to another A molecule to make an active A_2 reactive complex (a "dimer"), according to the simple chemical equilibrium in Equation 10.1,

$$A_2 \rightarrow\leftarrow 2A, \tag{10.1}$$

This means, that the concentration of A in solution must be high enough to bring about a sizable concentration of A_2, so that the diffusion forces do not destroy the dimer itself. What is the critical concentration of A which permits this?

By definition, 1 mole of any substance contains an Avogadro number of molecules – that is, 6.02×10^{23} (1 followed by 23 zeroes). The precise calculation of the critical minimal concentration of A is not easy, but a good approximation can show that there should be no A_2 formation below the picomole concentration of A. A picomolar solution of A (defined as 10^{-12} M) still contains 6×10^{11} molecules – that is, 600 billion. And even dealing with 1 microliter (one-thousandth of a milliliter, we would have in it 6×10^5 (i.e., 600 000) identical copies of A. This means that we would need several hundred thousand identical copies of A in order to start replication, even restricting the volume to 1 microliter.

In addition, in order to implement the replication, all monomers (the four bases in the case of RNA) must bind to the dimer A_2 – a very complex situation indeed. But even remaining for the sake of simplicity with only A_2, we must note that the formation of such an active dimer poses stringent limits to replication. To make this argument, we need to become a little more technical (see Box 10.1), but the consequence is very simple: with one single molecule we cannot do chemistry, neither replication nor evolution in solution.

The conclusion of the technical argument is simple: in order to do real chemistry we need a sizable concentration, and – abstract exercises of some theoretical biologists notwithstanding – with a single copy we cannot start any replication.

These considerations shed light on one of the most difficult problems facing the field of the origin of life: how to make in the laboratory with prebiotic chemistry a relatively long polypeptide (say, 30 residues) or polynucleotide, with a specific sequence in many identical copies. There are methods for the random polymerization of polypeptides or nucleic acid chains, yielding a wild mixture of all possible isomers, but random polymerization is not what we want here. We are looking for the prebiotic synthesis of specific sequences with a well-defined and unique functionality, and *in many identical copies* (since, as we have seen, chemistry cannot be put into action with only a single molecular copy). This is also a way to reiterate why the RNA world as a scenario for the origin of life is untenable at the present. For additional critical notes about the origin of life from the RNA world, see, for example, Benner (1993), Mills and Kenyon (1996), and Shapiro (1984, 1988).

Now let us return to the macromolecular conundrum. In nature, nucleic acids and proteins are locked into each other in a classical *operationally closed system*: proteins, in the form of enzymes, are specific catalysts that make nucleic acids; and nucleic acids, via the genetic code, make proteins. This is the kind of closed system we have encountered in the case of the cell's autopoietic network: the information does not come from the outside, but from the internal logic of the system – in this case, the apparatus of protein expression. We are clearly facing a chicken-and-egg problem: which came first, the proteins or the nucleic acids? How could they entangle each other, so that the one class would produce the other class? This is the problem of the origin of the genetic code. Or did the two classes of molecules evolve independently at the same time, collaborating only later on? (This hypothesis, known as the "two origins of life," was advocated by Dyson, 1985.)

Again and again, we do not have clear answers to all these important, fascinating questions. There are even international meetings on "Open questions about the origin of life," where scientists debate the still unanswered questions, and why they remain unanswered (Reine and Luisi, 2012; Stano and Luisi, 2007). Such a degree of ignorance may appear surprising. On the other hand, we need to remember that prebiotic chemistry is a field that started a little more than fifty years ago, quite a short period of time in comparison with other traditional branches of science. The researchers in this field are working with great intensity; continuous progress is being made, and there is every expectation that the macromolecular conundrum will be resolved some time soon, constituting a milestone in the quest for the origin of life.

10.3.3 Going upscale: self-organization and emergent properties

Oparin's idea that the origin of life derives from a spontaneous increase of molecular complexity (and order) seems to be, at first sight, against the common sense of thermodynamics. However, as we have discussed in Chapter 8, there are in fact spontaneous reactions that bring about greater complexity and structural order, being attended by an overall increase of entropy. Structural order alone would not be sufficient for molecular evolution to proceed along the pathway to life. We need another important element, the phenomenon we have called emergence – the arising of novel properties at various levels of increasing structural complexity, where "novel" means that these properties were not present in the single parts or components.

Thus, the progress from single molecular parts to functional biological entities is described in terms of two major concepts: self-organization and emergence. In Chapter 8 we discussed at length the interplay between these two key features, illustrating it with numerous examples. We saw in all these examples that, in the processes of the formation of biocomplexity, self-organization and emergent properties inextricably arise together, although it is often useful to distinguish them for heuristic purposes.

10.4 Laboratory approaches to minimal life

Since one of us (P.L.L.) has been involved in research on the origin of life for the past twenty-five years, we can give readers a first-hand account of how this research is carried out in various laboratories around the world. Implicitly, we all follow the assumptions listed in Section 10.1.1 as well as the precept of Kisakürek and Eschenmoser cited above. We tend to recreate in the laboratories those prebiotic conditions which permit the spontaneous synthesis of the complex structures of life – things like prebiotic self-reproducing RNA, folded enzymes, the genetic code, or the ribosomes, keeping in mind that the ultimate goal is then to assemble all these substructures to form a minimal cell.

Here the term "minimal" is a reminder that we are not looking for the reproduction in the laboratory of something as complex as *E. coli* or some other bacterium containing thousands of genes. We are looking instead for the cell structure having the minimal and sufficient components to be called alive. We have seen how complex the definition of life can be. However, at this level most biochemists and molecular biologists would agree that the quality of being alive could be attributed to a cellular structure possessing the following trilogy of properties: self-maintenance, self-reproduction, and evolvability.

It is not the aim of this section to discuss the various experimental approaches to the origin of life in detail. Instead, we limit ourselves to showing schematically one preferred way of thinking for the bottom-up approach (see Figure 10.4), starting from small initial components. Here the following six points are assumed:

1. that prebiotically formed, short peptides may act as the catalysts to make functional macromolecules like proteins;
2. that these macromolecules are then entrapped in vesicles, also prebiotically present (e.g., fatty acids capable of generating membranes);
3. that a primitive metabolism develops inside these vesicles, leading to
4. the production of enzymes that construct membranes made of lipids;
5. this construction leads to a higher form of metabolism and to the first form of the genetic code;
6. this leads to the DNA/RNA/protein living cell capable of self-reproduction.

This is just one hypothetical pathway, and, as already mentioned, research goes on for the various steps of this pathway. In particular, steps 4 to 6 are still obscure.

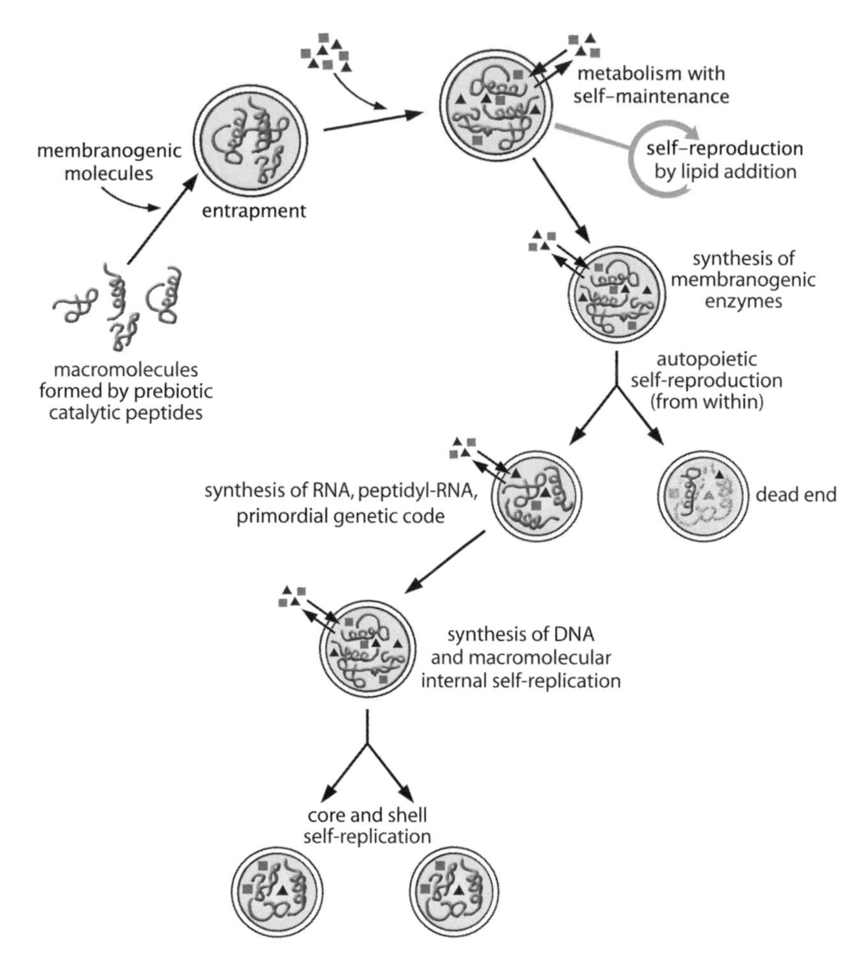

Figure 10.4 One hypothetical bottom-up approach to the origin of life. The basic macromolecules are synthesized first, and then entrapped in a vesicle which is semipermeable to low-molecular-weight compounds, but does not permit leaking of macromolecules. The increase of protocellular complexity takes place inside the vesicle, up to the possibility of self-reproduction.

Researchers pursuing the bottom-up approach include the proponents of the "prebiotic RNA world," who are hoping to discover a prebiotic chemistry that will lead to the self-construction of a self-reproducing RNA, or of an RNA polymerase. We have discussed this before. The basic conceptual scheme of the prebiotic RNA world is shown in Figure 10.7.

We have mentioned earlier, in connection with the macromolecular conundrum, the impossibility of implementing a scheme, such as that shown in Figure 10.5, starting from one single RNA molecule that originated by chance in the prebiotic soup. But even if that problem could be overcome, one should not assume that this reaction scheme is an easy, automatic thing. Consider that it is not enough that ribozymes might be endowed with the capability of catalyzing unspecifically the formation of peptide bonds: what is needed, as we

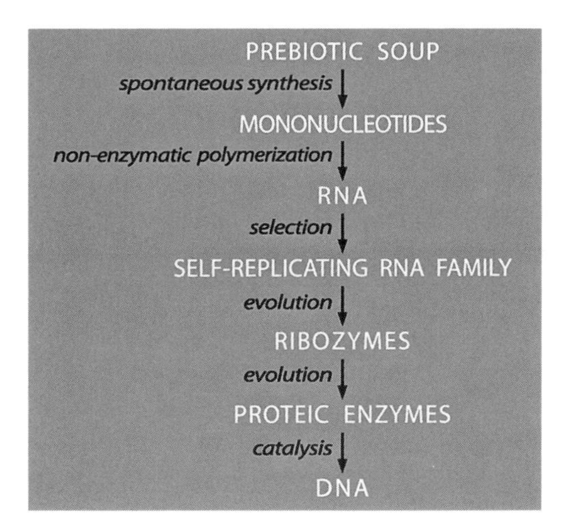

Figure 10.5 Schematic view of the RNA world vision of the origin of life, leading to proteins and then to DNA thanks to the catalytic action of special RNA molecules known as ribozymes.

have seen, are ordered sequences of proteins (or DNA) which must have a specific sequence, and in many identical copies. In addition, still unsolved are all problems of the genetic code and of the membrane formation, which are not even considered in such a scheme.

10.5 The synthetic-biology approach to the origin of life

Let us now turn to experimental approaches, starting from the attempts to reconstruct cellular life in the laboratory. The general framework is known as synthetic biology (SB). This relatively recent branch of life science is seen by many as advanced genetic engineering. In fact, SB has a rather focused, and ambitious, goal: to synthesize forms of life or biological structures alternative to those existing in nature. Why should we pursue this goal? Certainly, part of the answer lies in the old Faustian dream of creating life; we should immediately state, however, that our efforts to "create life" are limited at present to microbial life, and in particular to the genetic manipulation of extant bacteria to create new forms of bacterial life (Benner and Sismour, 2005; Forster and Church, 2006; Hutchison *et al.*, 1999; Luisi, 2002; Smith *et al.*, 2003). Making new forms of microbial life, or new biological structures such as new types of proteins or nucleic acids, can take two scientific directions. One is focused on applications in the biotechnological world; for example, making bacteria capable of producing hydrogen or biofuels (Lee *et al.*, 2008; Smith *et al.*, 2003; Waks and Silver, 2009), or capable of converting light into energy (Johnson and Schmidt-Dannert, 2008), producing specialized drugs (Benner *et al.*, 2008), or producing bioremediation (Pieper and Reineke, 2000). The other direction points more toward basic science, in trying to use the tools of SB for better understanding of the mechanisms of life.

In fact, these two different "souls" of SB were apparent within this new discipline from the very beginning. They correspond to different epistemic frameworks, as recently discussed (Luisi, 2011). The approach that is directed more toward basic science generally tries to tackle the question, "why this and not that?" In other words, why did nature do things in a certain way, and not in another? For example, why is only the sugar ribose present in DNA and RNA, and not the more stable and more widespread sugar glucose? With the techniques of SB, chemists can then synthesize DNA containing glucose instead of ribose, as was done by Eschenmoser's group (Bolli *et al.*, 1997); and from the comparison of the two forms of DNA we can perhaps derive a rationale for the choices of chemical natural selection. Likewise, we can synthesize nucleic acids with bases different from the canonical A, G, C, and T and study how these "alternative" biopolymers behave (Benner and Sismour, 2005; Breaker, 2004).

Another example: why are there in nature proteins with 20 different amino acids? Could nature have created enzymes having only a partial alphabet of amino acids? Proteins containing only 3, 5, or 7 different amino acids have been synthesized (Akanuma *et al.*, 2002; Doi *et al.*, 2005) and tested for their biological activity, often with quite surprising results.

In these more chemical examples, there is no need of altering or modifying the extant living structures. In fact, the term "chemical SB" has been proposed for this second kind of SB not based on genetic manipulation (Luisi, 2007). In the following sections we shall describe two projects within the framework of chemical SB, one directed toward the synthesis of minimal cellular life, including recent findings about the origin of cellular metabolism; the other toward the synthesis of proteins that do not exist in nature.

10.5.1 The construction of minimal cells

The main problem in our understanding of prebiotic chemical evolution lies in the fact that all intermediate steps to construct the products of this evolution, like the first proteins, or the first nucleic acids, are unknown to us. We do not know the conditions under which the turns, branching, and zigzagging of the contingency pathway took place – we only know the final result that emerged. Thus, as already mentioned, it is impossible to reconstruct the biogenesis of lysozyme or any other modern enzyme. If so, then the project of reconstructing life in the laboratory, even in the minimal form of the simplest possible cell, is simply impossible from the bottom-up approach. What we can do is to show that in principle such a pathway is possible, and we have previously cited in this regard the quotation from Eschenmoser and Kisakürek (Section 10.3).

However, there is another way of approaching the construction of minimal cells which is not the bottom-up approach. Let us first define the meaning of the term "minimal cell." A minimal cell is a cell that contains the minimal and sufficient number of components to be defined as alive. Clearly, this defines an entire family of possible structures, as there is not a single way to make such a minimal cell. "Alive" brings us again to the notion of life, and to the question, what is life? We have discussed this in Chapter 7 in terms of autopoiesis. We have also mentioned that, from an operational point of view, most biologists would use the term "alive" for cells that have a trilogy of properties: self-maintenance, self-reproduction, and evolvability. Drawing from autopoiesis, we have proposed an alternative trilogy, more consistent with the systems view of life, consisting of autopoietic structure, environment,

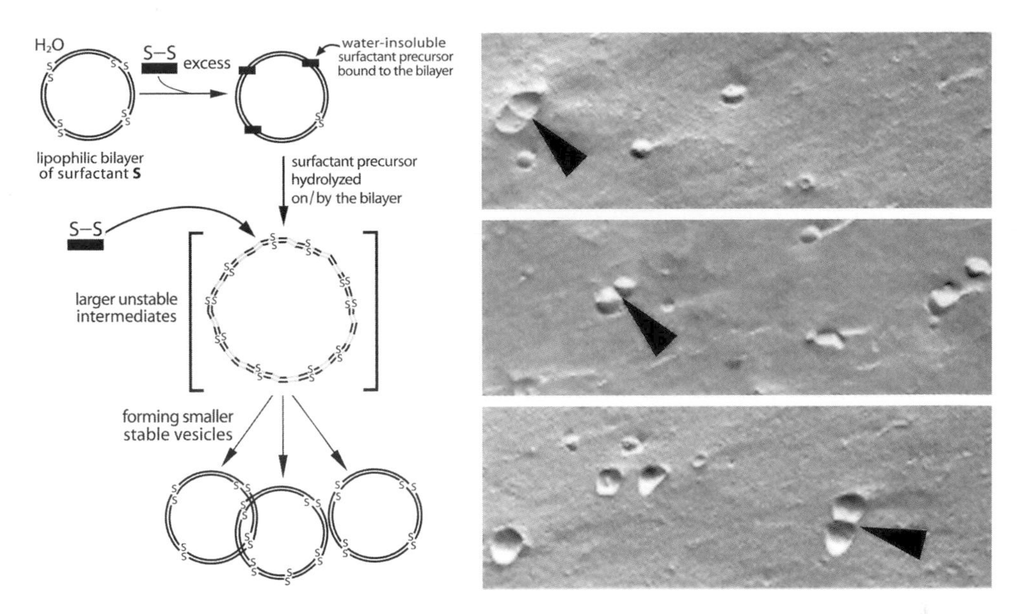

Figure 10.6 The left panel is a schematic representation of the self-reproduction of vesicles due to the addition of a precursor, which binds to the hydrophobic bilayer, and whose hydrolysis yields fresh surfactant molecules. The right panel shows the evidence of such self-reproduction as obtained by a special electron microscopy technique, called cryo-TEM (transmission electron microscopy). It is intriguing that in this kind of experiment only "twin" liposomes are produced (from Stano and Luisi, 2008).

and cognition. We should add that there are many types of approximations to a full-fledged living cell, so that the term "minimal cell" describes a family of compounds rather than a single, well-defined entity.

The staggering complexity of modern cells, encompassing thousands of genes and thousands of additional macromolecular components, elicits the question of whether such a complexity is really necessary for life, or whether cellular life can actually be achieved with a much lower degree of complexity. After all, it is reasonable to assume that at the origin of life the first cells could not have possessed this high degree of complexity.

How can we proceed to create such a minimal cell in a biochemical laboratory? We need first of all a compartment, and this can be provided by vesicles, which are indeed the best model of spherical cell membranes. Liposomes, as we show in Figure 10.7, are vesicles made out of lipids – namely, spherical aggregates in the form a bilayers. They contain an internal water pool in which nucleic acids, enzymes, and other biochemicals can be entrapped by special techniques.

The interest in liposomes as models for the minimal cell is also due to the fact that conditions have been found under which liposomes can self-reproduce; that is, they can induce the formation of copies of themselves at the expense of fresh lipids added to the solution. This is illustrated in Figure 10.6, which shows schematically the multiplication of

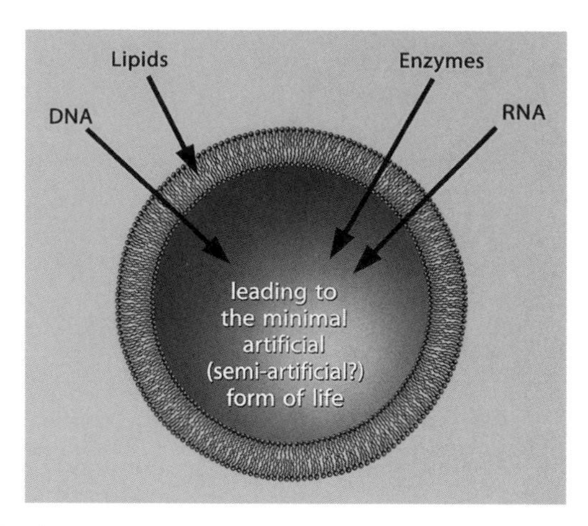

Figure 10.7 Diagrammatic approach to the construction of the minimal cell.

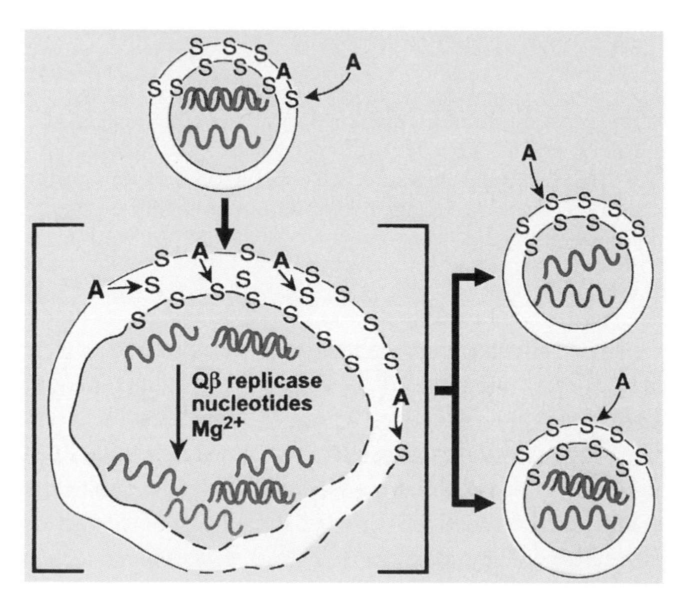

Figure 10.8 One of the first experiments on the construction of minimal cells in the laboratory. Inside the vesicle, the enzyme reproduces RNA, and simultaneously the vesicle self-reproduces. It is a core-and-shell self-reproduction, but the system is not living.

vesicles (formed by the surfactant *S*) upon addition of a hydrophobic precursor *S–S*. This binds to the surface, hydrolyzing and yielding more *S* compounds, which insert themselves into the membrane, provoking its enlargement and eventually the division.

The approach, then, is to insert into a liposome the minimal amount of molecular components so as to have a minimal living cell, as shown diagrammatically in Figure 10.7.

The first interesting experiment in the field is shown in Figure 10.8: an enzyme (known as Q-β replicase), which is able to make copies of a template RNA, has been entrapped in the liposome.

Simultaneously, the liposome self-reproduces via the mechanism depicted in Figure 10.8. Thus, we have a vesicular system that is self-reproducing, and which contains in its core the self-reproduction of nucleic acids – indeed a good model of a biological cell.

This system was first described in 1995 (Oberholzer *et al.*, 1995). The subsequent years saw more and more research groups getting involved in minimal cell projects until it was possible to incorporate inside the liposomes an entire ribosomal system with a minimal set of enzymes capable of expressing proteins (for a review, see Stano and Luisi, 2008). The best kit until now is the one developed by Ueda and collaborators in Tokyo (Shimizu *et al.*, 2001), consisting of only 37 enzymes. The entire set is composed, however, by a total of around 90 macromolecular components, still a sizable number, but one or two orders of magnitude smaller than that of typical bacterial cells. Actually, it has been possible to have protein expressions inside vesicles as small as 200 nm in diameter (2×10^{-5} cm, about one-tenth of the size of an average bacterial cell). This seems to suggest that cells of this size can exist and be viable (de Souza *et al.*, 2009).

All this suggests that life can indeed be produced in the laboratory by utilizing extant molecular components. Moreover, as we shall explain in the following section, there is something emerging from this research that sheds some light on the very origin of cell metabolism and therefore on the origin of cellular life.

10.5.2 On the origin of cellular metabolism

Before proceeding, it is important to mention how the incorporation of biomolecules into the vesicles actually takes place. The procedure is to start from an aqueous solution containing the biomolecules as solute, and then to produce the vesicles *in situ* – that is, in the same solution – so that during their formation and closure, they entrap part of the solution and hence part of the solute. In this way, the entire biochemical apparatus for expressing a protein is entrapped inside the liposomes. The protein chosen in the first experiments is the green fluorescence protein (GFP) for obvious analytical reasons: when the protein is produced, a characteristic green color appears, and the corresponding vesicles turn green.

As beautiful as these experiments are, there is something very odd in them. According to common-sense statistics, they should not work. The solution contains the minimal ensemble of about 90 different macromolecular compounds necessary to express that particular protein; and, in order to be viable, the vesicle must contain *all* those components. What is the probability that a relatively small vesicle, on closing, will entrap *all* of the different 90 macromolecular components in a rather diluted aqueous solution? According to standard statistics, the number of different entrapped components should follow a curve known as the Poisson distribution, in which the frequency of vesicles increases with the number of components entrapped, reaches a maximum at around 10 components, and then decreases again as the number of components increases. According to this curve (Figure 10.9), the probability for a vesicle containing all 90 components, for all practical purposes, should be zero.

Why, then, are these experiments successful? This question was already present at the beginning of this research (Luisi, 2006), and at that time no explanation could be given.

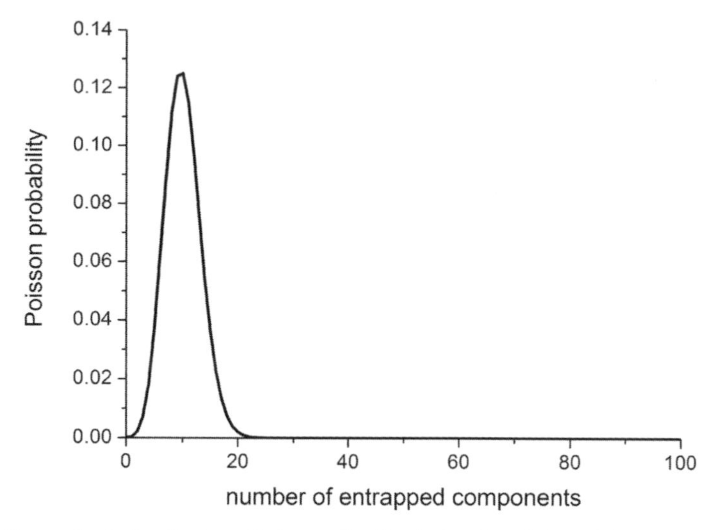

Figure 10.9 Theoretical distribution (Poisson distribution) of vesicles with different numbers of entrapped components. The bell-shaped curve shows the maximum probability for vesicles with the mean number of components and close to zero probability for vesicles with all 90 components.

Later on, the effects of the Poisson distribution were precisely calculated (Luisi *et al.*, 2010), and these calculations confirmed the "theoretical impossibility" that all 90 components would be entrapped in vesicles as small as 200 nm in diameter, according to the standard Poisson statistics. In other words, we were witnessing a situation in which some beautiful experiments were actually not supposed to give the observed results, on the basis of theory.

When something like that happens in science, the obvious conclusion is that there must be something wrong with the theory, or rather, that that particular theory does not apply to the system one is dealing with. Therefore, experiments were carried out in order to clarify which kind of entrapment statistics actually applied to this particular situation. This was done with electron microscopy and ferritin, a large protein containing over 3,000 iron ions. Because of the corresponding very high atomic density, this protein can serve as a useful label, as each protein molecule is visible as a black spot in electron microscopy (see Figure 10.10).

The results of these experiments were stunning. They showed that, in fact, the distribution of ferritin in the vesicles did not conform to a Poisson distribution. Instead, it was found that there was a large number of "empty" vesicles (no entrapped ferritin), and that there were a few vesicles which contained an extremely large number of entrapped ferritin molecules. A typical experiment which, in a first approximation, could be seen as a "all-or-nothing" situation is illustrated in Figure 10.10. The "overcrowded" vesicles have a concentration that is more than one order of magnitude higher than that of the bulk solution. The distribution of ferritin molecules actually follows what is known as a power law (see Figure 10.11).

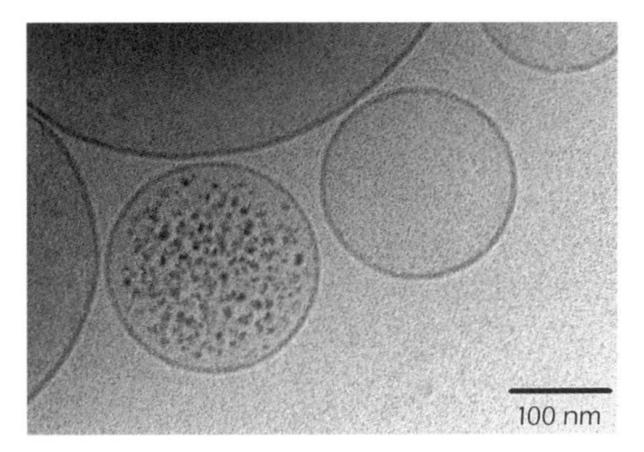

Figure 10.10 Cryo-TEM micrographs of ferritin entrapped in vesicles formed *in situ*. Each black spot is a protein molecule, and the "all-or-nothing" situation is apparent (de Souza *et al.*, 2011; Luisi *et al.*, 2010). The local concentration is much higher than the initial bulk solution.

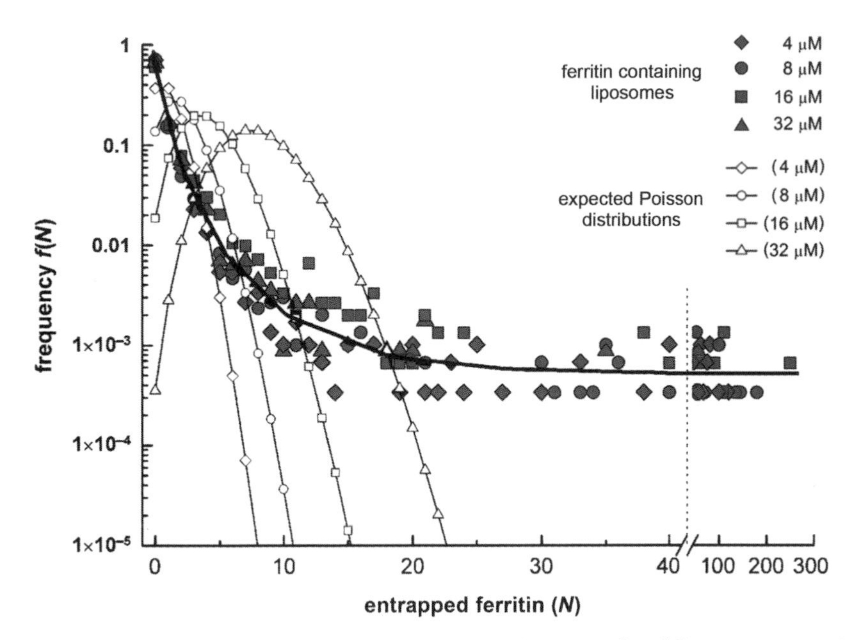

Figure 10.11 Distribution of ferritin in 100-nm diameter liposomes for different concentrations of ferritin (points around solid curve). For comparison, the bell-shaped curves are the calculated Poisson distribution curves. Note the large number of experimentally found "empty" liposomes, and the fact that the frequency of occupancy reaches a kind of plateau, even with a rather large number of entrapped ferritin molecules.

Several aspects of this intriguing phenomenon are still unclear. However, we may speculate about some intriguing similarities with the characteristics of complexity and self-organization discussed in previous chapters. The Poisson distribution applies to random events, independent of one another. The fact that it does not apply here may indicate that the entrapment of components at the moment of vesicle closure is not a random sequence of events but is guided by some emerging coherence, perhaps not unlike the coherent order of Bénard cells that emerges in certain types of heat convection (see Section 8.3.2). In the language of complexity theory, this would mean that the closure of the vesicles *in situ* produces a bifurcation point at which an attractor, representing the entrapment of all 90 components, can emerge (see Section 6.3).

The power-law distribution shown in Figure 10.11 may also be a telltale sign of some coherence among the entrapped components. Recent studies of metabolic networks have shown that the probability that a given node (chemical compound) has a certain number of links (participates in a certain number of reactions) does not follow the standard Poisson distribution but rather follows a power-law distribution (Jeong *et al.*, 2000). Such networks are known as "scale-free networks," because their connectivity does not peak around a characteristic "scale" for the number of links, as it would in a Poisson distribution. The authors of these studies emphasize that scale-free networks (in other words, power-law distributions) indicate a high degree of self-organization.

Indeed, we believe that these experiments could add some important details to our scenarios of the origin of life. In fact, the authors of the investigations see their results as pointing to a possible origin of cell metabolism (de Souza *et al.*, 2011; Luisi *et al.*, 2010; for a review paper, see also Luisi, 2012). While many questions still need to be clarified, it is already becoming apparent that research on the minimal cell as a particular section of SB may shed light on the general problem of the origin of cellular life. In the next section we shall discuss another remarkable development in SB.

10.5.3 The Never Born Proteins

Let us now return to the dichotomy between determinism and contingency and its relevance to the origin of life, which we discussed in Section 10.1.1. The central question is this: why did nature do things in one way and not in another? Was the way it happened the only way possible (determinism), or might it have happened in other ways (contingency)? In particular we can ask: why is there our particular set of proteins in nature instead of a different one?

The number of different existing proteins can be estimated to be of the order of thousands of billions – say, 10^{14} (100,000 billion) to be generous. These proteins are the product of billions of years of evolution, but we all accept the view that present-day proteins have the imprints and basically similar structures and functions as those of early life on Earth. The number of 10^{14} different proteins is certainly impressive, and yet it is ridiculously small compared to the theoretical number of possible proteins. If in each position of the chain we can have any one of the 20 amino acids (also referred to as "residues"), then, for a chain length of 100 residues, the number of mathematically possible different chains is 20^{100} (which is approximately equal to 10^{130}).

Probably, an extremely large fraction of them are energetically impossible, but even if only 1 in 10 billion were energetically allowed, the resulting number would still be a staggering 20^{90} (or 10^{120}). Some researchers have argued that in the course of evolution most of the mathematically possible structures would have been tested. But even if an immensely larger number of additional proteins had been tested, the fact is that our extant proteins are comparatively very few, and the question remains: how and why were these few selected?

To get a feeling for the magnitudes involved, we can represent the proteins by grains of sand. The ratio between our number of possible proteins with length 100 (10^{120}) and the number of actually existing proteins (10^{14}) would then correspond roughly to the ratio between all the sand of the Sahara and one single grain of sand (Luisi, 2006). Our human life thrives on that grain of sand. And all other grains of sand in the vast Sahara represent sets of "never born proteins" – proteins that are not with us presently. And the corresponding DNA codes represent, likewise, DNA stretches and genes which were never born.

How has that "grain of sand," our few existing proteins on Earth, been selected out of the immensity of possibilities? Why not a different grain? Why this and not that? Determinism or contingency?

A strictly deterministic view would require that our proteins had to be what they are as the result of the original selection of set features – for example, water solubility, stable folding capable of binding, hydrodynamic properties, etc. If this were so, then our proteins would be characterized by a series of particular, specific properties – for example, the combination of the above listed features – that all other grains of sand would not have, or have only sparingly.

In early 2000, the research team of one of us (P.L.L.) at the Swiss Federal Institute of Technology in Zurich, and later at the University of Rome (Chiarabelli *et al.*, 2006b), as well as a group led by Tetsuya Yomo at Osaka (Toyota *et al.*, 2008), decided to test this deterministic scenario experimentally. The main consideration was to make a different grain of sand and see what the corresponding proteins look like. In particular, to see how many of them have thermodynamically stable folding and solubility. We did that within a library of "never born proteins" (see Figure 10.12).

Such libraries have been produced by several teams of researchers (Chiarabelli *et al.*, 2006b). A limited but statistically significant number of clones have been randomly selected and investigated, and none of the corresponding 50-residue-long chains were found in the data bank of existing proteins, not even under the most stringent criteria of similarity (namely, they did not have in common even the shortest sequence of amino acids).

When the three-dimensional structures of these totally random *de novo* proteins were studied theoretically (unfortunately, only a very limited number), the results, within the limits of their statistical significance, were very interesting. The few investigated sequences displayed good folding, with a pronounced secondary and tertiary structure. And in the few cases investigated experimentally, these proteins appeared to be water-soluble and thermodynamically stable. Scientists familiar with proteins could easily recognize the usual structural features of "our" proteins, and the similarity in the physicochemical properties mentioned above permits us to say that "our" proteins appeared to be nothing special from the structural or thermodynamic point of view. Does that mean that they have not been selected because of some special features? The temptation is to answer this question

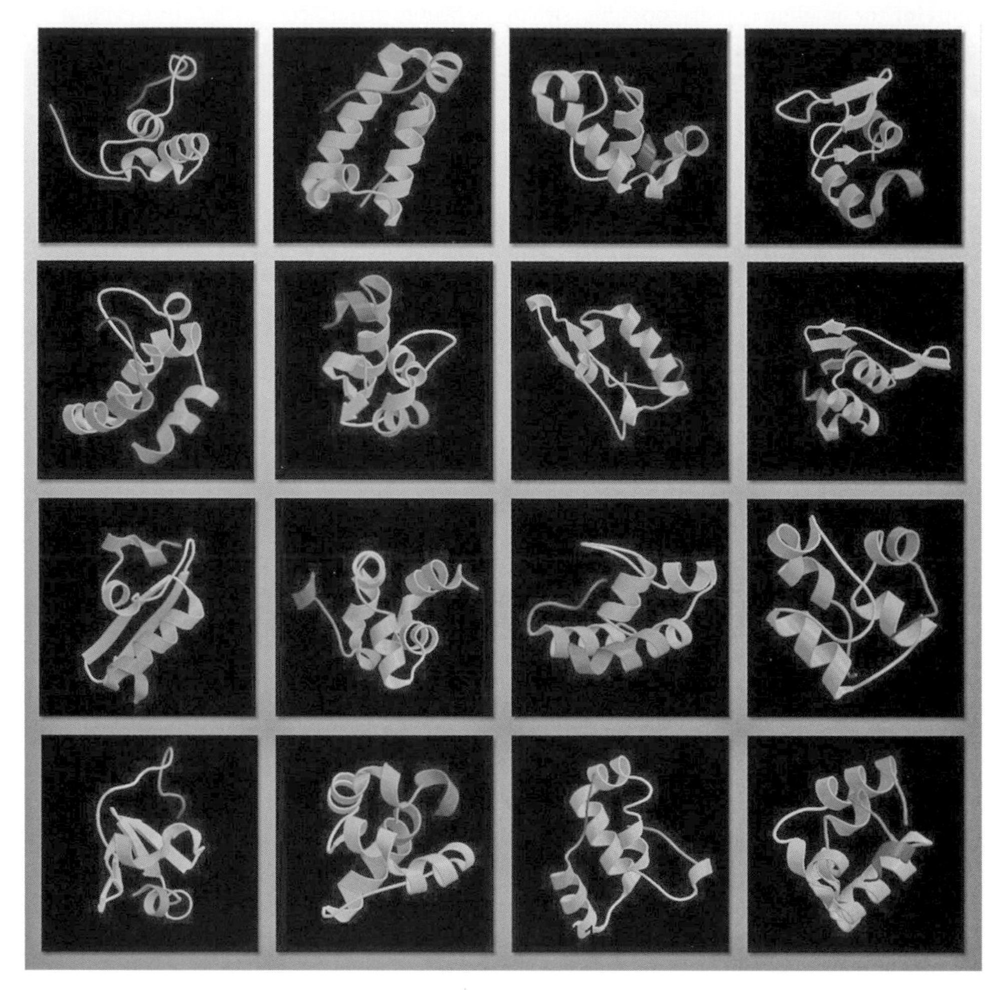

Figure 10.12 Computed three-dimensional structure of some of the Never Born Proteins (NBP). A few of them have been isolated and characterized spectroscopically. The few samples investigated until now have a thermodynamically stable folding, and are soluble in aqueous solution. About 20% of the NBP produced by phage display appear to have good folding (Chiarabelli *et al.*, 2006b).

affirmatively, an answer which is equivalent to asserting that "our" proteins, devoid of any distinctive signs of selection, were the products of contingency.

The point of this work was to show that experimental research in molecular biology can help to tackle subtle philosophical questions. In addition, of course, the libraries of the NBP (Never Born Proteins) are of great importance from the point of view of biotechnology, as they offer an immense arsenal of possibly novel therapeutic agents, as well as novel catalysts. Note that to make NBP proteins with the usual procedure of molecular biology, we need to have the corresponding intermediate RNA, which is then never born RNA. The study of the structure and properties of totally random RNA, which has not undergone the constraints of evolution, is another field of inquiry of chemical SB (de Lucrezia *et al.*, 2006), again with very surprising and most challenging findings (Anella *et al.*, 2011).

10.6 Concluding remarks

In this chapter we have outlined the basic ideas about the quest for the origin of life on Earth. We said at the start that we do not have an answer as yet on how life started on our planet. However we have seen that there is a large and rich spectrum of ideas and proposed scenarios – from the painstaking, stepwise reconstruction of prebiotic chemical reactions to the intriguing view of the "anthropic principle." We have also seen that these discussions touch upon deep philosophical questions that go beyond the current scientific framework and reach into the realm of mystery that always surrounds theoretical research at the forefront of science.

Within the chemical view, we have reviewed different operational frameworks and different philosophies. Again we had to deal with the dichotomy between contingency and determinism. We have used similar arguments as those used for discussing evolution in the preceding chapter. In the present case, the zigzagging of contingency concerns the stepwise pathway of the chemical reactions leading to biopolymers and their organizational complexity. And here, too, we have indicated a complementarity between contingency and determinism, in the sense that contingency cannot operate outside the laws of nature. But again, in the chemical pathway that eventually leads to the complexity of living structures, determinism implies that chemical processes should have a specific final outcome, predictable in principle, whereas contingency implies that such an outcome may not happen at all, or that things may have gone in quite a different way.

This led us to a challenging question, which is the basis of one important branch of modern SB: why did nature arrange things in one way and not in another? In this sense, SB may reveal itself in the coming years as a very important conceptual and experimental framework to better understand nature. In addition to this basic scientific perspective, SB has also a pragmatic bioengineering branch, aimed at producing in the laboratory new kinds of bacteria, and perhaps even higher organisms. SB has indeed a "double soul" (Luisi *et al.*, 2010).

The last three chapters were devoted to life in all its primordial steps up to the advent of biological cells and Darwinian evolution, which brought about the biodiversity enriching our environment today. Where is the human species in all this? And how is this very particular animal, the last guest in life's evolution, linked to and determined by its genes and the environment? It is time to dwell on these questions; they are the main theme of our next chapter.

11

The human adventure

11.1 The ages of life

To chart the unfolding of life on Earth, we have to use a geological time scale, on which periods are measured in billions of years. It begins with the formation of the planet Earth, a fireball of molten lava, around 4.5 billion years ago. We can distinguish three broad ages in the evolution of life on Earth, each extending for 1–2 billion years, and each containing several distinct stages of evolution. The first is the prebiotic age, in which the conditions for the emergence of life and the first protocells were formed. It lasted 1 billion years, from the formation of the Earth to the creation of the first cells, the beginning of life, around 3.5 billion years ago. The second age, extending for a full 2 billion years, is the age of the microcosm, in which bacteria and other microorganisms invented the basic processes of life and established the global feedback loops for the self-regulation of the Gaia system.

Around 1.5 billion years ago, the Earth's modern surface and atmosphere were largely established; microorganisms permeated the air, water, and soil, cycling gases and nutrients through their planetary network, as they do today; and the stage was set for the third age of life, the macrocosm, which saw the evolution of the visible forms of life, including ourselves.

The human evolutionary adventure is the most recent phase in the unfolding of life on Earth, and for us, naturally, it holds a special fascination. However, from the perspective of Gaia, the living planet as a whole, the evolution of human beings has been a very brief episode so far and may even come to an abrupt end in the near future.

To demonstrate how late the human species arrived on the planet, the Californian environmentalist David Brower (1995) devised a very ingenious narrative by compressing the age of the Earth into the six days of the biblical creation story.

In Brower's scenario, the Earth is created on Sunday at midnight. Life in the form of the first bacterial cells appears on Tuesday morning around 8 a.m. For the next two and a half days, the microcosm evolves, and by Thursday at midnight it is fully established, regulating the entire planetary system. On Friday around 4 p.m., the microorganisms invent sexual reproduction, and on Saturday, the last day of creation, all the visible forms of life evolve.

Around 1:30 a.m. on Saturday, the first marine animals are formed, and by 9:30 a.m. the first plants come ashore, followed 2 hours later by amphibians and insects. At 10 minutes to 5 p.m., the great reptiles appear, roam the Earth in lush tropical forests for 5 hours, and then suddenly die out around 9:45 p.m. In the meantime, the mammals have arrived on the Earth in the late afternoon, around 5:30, and the birds in the evening, around 7:15.

Shortly before 10 p.m., some tree-dwelling mammals in the tropics evolve into the first primates; 1 hour later some of those evolve into monkeys; and around 11:40 p.m. the great apes appear. Eight minutes before midnight, the first southern apes stand up and walk on two legs. Five minutes later, they disappear again. The first human species, *Homo habilis*, appears 4 minutes before midnight, evolves into *Homo erectus* half a minute later, and into the archaic forms of *Homo sapiens* 30 seconds before midnight. The Neanderthals command Europe and Asia from 15 to 4 seconds before midnight. The modern human species, finally, appears in Africa and Asia 11 seconds before midnight and in Europe 5 seconds before midnight. Written human history begins around two-thirds of a second before midnight.

11.2 The age of humans

The first step toward higher forms of life was the evolution of nucleated cells (eukaryotes), which became the fundamental components of all plants and animals. This decisive step occurred around 2.2 billion years ago as a result of long-term symbiosis, the permanent living together of various bacteria and other nonnucleated microorganisms (prokaryotes). This created a burst of evolutionary activity, generating the tremendous diversity of eukaryotic cells that we see today and preparing the next big step in evolution from unicellular to multicellular organisms.

When we examine these evolutionary milestones – the invention of photosynthesis, the production of oxygen, and the step from prokaryotes to eukaryotes, followed by the emergence of multicellular organisms – we need to realize (as we have pointed out in our previous chapter in connection with the notion of contingency) that none of them were bound to happen the way they did.

Still, when we follow the unfolding of life on Earth in detail from its very beginnings (see, e.g., Margulis and Sagan, 1986), we cannot help feeling a special sense of excitement when we arrive at the stage where the first apes stood up and walked on two legs. As we learn how reptiles evolved into warm-blooded vertebrates who care for their young; how the first primates developed flat fingernails, opposable thumbs, and the beginnings of vocal communication; and how the apes developed human-like chests and arms, complex brains, and tool-making capabilities, we can trace the gradual emergence of our human characteristics. And when we reach the stage of upright-walking apes with free hands, we feel that now the human evolutionary adventure begins in earnest.

The upright-walking apes, which became extinct around 1.4 million years ago, all belonged to the genus *Australopithecus*. The name, derived from the Latin *australis* ("southern") and the Greek *pithekos* ("ape"), means "southern ape" and is a tribute to the first discoveries of fossils belonging to this genus in South Africa. The oldest species of these southern apes is known as *A. afarensis*, named after fossil finds in the Afar region in Ethiopia that included the famous skeleton called "Lucy." They were lightly built primates, perhaps 140 cm (4.5 feet) in height, and probably as intelligent as present-day chimpanzees.

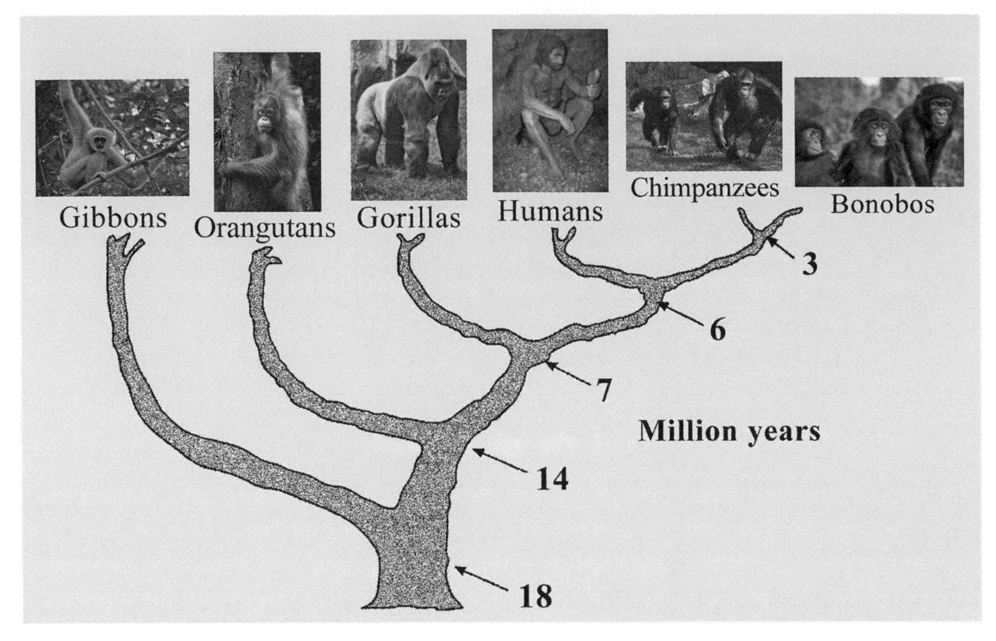

Figure 11.1 Branching and evolution of apes (from Bondi and Rickards, 2009 with modifications).

In the past, anthropologists used the term "hominids" to distinguish both the *Australopithecus* and the *Homo* species from the other apes. However, the present DNA evidence strongly indicates that chimpanzees and humans share a common ancestor which gorillas do not share (see Figure 11.1). The Smithsonian Institute changed its classification scheme accordingly. In the second edition of its standard reference work, *Mammal Species of the World* (Wilson and Reeder, 1993), all the members of the great ape family were moved into the family of hominids.

After almost 1 million years of genetic stability, from around 3–4 million years ago, the first species of southern apes evolved into several more heavily built species. These included two early human species that coexisted with the southern apes in Africa for several hundred thousand years, until the latter became extinct.

An important difference between human beings and the other primates is that human infants need much longer to pass into childhood, and human children need longer again to reach puberty and adulthood than any of the apes. Whereas the young of most other mammals develop fully in the womb and leave it ready for the outside world, our infants are incompletely formed at birth and utterly helpless. Compared with other animals, human infants seem to be born prematurely.

According to a widely accepted hypothesis, the helplessness of the prematurely born infants may have played a crucial role in the transition from apes to humans.

These newborns required supportive families, which may have formed the communities, nomadic tribes, and villages that became the foundations of human civilization. Females selected males who would take care of them while they nursed and protected their infants. Eventually, the females no longer went into heat at specific times, and since they could now be sexually receptive at any time, the males caring for their families may have changed their sexual habits as well, decreasing their promiscuity in favor of new social arrangements. At the same time, the freedom of the hands to make tools, wield weapons, and throw rocks stimulated the continuing brain growth that is characteristic of human evolution, and may even have contributed to the development of language.

The first human descendants of the southern apes emerged in east Africa around 2 million years ago. They were a small slender species with markedly expanded brains, which enabled them to develop tool-making skills far superior to any of their ape ancestors. This first human species was therefore given the name *Homo habilis* ("skillful human"). By 1.6 million years ago, *H. habilis* had evolved into a more robust and larger species, whose brain had expanded further. Known as *Homo erectus* ("upright human"), this species persisted well over a million years and became far more versatile than its predecessors, adapting its technologies and ways of life to a wide range of environmental conditions. There are indications that these early humans may have gained control of fire around 1.4 million years ago.

H. erectus was the first species to leave the comfortable African tropics and migrate into Asia, Indonesia, and Europe, settling in Asia around 1 million years ago and in Europe around 400,000 years ago. Far away from their African homeland, the early humans had to endure extremely harsh climatic conditions that had a strong impact on their further evolution. The entire evolutionary history of the human species, from the emergence of *H. habilis* to the agricultural revolution almost 2 million years later, coincided with the famous ice ages.

During these cold periods, when sheets of ice covered large parts of Europe and the Americas, as well as small areas in Asia, many animal species of tropical origin became extinct and were replaced by more robust, woolly species – oxen, mammoths, bisons, and the like – which could withstand the harsh conditions of the ice ages. The early humans hunted those animals with stone-tipped axes and spears, feasted on them by the fire in their caves, and used the animals' fur to protect themselves from the bitter cold. Hunting together, they also shared their food, and this sharing of food became another catalyst for human civilization and culture, eventually bringing forth the mythical, spiritual, and artistic dimensions of human consciousness.

Between 400,000 and 250,000 years ago, *H. erectus* began to evolve into *H. sapiens* ("wise human"), the species to which we modern humans belong. This evolution occurred gradually and included several transitional species. The transition to *H. sapiens* was complete around 100,000 years ago in Africa and Asia, and around 35,000 years ago in Europe. From that time on, fully modern humans have remained as the single surviving human species.

While *H. erectus* gradually evolved into *H. sapiens,* a different line branched off in Europe and evolved into the classic Neanderthal form around 125,000 years ago. Named after the Neander Valley in Germany, where the first specimen was found, this distinct species persisted until 35,000 years ago. The unique anatomical features of the Neanderthals – they were stocky and robust, with massive bones, low, sloping foreheads, heavy jaws, and long, protruding front teeth – were probably due to the fact that they were the first humans to spend long periods in extremely cold environments, having emerged at the onset of the most recent ice age. The Neanderthals settled in southern Europe and Asia, where they left behind signs of ritualized burials in caves decorated with a variety of symbols, and of cults involving the animals they hunted.

It is important to emphasize that we should not take this evolution of *Homo* as a linear progression. Different types of human species coexisted with each other, although often in different geographical habitats. For example, the Neanderthals coexisted with *H. sapiens*, and, according to Green *et al.* (2010), genetic evidence suggests that they actually interbred with *H. sapiens* (anatomically modern humans) between roughly 80,000 and 50,000 years ago in the Middle East, resulting in 1–4% of the genome of people from Eurasia containing genetic contributions from Neanderthals. According to certain authors, competition from *H. sapiens* may have contributed to Neanderthal extinction. Jared Diamond (1992) has even suggested a scenario of violent conflict and displacement.

By 35,000 years ago, the modern species of *H. sapiens* had replaced the Neanderthals in Europe and evolved into a subspecies known as Cro-Magnon – named after a cave in southern France – to which all modern humans belong. The Cro-Magnons were anatomically identical to us, had fully developed language, and brought forth a veritable explosion of technological innovations and artistic activities. Finely crafted tools of stone and bone, jewelry of shell and ivory, and magnificent paintings on the walls of damp, inaccessible caves are vivid testimonies to the cultural sophistication of those early members of the modern human race.

Until recently, archeologists believed that the Cro-Magnons developed their cave art gradually, beginning with rather crude and clumsy drawings and reaching their height with the famous paintings at Lascaux around 16,000 years ago. However, the sensational discovery of the Chauvet Cave in December 1994 forced scientists to radically revise their ideas. This large cave in the Ardèche region of southern France consists of a maze of underground chambers filled with over 300 highly accomplished paintings. The style is similar to the art at Lascaux, but careful radiocarbon dating has shown that the paintings at Chauvet are at least 30,000 years old (Chauvet *et al.*, 1996).

The figures, painted in ochre, hues of charcoal, and red hematite, are symbolic and mythological images of lions, mammoths, and other dangerous animals, many of them leaping or running across large panels (see Figure 11.2). Specialists in ancient rock art have been amazed by the sophisticated techniques – shading, special angles, staggering of figures, and so on – used by the cave artists to portray motion and perspective. In addition to the paintings, the Chauvet cave also contained a wealth of stone tools and ritualistic objects, including an altar-like stone slab with a bear skull placed on it. Perhaps the most intriguing find is a black drawing of a shamanistic creature, half human and half bison, in the innermost, darkest part of the cave.

Figure 11.2 Panel of horses and two fighting rhinos, Chauvet Cave, Pont-d'Arc, Ardèche, France (from Chauvet *et al.*, 1996).

The unexpectedly early date of those magnificent paintings means that high art was an integral part of the evolution of modern humans from the very beginning. As Margulis and Sagan (1986, pp. 223–4) point out,

> Such paintings alone clearly mark the presence of modern *Homo sapiens* on earth. Only people paint, only people plan expeditions to the rear ends of damp, dark caves in ceremony. Only people bury their dead with pomp. The search for the historical ancestor of man is the search for the story-teller and the artist.

11.3 The determinants of being human

During most of Western philosophy, human nature was believed to be unique and radically different from the nature of animals. Aristotle taught that the human soul shared certain characteristics with the animal soul, but that its principal and unique characteristic was reason. Christian medieval philosophers associated this faculty with the soul's divine origin, and they believed that it was uniquely human and immortal. And finally, Descartes postulated the fundamental division between mind and matter, which implied an even more radical difference between humans, inhabited by a rational soul, and animals, who were simply machines.

Charles Darwin challenged not only the traditional idea of the fixed nature of species but also the assumption of human uniqueness. In *The Descent of Man*, Darwin argued that our power of abstract thought was rooted in the cognitive abilities of our apelike ancestors, and he suggested that similar cognitive skills, including the use of tools, would be found in chimpanzees and other modern apes.

Contemporary studies in primatology have completely confirmed Darwin's revolutionary view. Today we know that the genomes of chimpanzees and humans differ by a mere 1.6%. As the primatologist Roger Fouts (1997, p. 57) explains, "Our skeleton is an upright version of the chimpanzee skeleton; our brain is an enlarged version of the chimpanzee brain; our vocal tract is an innovation on the chimpanzee vocal tract." In addition, it is well known that much of the chimpanzee facial repertoire is similar to our own.

The continuity between humans and chimpanzees does not end with anatomy but also extends to social and cultural characteristics. Like us, chimpanzees are social creatures. In captivity, they suffer most from loneliness and boredom. In the wild, they thrive on change, foraging in different fruit trees every day, building different sleeping nests every night, and socializing with various members of their community as they travel through the jungle. In addition, chimpanzees nurture family bonds, mourn the death of mothers and adopt orphans, struggle for power, and wage war. In short, there seems to be as much social and cultural continuity in the evolution of humans and chimpanzees as there is anatomical continuity.

Moreover, communication studies with chimpanzees, in particular with the help of sign language, have confirmed that the cognitive and emotional lives of animals and humans differ only by degree (see Fouts, 1997), that life is a great continuum in which differences between species are gradual and evolutionary. Cognitive scientists have fully confirmed this evolutionary conception of human nature. In the words of the cognitive linguists George Lakoff and Mark Johnson, "Reason, even in its most abstract form, makes use of, rather than transcends, our animal nature . . . Reason is thus not an essence that separates us from other animals; rather, it places us on a continuum with them." (Lakoff and Johnson, 1999, p. 4).

In view of the fact that our human genome is the result of a long, historical pathway through which we are related not only to our ape cousins but, ultimately, also to all living species, it is interesting to ask ourselves to what extent some of our outstanding human characteristics are genetically based, resulting from our animal instincts, and to what extent they have been acquired culturally.

11.3.1 The killing-ape instinct

One of the first questions that comes to mind concerns the aggressive nature of humans, as witnessed by thousands of years of bloody wars and killing. Is this the result of a genetic trait? Are we genetically determined to be aggressive, to make war and kill each other as a kind of genetic damnation of our being human? This question is made reasonable by the observation that many other animal species do not behave in such a way. They may kill each other to defend territory or their mates, but they do not band together in ferocious raiding groups for premeditated attacks on other groups of animals of the same species. There is also no doubt that the human species is the most belligerent and cruel of all species.

In an interesting book, *Demonic Males*, the anthropologists Wrangham and Peterson (1996), studying the aggressive behaviors of apes and humans, arrive at a grim conclusion: only humans and chimpanzees are "killing apes" with the habit or at least the capability of organizing themselves in male-bonding teams with the aim of killing in a cold-blooded way other individuals of the same species. Humans and chimpanzees diverged from each other only 5 million years ago, while the gorillas branched out 10 million years ago (see Figure 11.2). However, very interestingly, 2–3 million years later another kind of ape branched off from the chimpanzee: the bonobo (a kind of smaller chimpanzee, previously called pygmy chimp and correctly termed *Pan paniscus*).

The amazing thing is that bonobos are not aggressive at all; they are the friendliest and most peaceful of all animals (see, e.g., De Waals, 2006; Wrangham and Peterson, 1996). The bonobo branched off the chimpanzee about 3 million years ago (see Figure 11.2), and it is hypothesized that this diversification was due to the fact that some families of chimpanzees had to live in a different, more hospitable, environment (Wrangham and Peterson, 1996), an environmental difference that may eventually have caused a drastic change in genetic behavior. Might this kind of branching-off happen again, with the arising of a novel human species similar to the bonobos rather than to the chimpanzees?

For the time being, we have to accept the observations of the two already mentioned anthropologists, Wrangham and Peterson (1996, p. 82), who reject the notion that there are on our planet peaceful, idyllic places without violence, expressing this view thus:

> Neither in history nor around the globe today is there evidence of a truly peaceful society. But the suggestion that chimpanzees and humans have similar patterns of violence rests on more than the claims of universal human violence. It depends on something more specific – the idea that men in particular are systematically violent. Violent by temperament.

In other words, violence is not a general human characteristic, but rather a specifically *male* human characteristic.

We also need to take into account the secondary effects of human aggression; for example, the bloody bullfights and the barbarous practice of hunting and killing innocent animals for pleasure. Or think of competitive sports, which in a way can be seen as a sublimation of human aggression, but occasionally take the form of fierce animosity and violence. Another category that comes to mind is the field of economics and business. Are not the recent forms of colonialism (including missionary colonialism) and the present predatory capitalism expressions of this human aggressiveness? As we shall discuss in Chapter 17, our world today is dominated by a global economic system with disastrous social and environmental impacts. A tragic consequence of this form of violence is the fact that we are the only species on Earth who destroys its own habitat, threatening countless other species with extinction in the process. To this determinant of aggressiveness we also need to attribute negative emotions such as anger, hate, and jealousy. The whole issue of emotions has recently become an important topic of research in cognitive science, as we shall discuss in the following chapter.

Fortunately, killing and aggression are not the only determinant of our being human. There is also the opposite trait of love and altruism, to which we now turn.

11.3.2 Love and altruism

Love is certainly a very fundamental aspect of animal behavior. Obviously, love of the mother for her offspring can be seen as an instinct and is the main device provided by nature for the preservation of species, and love and sexual attraction between males and females can also be seen in this light, as the best way to ensure reproduction. In humans, love can also be seen as instinct, but it comes together with consciousness and moral codes.

To say that love is genetically determined is not to negate or diminish the beauty of love in all its many, wonderful manifestations, nor to diminish the cultural and artistic aspects brought about in our civilization by our conception of love. And the same is true for altruism. There is a vast literature on altruism and cooperation within the Darwinian scientific community, dealing with symbiosis and many other technical aspects upon which we cannot dwell here. Suffice it to say that altruism, cooperation, and love can be linked to natural selection: groups, tribes, or social structures that were characterized by altruism and cooperation had better chances of survival. This is certainly an important aspect of our being human. To the qualities of love and altruism, we should add positive emotions like empathy, joy, happiness, gratitude, euphoria, and hope – as well as positive feelings, such as feeling satisfied, sympathetic, or fulfilled. This is nowadays an important field of neurobiological inquiry.

The two determinants of being human discussed so far – the killing-ape instinct and love and altruism – are both linked inextricably to human consciousness, which is our next determinant.

11.3.3 Consciousness and spirituality

As we briefly indicated in Chapter 7 in connection with our discussion of autopoiesis, the cognitive dimension is an integral part of the systemic conception of life (see Section 7.8). During the past three decades, the study of mind and consciousness from this systemic perspective has grown into a richly interdisciplinary field of study known as cognitive science, which we shall discuss in detail in the following chapter. Here we want to limit ourselves to just a few comments regarding consciousness and spirituality as determinants of being human, leaving a deeper exploration of their nature and origin for later.

We have seen in the previous section that the paintings and artifacts found in Paleolithic caves strongly indicate that the Cro-Magnons expressed their sense of belonging in "religious" rituals. There is also evidence of burial ceremonies, indicating that these early humans were thinking about their own death, but this evidence is only about 25,000 years old, and is thus relatively "recent" history (Bondi and Rickards, 2009). The question is whether the early hominids had a consciousness of being. Was Lucy, over 3 million years ago, mentally aware of her own existence? And was this awareness, if present, connected to an evolutionary advantage?

This is a difficult question which has no clear-cut answer. One consideration that comes to mind is that a sense of awe and wonder may have arisen as soon as our hominid ancestors began to stand up and walk on two legs. Once they could walk erect and look up, they would have faced more directly the mysteries of nature – lightning in stormy weather, the starry night sky, the phases of the Moon, the sunrise, and so on. And at the same time this kind of spiritual perception may have induced a sense of self. In other words, the beginning of spirituality may well be attended by the consciousness of being.

The idea that some supernatural powers might be responsible for these phenomena must have arisen then, at the very beginning of human perception, and may well have been accompanied by the development of some religious rituals. It can also be argued that rituals of this sort may have had an impact on natural selection, since the groups or tribes who were involved in them would have had greater internal cohesion and strength, which would have helped their survival.

In this regard, one should mention one line of research in Darwinian evolution. Some authors have expressed the idea that humans are genetically characterized by being " born to believe" (see, e.g., Boyer, 2008; Girotto *et al.*, 2008). At this point, it is proper to refer to one other modern author, albeit not the most uncontroversial, Marc Hauser (2007). He argues that morality, which can be seen as a basic aspect of spirituality, is grounded in biology. According to him, there is an innate, universal, moral grammar that belongs to the human species as a product of evolution, while the specific expression of this morality varies among different places depending on the contingent constraints. This would imply that for humans the moral code comes from within human nature, without any need for religion.

11.3.4 Curiosity and the thirst for knowledge

Our list of the genetic determinants of being human would not be complete if we did not add another, very beautiful trait of humanity: the desire for knowledge, the search for understanding the nature around us, and the desire to conquer the difficulties presented to us by nature. We have mentioned the awe and wonder of our ancestor hominids facing the mysteries of nature – sunrise and sunset, the phases of the Moon, the colors of flowers, and the birth and growth of animals and plants. What they must have felt, right from the beginning, was not only awe and wonder but also curiosity and a desire to understand; and with that also the desire to master the environment with the help of tools, which became more and more sophisticated with time.

It is commonly accepted that this development was triggered by rapid brain growth at the dawn of human evolution, about 4 million years ago, when language, reflective consciousness, the ability to make and use tools, and organized social relations all evolved together (see Section 11.2). The size of the brain has been an important determinant for human development, since brain size is widely believed to be proportional to intelligence. The hominid brain has nearly quadrupled in size over the past 4 million years from the chimpanzee (400 cc) to *Australopithecus* (600 cc) to *H. erectus* (1,200 cc) to modern humans (1,400 cc).

However, the relation between brain size and intelligence is not straightforward. Intelligence as such – the capability of solving problems – is not necessarily hereditary; it can be retained indefinitely by an individual but cannot be genetically transmitted to descendants. Moreover, when we look at a possible direct relation between brain size and intelligence, it appears that human intelligence is not necessarily adaptive in an evolutionary sense.Indeed, large-headed babies are more difficult to give birth to, and large brains are costly in terms of nutrient and oxygen requirements. But the fact that the direct adaptive benefit of human intelligence may appear questionable makes it even clearer that cleverer humans may gain indirect selective benefits.

The relation between intelligence, thinking, and mind is a complex one, which we shall explore in our next chapter. For the purpose of this section, let us simply add human intelligence as a determinant that led to the rise of science – the desire to shed light on the darkness of ignorance – and of technology – the desire to apply this knowledge practically. These applications range from the invention of the alphabet and the wheel, to all forms of modern technology, including the invention of gunpowder, bombs, and other warfare devices that are linked to our first genetic determinant, the killer-ape instinct.

11.3.5 The search for beauty and harmony

Let us now consider the search for beauty and harmony, and the corresponding artistic creativity. A world without Greek statues, Chinese brush paintings, Indian Chola bronzes, or Renaissance frescoes; without the music of Mozart, Beethoven, or Bach, would not be our world. As we have seen, the artistic expressions of human consciousness began with magnificent paintings in Paleolithic caves 30,000 years ago, at the very birth of the modern human species (see Figure 11.2). From roughly the same era date the famous Paleolithic Venus figurines, often interpreted as fertility symbols, as well as musical instruments such as flutes.

Why do we create these monuments to beauty and harmony? Is this, too, connected with genes, meaning that making art has some reproductive advantage? Let us start with simpler animals. From the Darwinian point of view, beauty is one among many other biological properties. In certain animals, like birds, it plays an important role in sexual selection, orienting both males and females to make the best choice of their mates. Here, in nature, beauty is a symbol of youth, strength, and health. The peacock's tail is the emblematic example. But what about humans? Is our capability of appreciating beauty in nature – the display of colors of birds, the symmetry of flowers, the beauty of a painted Venus, or the harmony of Beethoven's symphonies – is this appreciation something inborn in our own nature? Or is it simply due to our education and therefore the product of culture?

According to the philosopher Denis Dutton (2009), an interest in art belongs to the list of evolutionary adaptations together with the enjoyment of sex, the response to facial expressions, the understanding of logic, and the spontaneous acquisition of language, all of which make it easier for us to survive and reproduce. He suggests that this appreciation for beauty may have been what pushed our hominid ancestors toward the beautiful savannas of Africa and other landscapes that would have appealed to them. Dutton uses arguments taken from evolutionary psychology to show that human perceptions undergo a kind of evolutionary development.

Darwinism is linked to beauty by Roger Scruton, in his book *Beauty* (2009). The idea is that contemplative appreciation is also instinctive, which permits the author to link high artistic values to our biology. It is perhaps interesting to recall that Immanuel Kant had already thought that our appreciation of nature is spontaneous, coming from our instinct.

11.4 Concluding remarks

The aesthetic sense in humans must be considered in conjunction with the development of consciousness, as well as spirituality, the determinants discussed earlier. Perhaps the sense of wonder before the beauty of a landscape – like our own wonder in looking at the Grand Canyon or the majestic peaks of the Alps – is also one of the preliminary emotions experienced by the first hominids, one that joins the appreciation of beauty with awe before the mysteries of nature, and eventually with the presence of some supernatural power.

The search for beauty and harmony, which we have identified as a key characteristic of human nature, is manifest not only in works of art but also in the persistent search for order in nature and in the cosmos. The harmony of the movements of the stars and planets became the foundation of astrology in ancient times, and then of the various attempts to interpret the universe, often on the basis of beautiful geometrical representations.

In concluding this chapter, let us note that in this discussion, we have presented the genetic determinants of being human separate from each other. This is, of course, valid only for the sake of simplicity. Consciousness, spirituality, artistic creativity, abstract thinking, and rationality intertwine with each other in an intricate maze. In most manifestations of our actions, and products of our civilization, it may be difficult to discriminate any one from the others. This reiterates the complexity of the species *H. sapiens* – the species capable of creating the splendors of St. Peter's Basilica, and also capable of dropping the atomic bomb.

12

Mind and consciousness

One of the most important philosophical implications of the new systemic understanding of life is a novel conception of the nature of mind and consciousness, which finally overcame the Cartesian division between mind and matter. In the seventeenth century (as we discussed in Chapter 1), René Descartes based his view of nature on the fundamental division between two independent and separate realms – that of mind, the "thinking thing" (*res cogitans*), and that of matter, the "extended thing" (*res extensa*). This conceptual split between mind and matter has haunted Western science and philosophy for more than 300 years.

Following Descartes, scientists and philosophers continued to think of the mind as an intangible entity and were unable to imagine how this "thinking thing" is related to the body. Although neuroscientists have known since the nineteenth century that brain structures and mental functions are intimately connected, the exact relationship between mind and brain remained a mystery. Less than ten years ago, the editors of an anthology titled *Consciousness in Philosophy and Cognitive Neuroscience*, Revonsuo and Kamppinen (1994, p. 5), stated frankly in their introduction: "Even though everybody agrees that mind has something to do with the brain, there is still no general agreement on the exact nature of this relationship."

12.1 Mind is a process!

The decisive advance of the systems view of life has been to abandon the Cartesian view of mind as a thing, and to realize that mind and consciousness are not things but processes. In biology, this novel concept of mind was developed during the 1960s by Gregory Bateson, who used the term "mental process," and independently by Humberto Maturana, who focused on cognition, the process of knowing. In the 1970s, Maturana and Francisco Varela, both working at the University of Chile in Santiago, expanded Maturana's initial work into a full theory, which has become known as the Santiago theory of cognition. During the past three decades, the study of mind from this systemic perspective has blossomed into a richly interdisciplinary field known as cognitive science, which transcends the traditional frameworks of biology, psychology, and epistemology.

Over the years, one of us (F.C.) developed a synthesis of the main ideas of some of the leading cognitive scientists with the aim of arriving at a coherent systemic understanding of life, mind, and consciousness (Capra, 1996, 2002). This chapter is based on that synthesis, which we summarize in Section 12.4 below. Its extension to the social dimension of life will be discussed in Chapter 14.

12.1.1 Bateson's "mental process"

Gregory Bateson, who was a regular participant in the legendary Macy Conferences during the early years of cybernetics (see Chapter 5), developed a concept of mind based on cybernetic principles, defining "mental process" as a systems phenomenon characteristic of all living organisms. Bateson (1972) listed a set of criteria that systems have to satisfy for mind to occur. Any system that satisfies those criteria will be able to develop the processes we associate with mind – learning, memory, decision-making, and so on. In Bateson's view, these mental processes are a necessary and inevitable consequence of a certain complexity that begins long before organisms develop brains and higher nervous systems. He also emphasized that mind is manifest not only in individual organisms but also in social systems and ecosystems.

Bateson presented his new concept of mental process for the first time in 1969 in Hawaii, in a paper he gave at a conference on mental health (reprinted in Bateson, 1972). This was the very year in which Maturana presented a different formulation of the same basic idea at a conference on cognition organized by Heinz von Foerster in Chicago (see Maturana, 1980/1970). Thus two scientists, both strongly influenced by cybernetics, had simultaneously arrived at the same revolutionary concept of mind. However, their methods were quite different, and so were the languages in which they described their groundbreaking discovery.

Bateson's whole thinking was in terms of patterns and relationships. His main aim, like Maturana's, was to discover the pattern of organization common to all living creatures. "What pattern," Bateson (1979, p. 8) asked, "connects the crab to the lobster and the orchid to the primrose and all four of them to me? And me to you?"

Bateson developed his criteria of mental process intuitively from his keen observation of the living world. It was clear to him that the phenomenon of mind was inseparably connected with the phenomenon of life. When he looked at the living world, he saw its organizing activity as being essentially mental. In his own words (personal communication to Capra, 1979), "mind is the essence of being alive."

In spite of his clear recognition of the unity of mind and life – or mind and nature, as he would put it – Bateson never asked, "What is life?" He never felt the need to develop a theory, or even a model, of living systems that would provide a conceptual framework for his criteria of mental process. To develop such a framework was precisely Maturana's approach.

12.1.2 Cognition – the process of life

When Maturana returned to the University of Chile in 1960 after six years of study and research in the UK and the USA, two major questions crystallized in his mind. As he remembered it later (Maturana, 1980/1970, p. xii), "I entered a situation in which my academic life was divided, and I oriented myself in search of the answers to two questions that seemed to lead in opposite directions – namely: 'What is the organization of the living being?' and 'What takes place in the phenomenon of perception?'"

Maturana struggled with these questions for almost a decade, and it was his genius to find a common answer to both of them. He discovered that the "organization of the living being" is a special network pattern, which he and Francisco Varela later called "autopoiesis" (see Chapter 7), and that the understanding of this pattern of self-generating networks provided him with the theoretical framework to understand perception and, more generally, cognition. The full theory of cognition, now known as the Santiago theory, was developed by Maturana and Varela in the 1970s and published in their pioneering monograph, *Autopoiesis and Cognition*, in 1980.

The central insight of the Santiago theory is the same as Bateson's – the identification of cognition, the process of knowing, with the process of life. Cognition, according to Maturana and Varela, is the activity involved in the self-generation and self-perpetuation of living networks. In other words, cognition is the very process of life. The organizing activity of living systems, at all levels of life, is mental activity. The interactions of a living organism – plant, animal, or human – with its environment are cognitive interactions. Thus life and cognition are inseparably connected. Mind – or, more accurately, mental activity – is immanent in matter at all levels of life.

As the complexity of species increased in the long history of evolution, so did their cognitive processes. Cognition, then, is a stratified notion (Luisi, 2003, 2006), whose sophistication increases with the increasing sophistication of the living organism's sensory apparatus – that is, from flagella to antennae, to photosensitive devices, to olfactory discrimination, the eye, and the brain. Thus, the interaction of each different living being with its environment, although exhibiting a common pattern of organization, is accomplished at each time through particular sensory organs, which are the product of its phylogeny.

This is a radical expansion of the concept of cognition and, implicitly, the concept of mind. In this new view, cognition involves the entire process of life – including perception, emotion, and behavior – and does not even necessarily require a brain and a nervous system. As Maturana (1980/1970, p. 13) put it in his original paper, "Biology of Cognition":

> Living systems are cognitive systems, and living as a process is a process of cognition. This statement is valid for all organisms, with and without a nervous system.

We should also mention that some cognitive scientists are reluctant to use "mind" for nonhuman cognition, preferring to use that term in the traditional sense, in which cognition is associated with reflective consciousness. In fact, while writing this book, we found ourselves in disagreement on this point. One of us (P.L.L.) prefers to reserve the term "mind" for forms of cognition involving living organisms with brains, while the other (F.C.), inspired by Gregory Bateson, prefers to extend "mind," or better "mental process," to all living systems, understanding the term "cognitive" as being synonymous with "mental."

Maturana studiously avoids using the term "mind," probably because he feels that, in view of its Cartesian association with a "thing," it is inappropriate to describe the process of cognition. Bateson, too, was aware of this dilemma. He used both "mind" and "mental process," but always emphasized that "mind" stands for a process, or rather a set of mental processes. At any rate, Maturana made it clear already in his first paper that, like Bateson, he sees no essential difference between the process of human cognition and the cognitive processes of other living beings. "Our cognitive process," Maturana (1980/1970, p. 49) wrote, "differs from the cognitive processes of other organisms only in the kinds of interactions into which we can enter, such as linguistic interactions, and not in the nature of the cognitive process itself."

12.2 The Santiago theory of cognition

12.2.1 Structural coupling

In the Santiago theory, cognition is closely linked to autopoiesis, the self-generation of living networks. The defining characteristic of an autopoietic system is that it undergoes continual structural changes while preserving its web-like pattern of organization. The components of the network continually produce and transform one another, and they do so in two distinct ways. One type of structural change is that of self-renewal. Every living organism continually renews itself, as its cells break down and build up structures, and its tissues and organs replace their cells in continual cycles. In spite of this ongoing change, the organism maintains its overall identity or pattern of organization.

The second type of structural change in a living system is that which creates new structures – new connections in the network. These changes, developmental rather than cyclical, also take place continually, either as a consequence of environmental influences or as a result of the system's internal dynamics.

According to the theory of autopoiesis, a living system couples to its environment *structurally* – that is, through recurrent interactions, each of which triggers structural changes in the system (see Section 7.4.1). Living systems are autonomous, however. The environment only triggers the structural changes; it does not specify or direct them.

As a living organism responds to environmental influences with structural changes, these changes will in turn alter its future response, because the organism responds to disturbances according to its structure, and that structure has now changed. But this process – a modification of behavior on the basis of previous experience – is what we mean by learning. In other words, a structurally coupled system is a learning system. Continual structural changes in response to the environment – and consequently continuing adaptation, learning, and development – are key characteristics of the behavior of all living beings. Because of this dynamic of structural coupling, we can call the behavior of an animal intelligent but would not apply that term to the behavior of a rock.

12.2.2 Bringing forth a world

Living systems, then, respond to disturbances from the environment autonomously with structural changes – that is, by rearranging their patterns of connectivity. According to Maturana and Varela, we can never direct a living system; we can only disturb it. More than that, the living system not only specifies its structural changes; it also specifies *which disturbances from the environment trigger them.* In other words, a living system has the autonomy to decide what to notice and what will disturb it. This is the key to the Santiago theory of cognition. The structural changes in the system constitute acts of cognition. By specifying which perturbations from the environment trigger changes, the system specifies the extent of its cognitive domain; it "brings forth a world," as Maturana and Varela put it.

Cognition, then, is not a representation of an independently existing world but rather a continual bringing forth of a world through the process of living. The interactions of a living system with its environment are cognitive interactions, and the process of living itself is a process of cognition. In the words of Maturana and Varela, "to live is to know." As a living organism goes through its individual pathway of structural changes, each of these changes corresponds to a cognitive act, which means that learning and development are merely two sides of the same coin.

12.2.3 Cognition and the soul

The identification of mind, or cognition, with the process of life is a novel idea in science, but it is one of the deepest and most archaic intuitions of humanity. In ancient times, the rational human mind was seen as merely one aspect of the immaterial soul or spirit. The basic distinction was not between body and mind, but between body and soul, or body and spirit.

While the conceptual boundaries between soul and spirit were often fluctuating in the philosophical schools of antiquity, both soul and spirit were described in the languages of ancient times with the metaphor of the breath of life. The words for "soul" in Sanskrit (*atman*), Greek (*psyche*), and Latin (*anima*) all mean "breath." The same is true of the words for "spirit" in Latin (*spiritus*), Greek (*pneuma*), and Hebrew (*ruah*). These, too, mean "breath."

The common ancient idea behind all these words is that of soul or spirit as the breath of life. Similarly, the concept of cognition in the Santiago theory goes far beyond the rational mind, as it includes the entire process of life. Describing cognition as the breath of life seems to be a perfect metaphor.

Among the ancient conceptions of the soul, the one that comes closest to the concept of cognition in the Santiago theory is that of Aristotle, which was held also by other early Greek philosophers (see Windelband, 2001/1901). As we mentioned in our Introduction, the soul was perceived in early Greek philosophy as the ultimate moving force and source of all life. Closely associated with that moving force, which leaves the body at death, was the idea of knowing. From the beginning of Greek philosophy, the concept of the soul had a cognitive dimension. The process of animation was also a process of knowing.

Aristotle, in particular, saw the soul both as the agent of perception and knowing, and as the force underlying the body's formation and movements, not unlike the Santiago theory sees cognition today. He conceived of the soul as being built up in successive levels, corresponding to levels of organic life, much like we think of levels of cognition. The first level is the "vegetative soul," which controls the organism's metabolic processes. The soul of plants is restricted to this metabolic level of a vital force. The next higher form is the "animal soul," characterized by autonomous motion in space and by feelings of pleasure and pain. The "human soul," finally, includes the vegetable and animal souls, but its main characteristic is reason.

12.2.4 Mind and brain

The conceptual advance of the Santiago theory is best appreciated by revisiting the thorny question of the relationship between mind and brain. In the Santiago theory, this relationship is simple and clear. The Cartesian characterization of mind as the "thinking thing" is abandoned. Mind is not a thing but a process – the process of cognition, which is identified with the process of life. The brain is a specific structure through which this process operates. The relationship between mind and brain, therefore, is one between process and structure. Moreover, the brain is not the only structure through which the process of cognition operates. The entire structure of the organism participates in the process of cognition, whether or not the organism has a brain and a higher nervous system.

The Santiago theory of cognition is the first scientific theory that overcomes the Cartesian division of mind and matter, and will thus have the most far-reaching implications. Mind and matter no longer appear to belong to two separate categories, but can be seen as representing two complementary aspects of the phenomenon of life – process and structure. At all levels of life, beginning with the simplest cell, mind and matter, process and structure, are inseparably connected. For the first time, we have a scientific theory that unifies mind, matter, and life.

12.3 Cognition and consciousness

Cognition, as understood in the Santiago theory, is associated with all levels of life and is thus a much broader phenomenon than consciousness. Consciousness – that is, conscious, lived experience – unfolds at certain levels of cognitive complexity that require a brain and a higher nervous system. In other words, consciousness is a special kind of cognitive process that emerges when cognition reaches a certain level of complexity.

The central characteristic of this special cognitive process is the experience of self-awareness – to be aware not only of one's environment but also of oneself. The literature on consciousness studies can be quite confusing because many authors use the term "consciousness" both for the broader phenomenon of cognition, which includes perception and awareness of the environment, and for the experience of self-awareness.

To distinguish between these two cognitive levels, the cognitive philosopher David Chalmers (1995) in an oft-quoted article labeled them as the "easy problem" and the "hard problem" of consciousness.

12.3.1 The "easy" and the "hard" problems of consciousness

It seems to us that there are two main reasons for the widespread confusion between cognition and consciousness in the literature. The first is that in everyday language, the term "consciousness" has a broad range of meanings, which do not all correspond to the terminology that has recently been developed in cognitive science.

Our common language is rich in expressions such as "acting consciously" (i.e., with critical awareness), "being conscious" rather than unconscious (i.e., being awake, in full possession of our cognitive faculties), or showing "social consciousness" (i.e., being aware of social problems). In all these phrases there is a subtle blending of cognition (including perception, emotions, and behavior) and consciousness in the sense of self-awareness.

In addition, the everyday meaning of " consciousness" is closely linked to "conscience" – the inner sense of what is right or wrong in one's motives and conduct – which has been examined by philosophers throughout the ages and, with its implications of ethics and morals, is an important part of religion (as we shall discuss in Chapter 13). In fact, in some languages the link between "consciousness" and "conscience" is so close that both are denoted by the same term, such as *conscience* in French.

These multiple, and often confusing meanings of "consciousness" in everyday language may be one reason why there is confusion also in the scientific and philosophical literature. A second reason is that many human acts of cognition are attended by subjective conscious experience – that is, by consciousness – as well as by thoughts and reflections on the experiences. And since most cognitive scientists and philosophers limit their research to the human mind, they often tend to neglect to distinguish between cognition, which is associated with all levels of life, and consciousness, which involves self-awareness and requires a brain and a higher nervous system. In this chapter, we shall continue to emphasize this important distinction, using the term "consciousness" to refer to what Chalmers and many others consider the hard problem.

The difference between the easy and the hard problem is profound. The easy problem (cognition) has to do with brain mechanisms; the hard problem has to do with the question, how and why personal experience arises. In fact, the physical reason why there should be a personal experience is rather elusive. Physical theories of consciousness based on brain behavior may give "correlates" of consciousness – for example, registered neuronal "wiggles" associated with the perception of certain colors – but these wiggles are silent on why and how a certain personal experience of the color red appears.

Thus, for some authors, there is an explanatory "gap" (Chalmers, 1995) between the brain mechanisms and the arising of personal conscious experience, which can only be bridged with the help of some additional assumptions. We shall come back to this problem below. Another problem, pointed out, for example, by Nicholas Humphrey (2006), is that "my red" is my personal experience, and there seems to be no way to communicate this sensation to others. How, then, can one build a science based on such first-person experience? We shall see in the following pages how the systemic approach to cognition and consciousness has allowed cognitive scientists in recent years to overcome these conceptual problems and make significant advances toward formulating a true science of consciousness.

12.3.2 The scientific study of consciousness

It is interesting that the notion of consciousness as a process appeared in science as early as the late nineteenth century in the writings of William James (1842–1910), whom many consider the greatest American psychologist. James was a fervent critic of the reductionist and materialist theories that dominated psychology in his time, and an enthusiastic advocate of the interdependence of mind and body. He pointed out that consciousness is not a thing, but an ever-changing stream, and he emphasized the personal, continuous, and highly integrated nature of this stream of consciousness.

In subsequent years, however, the exceptional views of William James were not able to break the Cartesian spell on psychologists and natural scientists, and his influence did not reemerge until the last few decades of the twentieth century. Even during the 1970s and 1980s, when new humanistic and transpersonal approaches were formulated by American psychologists, the study of consciousness as lived experience was still taboo in cognitive science.

During the 1990s, the situation changed dramatically. While cognitive science established itself as a broad, interdisciplinary field of study, new noninvasive techniques for analyzing brain functions were developed, making it possible to observe complex neural processes associated with mental imagery and other human experiences. And suddenly, the scientific study of consciousness became a respectable and lively field of research. Within a few years, several books about the nature of consciousness, authored by Nobel Laureates and other eminent scientists, were published (e.g., Crick, 1994; Dennett, 1991; Edelman, 1992; Penrose, 1994). In addition, dozens of articles by the leading cognitive scientists and philosophers appeared in the newly created *Journal of Consciousness Studies*; and "Toward a Science of Consciousness" became a popular theme for large scientific conferences.

Today, there is a bewildering variety of approaches to the study of consciousness pursued by quantum physicists, biologists, cognitive scientists, and philosophers. One of the best documentations of this tremendous intellectual diversity is provided in the recent hefty volume (more than 1,000 pages) edited by Roger Penrose *et al.* (2011), titled *Consciousness and the Universe*. It contains a collection of sixty-seven papers by scientists and philosophers on numerous aspects of consciousness studies – from quantum effects in the brain, to the biological origin of consciousness, the origin of life, animal consciousness, spirituality, near-death experiences, and other altered states of consciousness. It makes for fascinating, if rather overwhelming, reading. In this chapter, however, we shall limit ourselves to consciousness studies within cognitive science, and specifically to a systemic approach that is consistent with the overall theme of our book – the systems view of life.

12.3.3 Two types of consciousness

Although cognitive scientists have proposed many different approaches to the study of consciousness, and have sometimes engaged in heated debates, it seems that there is a growing consensus on two important points. The first, as mentioned above, is the recognition that consciousness is a cognitive process, emerging from complex neural activity. The second point is the distinction between two types of consciousness – in other words, two types of cognitive experiences – which emerge at different levels of neural complexity.

The first type, known as "primary consciousness" or "core consciousness," arises when cognitive processes are accompanied by basic perceptual, sensory, and emotional experience. "Core consciousness," writes the neurologist Antonio Damasio (1999, p. 16), "provides the organism with a sense of self about one moment – now – and about one place – here. The scope of core consciousness is the here and now." The biologist Gerald Edelman (1992) believes that this transient sense of self is probably experienced by most mammals and perhaps by some birds and other vertebrates.

The second type of consciousness, variously called "higher-order consciousness," "extended consciousness," or "reflective consciousness," involves more elaborate self-awareness – a concept of self, held by a thinking and reflecting subject. This extended experience of self-awareness, identity, and personhood is based on memories of the past and anticipation of the future. It emerged during the evolution of the great apes, or "hominids" (see Section 11.2), together with language, conceptual thought, and all the characteristics that fully unfolded in human consciousness. Because of the critical role of reflection in this extended conscious experience, we shall call it "reflective consciousness."

Reflective consciousness involves a level of cognitive abstraction that includes the ability to hold mental images, allowing us to formulate values, beliefs, goals, and strategies. This evolutionary stage established a fundamental link between consciousness and social phenomena, because with the evolution of language arose not only the inner world of concepts and ideas but also the social world of organized relationships and culture. We shall return to this important evolutionary link when we discuss the social dimension of life (Chapter 14).

12.3.4 The nature of conscious experience

The central challenge of a science of consciousness is to explain the experiences associated with cognitive events. Different states of conscious experience are sometimes called *qualia* by cognitive scientists, because each state is characterized by a special "qualitative feel." The challenge of explaining these *qualia* is the "hard problem" identified by Chalmers (1995). After reviewing the conventional attempts of cognitive science, Chalmers asserts that none of them can explain why certain neural processes give rise to experience. "To account for conscious experience," he concludes, "we need an *extra ingredient* in the explanation."

This statement is reminiscent of the debate between mechanists and vitalists about the nature of biological phenomena during the early decades of the twentieth century (see Section 4.1.1). Whereas the mechanists asserted that all biological phenomena can be explained in terms of the laws of physics and chemistry, the vitalists maintained that a "vital force" must be added to those laws as an additional, nonphysical "ingredient" to explain biological phenomena.

The insight that emerged from this debate, though not formulated until many decades later, is that in order to explain biological phenomena, we need to take into account not only the conventional laws of physics and chemistry but also the complex nonlinear dynamics of living networks. A full understanding of biological phenomena is reached only when we approach it through the interplay of three different levels of description – the biology of the observed phenomena, the laws of physics and biochemistry, and the nonlinear dynamics of complex systems.

It seems to us that cognitive scientists find themselves in a very similar situation, albeit at a different level of complexity, when they approach the study of consciousness. Conscious experience is an emergent phenomenon, meaning that it cannot be explained in terms of neural mechanisms alone. Experience emerges from the complex nonlinear dynamics of neural networks and can be explained only if our understanding of neurobiology is combined with an understanding of those dynamics.

To reach a full understanding of consciousness, we must approach it through the careful analysis of conscious experience; of the physics, biochemistry, and biology of the nervous system; and of the nonlinear dynamics of neural networks. A true science of consciousness will be formulated only when we understand how these three levels of description can be woven together into what Francisco Varela (1999) has called the "triple braid" of consciousness research.

When the study of consciousness is approached by braiding together experience, neurobiology, and nonlinear dynamics, the "hard problem" turns into the challenge of understanding and accepting two new scientific paradigms. The first is the paradigm of complexity theory. Since most scientists are used to working with linear models, they are often reluctant to adopt the nonlinear framework of complexity theory and find it difficult to fully appreciate the implications of nonlinear dynamics. This applies in particular to the phenomenon of emergence.

It seems quite mysterious that experience should emerge from neurophysiological processes. However, this is typical of emergent phenomena, as we discussed in Section 8.2. Emergence results in the creation of novelty, and this novelty is often qualitatively different from the phenomena out of which it emerged.

In addition to complexity theory, scientists will need to accept another new paradigm: the recognition that the analysis of lived experience – that is, of subjective phenomena– has to be an integral part of any science of consciousness. As Varela (1999) and Shear have argued, this amounts to a profound change of methodology, which many cognitive scientists are reluctant to embrace, and which lies at the very root of the "hard problem of consciousness."

The great reluctance of scientists to deal with subjective phenomena is part of our Cartesian heritage. Descartes' fundamental division between mind and matter, between the *I* and the world, made us believe that the world could be described objectively – that is, without ever mentioning the human observer. Such an objective description of nature became the ideal of all science. However, three centuries after Descartes, quantum theory showed us that this classical ideal of an objective science cannot be maintained when dealing with atomic phenomena (see Section 4.3.4). And more recently, the Santiago theory of cognition has made it clear that cognition itself is not a representation of an independently existing world, but rather a "bringing forth" of a world through the process of living.

We have come to realize that the subjective dimension is always implicit in the practice of science. However, in general it is not the explicit focus. In a science of consciousness, by contrast, some of the very data to be examined are subjective, inner experiences. To collect and analyze these data systematically requires a disciplined examination of "first-person," subjective experience. Only when such an examination becomes an integral part of the study of consciousness will it deserve to be called a "science of consciousness." As we have argued in Section 4.3.4, this does not mean that we have to give up scientific rigor. Even when the object of investigation consists of first-person accounts of conscious experience, the intersubjective validation that is standard practice in science need not be abandoned.

12.3.5 Schools of consciousness study

The use of complexity theory and the systematic analysis of first-person conscious experience will be crucial in formulating a proper science of consciousness. In the last decade, several significant steps have already been taken toward this goal. Indeed, the extent to which nonlinear dynamics and the analysis of first-person experience are utilized can be used to identify several broad schools of thought among the great variety of current approaches to the study of consciousness.

The first is the most traditional school of thought. It includes, among others, the neuroscientist Patricia Churchland and the molecular biologist Francis Crick. This school has been called "neuroreductionist" by Francisco Varela (1996), because it reduces consciousness to neural mechanisms. Thus, consciousness is "explained away," as Churchland puts it, much like heat in physics was explained away once it was recognized as the energy of molecules in motion. In the words of Francis Crick (1994, p. 3):

> "You," your joys and your sorrows, your memories and your ambitions, your sense of personal identity and free will, are in fact no more than the behavior of a vast assembly of nerve cells and their associated molecules. As Lewis Carroll's Alice might have phrased it: "You're nothing but a pack of neurons."

This statement certainly sounds like the classic reductionist position – conscious experience reduced to the firing of neurons – and in his book Crick describes the corresponding neurophysiology in considerable technical detail. However, in other parts of the book, somewhat inconsistently, he asserts that conscious experiences are emergent properties that "arise in the brain from the interactions of its many parts." He never addresses the nonlinear dynamics of these processes of emergence, and hence we do not consider his theory as truly systemic. However, we feel that Varela's (1996) categorical identification of Crick as a reductionist may be too harsh.

The second school of consciousness study, known as "functionalism," is the most popular among today's cognitive scientists and philosophers. Its proponents assert that mental states are defined by their "functional organization" – that is, by patterns of causal relations in the nervous system. The functionalists are not Cartesian reductionists, because they pay careful attention to nonlinear neural patterns. However, they deny that conscious experience is an irreducible, emergent phenomenon. It may seem an irreducible experience, but in their view a conscious state is defined completely by its functional organization and is therefore understood once that pattern of organization has been identified. Thus Daniel Dennett (1991), one of the leading functionalists, gave his book the catchy title *Consciousness Explained*. Many patterns of functional organization have been postulated by cognitive scientists, and consequently there are many different strands of functionalism today.

Finally, there is a small but growing school of consciousness studies that embraces both the use of complexity theory and the analysis of first-person experience. Francisco Varela (1996), who was one of the pioneers of this school of thought, gave it the name "neurophenomenology." Phenomenology is an important branch of modern philosophy, founded by Edmund Husserl at the beginning of the twentieth century and developed further by many European philosophers, including Martin Heidegger and Maurice Merleau-Ponty. The central concern of phenomenology is the disciplined examination of experience, and the hope of Husserl and his followers was, and is, that a true science of experience would eventually be established in partnership with the natural sciences.

Neurophenomenology is an approach to the study of consciousness that combines the disciplined examination of conscious experience with the analysis of corresponding neural patterns and processes. With this dual approach, neurophenomenologists explore various domains of experience and try to understand how they emerge from complex neural activities. In doing so, these cognitive scientists are indeed taking the first steps toward formulating a true science of experience.

12.3.6 The view from within

The basic premise of neurophenomenology is that brain physiology and conscious experience should be treated as two interdependent domains of research with equal status. The disciplined examination of experience and the analysis of the corresponding neural patterns and processes will generate reciprocal constraints, so that research activities in the two domains can guide one another in a systematic exploration of consciousness.

Today's neurophenomenologists are a very diverse group. They differ in the manner in which first-person experience is taken into account, and they have also proposed different models for the corresponding neural processes. At its inception, the whole field was presented in some detail in a special issue of the *Journal of Consciousness Studies* (vol. VI, nos. 2–3, 1999), titled "The View from Within" and edited by Francisco Varela and the philosopher Jonathan Shear.

As far as first-person experience is concerned, three main approaches are being pursued. The adherents of all three insist that they are not talking about a casual inspection of experience but about using strict methodologies that require special skills and sustained training, just like the methodologies in other areas of scientific observation. The first approach is introspection, a method developed at the very beginning of scientific psychology. The second is the phenomenological approach in the strict sense, as developed by Husserl and his followers. The third approach consists of using the wealth of evidence gathered from meditative practice in various spiritual traditions.

Throughout human history, the disciplined examination of experience has been used within widely differing philosophical and religious traditions, including Hinduism, Buddhism, Taoism, Sufism, and Christianity. We may therefore expect that some of the insights of these traditions will be valid beyond their particular metaphysical and cultural frameworks.

This applies especially to Buddhism, which has flourished in many different cultures, originating with the Buddha in India, then spreading to China and Southeast Asia, reaching Japan, and, many centuries later, crossing the Pacific to California. In these different cultural contexts, mind and consciousness have always been the primary objects of Buddhist contemplative investigations. Buddhists regard the undisciplined mind as an unreliable instrument for observing different states of consciousness, and, following the Buddha's initial instructions, they have developed a great variety of techniques for stabilizing and refining the attention.

Over the centuries, Buddhist scholars have formulated elaborate and sophisticated theories about many subtle aspects of conscious experience, which are likely to be fertile sources of inspiration for cognitive scientists. Indeed, the dialogue between cognitive science and Buddhist contemplative traditions has already begun, and the first results indicate that the evidence from meditative practices will be a valuable component of any future science of consciousness (see Luisi, 2009; Shear and Jevning, 1999; Siderits *et al.*, 2011).

12.3.7 Mind without biology?

The schools of consciousness study mentioned above all share the basic insight that consciousness is a cognitive process, emerging from complex neural activity. However, there are other attempts, mostly by physicists and mathematicians, to explain consciousness as a property of matter at the level of quantum physics, rather than viewing it as a phenomenon associated with life. An outstanding example of that position is the approach by the mathematician and cosmologist Roger Penrose (1994), who postulates that consciousness is a quantum phenomenon and claims that we do not understand it, because we do not know enough about the physical world.

According to Penrose, consciousness may arise from a deeper physical level within neurons, where certain small tube-like structures, known as "microtubules," may display (as yet mysterious) quantum effects that may play a crucial role in the functioning of synapses. Consciousness, in this highly speculative view, is not an emergent property of living neural networks, but is produced by quantum effects in their innermost structures.

Such views of "mind without biology," in the apt phrase of the neuroscientist Gerald Edelman (1992), also include the view of the brain as a complicated computer. Like many cognitive scientists (e.g., Edelman, 1992; Searle, 1984, 1995; Varela, 1996), we believe that these are extreme views that are fundamentally flawed and that conscious experience is an expression of life, emerging from complex neural activity.

Yet another approach posits an elementary form of consciousness, not as emerging from complex neural activity, but as a primary reality. (Note that "primary" is used here in a different sense from the "primary consciousness" discussed in Section 12.3.3). This view is widespread among spiritual traditions, many of which teach that the material world has emerged from such pure consciousness. We shall discuss the Buddhist conception of consciousness as an example of this spiritual viewpoint in Chapter 13.

Recently, some cognitive philosophers have developed a variation of this view, according to which some "pure experience," or "elementary consciousness," may not be a secondary feature emerging from neural activity, but may be primary in the strongest sense of the word, the very basis of all our observations. There is presently no scientific evidence for this hypothesis. However, since it has become a popular topic of discussion, we invited the quantum physicist and cognitive philosopher Michel Bitbol to explain it in a guest essay (p. 266).

12.3.8 The emergence of conscious experience

Let us now turn to the neural activity that underlies conscious experience. In recent years, cognitive scientists have made significant advances in identifying the links between neurophysiology and the emergence of experience. Two promising systemic models were developed in the 1990s by Francisco Varela (1995), and by Gerald Edelman in collaboration with Giulio Tononi (Tononi and Edelman, 1998); see also Edelman and Tononi, 2000). More recently, Antonio Damasio (1999) proposed a neurophysiological theory of consciousness that adds several important insights to these earlier models.

The core idea of the two models is the same: conscious experience is not located in a specific part of the brain, nor can it be identified in terms of special neural structures. It is an emergent property of a particular cognitive process – the formation of transient functional clusters of neurons. Varela calls such clusters "resonant cell assemblies," while Edelman and Tononi (2000) speak of a "dynamic core."

Tononi and Edelman (1998) also embrace the basic premise of neurophenomenology that brain physiology and conscious experience should be treated as two interdependent domains of research. "It is a central claim of this article," they write, "that analyzing the convergence between . . . phenomenological and neural properties can yield valuable insights into the kinds of neural processes that can account for the corresponding properties of conscious experience."

The detailed dynamics of the neural processes in these two models are different but not incompatible. They differ in part because the authors do not focus on the same characteristics of conscious experience, and hence emphasize different properties of the corresponding neural clusters.

Guest essay

On the primary nature of consciousness

Michel Bitbol

CREA (Centre de Recherche en Épistémologie Appliquée), CNRS (Centre national de la recherche scientifique) / École Polytechnique, Paris

Nobody can deny that complex features of consciousness, such as reflexivity (the awareness that there *is* awareness of something), or self-consciousness (the awareness of one's own identity) are late outcomes of a process of biological adaptation. But what about pure nonreflexive experience? What about the mere "feel" of sensing and being, irrespective of any second-order awareness of this feel? There are good reasons to think that pure experience, or elementary consciousness, or phenomenal consciousness, is no secondary feature of an objective item but plainly *here*, primary in the strongest sense of the word.

We start with this plain fact: the world as we found it (to borrow Wittgenstein's expression) is no collection of objects; it is indissolubly a perceptive-*experience*-of-objects, or an imaginative experience of these objects *qua* being out of reach of perceptive experience. In other words, conscious experience is self-evidently pervasive and *existentially primary*. Moreover, any scientific undertaking presupposes one's own experience and others' experiences as well. In history and on a day-to-day basis, the objective descriptions which are characteristic of science arise as an invariant structural focus for subjects endowed with conscious experience. In this sense, scientific findings, including results of neurophysiology and evolution theory, are methodologically secondary to experience. Experience, or elementary consciousness, can then be said to be *methodologically primary* for science. Consequently, the claim of the primary nature of elementary consciousness is not a scientific statement: it just expresses a most basic prerequisite of science.

But, conversely, this means that the objective science of nature has no real bearing on the pure experience that tacitly underpins it. The latter allegation sounds hard to swallow in view of the many momentous successes of the neurosciences. Yet, if one thinks a little harder, any sense of paradox vanishes. Actually, it is by virtue of the very efficience of the neurosciences that they can have no grip on phenomenal consciousness. Indeed, as soon as this efficience is fully put to use, nothing prevents us from offering a purely neurophysiological account of the chain of causes operating from a sensory input received by an organism to the elaborate behavior of this organism. At no point does one need to invoke the circumstance that this organism is perceiving and acting consciously (in the most elementary sense of "having a feel"). In mature cognitive neuroscience, the fact of *phenomenal* consciousness is bound to appear irrelevant or incidental.

As a result, any attempt at providing a scientific account of phenomenal consciousness, by way of neurological or evolutionary theories, is doomed to failure (not because of any deficiency of these sciences, but precisely as a side effect of their most fruitful methodological option). Modern neurological theories, such as global workspace theory or integrated information theory, have been remarkably successful in accounting for major features of higher levels of consciousness, such as the capacity of unifying the field of awareness and of elaborating self-mapping. They have also turned out to be excellent predictors of a subject's behavioral wakefulness and ability or inability to provide reports in clinical situations such as coma and epileptic seizure. But they have provided absolutely no clue about the origin of phenomenal consciousness. They have explained the *functions* of consciousness, but not the circumstance that there is *something it is like to be* an organism performing these functions. The same is true of evolutionist arguments. Evolution can select some useful functions ascribed to consciousness (such as the behavioral emotivity of the organism, integrated action planning, or self-monitoring), but not the mere fact that there is *something it is like to implement* these functions. Indeed, only the functions have adaptative value, not their being experienced.

Even the ability of neurophysiological inquiry to identify *correlates* of phenomenal consciousness can be challenged on that basis. After all, identifying such correlates relies heavily on the subject's ability to discriminate, to memorize, and to *report*, which is used as the ultimate experimental criterion of consciousness. Can we exclude the possibility that the large-scale synchronization of complex neural activity of the brain cortex often deemed indispensable for *consciousness* is in fact required only for *interconnecting* a number of cognitive functions, including those needed for *memorizing, self-reflecting* and *reporting*? Conversely, extrapolating Semir Zeki's suggestion, can we exclude that any (large or small) area of the brain or even of the body is associated with some sort of fleeting pure experience, although no report can be obtained from it?

Data from the administration of general anaesthesia feed this doubt. When the doses of certain classes of anaesthetic drugs are increased and coherent EEG frequency is decreased, mental abilities are lost step by step, one after another. At first, subjects lose some of their appreciation of pain, but can still have dialogue with doctors and remember every event. Then, they lose their ability to recall long-term explicit memories of what is going on, but they are still able to react and answer questions on a momentary basis. With higher doses of drugs, patients lose the ability to respond to requests, in addition to losing their explicit memory; but they still have "implicit memories" of the situation. To recapitulate, faculties that are usually taken *together* as necessary for consciousness are in fact dissociable from one another. And pure, instantaneous, unmemorized, nonreflective experience might well be the last item left. This looks like a scientific hint as to the ubiquity and primary nature of phenomenal consciousness. Of course, a scientific hint does not mean a scientific proof (at any rate, claiming that there exists a scientific proof of the primary nature of elementary consciousness would badly contradict our initial aknowledgment that objective science can have no real grip on pure experience). The former scientific hint is only an indirect indication coming from the very blind spot of science: the pure passing experience it presupposes, and of which it retains only a stabilized and intersubjectively shared structural residue.

Should we content ourselves with these negative remarks? As Francisco Varela has shown, one can overcome them by proposing a broadened definition of science. Instead of remaining stuck within the third-person attitude, the new science should include a "dance" of mutual definition taking place between first-person and third-person accounts, mediated by the second-person level of social exchange. As soon as this momentous turn is taken, elementary consciousness is no longer a mystery for a truncated science, but an acknowledged datum from which a fuller kind of science can unfold.

Bibliography

Bitbol M. (2002). "Science as if situation mattered", *Phenomenology and the Cognitive Science*, **1**, 181–224.
Bitbol, M. (2008). "Is consciousness primary?", *NeuroQuantology*, **6**, 53–71.
Bitbol, M. & P. L. Luisi. (2011). "Science and the self-referentiality of consciousness", *Journal of Cosmology*, **14**, 4728–4743.
Varela, F. (1998). "Neurophenomenology : a methodological remedy for the hard problem", in Shear J. (ed.), *Explaining Consciousness: the Hard Problem*, MIT Press.
Wittgenstein, L. (1968). "Notes for lectures on private experience and sense data", *Philosophical Review*, **77**, 275–320.
Zeki, S. (2008). "The disunity of consciousness", in R. Banerjee and B.K. Chakrabarti (Eds.), *Progress in Brain Research,* Vol. 168, Elsevier.

In spite of the differences in the detailed dynamics they describe, the two models of resonant cell assemblies and the dynamic core have much in common. Both view conscious experience as an emergent property of a transient process of integration, or synchronization, of widely distributed groups of neurons. Both offer concrete, testable proposals for the specific dynamics of that process, and thus are likely to lead to significant advances in the formulation of a proper science of consciousness in the years to come.

Core consciousness and the protoself

Antonio Damasio's approach seems to be complementary to those of Varela and of Edelman and Tononi. While these authors focused on the processes of synchronization of clusters of neurons without specifying the exact functions of those neural clusters, Damasio (1999) describes their functions in detail but does not really explain how conscious experience emerges from them. However, Damasio's theory provides a detailed account of the roots of consciousness in biological processes. In other words, he shows how consciousness grows out of cognition, the self-organizing process of life.

While Damasio's theory of consciousness does not use the conceptual framework of nonlinear dynamics to analyze neural networks, his views on introspection and neurophysiology as two parallel avenues of research with equal status are fully consistent with Varela's school of neurophenomenology:

> The idea that subjective experiences are not scientifically accessible is nonsense. Subjective entities require, as do objective ones, that enough observers undertake rigorous observations according to the same experimental design . . . Knowledge gathered from subjective observations, e.g. introspective insights, can inspire objective experiments, and, no less importantly, subjective experiences can be explained in terms of the available scientific knowledge.
>
> *(Damasio, 1999, p. 309)*

According to Damasio, the deep roots of consciousness and of the sense of self lie in a large ensemble of brain structures (located at various levels of the central nervous system, from the spinal cord and brainstem to the cerebral cortices) that continually and nonconsciously maintain the state of the body within the narrow range and relative stability required for survival – in other words, in homeostasis.

To maintain the body's homeostasis, the brain continually maps the state of the living body in structures that regulate the organism's life; and as the body's state changes, so does its neural map. Damasio calls this continually changing neural map of the organism the "protoself" and sees it as the nonconscious forerunner of the "core self" that is experienced with the emergence of primary or core consciousness. Since the mapping of the body as protoself is tied to the maintenance of the life process, it is evident that life and consciousness are indelibly interwoven. In the terminology of the Santiago theory, we may say that the mapping of the body as protoself is the cognitive activity out of which consciousness emerges.

Damasio's basic hypothesis about core consciousness is that it arises from the brain's nonverbal account of how the organism's own state is affected by the perception of an object (external or arising from memory). He explains that the brain maps not only the entire state of the organism in its many dimensions but also the perceived object in sensory and motor structures activated by the interaction of the organism with the object.

A key feature of Damasio's theory is the recognition of the critical role of emotions in the functioning of core consciousness. His detailed discussion of clinical cases shows that, when core consciousness is suspended because of brain damage, emotion is usually suspended as well. Patients whose core consciousness is impaired do not reveal emotions by facial expression, body expression, or vocalization. Damasio also argues that the reason why we so confidently attribute consciousness to the minds of some animals, especially domestic animals, comes from our observation of the emotions they exhibit.

Damasio distinguishes between emotions, which can be triggered and displayed nonconsciously, and feelings, which are emotions made conscious. Emotions are complex patterns of chemical and neural responses that have specific regulatory functions. Most emotional responses have a long evolutionary history; they automatically provide organisms with survival-oriented behaviors. A feeling, in Damasio's terminology, is the conscious experience, or "mental image," of an emotion.

Core consciousness arises, according to Damasio, when the neural maps of the protoself become mental images; and since these neural maps include the organism's emotional responses to perceived objects, the corresponding mental images are feelings. Core consciousness, then, is a feeling that accompanies the making of an image in the act of perception.

Damasio's core consciousness is created in pulses, each pulse triggered by an object that we interact with or recall. The continuous "stream of consciousness" arises from the steady generation of consciousness pulses that correspond to the endless processing of myriad objects, whose interactions, actually or recalled, modify the protoself.

12.3.9 Reflective consciousness

As human beings, we not only experience the transient states of primary consciousness; we also think and reflect, communicate through symbolic language, make value judgments, hold beliefs, and act intentionally with self-awareness and an experience of personal freedom. The fact that human consciousness is intextricably interwoven with thought and reflection has the interesting, and problematic, consequence that in cognitive science conscious experience is not only an object of investigation but also the precondition of any investigation, so that any questioning about consciousness is radically self-referential (see Bitbol and Luisi, 2011).

The "inner world" of our reflective consciousness emerged in evolution together with the evolution of language and of organized social relations. This means that human consciousness is inextricably linked to language and to our social world of interpersonal relationships and culture. In other words, our consciousness is not only a biological but also a social phenomenon.

Consciousness and language

Humberto Maturana was one of the first scientists to link the biology of human consciousness to language in a systematic way. He did so by approaching language through a careful analysis of communication. Communication, according to Maturana, is not primarily a transmission of information, but rather a coordination of behavior between living organisms. Such mutual coordination of behavior is the key characteristic of communication for all living organisms, with or without nervous systems, and it becomes more and more subtle and elaborate with nervous systems of increasing complexity.

Language arises when a level of abstraction is reached at which there is symbolic communication. This means that we use symbols – words, gestures, and other signs – as effective tools for the mutual coordination of our actions. In this process, the symbols become associated with abstract mental images of objects. The ability to form such mental images turns out to be a crucial characteristic of reflective consciousness. Abstract mental images are the basis of concepts, values, goals, and strategies. (Note that our use of "mental images," in the sense of abstract images created by reflective consciousness, is different from Damasio's use of them as conscious experiences of neural patterns.)

Maturana emphasizes that the phenomenon of language does not occur in the brain but in a continual flow of coordinations of behavior. As humans, we exist in language and we continually weave the linguistic web in which we are embedded. We coordinate our behavior in language, and together in language we bring forth our world. "The world everyone sees," write Maturana and Varela (1980, p. 245), "is not *the* world but *a* world, which we bring forth with others." This human world centrally includes our inner world of abstract thought, concepts, beliefs, mental images, intentions, and self-awareness. In a human conversation, our concepts and ideas, emotions, and body movements become tightly linked in a complex choreography of behavioral coordination.

The nature of the self

Damasio's theory has contributed significantly to our understanding of reflective consciousness. It enables us to form a concept of self that overcomes what Francisco Varela has called our "Cartesian anxiety": we are self-aware, aware of our individual identity – and yet when we look for an independent self within our world of experience we cannot find any such entity.

According to Damasio, there are two types of self, associated with the two types of consciousness, which he calls the core self and the autobiographical self. The core self is a transient experience of self that is continually recreated as we interact with objects in our environment. The autobiographical self, associated with reflective consciousness, is a collection of mental images that appears to remain constant (although it evolves over a person's lifetime). Damasio emphasizes that the autobiographical self requires the presence of a core self to begin its gradual development. The contents of the autobiographical self can only be known when there is a fresh construction of core self. In a nutshell, the core self is a feeling, while the autobiographical self is an idea. Both are real, but neither is a separate entity or structure.

12.4 Cognitive linguistics

The investigation of the connections between reflective consciousness and language, pioneered by Maturana, gave rise to the new scientific discipline of cognitive linguistics, which examines the nature of language from the perspective of cognitive science. In recent years, this new field has led to several significant advances in our understanding of the human mind. According to George Lakoff and Mark Johnson (1999), these can be summarized in terms of three major discoveries: thought is mostly unconscious; the mind is inherently embodied; and abstract concepts are largely metaphorical.

The first discovery means that most of our thought operates at a level that is inaccessible to ordinary, conscious awareness. This "cognitive unconscious" includes not only our automatic cognitive operations but also our tacit knowledge and beliefs. Without our awareness, the cognitive unconscious shapes and structures all conscious thought.

12.4.1 The embodied mind

The concept of the embodied mind was introduced by Francisco Varela in the early 1990s (see Varela *et al.*, 1991) and was expanded considerably during the subsequent years. When cognitive scientists say that the mind is embodied, they mean far more than the obvious fact that we need a brain in order to think. Recent studies in cognitive linguistics indicate strongly that human reason does not transcend the body, as much of Western philosophy has held, but is shaped crucially by the detailed nature of our bodies and brains and by our bodily experience. The very structure of reason arises from our bodies and brains.

This notion of the embodied mind is consistent with the hypothesis, advanced by the primatologist Roger Fouts (1997), that language was originally embodied in gesture and evolved from gesture together with human consciousness. According to Fouts, the early hominids communicated with their hands and developed the skill of precise hand movements both for gestures and for making tools. Speech would have evolved later from the capacity for "syntax" – an ability to follow complex patterned sequences in the making of tools, in gesturing, and in forming words.

The notion of the embodied mind is also consistent with Damasio's (1999, p. 284) assertion that all conscious cognitive processes "depend for their execution on representations of the organism. Their shared essence is the body." In other words, the mind is inherently embodied.

The mind's embodiment can easily be illustrated by our use of spatial relations, which are among our most basic concepts. As Lakoff and Johnson (1999, pp. 34–5) explain, when we perceive a cat "in front of" a tree, this spatial relationship does not exist objectively in the world, but is a projection from our bodily experience. We have bodies with inherent fronts and backs, and we project this distinction onto other objects. Thus, "our bodies define a set of fundamental spatial relations that we use not only in orienting ourselves, but in perceiving the relationship of one object to another."

Some of our embodied concepts are also the basis of certain forms of reasoning, meaning that the way we think is also embodied. For example, when we distinguish between "inside" and "outside," we tend to visualize this spatial relationship in terms of a container with an inside, a boundary, and an outside. This mental image, which is grounded in the experience of our body as a container, becomes the basis of a certain form of reasoning, as Lakoff and Johnson persuasively illustrate. Suppose we put a cup inside a bowl and a cherry inside the cup. We would know immediately, just by looking at it, that the cherry, being inside the cup, is also inside the bowl.

That inference corresponds to a well-known argument, or "syllogism," in classical Aristotelian logic. In its most familiar form, it goes: "All men are mortal. Socrates is a man. Therefore, Socrates is mortal." The argument seems conclusive because, like our cherry, Socrates is within the "container" (or category) of men, and men are within the "container" (or category) of mortals. We project the mental image of containers onto abstract categories, and then use our bodily experience of a container to reason about these categories.

In other words, the classical Aristotelian syllogism is not a form of disembodied reasoning but grows out of our bodily experience. Lakoff and Johnson argue that this is true for many other forms of reasoning as well. The structures of our bodies and brains determine the concepts we can form and the reasoning we can engage in.

12.4.2 Metaphors

When we project the mental image of a container onto the abstract concept of a category, we use it as a metaphor. This process of metaphorical projection turns out to be a crucial element in the formation of abstract thought. The discovery that most human thought is metaphorical has been another major advance in cognitive science. Metaphors make it possible to extend our basic embodied concepts into abstract theoretical domains. For example, when we say, "I don't seem to be able to grasp this idea," or "This is way over my head," we use our bodily experience of grasping an object to reason about understanding an idea. In the same way, we speak of a "warm welcome," or a "big day," projecting sensory and bodily experiences onto abstract domains.

These are examples of "primary metaphors" – the basic elements of metaphorical thought. Lakoff and Johnson theorize that we acquire most of our primary metaphors automatically and unconsciously in our early childhood. For example, for infants the experience of affection typically occurs together with that of warmth, of being held. Thus associations between the two experiential domains are built up, and corresponding pathways across neural networks are established. Later in life, these associations continue as metaphors when we speak of a "warm smile," or a "close friend."

Our thought and language contain hundreds of primary metaphors, most of which we use without even being aware of them; and since they originate in basic bodily experiences, they tend to be the same in most languages around the world. In our abstract thought processes, we combine primary metaphors into more complex ones, enabling us to use rich imagery and subtle conceptual structures when we reflect on our experience. For example, to think of life as a journey allows us to use our rich knowledge of journeys while reflecting on how to lead a purposeful life.

12.5 Concluding remarks

Let us now summarize the recent advances in cognitive science discussed in this chapter. The main achievement, in our view, has been the gradual but consistent healing of the Cartesian split between mind and matter. In the 1970s, a few cognitive scientists recognized that mind and consciousness are not "things" but cognitive processes, and they took the radical step of identifying these processes of cognition with the very process of life. Thus cognition became associated with all levels of life. This means that mind and body are not separate entities, as Descartes believed, but are two complementary aspects of life – its process and its structure.

More recent research in cognitive science has confirmed and refined this view by showing how the process of cognition evolved into forms of increasing complexity together with the evolution of corresponding biological structures. Consciousness – that is, conscious, lived experience – unfolded at certain levels of cognitive complexity that require a brain and a higher nervous system. The biological roots of consciousness lie in the unconscious protoself – a continually changing neural map of the organism, which is a characteristic feature of mammals and other higher vertebrates. Primary consciousness, or core consciousness, emerges with the emergence of mental images from these neural maps.

Primary consciousness provides the organism with a transient sense of self (the core self) in the act of perception. The stream of consciousness arises from the steady generation of consciousness pulses that correspond to the endless processing of myriad objects, actual or recalled. In the history of evolution, gestures evolved into spoken language when the hominid brain developed motor regions that control precise hand movements and precise movements of the tongue. And together with language, reflective consciousness and conceptual thought evolved in the early humans as parts of ever more complex processes of communication.

While primary consciousness is associated with a transient experience of self that is endlessly repeated, reflective consciousness is associated with an extended sense of self. This larger sense of self arises when the transient images of the core self are enriched by memorized and seemingly invariant autobiographical images.

At the level of reflective consciousness, the process of cognition is the continuous flow of mental images we experience as thought. Most of these images, and thus most of our thought processes, remain unconscious. This cognitive unconscious shapes and structures all conscious thought.

All mental images, and therefore all thought processes, whether conscious or not, emerge from the neural maps of the protoself. Thus the human mind is inherently embodied. The very structure of reason arises from our bodies and brains. The use of metaphors is fundamental to human thought because it allows us to project bodily experience onto abstract concepts. Indeed, our abstract concepts are largely metaphorical.

Many details of this science of mind and consciousness still remain to be clarified and integrated. However, we now have the outlines of a scientific theory that overcomes the Cartesian division of mind and matter that has haunted Western science and philosophy for more than 300 years. In this new science of cognition, mind and matter no longer appear to belong to two separate categories, but can be seen as representing two complementary aspects of the phenomenon of life – process and structure. At all levels of life, beginning with the simplest cell, mind and matter, process and structure, are inseparably connected. For the first time, we have a scientific theory that unifies mind, matter, and life.

13

Science and spirituality

13.1 Science and spirituality: a dialectic relationship

During its long evolutionary history, humanity has developed various pathways and methods for obtaining and expressing knowledge about the self and the world, including philosophy, science, religion, art, and literature. Among these, science and spirituality have been two major driving forces of civilization.

The power of science (and its applications in technology) is responsible for material and technological progress. Since the information technology revolution in the last century, in particular, we have been witnessing an incredible expansion of our capabilities of global communication and of our transport and travel facilities (even in outer space), while in medicine we have enjoyed the discovery of unimaginable surgical devices and techniques, which have had many beneficial impacts on our health. Spirituality (and its codification in religion), on the other hand, is responsible for the internal growth of individuals, as well as for ethical constraints on excessive consumption of the planet's resources.

Since the turn of the century, it has become abundantly evident that, even though the power of science and technology has brought us benefits never experienced before, the ways in which these benefits have been achieved, and how they have been distributed between and within countries, are now threatening the future well-being, and indeed the very existence of humanity. We only need to mention the continuing threat of nuclear weapons and the dangers of nuclear radiation; the many wars that seem to rage continually around the world; the dramatic crises of global climate collapse, resource depletion, and species extinction; and the severely unequal distribution of wealth and increase of poverty in so many countries – all of which contribute to an existential crisis of humanity (to be discussed in more detail in Chapter 17).

We have argued in Section 11.3.3 that these threatening aspects may be rooted in some basic genetic features of humanity – and in particular of the male gender – such as aggressiveness and the desire for power, which are manifest in today's predatory capitalism. However, we have seen that humans also have positive, contrasting features, which we may broadly associate with spirituality – the tendency to become a better human being, which encompasses the inner elevation toward the numinous and the mysteries of the cosmos; as well as love and respect for our fellow human beings. Religion and, more generally, religiosity are expressions of this second important force of humanity. Just as technical progress is the pragmatic side of science, so religion can be seen as the pragmatic side of spirituality.

It is evident from this brief, introductory overview that the fate and well-being of modern civilization will be shaped significantly by the balance (or lack thereof) between the two opposing developments of technological progress and spiritual wisdom. Clearly, a science "without a soul" would lead to disaster. Conversely, we cannot manage our complex modern world with a purely spiritual approach. In this chapter, we shall analyze the basis and implications of this dialectic relationship between science and spirituality.

13.2 Spirituality and religion

The view of science and religion as a dichotomy has a long history, especially in the Christian tradition, and has recently been revived in several books written by scientists (e.g., Dawkins, 2006; Gould, 1999; Hawking and Mlodinow, 2010), as we discuss briefly in Section 13.3 below. On the other hand, there are many scientists who see no intrinsic opposition between science and religion, or science and spirituality. At the very core of this confusing situation, in our opinion, lies the failure of many authors to distinguish clearly between spirituality and religion. In order to resolve the confusion, we shall carefully examine the meaning of both of these terms, as well as the relationship between religion and spirituality. We also want to remind our readers that we have discussed the meaning and nature of science in some detail in the Introduction, at the very outset of this book. In particular, we shall examine whether spiritual and religious worldviews are compatible with the systems view of life that we have been discussing.

13.2.1 Spirit and spirituality

Spirituality is a much broader and more basic human experience than religion. It has two dimensions: one going inward, or "upward," as it were; and the other going outward, embracing the world and our fellow human beings. Either of the two manifestations of spirituality may or may not be accompanied by religion. Thus, when we say that scientists like Einstein or Bohr were spiritual souls, we mean that they had a strong desire to come closer to, or perhaps even identify with, the mysteries of the cosmos.

On the other hand, when we see people like Gandhi or Martin Luther King as spiritual beings, we mean that they were expressing through their lives higher ideals of a better humanity. In these cases, there is a union of the inner and outer dimensions of spirituality, and we can talk about "lay spirituality," a form of being spiritual without the need of being associated with a particular religion.

Spirituality in this broad sense need not be in conflict with science. Indeed, as we show in Section 13.4 below, it is fully consistent with the systems view of life. However, it may be in opposition to certain forms of technology, such as genetically modified organisms

(GMOs), cloning, pollution, deforestation, or simply excessive consumerism. This is where the argument becomes political, and such forms of spirituality, or even religion, are by no means incompatible with political activism, as "spiritual activists" like Gandhi, Martin Luther King, Desmond Tutu, or Latin America's liberation theologians have impressively demonstrated.

For a deeper understanding of spirituality, it is useful to review the original meaning of the word "spirit." The Latin *spiritus* means "breath," as do the related Latin word *anima*, the Greek *psyche*, and the Sanskrit *atman*. The common meaning of these key terms indicates that the original meaning of spirit in many ancient philosophical and religious traditions, in the West as well as in the East, is that of the breath of life.

Since respiration is indeed a central aspect of the metabolism of all but the simplest forms of life, the breath of life seems to be a perfect metaphor for the network of metabolic processes that is the defining characteristic of all living systems. Spirit – the breath of life – is what we have in common with all living beings. It nourishes us and keeps us alive.

Spirituality, or the spiritual life, is usually understood as a way of being that flows from a certain profound experience of reality, which is known as "mystical," "religious," or "spiritual" experience. There are numerous descriptions of this experience in the literature of the world's religions, which tend to agree that it is a direct, nonintellectual experience of reality with some fundamental characteristics that are independent of cultural and historical contexts. One of the most beautiful contemporary descriptions can be found in a short essay titled "Spirituality as Common Sense," by the Benedictine monk, psychologist, and author David Steindl-Rast (1990).

In accordance with the original meaning of spirit as the breath of life, Brother David characterizes spiritual experience as a nonordinary experience of reality during moments of heightened aliveness. Our spiritual moments are moments when we feel intensely alive. The aliveness felt during such a "peak experience," as the psychologist Abraham Maslow (1964) called it, involves not only the body but also the mind. Buddhists refer to this heightened mental alertness as "mindfulness," and they emphasize, interestingly, that mindfulness is deeply rooted in the body. Spirituality, then, is always embodied. We experience our spirit, in the words of Brother David, as "the fullness of mind and body."

It is evident that this notion of spirituality is very consistent with the notion of the embodied mind that is now being developed in cognitive science (see Section 12.4.1). Spiritual experience is an experience of aliveness of mind and body as a unity. Moreover, this experience of unity transcends not only the separation of mind and body but also the separation of self and world. The central awareness in these spiritual moments is a profound sense of oneness with all, a sense of belonging to the universe as a whole.

This sense of oneness with the natural world is fully borne out by the new systemic conception of life. As we understand how the roots of life reach deep into basic physics and chemistry, how the unfolding of complexity began long before the formation of the first living cells, and how life has evolved for billions of years by using again and again the same basic patterns and processes, we realize how tightly we are connected with the entire fabric of life.

Spiritual experience – the direct, nonintellectual experience of reality in moments of heightened aliveness – is known as a mystical experience because it is an encounter with mystery. Spiritual teachers throughout the ages have insisted that the experience of a profound sense of connectedness, of belonging to the cosmos as a whole, which is the central characteristic of mystical experience, is ineffable – incapable of being adequately expressed in words or concepts – and they often describe it as being accompanied by a deep sense of awe and wonder together with a feeling of great humility.

Scientists, in their systematic observations of natural phenomena, do not consider their experience of reality as ineffable. On the contrary, we attempt to express it in technical language, including mathematics, as precisely as possible. However, the fundamental interconnectedness of all phenomena is a dominant theme also in modern science, and many of our great scientists have expressed their sense of awe and wonder when faced with the mystery that lies beyond the limits of their theories. Albert Einstein, for one, repeatedly expressed these feelings, as in the following celebrated passage (Einstein, 1949, p. 5):

> The fairest thing we can experience is the mysterious. It is the fundamental emotion which stands at the cradle of true art and true science . . . the mystery of the eternity of life, and the inkling of the marvellous structure of reality, together with the single-hearted endeavor to comprehend a portion, be it ever so tiny, of the reason that manifests itself in nature.

The sense of awe and wonder which lies at the core of spiritual, or mystical, experience may have developed very early on in human evolution. In fact, it may have been the original source of spirituality. In Section 11.3.3, we speculated that the sense of awe and wonder may have arisen as soon as our hominid ancestors began to stand up and walk on two legs, allowing them to freely look up and gaze at the Sun, the Moon, and the starry night sky.

Indeed, the experience of a strong feeling of reverence and humility when looking at the starry sky is a prevalent and ancient theme in literature and art. It is well expressed in the celebrated, nineteenth-century Flammarion engraving, (Figure 13.1), which depicts a pilgrim dressed in a long robe and carrying a staff, crawling under the edge of the starry sky and peering into a mysterious world beyond, known in Christian literature as the "empyrean." The engraving's caption reads: "A medieval missionary tells that he has found where Heaven and Earth meet."

The missionary's mystical quest for knowledge and his vision of secret worlds beyond ordinary reality is an apt representation of the medieval worldview. The existence and movements of the celestial bodies were ascribed to mysterious forces that humans could neither conceive of nor comprehend. Astrology was an important tool in most civilizations for the divination of human destiny from the positions and movements of the heavenly bodies. In fact, from antiquity to the Renaissance, astrology and astronomy were known in the West by the same name, *astrologia*. Thus, the two disciplines coexisted as one for centuries, becoming distinct endeavors only gradually during the Scientific Revolution.

With the rise of astronomy, first in its Ptolemaic and then in its Copernican form (see Section 1.1), the mystical vision of the medieval missionary of Figure 13.1 was transformed into rational scientific models, describing the movements of the Sun, the Moon, and the stars. Still, the prime mover of the heavenly bodies was God, who was also the divine

Figure 13.1 Flammarion engraving by unknown artist, first documented in 1888.
Source: Camille Flammarion, *L'Atmosphere: météorologie populaire* (Paris: 1888), pp. 163 (http://commons.wikimedia.org/wiki/File:Flammarion.jpg).

creator of life on Earth and its myriad species, including first and foremost "man," in the language of the time. Thus the sense of mystery persisted in the notion of a mysterious divine order whose ultimate explanation was hidden in the mind of God.

In the nineteenth century, the mysteries perceived by the missionary in our figure were reduced further by Charles Darwin, who demonstrated that the great diversity of life on Earth was not the result of a divine design, but was due to the interplay of natural forces. And in our time, scientists have begun to probe even the nature of the soul (see Section 12.3.3), which, in the Christian tradition, had always been thought to be of divine, and hence deeply mysterious, nature.

Does this mean that the spiritual outlook must diminish with continuing scientific progress? We do not think so. As we mentioned in connection with the quest for the origin of life on Earth (Chapter 10), there is always a realm of mystery surrounding theoretical research at the forefront of science. Even as the boundaries of the known world of our legendary pilgrim are being pushed further and further outward, the "empyrean" is still there, always to be felt when we reach the limits of scientific knowledge. As the great scientist and philosopher Blaise Pascal put it succinctly in the seventeenth century, "Knowledge is like a sphere; the greater its volume, the larger its contact with the unknown."

13.2.2 The nature of religion

To summarize our conclusions of the previous section, both the concept of spirituality and the essence of spiritual experience are fully consistent with the systems view of life. However, this is not necessarily true for religion, and here it becomes important to distinguish between the two. Spirituality is a way of being grounded in a certain experience of reality that is independent of cultural and historical contexts. Religion is the organized attempt to understand spiritual experience, to interpret it with words and concepts, and to use this interpretation as the source of moral guidelines for the religious community.

There are three basic aspects of religion: theology, morals, and ritual (see Capra and Steindl-Rast, 1991). In theistic religions, theology is the intellectual interpretation of the spiritual experience, of the sense of belonging, with God as the ultimate reference point. Morals, or ethics, is the rules of conduct derived from that sense of belonging; and ritual is the celebration of belonging by the religious community. All three of these aspects – theology, morals, and ritual – depend on the religious community's historical and cultural contexts.

13.2.3 Theology

Theology was originally understood as the intellectual interpretation of the theologians' own mystical experience. Indeed, according to the Benedictine scholar Thomas Matus (quoted in Capra and Steindl-Rast, 1991), during the first thousand years of Christianity virtually all of the leading theologians – the "Fathers of the Church" – were also mystics. Over the subsequent centuries, however, during the Scholastic period, theology became progressively fragmented and divorced from the spiritual experience that was originally at its core.

With the new emphasis on purely intellectual theological knowledge came a hardening of the language. Whereas the Church Fathers repeatedly asserted the ineffable nature of religious experience and expressed their interpretations in terms of symbols and metaphors, the Scholastic theologians formulated the Christian teachings in dogmatic language and required that the faithful accept these formulations as the literal truth. In other words, Christian theology (as far as the religious establishment was concerned) became more and more rigid and fundamentalist, devoid of authentic spirituality.

The awareness of these subtle relationships between religion and spirituality is important when we compare both of them with science. While scientists try to explain natural phenomena, the purpose of a spiritual discipline is not to provide a description of the world. Its purpose, rather, is to facilitate experiences that will change a person's self and way of life. However, in the interpretations of their experiences, mystics and spiritual teachers are often led to make statements also about the nature of reality, causal relationships, the nature of human consciousness, and the like. This allows us to compare their descriptions of reality with corresponding descriptions by scientists.

In these spiritual traditions – for example, in the various schools of Buddhism – the mystical experience is always primary; its descriptions and interpretations are considered secondary and tentative, insufficient to fully describe the spiritual experience. In a way, these descriptions are not unlike the limited and approximate models in science, which are always subject to further modifications and improvements.

In the history of Christianity, by contrast, theological statements about the nature of the world, or about human nature, were often considered literal truths, and any attempt to question or modify them was deemed heretical. This rigid position of the Church led to the well-known conflicts between science and fundamentalist Christianity, which have continued to the present day. In these conflicts, antagonistic positions are often taken by fundamentalists on both sides who fail to keep in mind the limited and approximate nature of all scientific theories, on the one hand, and the metaphorical and symbolic nature of the language in religious scriptures, on the other. In recent years, such fundamentalist debates have become especially problematic around the concept of a creator God.

In theistic religions, the sense of mystery that is at the core of spiritual experience is associated with the divine. In the Christian tradition, the encounter with mystery is an encounter with God, and the Christian mystics repeatedly emphasized that the experience of God transcends all words and concepts. Thus Dionysius the Areopagite, a highly influential mystic of the early sixth century, writes: "At the end of all our knowing, we shall know God as the unknown;" and St. John of Damascus writes in the early eighth century: "God is above all knowing and above all essence" (both quoted in Capra and Steindl-Rast, 1991, p. 47).

However, most Christian theologians do want to speak about their experience of God, and to do so the Fathers of the Church used poetic language, symbols, and metaphors. The central error of fundamentalist theologians in subsequent centuries has been to adopt a literal interpretation of these religious metaphors. Once this is done, any dialogue between religion and science becomes frustrating and unproductive.

13.2.4 Ethics, ritual, and the sacred

Religion not only involves the intellectual interpretation of spiritual experience but is also closely associated with morals and rituals. Morals or ethics means the rules of conduct derived from the sense of belonging that lies at the core of the spiritual experience, and ritual is the celebration of that belonging.

Both ethics and ritual develop within the context of a spiritual or religious community. According to David Steindl-Rast, ethical behavior is always related to the particular community to which we belong. When we belong to a community, we behave accordingly. In today's world, there are two relevant communities to which we all belong. We are all members of humanity, and we all belong to the global biosphere. We are members of *oikos*, the Earth Household, which is the Greek root of the word "ecology," and as such we should behave as the other members of the household behave – the plants, animals, and microorganisms that form the vast network of relationships that we call the web of life.

The outstanding characteristic of the Earth Household is its inherent ability to sustain life. As members of the global community of living beings, it behooves us to behave in such a way that we do not interfere with this inherent ability. This is the essential meaning of ecological sustainability. As members of the human community, our behavior should reflect a respect of human dignity and basic human rights. Since human life encompasses biological, cognitive, social, and ecological dimensions, human rights should be respected in all four of these dimensions. The political consequences of respecting these two core values – human dignity and ecological sustainability – will be discussed in more detail in the last part of this book (Chapters 16–17).

The original purpose of religious communities was to provide opportunities for their members to relive the mystical experiences of the religion's founders. For this purpose, religious leaders designed special rituals within their historical and cultural contexts. These rituals might involve special places, robes, music, psychedelic drugs, and various ritualistic objects. In many religions, these special means to facilitate mystical experience become closely associated with the religion itself and are considered sacred.

Sacred places, objects, or forms of art, however, are more than just tools or techniques to facilitate the experience of belonging. They are always integral parts of rituals, and it is those rituals that are the doorways to mystical experience.

13.3 Science versus religion: a "dialogue of the deaf"?

The debates between science and religion, especially in Christianity, have been going on for centuries – from the infamous trials of Galileo, Giordano Bruno, and Scopes (the high school biology teacher prosecuted in Tennessee in 1925 for teaching Darwin's theory) to the current attacks on evolutionary theory by Christian fundamentalists under the banner of creationism, or intelligent design (see Section 9.7). On the other hand, there has recently been a flourishing of books and essays on religion by well-known scientists who tend to contrast science and religion as two opposite ideologies. Other scientists, however, find the essential nature of religious experience to be in perfect harmony with the views of modern science, and in particular with the systems view of life.

We have argued in this chapter that this apparent confusion can be resolved if careful attention is given to the difference between religion and spirituality. When spirituality is understood as inner growth, associated with the experience of a profound sense of connectedness, of belonging to the universe as a whole, combined with a strong feeling of awe and wonder and with respect for a humanitarian and ecological ethics, then there cannot be any dichotomy between spirituality and science, nor between science and a religion that has such spiritual experience at its core.

Religion, as we explained, is the organized attempt to understand and interpret spiritual experience, and as such is often in danger of becoming dogmatic, requiring the faithful to accept its pronouncements, moral codes, and hierarchical structures as literal truths.

We shall see, however, that such fundamentalist attitudes are not limited to religious leaders. Scientists, too, can be fundamentalists, forgetting that all their models and theories are limited and approximate, and ignoring the important role of metaphors in religion, as well as in science. When that happens, the debate between scientists and religious leaders soon turns into a *dialogue de sourds* ("dialogue of the deaf").

With these caveats, let us now review the pronouncements of some representative scientists on the relationship between science and religion. Because of the large number of recent publications on this subject, we have to limit ourselves to mentioning just a few authors; and we shall also have to restrict ourselves to debates between science and Christianity, leaving aside all other religions.

Central to most of these debates is the concept of a monotheistic God, creator of the world, nature, human life, and human consciousness. As we have mentioned, the failure to understand all attributes of such a God as metaphors, and thus either to accept a literal concept of God as an act of faith, or reject it as unscientific, is the reason why most of these debates remain at a fruitless, fundamentalist level.

One idea that has been widely discussed and has gained much acceptance is that of "non-overlapping magisteria (NOMA)" by the evolutionary biologist Stephen Jay Gould (1999). According to Gould, there should be a clear demarcation between science and religion, both a methodological separation and a clear distinction between their final aims. The "magisterium" (Church Latin for "teaching authority") of the natural sciences should deal with the interpretation of the functioning of the world, based on natural laws, whereas the magisterium of religion should deal with the spiritual and humanistic world. For Gould, this separation does not mean that the two realms cannot communicate with each other. There should be enough questions along the border between the two, and discussions should go on based on mutual respect and tolerance; but there should be no interference, or encroachment of one upon the other.

This sounds like common sense, easy to accept, and this is why the idea of NOMA is relatively popular among scientists. But things are not that simple. In fact, common experience shows that the strict separation advocated by Gould is rarely maintained (see Pievani, 2011). For example, consider creationism and its offshoot intelligent design (ID). Here, clearly, there is a strong encroachment of religion upon science, since the advocates of ID claim to possess a "scientific" view that is more valid than the Darwinian view. Or, consider the interference of the Vatican with research on stem cells, contraceptives, genetic identity, end-of-life problems, and even the very definition of life and consciousness.

We are witnessing almost daily an invasion of religious thought into the "magisterium" of science, and with considerable political impact. Not to mention the infamous examples of the past: the trial of Galileo or the burning alive of philosophers like Giordano Bruno and of "witches," who were natural healers practicing folk medicine; this was certainly something much more than benign interference! Conversely, there are a fair number of fundamentalist scientists who claim to have found a rational demonstration of the nonexistence of God.

In addition, NOMA does not work well at the personal level for a scientist who is also a believer. Should he or she accept the idea that life on Earth originated from inanimate matter, or accept the biblical account of creation as literal truth. To accept both would correspond to a kind of schizophrenia; or should the scientist investigate one in the laboratory and believe in the other one in church? As we have argued, this dilemma can only be resolved from a nonfundamentalist perspective, when the language of the Bible is accepted as metaphorical narrative, and the theories of science as limited and approximate models.

Thus, NOMA is a nice theoretical principle, but it does not work at the everyday practical level. As Pievani (2011) has pointed out, most scientists, even if they are not familiar with the idea of NOMA, are either agnostics (e.g., they do not take a position about God's existence, as the question of God does not belong to the realm of science), or they profess a methodological atheism restricted to science (the hypothesis of God is ruled out as long as we consider questions belonging to the natural world).

Another well-known biologist who has passionately discussed the question of God from a fundamentalist point of view is Richard Dawkins, the author of *The Selfish Gene*, whose genetic determinism we critically discussed in Section 9.6. As pointed out in Pievani (2011), the title of Dawkins' book, *The God Delusion* (Dawkins, 2006), already summarizes the polemic nature of his campaign. It is certainly a good example of a scientist's deliberate interference with the magisterium of religion. In fact, Dawkins attempts to justify atheism in a rational, scientific way. The existence, or nonexistence, of God becomes for him a scientific hypothesis which can be dealt with just like any other scientific hypothesis.

Dawkins recognizes that there is no way to conclusively demonstrate the nonexistence of God. However, he develops arguments based on probability and common sense. He concludes that the idea of God is highly improbable. For example, to explain the numerical combination of physical constants that make life on Earth possible as an act of God, would be to explain something improbable with something even more improbable. God, for Dawkins, is statistically improbable because this hypothesis needs more assumptions and explanations than we can furnish. We agree with the final comment of Pievani that Dawkins, with this kind of argumentation, is the perfect contributor to a "dialog among the deaf."

Biologists are not the only scientists engaging in fundamentalist discussions about the existence of God. There are several physicists, too, who have contributed to the debate, first and foremost among them the celebrated astrophysicist and cosmologist Stephen Hawking. In his *A Brief History of Time*, Hawking (1988, p. 6), writes:

> So long as the universe had a beginning, we could suppose it had a creator. But if the universe is really completely self-contained, having no boundary or edge, it would have neither beginning nor end: it would simply be. What place, then, for a creator?

Thus, it is all very simple: as soon as physicists can show that there was no big bang, the notion of a creator God will disappear. It is astonishing to us that such a simplistic, linear concept of God is entertained by one of the most brilliant scientists.

Physicists seem to be especially fond of using God as a metaphor, usually in rather fundamentalist ways. Best known, perhaps, is Albert Einstein's statement that "God does not play dice." More recently, the physicist and cosmologist Paul Davies has expressed his belief that it is by the means of science that we can truly see into "the mind of God." In fact, God appears in the titles of two books by Davies: *God and the New Physics* (1983) and *The Mind of God* (1992). Another well-known scientist and author, Stuart Kauffman (2008), equates God with the creativity of the universe, a metaphor we can certainly agree with.

We have provided only a short analysis of the thinking of some brilliant scientists on the relation between science and religion. Again and again, we have seen that fundamentalist positions on both sides of the debate are the main stumbling block. From our point of view, the apparent dichotomy dissolves when we move from organized religion to the broader realm of spirituality, and when we recognize that both spiritual experience and the mystery we find at the edge of every scientific theory transcend all words and concepts. With this attitude we can marvel at the scientific narrative of how matter condensed into the first forms of life and evolved ever more complex structures and cognitive processes, all the way to the emergence of consciousness, while we can enjoy the richness of spiritual teachings gleaned from the world's religious traditions at each stage of this unfolding. This is a dialogue of mutual tolerance and respect, and one which can foster a dialectic interplay that can be instructive and inspiring to scientists and religious leaders alike.

Having indicated a possible resolution of the dichotomy between science and religion, we shall now review the fascinating similarities between the worldviews of scientists and mystics that have recently been discovered.

13.4 Parallels between science and mysticism

As we have mentioned, scientists and spiritual teachers pursue very different goals. While the purpose of the former is to find explanations of natural phenomena, that of the latter is to change a person's self and way of life. However, in their different pursuits, both are led to make statements about the nature of reality that can be compared.

Among the first modern scientists to make such comparisons were some of the leading physicists of the twentieth century who had struggled to understand the strange and unexpected reality revealed to them in their explorations of atomic and subatomic phenomena (see Section 4.2). In the 1950s, several of these scientists published popular books about the history and philosophy of quantum physics, in which they hinted at remarkable parallels between the worldview implied by modern physics and the views of Eastern spiritual and philosophical traditions. The following three quotations are examples of such early comparisons.

> The general notions about human understanding . . . which are illustrated by discoveries in atomic physics are not in the nature of things wholly unfamiliar, wholly unheard of, or new. Even in our own culture they have a history, and in Buddhist and Hindu thought a more considerable and central place.
>
> *(J. Robert Oppenheimer, 1954, pp. 8–9)*

> For a parallel to the lesson of atomic theory . . . [we must turn] to those kinds of epistemological problems with which already thinkers like the Buddha and Lao Tzu have been confronted.
>
> *(Niels Bohr, 1958, p. 20)*

> The great scientific contribution in theoretical physics that has come from Japan since the last war may be an indication of a certain relationship between philosophical ideas in the tradition of the Far East and the philosophical substance of quantum theory.
>
> *(Werner Heisenberg, 1958, p. 202)*

During the 1960s, there was a strong interest in Eastern spiritual traditions in Europe and North America, and many scholarly books on Hinduism, Buddhism, and Taoism were published by Eastern and Western authors (e.g., Rahula, 1967; Ross, 1966; Suzuki, 1963; Watts, 1957). At that time, the parallels between these Eastern traditions and modern physics were discussed more frequently (see, e.g., LeShan, 1969), and a few years later they were explored systematically by Capra (2010/1975) in a book titled *The Tao of Physics*.

The main thesis in *The Tao of Physics* is that the approaches of physicists and mystics, even though they seem at first quite different, share some important characteristics. To begin with, their method is thoroughly empirical. Physicists derive their knowledge from experiments; mystics from meditative insights. Both are observations, and in both fields these observations are acknowledged as the only source of knowledge. The objects of observation are, of course, very different in the two cases. The mystic looks within and explores his consciousness at various levels, including the physical phenomena associated with the mind's embodiment. The physicist, by contrast, begins his inquiry into the essential nature of things by studying the material world. Exploring ever deeper realms of matter, he becomes aware of the essential unity of all natural phenomena. More than that, he also realizes that he himself and his consciousness are an integral part of this unity. Thus the mystic and the physicist arrive at the same conclusion; one starting from the inner realm, and the other from the outer world. The harmony between their views confirms the ancient Indian wisdom that *brahman*, the ultimate reality without, is identical to *atman*, the reality within.

A further important similarity between the ways of the physicist and the mystic is the fact that their observations take place in realms that are inaccessible to the ordinary senses. In modern physics, these are the realms of the atomic and subatomic world; in mysticism, they are nonordinary states of consciousness in which the everyday sensory world is transcended. In both cases, access to these nonordinary levels of experience is possible only after long years of training within a rigorous discipline, and in both fields the "experts" assert that their observations often defy expressions in ordinary language.

Twentieth-century physics was the first discipline in which scientists experienced dramatic changes in their basic concepts and ideas – a paradigm shift from the mechanistic worldview of Descartes and Newton to a holistic and systemic conception of reality. Subsequently, the same change of paradigms occurred in the life sciences with the gradual emergence of the systems view of life, the subject of this book. It should therefore not come as a surprise that the similarities between the worldviews of physicists and Eastern mystics are relevant not only to physics but also to science as a whole.

After the publication of *The Tao of Physics* in 1975, numerous books appeared in which physicists and other scientists presented similar explorations of the parallels between physics and mysticism (e.g., Davies, 1983; Talbot, 1980; Zukav, 1979). Other authors extended their inquiries beyond physics, finding similarities between Eastern thought and certain ideas about free will; death and birth; and the nature of life, mind, consciousness, and evolution (see Mansfield, 2008). Moreover, the same kinds of parallels have been drawn also to Western mystical traditions (see Capra and Steindl-Rast, 1991).

Some of the explorations of parallels between modern science and Eastern thought were initiated by Eastern spiritual teachers. The Dalai Lama, in particular, hosted a series of dialogues with Western scientists on "Mind and Life" at his home in Dharamsala, India, which we shall briefly describe.

13.4.1 The Mind and Life Institute: science and Buddhism

Tenzin Gyatso, the 14th Dalai Lama, has not only been one of the most charismatic and authentic spiritual teachers to come from the East; he has also had, since his youth, a keen interest in science and technology (see Dalai Lama, 2005). The idea of hosting a series of dialogues between scientists and Buddhists originated in 1983 at an international conference in Alpbach, Austria, on "Alternative Realities: Convergence Between New Sciences and Ancient Spiritual Traditions." At this memorable gathering (which was attended by both P.L.L. and F.C.), the Dalai Lama met Francisco Varela, who was inspired by this encounter to found the Mind & Life (M&L) Institute a few years later.

Since 1987, the institute has organized a series of dialogues between scientists and Buddhists, which take place every other year in Dharamsala under the patronage of the Dalai Lama. One of us (P.L.L.) has participated in all of these meetings (see Luisi, 2009). To give our readers a flavor of the conversations between scientists and Buddhists, we reproduce below a brief exchange with the Dalai Lama on the Buddhist view of consciousness (Box 13.1).

The M&L dialogues last for one week. On each day, a leading scientist, generally from the West, presents to the Dalai Lama and his fellow monks for discussion some of the key ideas he or she is pursuing in a particular discipline. The subject has been mostly cognitive science, but there have also been discussions on cosmology, quantum physics, and biology. The meetings are restricted to 40–50 invited participants. However, the institute also organizes larger meetings, open to the public, at American and European universities (see www.mindandlife.org).

Since 1987, a large number of scientists have accepted the invitation by Francisco Varela to attend the M&L meetings, which have acquired a considerable reputation also in established scientific institutions, some of which have collaborated in the organization of the meetings. These institutions include the Massachusetts Institute of Technology, the Swiss Federal Institute of Technology, the University of Innsbruck (Austria), the University of Wisconsin, and Johns Hopkins Medical University. Several books have been published on the results of these meetings (see Luisi, 2009, for references); and many of the participating scientists, for whom this was often their first encounter with Buddhism, have remained fascinated by and personally engaged with this spiritual tradition. Conversely, the Dalai Lama was invited to open the 2005 international meeting of the Society for Neuroscience in Washington, DC – another sign of recognition of the M&L Institute.

Important is the fact that these contacts of Buddhists with the academic world have led to new experimental research. Of particular resonance in the scientific community were the neurobiological experiments conducted by the team of Richie Davidson at the University

Box 13.1
The Buddhist view of consciousness

This brief exchange between Pier Luigi Luisi and the Dalai Lama (see Luisi, 2009) is about the nature of consciousness. In most of the current Buddhist literature, in particular in the Tibetan tradition, the term "consciousness" is used on two levels: the first one concerns perception, intentionality, and the conscious (rather than unconscious) waking state, all of which are part of the "easy problem", and which we have associated with cognition in this chapter. The second level is what the Tibetan Buddhists refer to as "subtle consciousness," which is seen as the very source of the human capability of conscious experience. According to them, this subtle consciousness is not based on matter. In the following exchange the Dalai Lama, in his colorful and forceful language, conveys a clear idea of the Buddhist position on this subject.

P.L.L.: Does your Holiness accept the view that consciousness arises naturally as an emergent property at a certain level of brain and neuronal complexity?

D.L.: It is very clear that specific modes of embodied consciousness, such as the human psyche, or human visual perception, do not arise in the absence of the brain or the appropriate faculty. But if we examine the clear, luminous, and cognizant aspect of these mental processes – in other words, consciousness itself – then the Buddhist perspective is that the event of consciousness does not emerge from the brain or from matter.

P.L.L.: This is an important difference. Many scientists accept the idea that all properties of mankind come from within, even consciousness and the idea of God, as self-generated values. This is not so for Buddhism?

D.L.: That's correct.

of Wisconsin on long-term Buddhist practitioners. Davidson's team were able to show that meditation induces a particular set of neuronal oscillations, which differ significantly from those of control groups. Importantly, the data suggest that mental training involves temporal integrative mechanisms and may induce short-term and long-term neural changes. These results were published in one of the most prestigious scientific journals, *Proceedings of the National Academy of Sciences of the USA* (Lutz *et al.*, 2004).

This kind of work started a new scientific discipline, now called "contemplative neuroscience": the study of the effects of contemplative practices and purposeful mental training on the human brain and behavior. And this in turn gave rise to "contemplative clinical science," concerned with the prevention and treatment of disease by meditation practices. Also to be mentioned is the success of the M&L Summer Research Institute, an annual retreat for about 200 participants, who work together for one week to investigate the effects of contemplative practice.

With these programs, the M&L Institute has established an important channel of communication between academic institutions and key representatives of Buddhist spirituality, advancing both conceptual and organizational links between ancient spirituality and modern science.

13.5 Spiritual practice today

The strong interest in Eastern spiritual traditions that emerged in Europe and North America in the 1960s was generated by the so-called counterculture of the time (see Roszak, 1969). During the subsequent decades, many of the values of this subculture were embraced by mainstream society, including in particular the interest in Eastern meditation practices. Today, there are countless meditation centers throughout the Western hemisphere where various techniques of yoga, tai ji, qigong, and many forms of sitting meditation are being taught and practiced. While many people practice these techniques for reasons of health and relaxation, many others are engaged in the spiritual life that flows from the nonordinary states of consciousness experienced in deep meditation.

13.5.1 The spread of Buddhism in the West

Among the spiritual or religious philosophies taught in these Eastern traditions, that of Buddhism has been by far the most popular in the West, probably because Buddhism is a nontheistic, nondogmatic religion with a very pragmatic approach (see Rahula, 1967; Suzuki, 1970). Today there are numerous Buddhist centers in Europe, North America, Latin America, and Australia, representing a wide variety of different schools, including Zen, Vipassana, and various schools of Tibetan Buddhism. In fact, in many of these regions, Buddhism has been the fastest growing religious philosophy in recent years.

Throughout its history, Buddhism has shown that it can easily adapt to a variety of cultural situations. It originated with the Buddha in India in the fifth century BC, spread to China and Southeast Asia, reached Japan in the first century AD and, almost two millennia later, jumped across the Pacific Ocean to California. There, the Japanese school of Zen Buddhism was embraced first in the 1950s by the "beat poets" – Jack Kerouac, Allen Ginsberg, Gary Snyder, and others (see Cook, 1971; Watson, 1995). At the end of that decade, the first Zen master, Suzuki Roshi, arrived in San Francisco, where he established the San Francisco Zen Center, which is still flourishing today.

Similar Zen centers were founded in other American cities and in Europe, where the Vietnamese Buddhist monk and peace activist Thich Nhat Hanh became a very influential teacher. The early 1970s saw a dramatic growth of interest in Tibetan Buddhism, mainly due to the arrival of exiled lamas from various Tibetan lineages, who established meditation centers in many parts of Europe and the USA. More than any of them, however, the charismatic figure of the Dalai Lama introduced a worldwide audience to the basic principles of Buddhist philosophy.

13.5.2 Buddhist philosophy and science

With its pragmatic, empirical approach, Buddhist philosophy seems to have a special affinity with modern science. One aspect of its philosophy that is especially attractive to modern scientists is that it denies the duality of mind and matter. In the Buddhist view, mental activity is one of the physical senses (much like in cognitive science), so that there is no opposition between subject and object, self and world.

Moreover, for Buddhists no single phenomenon in the world has an independent, intrinsic reality; all phenomena arise in mutual dependence and are dependent on contextual causes and conditions. This is the celebrated Buddhist doctrine of "codependent arising." Nagarjuna, perhaps the most intellectual among the early Buddhist philosophers, introduced the term "emptiness" (*sunyata*) to indicate this lack of any intrinsic reality in the phenomena we perceive:

> Things derive their being and nature by mutual dependence and are nothing in themselves.

This remarkable statement can be seen, from our modern perspective, as a quintessential expression of the systemic conception of reality. Buddhists apply their conception of phenomena in terms of processes and relationships also to the structures of the mind. In accordance with the principle of emptiness, they hold that there is no independently existing, immutable self (see Siderits *et al.*, 2011). Rather, the self is an emergent property that changes from moment to moment – a notion dear to most modern cognitive scientists.

Another point of attraction in Buddhism is its great tolerance: everybody is accepted without question, regardless of political or religious affiliation, caste, or personal preferences. In terms of morality, this means recognition of the diversity and relativity of ethical norms. Rather than judging unethical behavior as bad in an absolute sense, Buddhists consider it "unskillful," because it is a hindrance to one's self-realization.

Many scientists have been attracted to Buddhist philosophy because of its intellectual, speculative nature. However, this is only one side of Buddhism. Complementary to it is a strong emphasis on love and compassion (see Luisi, 2008). Enlightened wisdom (*bodhi*) is seen as being composed of two key elements: intuitive intelligence (*prajna*) and compassion (*karuna*) for all sentient beings. To the conscientious scientist this means that our science is of little value unless it is accompanied by social and ecological concern.

13.6 Spirituality, ecology, and education

13.6.1 Deep ecology and spirituality

The extensive explorations of the relationships between science and spirituality over the past three decades have made it evident that the sense of oneness, which is the key characteristic of spiritual experience, is fully confirmed by the understanding of reality in contemporary science. Hence, there are numerous similarities between the worldviews of mystics and spiritual teachers – both Eastern and Western – and the systemic conception of nature that is now being developed in several scientific disciplines.

The awareness of being connected with all of nature is particularly strong in ecology. Connectedness, relationship, and interdependence are fundamental concepts of ecology (as we discuss in Chapter 16); and connectedness, relationship, and belonging are also the essence of spiritual experience. We believe therefore that ecology – and in particular the philosophical school of deep ecology (see Introduction) – is the ideal bridge between science and spirituality.

When we look at the world around us, we find that we are not thrown into chaos and randomness but are part of a great order, a grand symphony of life. Every molecule in our body was once a part of previous bodies – living or nonliving – and will be a part of future bodies. In this sense, our body will not die but will live on, again and again, because life lives on. Moreover, we share not only life's molecules, but also its basic principles of organization with the rest of the living world. And since our mind, too, is embodied, our concepts and metaphors are embedded in the web of life together with our bodies and brains. Indeed, we belong to the universe, and this experience of belonging can make our lives profoundly meaningful.

13.6.2 The spiritual dimension of education

We shall argue in Chapter 16 that in the twenty-first century the well-being, and even survival, of humanity will depend crucially" on our "ecological literacy" – our ability to understand the basic principles of ecology, or principles of sustainability, and to live accordingly. This means that ecological literacy must become a critical skill for politicians, business leaders, and professionals in all spheres, and should be the most important part of education, especially at the university level where certain kinds of knowledge and certain values are taught to the leaders of tomorrow.

Ecological literacy involves not only the intellectual understanding of the basic principles of ecology but also the deep ecological awareness of the fundamental interdependence of all phenomena and of the fact that, as individuals and societies, we are embedded in, and dependent upon, the cyclical processes of nature. And since this awareness, ultimately, is grounded in spiritual awareness (see Section 13.6.1), it is evident that ecological literacy has an important spiritual dimension.

In today's academic world, it is very difficult to explore the spiritual dimension of education. Our classic academic institutions produce technological or humanistic specialists in one particular discipline at a time and are only very rarely capable of pursuing the interdisciplinary approach to knowledge that we are advocating in this book. Such an approach, however, is urgently needed today, since none of the major problems of our time can be understood in isolation (as we shall discuss in Chapter 17). All of them are systemic problems – interconnected and interdependent – and hence are in need of systemic solutions.

To include the spiritual dimension in education is even more difficult than to pursue a systemic approach, because of the widespread confusion of spirituality and religion (see Section 13.2). Thus, most of our graduate students – the world leaders of tomorrow – are deprived of the stimulating experience of interdisciplinary dialogues; and most future scientists are kept from examining the values of ethics, art, music, poetry, and personal introspection. Consequently, there is a great danger that we are educating leaders in various fields who do not know each other, and who are not sensitive to the values of the human spirit.

Fortunately, the last two decades have seen the creation of many research institutes and centers of learning where knowledge is pursued in an interdisciplinary way, and where the ecological and spiritual dimensions of education are explicitly emphasized. As we shall discuss in more detail in Chapter 18, these new institutions are an integral part of the global civil society that has emerged over the past twenty years. Some of them have links to traditional academic institutions while others do not. Most of these research institutes and centers of learning are communities of thinkers and activists, and they all pursue their research and teaching within an explicit framework of shared values.

The spiritual dimension is emphasized in these alternative institutions to varying degrees. To conclude this chapter, we shall profile two European organizations of this kind in which both of us have been centrally involved: the Cortona Week in Italy and Schumacher College in the UK. In both institutions, the relationship between science and spirituality is a frequent topic of discussion, and in both spirituality is not only discussed but also experienced in various forms of practice.

However, we should add a word of caution. The founders of both the Cortona Week and Schumacher College are no longer involved in the direction and programming of these courses, seminars, and workshops; and it may well be that the character of both institutions may change considerably over the coming years. What we are describing in the following pages is the way these centers of learning have been functioning over the past 20–25 years.

13.6.3 The Cortona Week experiment

The Cortona Week is a summer school for graduate students of the Swiss Federal Institute of Technology (*Eidgenössische Technische Hochschule* – ETH). It was conceived and established in 1985 by one of us (P.L.L.) while he was professor of chemistry at ETH. He directed the Cortona Week from its inception to 2003, his year of retirement from ETH. Interestingly, the inspiration for the Cortona Week came from the same Alpbach meeting in 1983 that inspired Francisco Varela to found the M&L Institute (see Section 13.4.7). At first the university did not approve the project, which was funded by a private sponsor (Dr. Branco Weiss) during its first five years; but, eventually, ETH came on board, and to this day it remains the only school of higher education in the world that supports such a holistic, interdisciplinary initiative.

In this summer school, which takes place at Cortona, a beautiful Etruscan and medieval town in Tuscany, graduate students from the sciences and other faculties mix with artists, musicians, and religious leaders, working and living together for one whole week (partial participation is not allowed for either teachers or students). Each morning there are either lectures or panel discussions in an environment that honors the full range of both human experience and intellectual analysis. The panellists and speakers are high-level professionals who present examples of integration and holistic views from their own lives.

In addition to the plenary lectures, an important characteristic of the Cortona Week is the "experiential workshop," which takes place in the afternoon. The 150 or so participants choose to join small groups in which they can paint, sculpt, or practice improvisation theatre, dancing, psychological training, or body-oriented disciplines. There are also meditative practices in the early morning – yoga, tai ji, and the like. The idea is that the participants may discover, or rediscover, some practices they feel they need in their lives.

Since 1985, many important scientists, philosophers, and writers, as well as political and religious leaders, have taught at the Cortona Weeks; and a total of over 1,000 graduate students have now participated in one or more Cortona Weeks. The feedback received from them has generally been very positive and encouraging. It seems that they have the potential to become a new generation of leaders well aware of the ecological difficulties of our time and of the importance of a systemic approach to solve the world's problems. Each Cortona Week is devoted to a different theme; subjects related to neuroscience, cognitive science, spirituality, and ethics have been central.

While the Cortona Week is continuing its course with the support of ETH, Pier Luigi Luisi has recently devoted his efforts to export this experiment into the world under the headline "International Cortona Week on Science and Spirituality." One of these international gatherings was held in 2009 at Cortona (see www.cortona-week.org); and another, "Cortona-India," in Hyderabad (see www.cortona.ethz.ch). Luisi believes that this is one of the most effective means to bring forth a new generation of world leaders, where high professional capability is in harmony with a high degree of spirituality.

To conclude this brief introduction to the Cortona Week experiment, we list below a selection of questions collected by the students and discussed in the last two sessions (2009–10) of the International Cortona Week devoted to science and spirituality. We feel that it might be interesting to see which questions about the relationship between science and spirituality trouble the minds of these graduate students. Their questions include the following.

- Can all reality be explained in terms of atoms and their interactions?
- Is nature "reasonable and rational"; that is, can the cosmos ever be understood as it really is?
- Who is the final judge of reality: science or spirituality?
- Do we need spirituality to give meaning to life?
- Will spiritual insights be critical to future science? Will science be critical to understanding the spiritual dimensions of the human mind?
- The mystery of order: is the order of the universe spiritual, or natural?
- Is the brain the only responsible organ for extraordinary states of experience, such as out-of-body states, near-death encounters, ecstasy, etc.?
- Is consciousness based on matter (the brain), or does it have a transpersonal dimension as well?
- Can mind have healing powers?
- How can we ensure personal dignity and freedom in a society dominated by science and technology?

Some of these questions have been considered in this chapter; others will be discussed in subsequent chapters. Notice that most of them refer to the inner, or "vertical," dimension of spirituality. The answers are, of course, not always easy, and at any rate are highly personal, bound to haunt each person for many years, perhaps for life.

13.6.4 Schumacher College: a unique learning experience

Schumacher College (www.schumachercollege.org.uk), an international center for ecological studies in Devon, England, is a unique institution of learning. It is not a traditional college with a well-defined faculty and student body, and, unlike most colleges and universities, it was not founded by any government agency or individual, and its foundation was not associated with business. The college grew out of the global civil society that emerged during the 1990s. Thus, from the beginning its faculty have been part of an international network of scholars and activists, a network of friends and colleagues that has existed for several decades.

The college was founded by Satish Kumar, an Indian spiritual teacher and Gandhian activist, who lives in the UK, where he publishes *Resurgence*, one of the most important and beautiful ecological magazines (www.resurgence.org). Satish (as he is known to his friends and disciples around the world) was a close associate of the environmental pioneer and author of the classic book *Small Is Beautiful*, E.F. Schumacher (1975), after whom the college was named.

Before its foundation in 1991, there was no center of learning where ecology could be studied in a rigorous, in-depth way from many different perspectives. During the subsequent years, the situation changed significantly when the global coalition of nongovernmental organizations (NGOs), now known as the global civil society, formed around the core values of human dignity and ecological sustainability (see Chapter 17). To place their political discourse within a systemic and ecological perspective, the global civil society developed a network of scholars, research institutes, think tanks, and centers of learning that operate largely outside our leading academic institutions, business organizations, and government agencies. Today, there are dozens of these institutions of research and learning in all parts of the world. Schumacher College was one of the first and continues to play a leading role. These research institutes are communities of scholars and activists engaged in a wide variety of projects and campaigns. As their scope grew and diversified during the past two decades, so did the faculty and curriculum of Schumacher College.

From the very beginning, Satish had the vision that the college should not represent a Eurocentric view but should give voice to a broad range of opinions – that it should be international. When Americans and Europeans discuss science, technology, and philosophy here, they are joined by voices from Africa, India, Japan, and other parts of the world.

The same ethnic, cultural, and intellectual diversity exists among the students. One of us (F.C.) has taught at Schumacher College since the early 1990s, and during those years it has not been unusual for him to have twenty-four course participants (the limit that was established) from ten or more different countries. Participants are usually highly educated.

They are professionals in various fields. Some of them are young students, but there are also older people; and thus they contribute to the discussions from a multitude of perspectives.

The level of education and experience of the course participants, who come from all over the world and engage one another in intensive discussions, is truly amazing. In a way, these diverse perspectives mirror the richness of the field of ecology, which is the central focus of the college. There is ecology as science, as politics, as technology, and as a philosophy grounded in spirituality. This great diversity of ecology is embodied in the very structure and curriculum of the college.

Another key characteristic of Schumacher College is the strong sense of community it engenders. Participants come here for several weeks to live together, to learn together, and to work together to sustain the learning community. They are divided into working groups that cook, clean, garden – doing all the work that is needed to maintain the college in a practice of Gandhian spirituality.

In these groups, conversations go on virtually round the clock. While they are cutting vegetables in the kitchen, they talk; while they are mopping the floor, or rearranging chairs for a special event, they talk. Everybody here is immersed in a continual experience of community and in exciting intellectual dialogues and discussions.

All this stimulates tremendous creativity. At Schumacher College, many things are created collectively, from meals in the kitchen to ideas in the classroom. Creativity flourishes because there is total trust in the community. Satish has created here a unique learning environment where discussions take place in an atmosphere that is intellectually intense and challenging, but is emotionally very safe. To the faculty who teach at the college, it feels almost like being among family, and this strong feeling of community emerges after being together for not more than a week or two. To most scholars such a situation is extremely attractive, as it offers them a unique opportunity to examine their work in depth, and to try out new ideas in a safe environment. Hence, Schumacher College is a unique place not only for course participants to learn but also for the teaching faculty to deeply engage over a relatively long period with a group of highly educated and motivated students, and to pursue a process of sustained self-exploration.

13.7 Concluding remarks

In this chapter, we have argued that the age-old tensions between science and spirituality have usually arisen when spirituality is confused with religion and when antagonistic positions are taken up by fundamentalists on both sides. We have tried to clarify the nature of true spirituality and have shown that both the concept of spirituality and the nature of spiritual experiences of reality are fully consistent with modern science, and in particular with the systems view of life. Moreover, we have argued that deep ecology seems to provide an ideal bridge between science and spirituality.

In our view, it is of the utmost importance today to introduce the ecological and spiritual dimensions into education at all levels. We have profiled two institutions that do so in exemplary ways: the Cortona Week, a summer school for graduate students of the Swiss

Federal Institute of Technology, and Schumacher College, an international center for ecological studies that offers courses year round. We believe that both of these institutions can serve as valuable models for centers of learning in other parts of the world, dedicated to interdisciplinary approaches to education that explicitly include ecological and spiritual dimensions.

We see this broadening of education as essential for bringing forth a new generation of world leaders to succeed our current politicians, many of whom are corrupt, pursue narrow-minded goals, and lack a "moral compass," as the Czech playwright and statesman Václav Havel famously put it. In his opening address to Forum 2000, a series of international symposia held in Prague, Havel (1997) bemoaned the paradoxical fact that humanity today is well aware of the multiple threats of global crisis, and yet does almost nothing to confront or avert them. And he concluded with the following memorable words:

> It is my deep conviction that the only option is for something to change in the sphere of the spirit, in the sphere of human conscience, in the actual attitude of man towards the world and his understanding of himself and his place in the overall order of existence.

It is evident that, for such a new direction to take hold in education, we shall need a profound transformation of our academic institutions and, more generally, of the values that currently dominate industrial societies. This brings us to the social dimension of the systems view of life, the subject of our next chapter.

14

Life, mind, and society

14.1 The evolutionary link between consciousness and social phenomena

Around 4 million years ago, an extraordinary confluence of events occurred in the evolution of primates with the appearance of the first upright-walking apes of the genus *Australopithecus* (see Section 11.2). The new freedom of their hands allowed these early hominids to make tools, wield weapons, and throw rocks, which stimulated the rapid brain growth that became characteristic of human evolution, leading eventually to the development of language and reflective consciousness. While they developed complex brains, tool-making skills, and language, the helplessness of their prematurely born infants led to the formation of the supportive families and communities that became the foundation of human social life. Thus, the evolution of language and human consciousness was inextricably connected with that of technology and of organized social relations from the very beginning of human life. In particular, the evolutionary stage of the Australopithecine hominids established a fundamental link between consciousness and social phenomena. With the evolution of language arose not only the inner world of concepts and ideas but also the social world of organized relationships and culture, which is the subject of this chapter.

From an evolutionary perspective, then, it is very natural to ground the understanding of social phenomena in a unified conception of life and consciousness. Indeed, the systems view of life attempts to integrate life's biological, cognitive, and social dimensions (see Capra, 2002). As we discussed in Section 12.3.3, our ability to hold abstract mental images – a critical property of reflective consciousness – is of special relevance to such an integrative approach. Being able to hold mental images enables us to choose among several alternatives, an ability which is necessary to formulate values and social rules of behavior. On the other hand, differences of values give rise to conflicts of interest, which are the origin of relationships of power, as we shall discuss below. Thus, the ability of human consciousness to form abstract mental images of material objects and events lies at the roots of the main characteristics of social life.

14.2 Sociology and the social sciences

The study of society is the domain of sociology, also known as social science or social theory. It is a very broad field including a variety of academic disciplines, which are commonly referred to as the social sciences and are contrasted with the natural sciences. In addition to the core discipline of sociology, the social sciences include fields of study such as anthropology, economics, political science, management science, history, and law, to name just a few.

14.2.1 Social theory in the twentieth century

As we discussed in Chapter 3, social thought in the late nineteenth and early twentieth centuries was greatly influenced by positivism, a doctrine formulated by the social philosopher Auguste Comte, who asserted that the social sciences should search for general laws of human behavior, emphasize quantification, and reject all explanations in terms of subjective phenomena, such as intentions or purpose. The positivist framework was clearly patterned after Newtonian physics. Indeed, Comte called the scientific study of society at first "social physics" before introducing the term "sociology."

The major schools of thought in early twentieth-century sociology were different attempts at emancipation from the narrow confines of positivism (see Baert, 1998). In these attempts several sociologists were influenced by the new school of systems thinking that was being developed around the same time (see Chapter 4). However, none of them managed to truly integrate their understanding of social phenomena with the basic ideas about the relevant biological and cognitive phenomena in the natural sciences of the time.

In our opinion, this was due to the fact that, even though the fundamental connections between consciousness, culture, and the social domain were recognized by many social theorists, they lacked a conceptual framework that would allow them to truly integrate the biological, cognitive, and social dimensions of life. Even those social scientists who were strongly influenced by the new systems theories were unable to bridge the gap between the natural and the social sciences (which still exists in most of today's academic world).

A true integration of these "two cultures," to use the memorable phrase of C.P. Snow (1960), had to wait until the late twentieth century when the development of complexity theory, the concept of cognitive autopoietic networks, and the wide appreciation of the importance of networks of communications (social networks) provided an appropriate conceptual framework. We shall discuss this integrative framework in detail in Section 14.3 below.

But now let us return to the beginnings of social theory in the early twentieth century. One inheritance of positivism during these early decades of sociology was the focus on a narrow notion of "social causation," which linked social theory conceptually to physics rather than to the life sciences. Émile Durkheim (1858–1917), who, along with Max Weber (1864–1920), is considered one of the principal founders of modern sociology, identified "social facts," such as beliefs or practices, as the causes of social phenomena. Even though these social facts are clearly nonmaterial, Durkheim insisted that they should be treated like material objects. He saw social facts as being caused by other social facts, in analogy to the operations of physical forces.

14.2.2 Structuralism and functionalism

Durkheim's ideas exerted a major influence on both structuralism and functionalism, the two dominant schools of early twentieth-century sociology. Both of these schools of thought assumed that the task of social scientists is to unravel a hidden causative reality beneath the surface level of observed phenomena. Such attempts to identify some hidden phenomena – vital forces or other "extra ingredients" – have occurred repeatedly in the life sciences when scientists struggled to understand the emergence of novelty that is characteristic of all life (see Chapter 8) and cannot be explained in terms of linear relations of cause and effect.

For structuralists, the hidden realm consists of underlying "social structures." Although the early structuralists treated those social structures like material objects, they also understood them as integrated wholes and used the term "structure" not unlike the ways in which the early systems thinkers used "pattern of organization."

Whereas the structuralists searched for hidden social structures, the functionalists postulated that there is an underlying social rationality that causes individuals to act according to the "social functions" of their actions – that is, to act in such a way that their actions fulfill society's needs. Durkheim insisted that a full explanation of social phenomena must combine both causal and functional analyses, and he also emphasized that one should distinguish between functions and intentions. It seems that, somehow, he attempted to take into account the cognitive phenomena of intentions and purposes without abandoning the conceptual framework of classical physics with its material structures, forces, and linear cause-and-effect relationships. The emphasis on intention and purpose was identified by subsequent social theorists with the focus on "human agency" or purposeful action.

Several of the early structuralists also recognized the connections between social reality, consciousness, and language. The linguist Ferdinand de Saussure (1857–1913) was one of the founders of structuralism, and the anthropologist Claude Lévi-Strauss (1908–2009), whose name is closely associated with the structuralist tradition, was one of the first to analyze social life by systematically employing analogies with linguistic systems. The focus on language intensified around the 1960s with the advent of the so-called interpretative sociologies, which emphasize that individuals interpret their surrounding reality and act accordingly.

14.2.3 Giddens and Habermas: two integrative theories

During the second half of the twentieth century, social theory was shaped significantly by several attempts to transcend the opposing schools of the earlier decades and to integrate the notions of social structure and human agency with an explicit analysis of meaning. Perhaps the most influential of those integrative theoretical frameworks have been the theories of two eminent European sociologists: the structuration theory of Anthony Giddens and the critical theory of Jürgen Habermas.

Anthony Giddens has been a leading contributor to social theory since the early 1970s. His structuration theory is designed to explore the interaction between social structures and human agency in such a way that it integrates insights from structuralism and functionalism on the one hand, and from interpretative sociologies on the other hand. To do so, Giddens employs two different but complementary methods of investigation. Institutional analysis is his method for studying social structures and institutions, while strategic analysis is used to study how people draw upon social structures in their pursuit of strategic goals.

Giddens emphasizes that people's strategic conduct is based largely on how they interpret their environment. In fact, he points out that social scientists have to deal with a "double hermeneutic" (from Greek *hermeneuein*, "to interpret"). They interpret their subject matter, which itself is engaged in interpretations. Consequently, Giddens believes that subjective phenomenological insights must be taken seriously if we are to understand human conduct.

As would be expected from an integrative theory that attempts to transcend traditional opposites, Giddens' concept of social structure is rather complex. As in most contemporary social theory, it is defined as a set of rules enacted in social practices, and Giddens also includes resources in his definition of social structure.

The interaction between social structures and human agency is cyclical, according to Giddens. Social structures are both the precondition and the unintended outcome of people's agency. People draw upon them in order to engage in their daily social practices, and in so doing they cannot help but reproduce the very same structures.

For example, when we speak we necessarily draw upon the rules of our language, and as we use language we continually reproduce and transform the very same semantic structures. Thus social structures enable us to interact and are also reproduced by our interactions. Giddens calls this the "duality of structure," and he acknowledges the similarity to the circular nature of autopoietic networks in biology (cited in Baert, 1998).

The conceptual links with the theory of autopoiesis are even more evident when we turn to Giddens' view of human agency. He insists that agency does not consist of discrete acts but is a continuous flow of conduct. Similarly, a living metabolic network embodies an ongoing process of life. And as the components of the living network continually transform or replace other components, so the actions in the flow of human conduct have a "transformative capacity" in Giddens' theory.

During the 1970s, while Anthony Giddens developed his structuration theory at Cambridge University, Jürgen Habermas formulated a theory of equal scope and depth, which he called the "theory of communicative action" at the University of Frankfurt. By integrating numerous philosophical strands, Habermas has become a leading intellectual force and a major influence on philosophy and social theory. He is the most prominent contemporary exponent of critical theory, the social theory with Marxian roots that was developed by the Frankfurt School in the 1930s (see, e.g., Held, 1990). True to their Marxian origins (see Section 3.3.2), critical theorists do not want simply to explain the world. Their ultimate task, according to Habermas, is to uncover the structural conditions of people's actions and to help them transcend these conditions. Critical theory deals with power and is aimed at emancipation.

Like Giddens, Habermas asserts that two different but complementary perspectives are needed to fully understand social phenomena. One perspective is that of the social system, which corresponds to the focus on institutions in Giddens' theory; the other is the perspective of the "life world" (*Lebenswelt*), corresponding to Giddens' focus on human conduct.

For Habermas, the social system has to do with the ways social structures constrain people's actions, which includes issues of power and specifically the class relationships involved in production. The life world, on the other hand, raises issues of meaning and communication. Accordingly, Habermas sees critical theory as the integration of two different types of knowledge. Empirical-analytical knowledge is associated with the external world and is concerned with causal explanations. Hermeneutics, the understanding of meaning, is associated with the inner world, and is concerned with language and communication.

Like Giddens, Habermas recognizes that hermeneutic insights are relevant to the workings of the social world because people attribute meaning to their surroundings and act accordingly. However, he points out that people's interpretations always rely on a number of implicit assumptions that are embedded in history and tradition, and he argues that this means that all assumptions are not equally valid. According to Habermas, social scientists should evaluate different traditions critically, identify ideological distortions, and uncover their connections with power relations. Emancipation takes place whenever people are able to overcome past restrictions that resulted from distorted communication.

14.3 Extending the systems approach

The theories of Giddens and Habermas are outstanding attempts to integrate studies of the external world of cause and effect, the social world of human relationships, and the inner world of values and meaning. Both social theorists integrate insights from the natural sciences, the social sciences, and the cognitive philosophies, while rejecting the limitations of positivism.

We believe that this integration can be advanced significantly by extending the systems view of life that we have presented in the preceding chapters to the social domain. To do so, we shall use a conceptual synthesis developed by one of us several years ago (Capra, 1996, 2002).

14.3.1 Three perspectives on life

Let us return once more to autopoiesis, the defining characteristic of biological life. In their theory, Maturana and Varela (1980) distinguish between two fundamental aspects of living systems: organization and structure. The *pattern of organization* of any system, living or nonliving, is the configuration of relationships among the system's components that determines the system's essential characteristics. In other words, certain relationships must be present for something to be recognized as, say, a chair, a bicycle, or a tree. The configuration of relationships that gives a system its essential characteristics is what we mean by its pattern of organization.

The *structure* of a system is the physical embodiment of its pattern of organization. Whereas the description of the system's organization involves an abstract mapping of relationships, the description of its structure involves describing the system's actual physical components – their shapes, chemical compositions, and so forth.

To illustrate the difference between organization and structure, let us look at a well-known nonliving system, a bicycle. In order for something to be called a bicycle, there must be a number of functional relationships between components known as frame, pedals, handlebars, wheels, chain, sprocket, and so on. The complete configuration of these functional relationships constitutes the bicycle's pattern of organization. All of those relationships must be present to give the system the essential characteristics of a bicycle.

The structure of the bicycle is the physical embodiment of its pattern of organization in terms of components of specific shapes, made of specific materials. The same pattern "bicycle" can be embodied in many different structures. The handlebars will be shaped differently for a touring bike, a racing bike, or a mountain bike; the frame may be heavy and solid, or light and delicate; the tires may be narrow or wide, tubes or solid rubber. All these combinations and many more will easily be recognized as different embodiments of the same pattern of relationships that defines a bicycle.

In a machine such as a bicycle the parts have been designed, manufactured, and then put together to form a structure with fixed components. In a living system, by contrast, the components change continually. There is a ceaseless flux of matter through a living organism. Each cell continually synthesizes and dissolves structures, and eliminates waste products. Tissues and organs replace their cells in continual cycles. There is growth, development, and evolution. From the very beginning of biology, the understanding of living structure has been inseparable from the understanding of metabolic and developmental processes.

This striking property of living systems suggests *process* as a third perspective for a comprehensive description of the nature of life. The process of life is the activity involved in the continual embodiment of the system's pattern of organization. Thus the process perspective is the link between organization and structure. In the case of the bicycle, the pattern of organization is represented by the design sketches that are used to build the bicycle, the structure is a specific physical bicycle, and the link between pattern and structure is in the mind of the designer. In the case of a living organism, by contrast, the pattern of organization is always embodied in the organism's structure, and the link between pattern and structure lies in the process of continual embodiment.

Drawn from autopoiesis, the three perspectives of *organization*, *structure*, and *process* provide an integrative conceptual framework for the understanding of biological life (Capra, 1996). All three are totally interdependent. The pattern of organization can be recognized only if it is embodied in a physical structure, and in living systems this embodiment is an ongoing process. The three aspects – organization, structure, and process – are three different but inseparable perspectives on the phenomenon of life.

When we study living systems from the perspective of organization, we find that their pattern of organization is that of a self-generating (autopoietic) network. From the structure perspective, the material structure of a living system is a dissipative structure – that is, an open system operating far from equilibrium (as we discussed in Section 8.3). From the process perspective, finally, living systems are cognitive systems in which the process of cognition is closely linked to the pattern of autopoiesis. In a nutshell, this is our synthesis of the new systemic understanding of biological life.

It is interesting to note that this synthesis is also implicit in the "trilogy of life" pictured in Figure 7.5. There, the "autopoietic unit" represents the system's pattern of organization, while "cognition" represents the process perspective, and the structure perspective is implicit in the "environment" domain, since the autopoietic system is linked to its environment through structural coupling (see Section 7.4.1).

We should also remember that, although the structure of a living system is always a dissipative structure, not all dissipative structures are autopoietic networks. Thus a dissipative structure may be a living or a nonliving system. For example, the Bénard cells and chemical clocks studied extensively by Prigogine (see Section 8.3.2) are dissipative structures but not living systems.

Let us illustrate the fundamental interdependence of our three perspectives with a living cell. It consists of a network (*pattern*) of chemical reactions (*processes*), which involve the production of the cell's components (*structures*), and which respond cognitively – that is, through self-directed structural changes (*processes*) – to disturbances from the environment. Similarly, the phenomenon of emergence in dynamic systems (as discussed in Section 8.3) is a *process* characteristic of a dissipative *structure*, which involves multiple feedback loops (*pattern*).

To give equal importance to each of these three perspectives is difficult for most scientists because of the persistent influence of our Cartesian heritage. The natural sciences are supposed to deal with material phenomena, but only the structure perspective is concerned with the study of matter. The other two deal with relationships, qualities, patterns, and processes, all of which are nonmaterial (see Section 4.3). Of course, no scientist would deny the existence of patterns and processes, but most scientists tend to think of a pattern of organization as an idea abstracted from matter, rather than a generative force.

To focus on material structures and the forces between them, and to view the patterns of organization resulting from these forces as secondary phenomena have been very effective in physics and chemistry, but when we come to living systems this approach is no longer adequate. The essential characteristic that distinguishes living from nonliving systems – autopoiesis – is not a property of matter, nor a special "vital force." It is a specific pattern of relationships among chemical processes. Although these processes produce material components, the network pattern itself is nonmaterial.

The processes of self-organization in this network pattern are understood as cognitive processes that eventually give rise to conscious experience and conceptual thought. All these cognitive phenomena are nonmaterial, but they are *embodied*; they arise from and are shaped by the body. Thus, life is never divorced from matter, even though its essential characteristics – organization, complexity, processes, and so on – are nonmaterial.

14.3.2 *Meaning – the fourth perspective*

The extension of our integrative conceptual framework to the social domain is not trivial, because social theorists have traditionally used the term "structure" in a sense that is quite different from that in the natural sciences. As we have mentioned, the structuralists used "social structure" somewhat in the way the early systems thinkers used "pattern of organization;" and today sociologists generally define social structure as a set of rules enacted in social practices.

To accommodate these different usages of the term "structure," we shall slightly modify our terminology by using the more general concepts of "form" and "matter" instead of "pattern" and "structure." In this more general terminology, the three perspectives on the nature of life correspond to the study of *form* (or pattern of organization), the study of *matter* (or material structure), and the study of *process*.

When we try to extend these three perspectives to the social domain, we are confronted with a bewildering multitude of phenomena – rules of behavior, values, intentions, goals, strategies, designs, power relations – that play no role in most of the nonhuman world but are essential to human social life. However, as we mentioned at the beginning of this chapter, these diverse characteristics of social reality, ultimately, are all based on the ability of human consciousness to form abstract mental images.

This fundamental link between the cognitive and social dimensions of life provides a natural way of extending the systems approach, as it makes evident that our inner world of concepts and ideas, images, and symbols is a critical dimension of social reality. Social scientists have often referred to it as the "hermeneutic" dimension to express the view that human language, being of symbolic nature, centrally involves the communication of meaning, and that human action flows from the meaning that we attribute to our surroundings.

Accordingly, we shall extend the systemic understanding of life to the social domain by adding the perspective of *meaning* to the other three perspectives. In doing so, we are using "meaning" as a shorthand notation for the inner world of reflective consciousness, which contains a multitude of interrelated characteristics. A full understanding of social phenomena, then, must involve the integration of four perspectives – form, matter, process, and meaning. The fundamental interconnectedness of these four perspectives on life is indicated in the diagram in Figure 14.1 by representing them as the corners of a geometric figure. The first three perspectives form a triangle. The perspective of meaning is represented as lying outside the plane of this triangle to indicate that it opens up a new "inner" dimension, so that the entire conceptual structure forms a tetrahedron.

Integrating the four perspectives means recognizing that each contributes significantly to the understanding of a social phenomenon. For example, we shall see that culture is created and sustained by a network (*form*) of communications (*processes*) in which *meaning* is generated. The culture's material embodiments (*matter*) include artifacts and written texts, through which meaning is passed on from generation to generation.

It is interesting to note that this conceptual framework of four interdependent perspectives on life shows some similarities with the four principles, or "causes," postulated by Aristotle

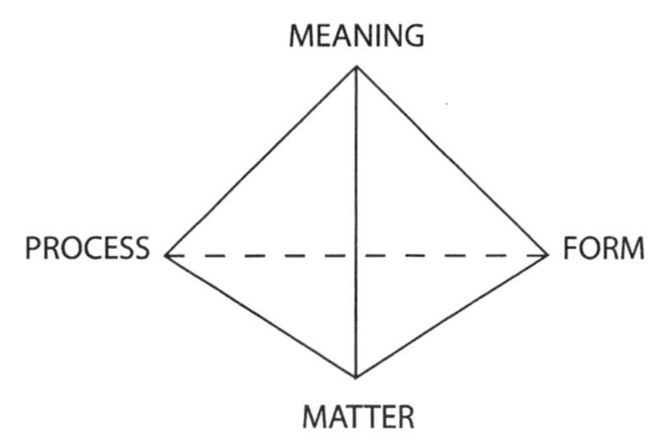

Figure 14.1 The interconnectedness of the four perspectives on life, represented as a tetrahedron (from Capra, 2002).

as the interdependent sources of all phenomena. Aristotle distinguished between internal and external causes. The two internal causes are matter and form. The external causes are the efficient cause, which generates the phenomenon through its action, and the final cause, which determines the action of the efficient cause by giving it a goal or purpose.

Aristotle's detailed description of the four causes and their interrelations is quite different from our conceptual scheme (see, e.g., Windelband, 2001/1901). In particular, the final cause, which corresponds to the perspective we have associated with meaning, operates throughout the material world, according to Aristotle, whereas contemporary science asserts that it plays no role in nonhuman nature. Nevertheless, we find it fascinating that after more than 2,000 years of philosophy, we still analyze reality within the four perspectives identified by Aristotle.

14.3.3 Living networks

Our extension of the systems view of life to social reality is based on the assumption that there is a fundamental unity to life, that different living systems exhibit similar patterns of organization. This assumption is supported by the observation that evolution has proceeded for billions of years by using the same patterns again and again. As life evolves, these patterns tend to become more and more elaborate, but they are always variations on the same basic themes.

The network pattern, in particular, is one of the very basic patterns of organization in all living systems. At all levels of life – from the metabolic networks of cells to the food webs of ecosystems – the components and processes of living systems are interlinked in network fashion. Extending the systemic understanding of life to the social domain, therefore, means applying our knowledge of life's basic patterns and principles of organization, and specifically our understanding of living networks, to social reality.

However, while insights into the *organization* of biological networks may help us understand social networks, we should not expect to transfer our understanding of the networks' material *structures* from the biological to the social domain. Let us take the metabolic network of cells as an example to illustrate this point. A cellular network is a nonlinear pattern of organization, and we need complexity theory (nonlinear dynamics) to understand its intricacies. The cell, moreover, is a chemical system, and we need molecular biology and biochemistry to understand the nature of the structures and processes that form the network's nodes and links. If we do not know what an enzyme is and how it catalyzes the synthesis of a protein, we cannot expect to understand the cell's metabolic network.

A social network, too, is a nonlinear pattern of organization, and concepts developed in complexity theory, such as feedback or emergence, are likely to be relevant in a social context as well. However, the nodes and links of the network are not merely biochemical. Social networks are first and foremost networks of communication involving symbolic language, cultural constraints, relationships of power, and so on. To understand the structures of such networks, we need to use insights from social theory, philosophy, cognitive science, anthropology, and other disciplines. A unified systemic framework for the understanding of biological and social phenomena will emerge only when the concepts of nonlinear dynamics are combined with insights from these fields of study.

To apply our knowledge of living networks to social phenomena, we need to find out, among other things, whether the concept of autopoiesis is valid in the social domain. There has been considerable discussion of this point in recent years, but the situation is still far from clear (see Mingers, 1992, 1995, 1997). Most of the research in the theory of autopoiesis, so far, has been concerned with minimal autopoietic systems – simple cells, computer simulations, and the autopoietic chemical structures, or "minimal cells," created recently in the laboratory (see Section 10.5.1). Much less work has been done on studying the autopoiesis of multicellular organisms, ecosystems, and social systems. Current ideas about the network patterns in those living systems are therefore still rather speculative.

All living systems are networks of smaller components, and the web of life as a whole is a multilayered structure of living systems nesting within other living systems – networks within networks. Organisms are aggregates of autonomous but closely coupled cells; populations are networks of autonomous organisms belonging to a single species; and ecosystems are webs of organisms, both single-celled and multicellular, belonging to many different species.

What is common to all these living systems is that their smallest living components are always cells, and therefore we can confidently say that all living systems, ultimately, are autopoietic. However, it is also interesting to ask whether the larger systems formed by those autopoietic cells – the organisms, societies, and ecosystems – are in themselves autopoietic networks.

In their book *The Tree of Knowledge* (1998), Maturana and Varela proposed that the concept of autopoiesis should be restricted to the description of cellular networks, and that the broader concept of "operational closure," which does not specify production processes, should be applied to all other living systems. The authors also pointed out that the three types of multicellular living systems – organisms, ecosystems, and societies – differ greatly in the degrees of autonomy of their components. In organisms, the cellular components have a minimal degree of independent existence, while the components of human societies, individual human beings, have a maximum degree of autonomy, enjoying many dimensions of independent existence. Animal societies and ecosystems occupy various places between those two extremes.

Organisms and human societies are therefore very different types of living systems. Totalitarian political regimes have often severely restricted the autonomy of their members and, in doing so, have depersonalized and dehumanized them. Thus fascist societies function more like organisms, and it is not a coincidence that dictatorships have often been fond of using the metaphor of society as a living organism.

14.3.4 Autopoiesis in the social domain

The central problem with all attempts to extend the concept of autopoiesis to the social domain is that it has been defined precisely only for systems in physical space and for computer simulations in mathematical spaces. Human social systems, however, exist not only in the physical domain but also in a symbolic social domain, shaped by the "inner world" of concepts, ideas, and symbols that arises with human thought, consciousness, and language.

Thus a human family can be described as a biological system, defined by certain blood relations, but also as a "conceptual system," defined by certain roles and relationships that may or may not coincide with any blood relationships among its members. These roles depend on social convention and may vary considerably in different periods of time and different cultures. For example, as Mingers (1995) points out, in contemporary Western culture, the role of "father" may be fulfilled by the biological father, a foster father, a stepfather, an uncle, or an older brother. In other words, these roles are not objective features of the family system but are flexible and continually renegotiated social constructs.

While behavior in the physical domain is governed by cause and effect, the so-called "laws of nature," behavior in the social domain is governed by rules generated by the social system that are often codified into law. The crucial difference is that social rules can be broken, but natural laws cannot. Human beings can choose whether and how to obey a social rule; molecules cannot choose whether or not they should interact.

Given the simultaneous existence of social systems in two domains, the physical and the social, is it meaningful to apply the concept of autopoiesis to them at all, and, if so, in which domain should it be applied?

While Maturana and Varela left this question open in their original work, other authors have asserted that an autopoietic social network *can* be defined if the description of human social systems remains entirely within the social domain. This school of thought was pioneered in Germany by the sociologist Niklas Luhmann (1990, p. 3), who has developed the concept of social autopoiesis in considerable detail. Luhmann's central point is to identify the social processes of the autopoietic network as processes of communication:

> Social systems use communication as their particular mode of autopoietic reproduction. Their elements are communications that are . . . produced and reproduced by a network of communications and that cannot exist outside of such a network.

A family system, for example, can be defined as a network of conversations exhibiting inherent circularities. Each conversation creates thoughts and meaning, which give rise to further conversations, and thus the entire network generates itself – it is autopoietic. The communicative acts of the network of conversations include the "self-production" of the roles by which the various family members are defined and of the family system's boundary.

According to Luhmann, this is a general pattern in social networks. As communications recur in multiple feedback loops, they produce a shared system of beliefs, explanations, and values – a common context of meaning – that is continually sustained by further communications. Through this shared context of meaning individuals acquire identities as members of the social network, and in this way the network generates its own boundary. It is not a physical boundary but a boundary of expectations, of confidentiality and loyalty, which is continually maintained and renegotiated by the network itself.

To explore the implications of viewing social systems as networks of communications, it is helpful to remember the dual nature of human communication, discussed in Section 12.3.9. Like all communication among living organisms, it involves a continual coordination of behavior, and because it involves conceptual thinking and symbolic language, it also generates mental images, thoughts, and meaning. Accordingly, we can expect networks of communications to have a dual effect. They will generate, on the one hand, ideas and contexts of meaning, and, on the other hand, rules of behavior or, in the language of social theorists, social structures.

14.4 Networks of communications

14.4.1 Meaning, purpose, and human freedom

Having identified the organization of social systems as self-generating networks of communications, we now need to turn our attention to the structures that are produced by these networks and to the nature of the relationships that are engendered by them. A comparison with biological networks will again be useful. The metabolic network of a cell, for example, generates various molecular structures. Some of them become structural components of the network, forming parts of the cell membrane or of other cellular components. Others are exchanged between the network's nodes as carriers of energy or information, or as catalysts of metabolic processes.

Social networks, too, generate material structures – buildings, roads, technologies, etc. – that become structural components of the network; and they also produce material goods and artifacts that are exchanged between the network's nodes. However, the production of material structures in social networks is quite different from that in biological and ecological networks. The structures are created for a purpose, according to some design, and they embody some meaning. To understand the activities of social systems, it is therefore crucial to study them from that perspective.

The perspective of meaning includes a multitude of interrelated characteristics that are essential to understanding social reality. Meaning itself is a systemic phenomenon; it always has to do with context. A general dictionary definition of *meaning* is "an idea conveyed to the mind that requires or allows of interpretation," while *interpretation* may be defined as "conceiving in the light of individual belief, judgment, or circumstance." In other words, we interpret something by putting it into a particular context of concepts, values, beliefs, or circumstances. To understand the meaning of anything we need to relate it to other things in its environment, in its past, or in its future. Nothing is meaningful in itself.

For example, to understand the meaning of a literary text, one needs to establish the multiple contexts of its words and phrases. This can be a purely intellectual endeavor, but it may also reach a deeper level. If the context of an idea or expression includes relationships involving our own selves, it becomes meaningful to us in a personal way. This deeper sense of meaning includes an emotional dimension and may even bypass reason altogether. Something may be profoundly meaningful to us through a direct experience of context.

Meaning is essential to human beings. We continually need to make sense of our outer and inner worlds, find meaning in our environment and in our relationships with other human beings, and act according to that meaning. This includes in particular our need to act with a purpose or goal in mind. Because of our ability to project mental images into the future we act with the conviction, valid or invalid, that our actions are voluntary, intentional, and purposeful.

As human beings we are capable of two kinds of actions. Like all living organisms we engage in involuntary, unconscious activities, such as digesting our food or circulating our blood, which are part of the process of life and therefore cognitive in the sense of the Santiago theory of cognition (see Section 12.2). In addition we engage in voluntary, intentional activities, and it is in acting with intention and purpose that we experience human freedom (see Searle, 1984).

As we mentioned in Section 7.4.2, the systemic understanding of life sheds new light on the age-old philosophical debate about freedom and determinism. The key point is that the behavior of a living organism is constrained but not determined by outside forces. Living organisms are self-organizing, meaning that their behavior is not imposed by the environment but is established by the system itself. More specifically, the organism's behavior is determined by its own structure, a structure formed by a succession of autonomous structural changes.

The autonomy of living systems must not be confused with independence. Living organisms are not isolated from their environment. They interact with it continually, but the environment does not determine their organization. At the human level, we experience this self-determination as the freedom to act according to our own choices and decisions. To experience these as "our own" means that they are determined by our nature, including our past experiences and genetic heritage. To the extent that we are not constrained by human relationships of power, our behavior is self-determined and therefore free.

14.4.2 The dynamics of culture

Our ability to hold mental images and project them into the future not only allows us to identify goals and purposes and develop strategies and designs; it also enables us to choose among several alternatives and hence to formulate values and social rules of behavior. All of these social phenomena are generated by networks of communications as a consequence of the dual role of human communication. On the one hand, the network continually generates mental images, thoughts, and meaning; on the other hand, it continually coordinates the behavior of its members. From the complex dynamics and interdependence of these processes emerges the integrated system of values, beliefs, and rules of conduct that we associate with the phenomenon of culture.

The term "culture" has a long and intricate history and is now used in different intellectual disciplines with diverse and sometimes confusing meanings. In his classic text, *Culture*, historian Raymond Williams (1981) traces the meaning of the word back to its early use as a noun denoting a process: the culture (i.e., cultivation) of crops, or the culture (i.e., rearing and breeding) of animals. In the sixteenth century this meaning was extended metaphorically to the active cultivation of the human mind; and in the late eighteenth century, when the word was borrowed from the French by German writers (who first spelled it *Cultur* and subsequently *Kultur*), it acquired the meaning of a distinctive way of life of a people. In the nineteenth century the plural "cultures" became especially important in the development of comparative anthropology, where it has continued to designate distinctive ways of life.

In the meantime, the older use of "culture" as the active cultivation of the mind continued. Indeed, it expanded and diversified, covering a range of meanings from a developed state of mind ("a cultured person") to the process of this development ("cultural activities") to the means of these processes (administered, for example, by a "Ministry of Culture"). In our time, the different meanings of "culture" that are associated with the active cultivation of the mind coexist – often uneasily, as Williams notes – with the anthropological use as a distinctive way of life of a people or social group (as in "aboriginal culture" or "corporate culture"). In addition, the original biological meaning of "culture" as cultivation continues to be used, as, for example, in "agriculture," "monoculture," or "germ culture."

For our systemic analysis of social reality we need to focus on the anthropological meaning of culture, which the *Columbia Encyclopedia* defines as "the integrated system of socially acquired values, beliefs, and rules of conduct that delimit the range of accepted behaviors in any given society." When we explore the details of this definition, we discover that culture arises from a complex, highly nonlinear dynamics. It is created by a social network involving multiple feedback loops through which values, beliefs, and rules of conduct are continually communicated, modified, and sustained. It emerges from a network of communications among individuals; and as it emerges, it produces constraints on their actions. In other words, the social structures, or rules of behavior, that constrain the actions of individuals are produced and continually reinforced by their own network of communications.

The social network also produces a shared body of knowledge – including information, ideas, and skills – that shapes the culture's distinctive way of life in addition to its values and beliefs. Moreover, the culture's values and beliefs affect its body of knowledge. They are part of the lens through which we see the world. They help us to interpret our experiences and to decide what kind of knowledge is meaningful. This meaningful knowledge, continually modified by the network of communications, is passed on from generation to generation together with the culture's values, beliefs, and rules of conduct.

The system of shared values and beliefs creates an identity among the members of the social network, based on a sense of belonging. People in different cultures have different identities because they share different sets of values and beliefs. At the same time, an individual may belong to several different cultures. People's behavior is informed and restricted by their cultural identities, and this in turn reinforces their sense of belonging. Culture is embedded in people's way of life, and it tends to be so pervasive that it escapes our everyday awareness.

Cultural identity also reinforces the closure of the network by creating a boundary of meaning and expectations that limits the access of people and information to the network. Thus the social network is engaged in communications within a cultural boundary, which its members continually recreate and renegotiate. This situation is not unlike that of the metabolic network of a cell, which continually produces and recreates a boundary – the cell membrane – that confines it and gives it its identity.

However, there are some crucial differences between cellular and social boundaries. Social boundaries, as we have emphasized, are not necessarily physical boundaries but boundaries of meaning and expectations. They do not literally surround the network but exist in a mental realm that does not have the topological properties of physical space. The similarities and differences between biological and social boundaries are a fascinating subject which, in our view, deserves to be explored in much greater depth.

14.4.3 The origin of power

One of the most striking characteristics of social reality is the phenomenon of power. In the words of the economist John Kenneth Galbraith (1984), "The exercise of power, the submission of some to the will of others, is inevitable in modern society; nothing whatever is accomplished without it. . . . Power can be socially malign; it is also socially essential (quoted in Lukes, 1986) ." The essential role of power in social organization is linked to inevitable conflicts of interest. Because of our ability to affirm preferences and make choices accordingly, conflicts of interest will appear in any human community, and power is the means by which these conflicts are resolved.

This does not necessarily imply the threat or use of violence. In his lucid book, Galbraith distinguishes three kinds of power, depending on the means that are employed. Coercive power wins submission by inflicting or threatening sanctions; compensatory power by offering incentives or rewards; and conditioned power by changing beliefs through persuasion or education. To find the right mixture of these three kinds of power in order to resolve conflicts and balance competing interests is the art of politics.

The association of power with the advancement of one's own interests is the basis of most contemporary analyses of power. In the words of Galbraith (1984), quoted in Lukes (1986), "Individuals and groups seek power to advance their own interests and to extend to others their personal, religious, or social values." This is when power becomes linked to exploitation. A further stage is reached when power is pursued not only to advance one's own personal interests, values, or social perceptions but also for its own sake. It is well known that for most people the exercise of power brings high emotional and material rewards, conveyed by elaborate symbols and rituals of obeisance – from standing ovations, fanfares, and military salutes to office suites, limousines, corporate jets, and motorcades.

As a community grows and increases in complexity, its positions of power will also increase. In complex societies, resolutions of conflicts and decisions about how to act will be effective only if authority and power are organized within administrative structures. In the long history of human civilization, numerous forms of social organization have been generated by this need to organize the distribution of power.

Thus, power plays a central role in the emergence of social structures. In social theory, all rules of conduct are included in the concept of social structures, whether they are informal, resulting from continual coordinations of behavior, or formalized, documented, and enforced by laws. All such formal structures, or social institutions, are ultimately rules of behavior that facilitate decision-making and embody relationships of power. This crucial link between power and social structure has been discussed extensively in the classic texts on power. The sociologist and economist Max Weber states: "Domination has played the decisive role . . . in the economically most important social structures of the past and present"; and, according to the political theorist Hannah Arendt, "All political institutions are manifestations and materializations of power" (both quoted, respectively, in Lukes, 1986, p. 28, p. 62).

Power in social networks

In recent years, social networks have become a major focus of attention not only in science but also in society at large and throughout the newly emerging global culture. Social networks have existed in all human communities throughout history, but the recent information technology revolution has given them unprecedented flexibility and global scale, and has made them the dominant feature of our age. According to the sociologist Manuel Castells (2000), society in the early twenty-first century is characterized by a social structure that he calls the "network society."

In this network society the nature of power in a social network is an important, and intriguing, issue, both from a theoretical and from a practical point of view. It is explored in considerable detail by Castells (2009) in a book, titled *Communication Power*. Castells distinguishes between several distinct kinds of power in social networks. He argues that the paramount form of power in the network society is the power to constitute networks – to connect individuals and institutions to these networks, or to exclude them, and to interconnect different networks.

A detailed discussion of Castells' analysis is beyond the scope of this book. However, we want to add a general comment about power in networks. In addition to power as domination of others, which arises from conflicts of interest, as we have seen, there is another kind of power – power as empowerment of others. Whereas power as domination is most effectively exercised through a hierarchy, the most effective social structure for power as empowerment is the network. In a social network, people are empowered by being connected to the network.

In such a network the success of the whole community depends on the success of its individual members, while the success of each member depends on the success of the community as a whole. Any enrichment of individuals, due to increased connectedness in the network, will therefore also enrich the entire network. In social networks, the hubs with the richest connections become centers of power. Because they connect large numbers of people to the network they are sought out as authorities in various fields. Thus, in a social network the centers of power are centers of both empowerment and of authority.

14.4.4 Structure in biological and social systems

As we explored the dynamics of social networks, culture, and power in the preceding pages, we saw repeatedly that the generation of structures, both material and social, is a key characteristic of those dynamics. At this point, it will be useful to review the role of structure in social systems in a systematic way.

A central focus of the systemic understanding of life is the concept of organization, or "pattern of organization." Living systems are self-generating networks, meaning that their pattern of organization is a network pattern in which each component contributes to the production of the other components. This idea is extended to the social domain by identifying the relevant living networks as networks of communications.

In the social realm, however, the concept of organization takes on an additional meaning. Social organizations, such as businesses or political institutions, are systems whose patterns of organization are designed specifically to organize the distribution of power. These formally designed patterns are known as organizational structures and are visually represented by the standard organizational charts. They are ultimately rules of behavior that facilitate decision-making and embody relationships of power. (The complex interactions between these formal organizational structures and informal networks of communications, which exist within all organizations, are discussed in Section 14.5.3 below.)

In biological systems, all structures are material structures. The processes in a biological network are production processes of the network's material components, and the resulting structures are the material embodiments of the system's pattern of organization. All biological structures change continually; hence, the process of material embodiment is a continual process.

Social systems produce nonmaterial as well as material structures. The processes that sustain a social network are processes of communication, which generate shared meaning and rules of behavior (the network's culture), as well as a shared body of knowledge. The rules of behavior, whether formal or informal, are known in sociology as social structures.

The ideas, values, beliefs, and other forms of knowledge generated by social systems constitute structures of meaning, or semantic structures. In modern societies, the culture's semantic structures are documented – that is, materially embodied – in written and digital texts. They are also embodied in artifacts, works of art, and other material structures, as they are in traditional, nonliterate cultures. Indeed, the activities of individuals in social networks specifically include the organized production of material goods. All these material structures – texts, works of art, technologies, and material goods – are created for a purpose and according to some design. They are embodiments of the shared meaning generated by the society's networks of communications.

14.4.5 Technology and culture

In biology, the behavior of a living organism is shaped by its structure (see Section 14.4.1). As the structure changes during the organism's development and during the evolution of its species, so does its behavior. A similar dynamic can be observed in social systems. The biological structure of an organism corresponds to the material infrastructure of a society, which embodies the society's culture. As the culture evolves, so does its infrastructure – they coevolve through continual mutual influences. The influences of the material infrastructure on people's behavior and culture are especially significant in the case of technology, and hence the analysis of technology has become an important subject in social theory (see, e.g., Fischer, 1985).

The meaning of "technology," like that of "science," has changed considerably over the centuries. The original Greek *technologia*, derived from *techne* ("art"), meant a discourse on the arts. When the term was first used in English in the seventeenth century, it meant a systematic discussion of the "applied arts," or crafts, and gradually it came to denote the crafts themselves. In the early twentieth century, the meaning was extended to include not only tools and machines but also nonmaterial methods and techniques, with the understanding of a systematic application of any such techniques. Thus, we speak of "the technology of management," or of "simulation technologies." Today, most definitions of technology emphasize its connection with science. The sociologist Manuel Castells (2000) defines technology as "the set of tools, rules, and procedures through which scientific knowledge is applied to a given task in a reproducible manner."

Technology, however, is much older than science. Indeed, its origins in tool making go back to the very dawn of the human species when language, reflective consciousness, and the ability to make tools evolved together (see Section 9.4.1). Accordingly, the first human species was given the name *Homo habilis* ("skillful human") to denote its ability to make sophisticated tools. Thus technology is a defining characteristic of human nature: its history encompasses the entire history of human evolution.

As a fundamental aspect of human nature, technology has crucially shaped successive epochs of civilization. Indeed, we characterize the great periods of human civilization in terms of their technologies – from the Stone Age, Bronze Age, and Iron Age to the Industrial Age and the Information Age. Throughout the millennia, but especially since the Industrial Revolution, critical voices have pointed out that the influences of technology on human life and culture are not always beneficial. In the early nineteenth century, William Blake decried the "dark Satanic mills" of Great Britain's growing industrialism, and several decades later

Karl Marx described the horrendous exploitation of workers in the British lace and pottery industries in one of the most vivid and moving chapters of *Das Kapital*.

More recently, critics have emphasized the increasing tensions between cultural values and high technology (see Ellul, 1964; Mander, 1991; Postman, 1992; Winner, 1977). Technology advocates often discount those critical voices by claiming that technology is neutral; that it can have beneficial or harmful effects depending on how it is used. However, these promoters of technology do not realize that a specific technology will always shape human nature in specific ways, because the use of technology is such a fundamental aspect of being human.

14.5 Life and leadership in organizations

14.5.1 Complexity and change

In the social sciences, the systems view of life has found its greatest advocates in management science. In recent years, the nature of human organizations has been discussed extensively in business and management circles in response to a widespread feeling that today's businesses need to undergo a fundamental transformation to adapt to a new global business and organizational environment that is almost unrecognizable from the point of view of traditional management theory and practice (see, e.g., Beerel, 2009; Wheatley and Kellner-Rogers, 1998). Moreover, a growing number of business leaders are becoming aware that our complex industrial systems, both organizational and technological, are the main driving force of global environmental destruction, and need to be fundamentally reorganized to become ecologically sustainable (as we shall discuss in Chapter 17).

This double challenge – the complexity of today's business environment and the need to become ecologically sustainable – is urgent and real, and the recent extensive discussions of organizational change, or "change management," are fully justified. However, in spite of these extensive discussions and some anecdotal evidence of successful attempts to transform organizations, the overall track record seems to be very poor. In recent surveys, CEOs (chief executive officers) reported again and again that their efforts at organizational change did not yield the promised results. Instead of managing new organizations, they ended up managing the unwanted side effects of their efforts (see Wheatley and Kellner-Rogers, 1998).

From the systems point of view, it is evident that one of the main obstacles to organizational change today is the – largely unconscious – embrace by business leaders of the mechanistic approach to management (discussed in Section 3.6). The principles of classical management theory are so deeply ingrained in the ways we think about organizations that for most managers the design of formal structures, linked by clear lines of communication, coordination, and control, has become almost second nature.

The core problem seems to be a confusion arising from the dual nature of all human organizations (see Capra, 2002). On the one hand, they are social institutions designed for specific purposes, such as making money for their shareholders, managing the distribution of political power, transmitting knowledge, or spreading religious faith. At the same time, organizations are communities of people who interact with one another to build relationships, help each other, and make their daily activities meaningful at a personal level.

These two aspects of organizations correspond to two very different types of change. Many CEOs are disappointed about their efforts to achieve change, in large part because they see their company as a well-designed tool for achieving specific purposes, and when they attempt to change its design they want predictable, quantifiable change in the entire structure. However, the designed structure always intersects with the organization's living individuals and communities, for whom change cannot be designed.

It is common to hear that people in organizations resist change. In reality, people do not resist change; they resist having change *imposed on them*. Being alive, individuals and their communities are both stable *and* subject to change and development, but their natural change processes are very different from the organizational changes designed by "re-engineering" experts and mandated from the top.

From our perspective of the systems view of life, it seems that, in order to resolve the problem of organizational change, we first need to understand the natural change processes that are embedded in all living systems. Once we have that understanding, we can begin to design processes of organizational change accordingly and to create human organizations that mirror life's adaptability, diversity, and creativity.

As we have learned from the theory of autopoiesis (Chapter 7), living systems continually create, or recreate, themselves by transforming or replacing their components. They undergo continual structural changes while preserving their web-like patterns of organization. Understanding life means understanding its inherent change processes. It seems, therefore, that organizational change will appear in a new light when we understand clearly to what extent and in what ways human organizations are alive. Indeed, a number of organizational theorists have taken this approach in recent years (De Geus, 1997; Senge, 1990; Wheatley, 1999; Wheatley and Kellner-Rogers, 1998).

14.5.2 Communities of practice

Living social systems, as we have seen (in Section 14.3.4), are self-generating networks of communications. This means that a human organization will be a living system only if it is organized as a network or contains smaller networks within its boundaries. Indeed, organizational theorists have come to realize that informal social networks exist within every organization. They arise from various alliances and friendships, informal channels of communication, and other tangled webs of relationships that continually grow, change, and adapt to new situations.

The social learning theorist Étienne Wenger (1998) has coined the term "communities of practice" for these informal, self-generating networks within organizations. Wenger points out that, in our daily activities, most of us belong to several communities of practice – at work, in schools, in sports and hobbies, or in civic life. Some of them may have explicit names and formal structures; others may be so informal that they are not even identified as communities. Whatever their status, communities of practice are an integral part of our lives.

As far as human organizations are concerned, we can now see that their dual nature as legal and economic entities, on the one hand, and communities of people on the other hand, derives from the fact that various communities of practice invariably arise and develop within the organization's formal structures. Within every organization, there is a cluster of interconnected communities of practice. The more people are engaged in these informal networks, and the more developed and sophisticated the networks are, the better will the organization be able to learn, respond creatively to unexpected new circumstances, change, and evolve. In other words, the organization's aliveness resides in its communities of practice.

14.5.3 The living organization

In order to maximize a company's creative potential and learning capabilities, it is crucial for managers and business leaders to understand the interplay between the organization's formal, designed structures and its informal, self-generating networks. The formal structures are sets of rules and regulations that define relationships between people and tasks, and determine the distribution of power. Boundaries are established by contractual agreements that delineate well-defined subsystems (departments) and functions. The formal structures are depicted in the organization's official documents – the organizational charts, bylaws, manuals, and budgets that describe the organization's formal policies, strategies, and procedures.

The informal structures, by contrast, are fluid and fluctuating networks of communications. These communications include nonverbal forms of mutual engagement in a joint enterprise through which skills are exchanged and shared tacit knowledge is generated. The shared practice creates flexible boundaries of meaning that are often unspoken. The distinction of belonging to a network may be as simple as being able to follow certain conversations, or knowing the latest gossip.

Informal networks of communications are embodied in the people who engage in the common practice. When new people join, the entire network may reconfigure itself; when people leave, the network will change again, or may even break down. In the formal organization, by contrast, functions and power relations are more important than people, persisting over the years while people come and go. We should also note that not all informal networks are fluid and self-generating. For example, the well-known "old boys' networks" are informal patriarchal structures that can be very rigid and may exert considerable power. When we speak of "informal structures" in the following paragraphs, we refer to continually self-generating networks of communications, or communities of practice.

In every organization, there is a continuous interplay between its informal networks and its formal structures. Formal policies and procedures are always filtered and modified by the informal networks, which allow workers to use their creativity when faced with unexpected and novel situations. The power of this interplay becomes strikingly apparent when employees engage in a "work-to-rule" protest. By working strictly according to the official manuals and procedures, they seriously impair the organization's functioning. Ideally, the formal organization recognizes and supports its informal networks of relationships and incorporates their innovations into its structures.

To repeat, the aliveness of an organization – its flexibility, creative potential, and learning capability – resides in its informal communities of practice. The formal parts of the organization may be "alive" to varying degrees, depending on how closely they are in touch with their informal networks. Experienced managers know how to work with the informal organization. They will typically let the formal structures handle the routine work and rely on the informal organization to help with tasks that go beyond the usual routine. They may also communicate critical information to certain people, knowing that it will be passed around and discussed through the informal channels. These considerations imply that the most effective way to enhance an organization's potential for creativity and learning, to keep it vibrant and alive, is to support and strengthen its communities of practice. The first step in this endeavor will be to provide the social space for informal communications to flourish.

The more managers know about the detailed processes involved in self-generating social networks, the more effective they will be in working with the organization's communities of practice. Let us see, then, what kinds of lessons for management can be derived from the systemic understanding of life.

According to the Santiago theory of cognition, a living network responds to disturbances with structural changes, and it chooses both *which* disturbances to notice and *how* to respond (see Section 12.2.2). What people notice depends on who they are as individuals, and on the cultural characteristics of their communities of practice. A message will get through to them not only because of its volume or frequency but also because it is meaningful to them.

We are dealing here with a crucial difference between a living system and a machine. A machine can be controlled; a living system can only be disturbed. This implies that human organizations cannot be controlled through direct interventions, but they can be influenced by giving impulses rather than instructions. To change the conventional style of management accordingly requires a shift of perception that is anything but easy, but it also brings great rewards. Working with the processes inherent in living systems means that we do not need to spend a lot of energy to move an organization. There is no need to push, pull, or bully it to make it change. Force, or energy, is not the issue; the issue is meaning. Meaningful disturbances will get the organization's attention and will trigger structural changes.

Offering impulses and guiding principles rather than strict instructions evidently amounts to significant changes in power relations, changes from domination and control to cooperation and partnerships. This, too, is a fundamental implication of the new understanding of life. As we mentioned in our chapter on evolution (in Section 9.6.3), biologists and ecologists have come to realize that most relationships between organisms in nature are essentially cooperative ones. The tendency to associate, establish links, cooperate, and maintain symbiotic relationships is one of the hallmarks of life.

In terms of our previous discussion of power in Section 14.4.3, we could say that the shift from domination to partnership corresponds to a shift from coercive power, which uses threats of sanctions to ensure adherence to orders, and compensatory power, which offers financial incentives and rewards, to conditioned power, which tries to make instructions meaningful through persuasion and education.

14.5.4 Emergence and design

If the aliveness of an organization resides in its communities of practice, and if creativity, learning, change, and development are inherent in all living systems, how do these processes actually manifest themselves in the organization's living networks and communities? To answer this question, we need to turn to a key characteristic of life that we discussed in Chapter 8 – the spontaneous emergence of new order. As we showed with many examples, emergent properties are ubiquitous in chemistry (see Section 8.2.2), and emergence acquires its full potential in dynamical systems that operate far from equilibrium (see Section 8.3). There, the phenomenon of emergence takes place at critical points of instability that arise from fluctuations in the environment, amplified by feedback loops. Emergence results in the creation of novelty that is often qualitatively different from the phenomena out of which it emerged. The constant generation of novelty – "nature's creative advance," as the philosopher Alfred North Whitehead called it – is a key property of all living systems.

In a human organization, the event triggering the process of emergence may be an offhand comment, which may not even seem important to the person who made it but is meaningful to some people in a community of practice. Because it is meaningful to them, they "choose to be disturbed" and circulate the information rapidly through the organization's networks. As it circulates through various feedback loops, the information may get amplified and expanded, even to such an extent that the organization can no longer absorb it in its present state. When that happens, a point of instability has been reached. The system cannot integrate the new information into its existing order; it is forced to abandon some of its structures, behaviors, or beliefs. The result is a state of chaos, confusion, uncertainty, and doubt; and out of that chaotic state a new form of order, organized around a new meaning, emerges. The new order was not designed by any individual but emerged as a result of the organization's collective creativity.

This process of emergence involves several distinct stages. To begin with, there must be a certain openness within the organization, a willingness to be disturbed, in order to set the process in motion; and there has to be an active network of communications with multiple feedback loops to amplify the triggering event. The next stage is the point of instability, which may be experienced as tension, chaos, uncertainty, or crisis. At this stage, the system may either break *down*, or it may break *through* to a new state of order, which is characterized by novelty and involves an experience of creativity that often feels like magic. Since the process of emergence is thoroughly nonlinear, involving multiple feedback loops, it cannot be fully analyzed with our conventional, linear ways of reasoning, and hence we tend to experience it with a sense of mystery.

Throughout the living world, the creativity of life expresses itself through the process of emergence. The structures that are created in this process – biological structures of living organisms as well as social structures in human communities – may appropriately be called "emergent structures." Before the evolution of humans, all living structures on the planet were emergent structures. With human evolution, language, conceptual thought, and all the other characteristics of reflective consciousness came into play. This enabled us to form mental images of physical objects, to formulate goals and strategies, and thus to create structures by design.

We sometimes speak of the structural "design" of a blade of grass or an insect's wing, but in doing so we use metaphorical language. These structures were not designed; rather, they were formed during the evolution of life and survived through natural selection. They are emergent structures. Design requires the ability to form mental images, and since this ability, as far as we know, is limited to humans and the great apes, there is no design in nature at large.

Human organizations always contain both designed and emergent structures. The designed structures are the formal structures of the organization, as described in its official documents. The emergent structures are created by the organization's informal networks and communities of practice. The two types of structures are very different, as we have seen, and every organization needs both kinds. Designed structures provide the rules and routines that are necessary for the effective functioning of the organization. Designed structures provide stability.

Emergent structures, on the other hand, provide novelty, creativity, and flexibility. They are adaptive, capable of changing and evolving. In today's complex business environment, purely designed structures do not have the necessary responsiveness and learning capability. The issue is not one of discarding designed structures in favor of emergent ones. We need both. In every human organization there is a tension between its designed structures, which embody relationships of power, and its emergent structures, which represent the organization's aliveness and creativity. Skillful managers understand the interdependence of design and emergence. They know that in today's turbulent business environment, their challenge is to find the right balance between the creativity of emergence and the stability of design.

14.6 Concluding remarks

Bringing life into human organizations by empowering their communities of practice not only increases their flexibility, creativity, and learning potential but also enhances the dignity and humanity of the organization's individuals, as they connect with those qualities in themselves. In other words, the focus on life and self-organization empowers the self. It creates mentally and emotionally healthy working environments in which people feel that they are supported in striving to achieve their own goals and do not have to sacrifice their integrity to meet the goals of the organization.

The problem is that human organizations are not only living communities but also social institutions designed for specific purposes and functioning in a specific economic environment. Today that environment is not life-enhancing but is increasingly life-destroying. The more we understand the nature of life and become aware of how alive an organization can be, the more painfully we notice the life-draining nature of our current economic system.

When shareholders and other outside bodies assess the "health" of a business organization, they generally do not inquire about the aliveness of its communities, the integrity and well-being of its employees, or the ecological sustainability of its products. They ask about profits, shareholder value, market share, and other economic parameters; and they will apply any pressure they can to ensure quick returns on their investments, irrespective of the long-term consequences for the organization, the well-being of its employees, or of the broader social and environmental impacts.

It is evident that the key characteristics of today's business environment – global competition, turbulent markets, corporate mergers with rapid structural changes, increasing workloads, and demands for "24–7" accessibility – combine to create a situation that is highly stressful and profoundly unhealthy. In this business climate it is often difficult to hold on to the vision of an organization that is alive, creative, and concerned about the well-being of its members and of the living world at large.

Paradoxically, the current business environment, with its turbulences and complexities, is also one in which the flexibility, creativity, and learning capability that come with the organization's aliveness are most needed. This is now being recognized by a growing number of visionary business leaders who are shifting their priorities toward developing the creative potential of their employees, enhancing the quality of the company's internal communities, and integrating the challenges of ecological sustainability into their strategies (see Petzinger, 1999).

In the long run, organizations that are truly alive will be able to flourish only when we change our economic system so that it becomes life-enhancing rather than life-destroying. The systemic understanding of life makes it clear that in the coming years such a change will be imperative not only for the well-being of human organizations but also for the survival and sustainability of humanity as a whole. This is the subject of the last three chapters of this book.

15

The systems view of health

In the last eight chapters we have discussed the biological, cognitive, and social dimensions of the systems view of life, emphasizing the fundamental patterns of self-organization and emergence that involve all three dimensions and allow us to integrate them into a unifying vision. No less important is the ecological dimension of life, which we shall explore in Chapter 16.

This new unifying vision of life, which has emerged in science over the last three decades, has important implications for almost every field of study and every human endeavor. This should not come as a surprise, since most phenomena we deal with in our professional and personal lives have to do with living systems. Whether we talk about economics, the environment, education, healthcare, law, or management, we are dealing with living organisms, social systems, or ecosystems. And consequently, the fundamental shift of perception from the mechanistic to the systemic view of life is relevant to all these areas.

Within the limited scope of this book, we can review only a few of these fields. In the previous chapter we discussed the influence of the systems view of life on the science and practice of management (Section 14.5). In this chapter we shall concentrate on the important field of health and healthcare, perhaps the field in which the limits of the mechanistic view of life are most clearly visible.

The critique of the conventional mechanistic approach to health and healing, and the outlines of an alternative systemic conception had already been developed during the late 1970s (see Capra, 1982, and references given therein). However, the systemic approach has still not found broad acceptance in today's scientific medicine. In fact, we can observe two parallel streams in contemporary healthcare. On the one hand, the biomedical approach (see Section 15.1.1) has become ever more efficient within its own terms, with improved computer-assisted diagnostics, a vastly expanded pharmacology, and novel, minimally intrusive surgical techniques – all of which, however, have enormously increased the costs of healthcare and are often criticized for being applied without sufficient discrimination.

On the other hand, the systemic and integrative vision of health and healing (see Section 15.2) is now widely accepted by the public and the media; and yet, it exists largely as a parallel structure outside the "official" healthcare system. We believe, therefore, that it is important to review the systemic critique of conventional medicine, as well as the outlines of an alternative, integrative system of healthcare. The full implementation of such a system of "integrative medicine" (see Section 15.3) is now essentially a question of financial power and political will – issues to which we shall return in the last part of our book.

15.1 Crisis in healthcare

In spite of the great advances of medical science in the last century, we are witnessing a widespread dissatisfaction with medical institutions. Many reasons are given for this discontent – inaccessibility of services, lack of sympathy and care, malpractice – but the central theme of all criticism is the striking disproportion between the cost and overall effectiveness of modern medicine. The manifestations of this disparity are different from country to country, due to their different healthcare systems, but the overall picture is the same. Despite a staggering increase in healthcare costs over the past decades, and amidst astonishing advances in diagnostic techniques and surgical procedures, the overall health of the population does not seem to have improved significantly (see World Health Statistics published annually by the World Health Organization (www.who.int); see also McMichael, 2001).

The causes of our health crisis are manifold; they can be found both within and without medical science, and are inextricably linked to the larger, multifaceted global crisis (to be discussed in Chapter 17). Still, increasing numbers of people, both within and outside the medical field, perceive many shortcomings of the current healthcare system as being rooted in the conceptual framework that supports medical theory and practice.

15.1.1 The biomedical model

The conceptual foundation of modern scientific medicine is the so-called biomedical model, which is firmly grounded in Cartesian thought (see Section 2.4). As we have discussed, the conceptual problem at the center of contemporary healthcare is the confusion between the origins of disease and the processes through which it manifests itself. Rather than asking why an illness occurs and trying to remove the conditions that led to it, medical researchers and practitioners often limit themselves to understanding the mechanisms through which the disease operates, so that they can then interfere with them.

A systemic approach, by contrast, would broaden the scope from the levels of organs and cells to the whole person – to the patient's body and mind, as well as his or her interactions with a particular natural and social environment. Such a broad, systemic perspective will enable health professionals to better understand the phenomenon of healing, which today is often considered outside the scientific framework. Although every practicing physician knows that healing is an essential part of all medical care, the phenomenon is presently not part of scientific medicine. The reason is evident: it is a phenomenon that cannot be understood when health is reduced to mechanical functioning.

15.1.2 Genes and disease

In the process of reducing an illness of the whole patient to the disease of a particular organ, physicians have focused their attention on smaller and smaller parts of the body – shifting their perspective from the study of bodily organs and their functions to that of cells, and finally, with the development of genetic engineering, to the study of molecules. Indeed, when the techniques of DNA sequencing and gene splicing were developed in the 1970s, geneticists first turned to the medical applications of genetic engineering. Since genes were thought to determine biological functions, it was natural for geneticists to set themselves the task of precisely identifying the genes that caused specific diseases. If they were successful in doing so, they thought, they might be able to prevent or cure these "genetic" diseases by correcting or replacing the defective genes.

This was, in large part, the dream behind the Human Genome Project (see Section 9.5), that specific faulty genes would be recognized as the root causes of most diseases – a dream which has not been realized. In a recent review of medical genetics, published in the journal *Nature*, T.A. Manolio (2009) and 26 other prominent geneticists came to the conclusion that, in spite of more than 700 published papers on genome scanning, geneticists had still not found more than a fractional genetic basis for human disease.

In fact, it was soon discovered that there is a huge gap between the ability to identify genes that are involved in the development of disease and the understanding of their precise function, let alone their manipulation to obtain a desired outcome. This gap is a direct consequence of the mismatch between the linear causal chains of genetic determinism and the nonlinear genetic and epigenetic networks of biological reality, as we discussed in Section 9.6.4.

Initially, there was the idea to associate specific diseases with single genes, but it turned out that single-gene disorders are extremely rare, accounting for less than 2% of all human diseases. Even in these clear-cut cases – for example, sickle-cell anemia, muscular dystrophy, or cystic fibrosis – where a mutation causes a malfunction in a single protein of crucial importance, the links between the defective gene and the onset and course of the disease are still poorly understood.

The problems encountered in the rare single-gene disorders are compounded when geneticists study common diseases like cancer and heart disease, which involve networks of multiple genes. As the molecular biologist Richard Strohman (1997) explained,

> In the case of coronary artery disease [for example], there are more than 100 genes identified as having some interactive contribution. With networks of 100 genes and their products interacting with subtle environments to affect [biological functions], it is naive to think that some kind of nonlinear networking theory could be omitted from a diagnostic analysis.

Another problem is that the defective genes in single-gene diseases are often very, very large. For example, the gene that is critical to cystic fibrosis, a disease affecting mainly infants and children, consists of some 230,000 base pairs and codes for a protein composed of almost 1,500 amino acids. More than 400 different mutations have been observed in this gene. Only one of them results in the disease, and identical mutations may lead to different symptoms in different individuals. All this makes screening for the "cystic fibrosis defect" highly problematic.

15.1.3 Genetic therapy?

What we have outlined above should make clear that the pathway from genes to diseases is far from being understood sufficiently for clinical applications. Nevertheless, with the advent of synthetic biology (see Section 10.5), and with the remarkable improvements of analytical tools within genetics, "medical genetics" has been developed as a new, and initially very promising, field of medicine, dedicated to the study of the relation between genetic structure and health disorders.

In principle, different approaches can be used toward this aim. First, one might think of using the techniques of genetic engineering. Accordingly, once the faulty gene is identified as the cause of the disorder, it can be cut away and replaced with a new, correct one. There is also, in principle, an alternative and less drastic procedure of genetic therapy: instead of cutting away the faulty gene, one might simply add the healthy gene system to the organism, to work in parallel, producing the correct proteins and correct functions. And finally, one can try to "knock down" the ill-functioning genes in a kind of molecular surgery.

It is important to understand that in the first two cases the new gene must first be transported into the identified cells. This is not easy. If the gene is administered in an intravenous solution, its DNA sequence is going to be split and destroyed (hydrolyzed) before it reaches its target. Thus, the foreign gene must be protected during its pathway to the target cells. This can be done with special "vectors." For example, the gene, a negatively charged DNA macromolecule, can be associated with positively charged vesicles, an association which may considerably increase the circulation time in the organism. Theoretically, then, the "sick" cells will eventually be hit by the vectors, and the gene will be incorporated into the cells, where it can, in principle, perform its function.

There are, however, several problems associated with this kind of DNA delivery. One is that healthy cells, too, are hit by these vectors; the other is that the efficiency of the yield is generally extremely modest with respect to the necessary therapeutic doses. In fact, despite many studies and attempts over the past years, this method has never reached the threshold of a solid therapy. By contrast, the delivery with vesicles is well accepted, and it is more successful when simple drugs are used instead of genes – for example, water-insoluble anticancer drugs. This was shown in a research project in which one of us (P.L.L.) was involved (Stano *et al.*, 2004).

The use of viruses as vectors yields, in principle, a more efficient way of delivering the gene into the cell. A virus has the capability of infecting cells, and of transferring its own genetic material to them. This can be turned into a positive trait if part of the virus structure – duly modified if necessary – contains the genetic information that is defective in the cell. Of course, in this case it is important that the virus first be inactivated so that it is no longer capable of self-reproducing inside the cell.

It must be clearly said that scientific advances in this entire field are not yet at the stage of warranting a safe therapeutic procedure for humans. In fact, the US Food and Drug Administration (FDA) has not yet approved any human gene therapy product for sale. Actually, very little progress has been made since the first clinical trial of a gene therapy began in the 1990s. In fact, there have been a few quite negative results, which have caused a major setback in gene therapy. In 1999, there was the death of an 18-year-old volunteer who participated in a gene therapy trial; and four years later, children treated in a French gene therapy trial reportedly developed a leukemia-like condition (see Johnston and Baylis, 2004).

As David Weatherall (1998), director of Oxford University's Institute of Molecular Medicine, sums up, "Transferring genes into a new environment and enticing them to . . . do their jobs, with all the sophisticated regulatory mechanisms that are involved, has, so far, proved too difficult a task for molecular geneticists."

In response to these difficulties, there has recently been a major shift of perspective among leading geneticists from genes to epigenetic networks, as we discussed in Section 9.6. Nevertheless, the central assertion of genetic determinism, that genes determine behavior, continues to be promoted vigorously by the biotechnology industry and repeated constantly in the popular media. Once we know the exact sequence of the genetic bases in DNA, the public is told, we will understand how genes cause cancer, alcoholism, or criminality.

The primary interest of the biotech companies, of course, is not the advancement of science, the cure of disease, or the feeding of the hungry (even though they publicly profess all of these noble goals), but rather their own financial gain. One of the most effective ways of ensuring that the shareholder values of their ventures remain high, even though any tangible benefits lie far in the future, is to perpetuate the perception among the general public that genes determine behavior. Thus, year after year, bold headlines in newspapers and cover stories in magazines have excitedly reported discoveries of new "disease-causing" genes and corresponding new potential therapies, usually with serious scientific caveats appearing a few weeks later but published as small notices among the bulk of other news.

In our view, this is one more indication that medical scientists urgently need to shift their attention from genes to genetic and epigenetic networks, from parts of the body to the whole person, and from a mechanistic to a systemic view of health.

15.2 What is health?

In the biomedical model, health is defined as the absence of disease, and disease as the malfunctioning of biological mechanisms. An alternative conception of health, based on the systems view of life, begins with the realization that it is impossible to give a precise definition of health. The reason is that health is largely a subjective experience whose quality can be known intuitively but can never be exhaustively described or quantified. Health is a state of well-being that arises when the organism functions in a certain way.

The description of this way of functioning will depend on how we describe the organism and its interactions with its environment. In other words, our understanding of health will always be linked to our understanding of life. Different models of living organisms will lead to different definitions of health. The concept of health, therefore, and the related concepts of illness, disease, and pathology, do not refer to well-defined entities but are integral parts of limited and approximate models that mirror the complex and fluid phenomenon of life.

Once we recognize the relative and subjective nature of the concept of health, we can begin to explore how the systems view of life can help us develop a corresponding systems view of health. Systems thinking is process thinking (see Section 4.3.3), and hence the systems view sees health as an ongoing process. Rather than defining health as a static state of perfect well-being, the systemic conception of health implies continual activity and change, reflecting the organism's creative response to environmental challenges. Since a person's condition will always depend on the natural and social environment, there can be no absolute level of health independent of this environment. The continual changes of one's body in relation to the changing environment will naturally include temporary phases of ill health, and it will often be impossible to draw a sharp line between health and illness.

Moreover, health is a multidimensional process. From the systems point of view, the experience of illness results from patterns of disorder that may become manifest at various levels of the organism – biological as well as psychological – and also in the various interactions between the organism and the larger systems in which it is embedded. This means that the biological, cognitive, social, and ecological dimensions of life, which we have emphasized throughout this book, correspond to similar dimensions of health.

The systems view of life recognizes that living systems in nature include individual organisms, parts of organisms, and communities of organisms, and that they all share a set of common properties and principles of organization (see Section 4.3). Accordingly, the systems view of health can be applied to different systems levels, with corresponding levels of health being mutually interconnected. In particular, we can discern three interdependent levels of health – individual, social, and ecological.

In summary, the systems view of life leads us to see health as a process, and as a multidimensional and multileveled phenomenon. With these general characteristics in mind, we can now see how health can actually be defined from a systemic perspective.

15.2.1 The systemic conception of health

As we have discussed, a living system is understood as a self-generating, self-organizing network that displays a high degree of stability. This stability is utterly dynamic and is characterized by continual, multiple, and interdependent fluctuations (see Section 8.3). All the variables of a living system fluctuate continually between tolerance limits: the more dynamic the state of the system, the greater is its flexibility. Whatever the nature of this flexibility – physical, mental, social, technological, or economic – it is essential to the system's ability to adapt to environmental changes. Loss of flexibility means loss of health.

Moreover, the living system responds to disturbances from its environment autonomously and cognitively (see Section 12.3). The resulting structural changes can be either changes of self-renewal or developmental changes in which new forms of order emerge.

This view of living systems suggests the notion of dynamic balance as a useful concept to define health. Such a state of balance is not a static equilibrium but rather a flexible pattern of fluctuations. With this meaning of "dynamic balance" in mind, health can be defined as

> "a state of well-being, resulting from a dynamic balance that involves the physical and psychological aspects of the organism, as well as its interactions with its natural and social environment".
>
> *(Capra, 1982, p. 323)*

An important part of this systemic definition of health is that the dynamic balance of the healthy system involves both the physical and mental, or psychological, aspects of the organism. As we discussed in Chapter 12, the systemic conception of life includes a radically new concept of mind. When this new concept of mind, or cognition, is adopted, it becomes evident that every illness has mental aspects. Getting sick (moving out of balance) and healing (regaining balance) are both integral parts of the life process, and if that process is identified with cognition (see Section 12.1.2), the processes of getting sick and of healing can be seen as cognitive processes. This means that there is a mental dimension in every illness, even if it often lies in the realm of the unconscious.

This amounts to a radical redefinition of the term "psychosomatic." In the biomedical model, this term has been used to refer to disorders without a clearly diagnosed organic basis, and because of the mechanistic bias of that model, such "psychosomatic disorders" often tended to be regarded as imagined, not real. The systemic use of the term is quite different. Researchers and clinicians have become increasingly aware that all disorders are psychosomatic in the sense that they involve the continual interplay of mind and body in their origin, development, and cure. This refined notion of "psychosomatic" has recently led to many important new insights (see Borysenko, 2007; Harrington, 1997; Pelletier, 2000; Pert, 1997; Pert *et al.*, 1998; Siegel, 2010; Weil, 1995).

The new systemic understanding of "psychosomatic" makes it clear why mental attitudes and psychological techniques are important means for the healing of illness. A positive attitude will have a strong positive impact on the mind/body system and will often be able to reverse the disease process, and even to heal severe biological disorders. An impressive proof of the healing power of positive expectations alone is provided by the well-known placebo effect (see guest essay by Fabrizio Benedetti, MD, on p. 329). A placebo is an imitation medicine, dressed up like an authentic pill and given to patients who think they are receiving the real thing. Studies have shown that 35% of patients consistently experience

Guest essay

Placebo and nocebo responses

Fabrizio Benedetti, MD

Department of Neuroscience, University of Turin Medical School, and National Institute of Neuroscience, Turin, Italy

Definition

A placebo response is a psychological and biological phenomenon occurring in the patient's brain following the administration of an inert substance or of a sham physical treatment, along with verbal suggestions, or any other cue, of clinical benefit. The effect that follows the administration of a placebo is not due to the inert treatment *per se*, for saline solutions or sugar pills will never acquire therapeutic properties, but to the psychosocial context that surrounds the inert substance and the patient. In this sense, to the clinical trialist and to the neurobiologist, the term "placebo effect" or "placebo response" has different meanings. The former is interested in any improvement that may occur in the group of patients who receive the inert treatment, regardless of its cause, because he only wants to establish whether the patients who take the true treatment are better off than those who take the placebo, irrespective of whether such observed improvement is due to spontaneous remission of the disease, doctor's biases in selecting patients, or patient's biases in reporting symptoms. By contrast, the psychologist and the neuroscientist are only interested in the improvement that derives from active processes occurring in the patient's brain, such as the patient's expectations of clinical improvement. The same concepts hold true for the nocebo response, a phenomenon that is opposite to the placebo response, whereby an inert treatment is administered along with negative verbal suggestions of worsening. Taking these considerations into account, the placebo and nocebo are not the inert treatments alone, but rather their administration within a set of sensory and social stimuli that tell the patient that either a beneficial or a harmful treatment, respectively, is being given.

Mechanisms

There is not a single placebo effect, but many, with different mechanisms in different diseases and in different systems. Most of the research on placebos has focused on expectations as the main factor involved in placebo responsiveness. In the many studies in the literature where expectations are analyzed, the terms "effects of placebos" and "effects of expectations" are frequently used interchangeably. There are several mechanisms through which expectation of a future event may affect different physiological functions. First, anxiety has been found to be reduced after placebo administration and increased after nocebo administration. Therefore, expecting either improvement or worsening of a symptom may either decrease or increase anxiety, respectively. Second, expectations of future events may also induce physiological changes through reward mechanisms. These mechanisms are mediated by specific neuronal circuits linking cognitive, emotional, and motor responses, and are traditionally studied in the context of the pursuit of natural (e.g., food), monetary, and drug rewards. There is compelling experimental evidence that the so-called "mesolimbic dopaminergic system" (a pathway in the brain containing dopamine, which controls the brain's pleasure and reward centers) may be activated in some circumstances when a subject expects clinical improvement after placebo administration. In this case, the reward is represented by the therapeutic benefit itself. Third, learning is another mechanism that is central to placebo responsiveness. Subjects who suffer from a painful condition, such as headache, and who regularly consume aspirin, can associate the shape, color, and taste of the pillwith pain decrease. After repeated associations, if they are given a sugar pill resembling aspirin, they will experience pain decrease. Needless to say, besides shape, color, and taste of pills, countless other stimuli can be associated with therapeutic benefit, such as hospitals, diagnostic and therapeutic equipment, and medical personnel features. Fourth, there is some experimental evidence supporting the role of genes in some types of placebo responses – for example, a genetically controlled modulation of amygdala activity by the neurotransmitter serotonin, which is linked to placebo-induced anxiety relief.

Neurobiology

Besides these general mechanisms, recent neuroscientific research has shed light on the neurotransmitters and brain areas that are involved in the placebo and nocebo responses. For example, pain research has shown that placebo administration activates regulatory systems involving endogenous opiates and endocannabinoids. These systems represent a top-down regulatory network extending from cognitive and affective cortical brain regions to the brainstem and spinal cord dorsal horns, with the ability to negatively modulate the incoming pain-causing signals. Another neurotransmitter identified in placebo analgesia is cholecystokinin (CCK), which plays an inhibitory role in placebo analgesia. Interestingly, CCK is also the neurotransmitter that induces the pain worsening known as nocebo hyperalgesia. Dopamine is involved in the placebo analgesic response as well, but its role is better understood in Parkinson's disease. In fact, it has been shown that placebo administration in Parkinson patients leads to the activation of dopamine in the subcortical forebrain known as the striatum, along with changes in neuronal activity in different regions of the basal ganglia.

The crucial role of the prefrontal lobes in placebo responsiveness is shown by several studies on the impairment of prefrontal cortex functioning. First, Alzheimer patients with reduced cognitive scores and reduced connectivity of the prefrontal lobes show reduced placebo responses. Second, placebo analgesic responses are disrupted by the pharmacological blockade of the prefrontal opiate network. Third, the inactivation of prefrontal regions by transcranial magnetic stimulation abolishes placebo responses. Therefore, in the presence of a loss of prefrontal control, we also witness a loss of placebo response.

Physical performance

As in the clinic, also in the sport world there is a zone where placebos and nocebos can exert their influence. Here, too, chemicals such as vitamins, performance-enhancing aids, or diet supplements are handed out, or physical treatments and manipulations of different kinds are delivered, and expectations about their effects are set in motion in the athlete's brain. In spite of very different experimental conditions, ranging from short anaerobic sprints to long aerobic endurance cycling, and across many different outcome measures, such as mean power output, time, speed, weight lifted, or perceived exertion, all the data indicate athletes' expectations as important factors in physical performance, to be taken into account in training strategies.

Apart from a role for endogenous opiates in placebo pain endurance, not much is known at present in this context. In many studies, athletes are asked to perform at their limit, in an all-out effort. Placebos apparently act by pushing this limit forward, possibly impacting on a central governor of fatigue which, although not identified, has been proposed as a brain center integrating peripheral signals and central control processes, so as to continuously regulate exercise performance and avoid reaching maximal physiological capacity. This would provide protection against damage on the one hand, and constant availability of a reserve capacity on the other hand. By altering expectations, placebos could then represent a psychological means to signal the central governor to release the brake, allowing an increase in performance.

References

Benedetti, F. (2008). *Placebo Effects: Understanding the Mechanisms in Health and Disease.* Oxford University Press.

(2010). *The Patient's Brain: The Neuroscience Behind the Doctor–Patient Relationship.* Oxford University Press.

"satisfactory relief" when placebos are used instead of regular medication for a wide range of medical problems (see Benedetti, 2009; Harrington, 1997).

Placebos have been strikingly successful in reducing or eliminating physical symptoms, and have produced dramatic recoveries from illnesses for which there are no known medical cures. The only active ingredient in these treatments appears to be the power of the patient's positive expectations, supported by interaction with the therapist. The placebo effect is not limited to the administration of pills but can be associated with any form of treatment. Indeed, the patient's will to get well and confidence in the treatment are crucial aspects of any therapy, from shamanistic healing rituals to modern medical procedures.

15.2.2 Illness as imbalance

The view of health as a state of dynamic balance implies that illness is a consequence of imbalance and disharmony, which may arise at various levels of the organism and may generate symptoms of a physical, psychological, or social nature. To describe an organism's imbalance, the concept of stress has proven very useful. Temporary stress is an essential aspect of life, but prolonged or chronic stress can be harmful and plays a significant role in the origin and development of most illnesses. From the systemic perspective, stress is an imbalance in the organism that occurs when one or several of its fluctuating variables are pushed to their extreme values, which induces increased rigidity throughout the system, and hence loss of flexibility. A key element in the link between stress and illness is the fact that prolonged stress suppresses the body's immune system, which plays an important role in the organism's self-organization.

The recognition of the role of stress in the development of illness leads to the important notion of illness as a "problem solver." Because of social and cultural conditioning, people often find it impossible to release their stresses in healthy ways and therefore choose – consciously or unconsciously – to get sick as a way out. Their illness may be physical or mental, or it may manifest itself as violent and reckless behavior, which may appropriately be called social illnesses. All these "escape routes" are forms of ill health, physical disease being only one of several unhealthy ways of dealing with stressful life situations. Hence, curing a disease will not necessarily make the patient healthy. If the escape into a particular disease is blocked effectively by medical intervention while the stressful situation persists, this may merely shift the patient's response to a different mode, such as mental illness or asocial behavior, which will be just as unhealthy.

The idea of illness as a way to cope with stressful life situations naturally leads to the notion of the "meaning" of an illness, or the "message" that is transmitted by a particular disease. To understand this message, ill health should be taken as an opportunity for introspection, so that the original problem can be brought to a conscious level, where it may be resolved. This is where psychological counseling and psychotherapy can play an important role, even in the treatment of physical illnesses. To integrate physical and psychological therapies in such a way amounts to a major reconceptualization of healthcare, since it requires the full recognition of the interdependence of mind and body in health and illness.

15.2.3 The nature of healing

The systemic understanding of health and illness, which we have outlined in the previous sections, implies a corresponding systemic understanding of healing. Many traditional models of health and healing acknowledge the self-healing capacities inherent in every living organism – that is, the organism's innate tendency to re-establish itself in a balanced state when it has been disturbed. In systemic language, this tendency is associated with the self-balancing feedback loops inherent in living systems (see Section 8.3), through which the self-organizing system returns, more or less, to the original fluctuating state. Examples of this phenomenon would be periods of ill health involving minor symptoms. These are normal and natural stages in the organism's process of restoring balance by interrupting our usual activities and forcing a change of pace. As a consequence, the symptoms of these minor illnesses usually disappear after a few days, whether or not any treatment is received.

More serious illnesses will require greater efforts of regaining our balance, generally including the help of a doctor or therapist, and they may include stages of crisis and transformation, leading to the emergence of an entirely new state of balance.

Again, feedback loops are relevant here, but this time the feedback is self-amplifying, and the whole process is the process of emergence (see Section 8.3). Major changes in a person's lifestyle, induced by severe illness, are often examples of such creative responses, which may even leave the person at a higher level of health than the one enjoyed before the challenge.

15.3 A systemic approach to healthcare

The systems view of health that we are discussing in this chapter implies a number of guidelines for a corresponding systemic approach to healthcare. Indeed, such a systemic approach has been developed over the past four decades by numerous physicians, biologists, public health professionals, nurses, psychologists, and scientists from other disciplines. For a short selection of representative literature, see Micozzi, 2006; Moyers *et al.*, 1993; Ornish, 1998; Spencer and Jacobs, 1999; Sternberg, 2000; Weil, 2009; see also the US Public Broadcasting Service (PBS) documentary, "The New Medicine," 2006 (www.thenewmedicine.org).

In the late 1970s and early 1980s, the leading catchphrases of this new approach to health and healing were "holistic healthcare," "holistic medicine," "alternative medicine," and "wellness." In the subsequent decades, the phrase "integrative medicine" established itself as the unifying term. Today integrative medicine is a well-established discipline with its own Consortium of Academic Health Centers, professional journals, and international congresses (see guest essay by Helmut Milz, MD, on p. 334). It is understood by its practitioners as a healing-oriented approach that attempts to combine the best of conventional and alternative, or "complementary," therapies. In the following pages, we shall briefly outline the vision of a future healthcare system, based on the systemic approach of integrative medicine.

15.3.1 Individual and social healthcare

An integrative system of healthcare will consist, first and foremost, of a comprehensive, effective, and well-integrated system of health promotion and preventive care. Healthcare will consist of maintaining and restoring the dynamic balance of individuals, families, and other social groups. Health maintenance will be partly an individual matter and partly a collective matter, and most of the time the two will be closely interrelated.

Individual healthcare is based on the recognition that the health of human beings is determined, above all, by their behavior, their food, and the nature of their environment. As individuals, we have the power and the responsibility to keep our organism in balance by observing a number of simple rules of behavior relating to sleep, food, exercise, and drugs.

The role of therapists and health professionals will be mainly to assist us in doing so. Indeed, this is the original meaning of the word "therapist," from the Greek *therapeuein* ("to attend"). The Hippocratic writings, which constitute the foundation of Western medicine, define the role of the therapist as that of an attendant, or assistant, to the natural healing forces.

While individual responsibility will be crucial to a future integrative healthcare system, it will be equally crucial to recognize that this responsibility is subject to severe constraints. Many health problems arise from economic and political forces that can be modified only by collective action. Individual responsibility, therefore, has to be accompanied by social responsibility, and individual healthcare by social actions and policies. "Social healthcare"

Guest essay

Integrative practice in healthcare and healing

Helmut Milz, MD

Honorary Professor of Public Health, University of Bremen

Biomedicine has made great progress with respect to diagnostic imaging technologies, emergency interventions, high-tech surgical procedures, and gene-related diagnostics. But it has great problems with public recognition, specifically when it comes to its credibility and results for the majority of the complaints that are presented in ambulant care. Biomedicine's narrow focus on the specific treatment of diseases, mainly with drugs or surgery, has lost its efficiency. This is a particular problem with respect to the growing spectrum of chronic health problems like diabetes, obesity, heart problems, musculoskeletal pains, abdominal pains, depression, Alzheimer's disease, or the many other disorders that show no clear somatic origin. They now account for three-quarters of our healthcare spending. Many of them are lifestyle-related problems, which need a broader, cooperative assessment. Many might be prevented, or even reversed, through lifestyle changes.

Lifestyle-related health problems need to be looked at from a participatory perspective, which puts the options of active behavioral changes back on the agenda of patients, health professionals, communities, and public health policies. They require healthcare strategies that reconsider some lost skills in biomedicine, such as longer initial meetings, covering the many details of the patient's history and situation; a more calming atmosphere; a focus on the general health hazards in a person's life, on reducing everyday stress, on diet changes and exercise, and on treatment options that consider the living organism's ability to heal itself; and more frequent follow-up consultations. All this is barely on the agenda of today's speedy medical care.

Modern physicians do not get enough training in understanding illness in context. They are paid for short interventions, which avoid the exploration of emotional issues or anxieties. They tend to interrupt patients' statements after a few moments and provide little information on their prescriptions and treatments. They have almost no time to listen carefully, nor to help and advise their patients on healthier attitudes and behaviors.

This is where some of the strengths of alternative, or complementary, care fill a perceived gap. Many alternative practitioners spend more time exploring the problems and co-developing options for lifestyle changes. Another, sometimes unproven, claim is that, in general, alternative methods have less serious side effects. Critics say that alternative methods often have "only subjective effects," which may not be more than placebo. The placebo effect, which had been acknowledged in biomedicine only at the periphery of drug testing, has in recent years received much more rigidly scientific interest [see guest essay by Fabrizio Benedetti, MD, on p. 329]. The realm of consciousness, which was for decades considered to be only a philosophical issue, has earned very serious scientific interest in the neurosciences, particularly with the new imaging technologies [see Chapter 12]; and consequently visualization techniques, biofeedback, and mental and motor imagery strategies are now the focus of mainstream medical research.

About 40% of Americans have tried some form of alternative medicine, and some $35 billion is spent on it annually. Some methods that are considered "alternative" in the USA, such as homeopathy or massage therapy, are mainstream and covered by insurance in many

European countries. Some allied health professionals, such as the *Heilpraktiker* in Germany, would be considered alternative in the USA.

At the moment we often have two parallel healthcare systems: a mainstream, insurance-covered, biomedical system, and a broad, unregulated, and mainly privately paid alternative system. Their borders, most of the time, are strict. Nevertheless, many medical professionals are individually open to and interested in learning from the other system. On the other hand, the biological principles of conventional medicine are strongly influencing the practice of Western-born alternatives. In addition, many culture-bound, traditional healing practices continue to exist for immigrants as alternative, often hidden, healthcare systems.

Academic teaching centers have become more sensitive to the dilemma of this split. Anne Harrington, a historian of science at Harvard, has published an excellent review of the history of mind–body medicine, titled *The Cure Within* (Harrington, 2008). Today education and training in "mind–body medicine" is slowly being reintroduced into the curricula of medical schools. Alternative and complementary forms are being used increasingly in integrative ways with biomedical knowledge in an approach that has become known as "integrative medicine."

In the last two decades, many international professional organizations and journals promoting complementary, alternative, and integrative medicine have been established. The following list is a representative sample:

- The National Center for Complementary and Alternative Medicine (NCCAM) in the USA (www.nccam.nih.gov/) describes itself as "the Federal Government's lead agency for scientific research on the diverse medical and healthcare systems, practices, and products that are not generally considered part of conventional medicine." Founded originally in 1992 as the "Office of Alternative Medicine," with a budget of $2 million, the NCCAM became one of the twenty-seven National Institutes of Health in 1998. Its current budget is $127 million. The advisory council of the institute includes members from respected academic and professional boards. The institute's extensive website provides detailed information on research findings relevant to its mission, which are constantly updated, with the aim of helping informed consumers in their choices of complementary and alternative medicine (CAM) options.
- The Consortium of Academic Health Centers for Integrative Medicine (www.imconsortium.org), supported by the Fetzer Foundation, was convened first in 1999. Its idea was to set up reliable mechanisms that could help major medical centers to develop "clinical, educational, and research opportunities arising in this rapidly growing field" of "complementary, integrative, mind/body perspectives", which "can positively influence the medicine of the future." Today the consortium includes fifty-one highly esteemed academic medical centers and affiliate institutions in the USA.
- CAMbrella (www.cambrella.eu), founded in 2009, is a European research network for complementary and alternative medicine. The goal of this collaboration is to develop a road map for future European CAM research that is appropriate for the healthcare needs of European Union (EU) citizens, and acceptable to the EU Parliament as well as national research funders and healthcare providers.
- The European Information Center for Complementary and Alternative Medicine (EICCAM) has been created with the aim of providing and disseminating understandable, objective, and high-quality information on the safety, effectiveness, and efficiency of complementary and alternative medicine. EICCAM (www.eiccam.eu) has been set up as a Public Utility Foundation under Belgian law, with a management board and a scientific board.

- The European Federation for Complementary and Alternative Medicine (EFCAM), formed in December 2004 as a federation where representatives of organizations of patients, CAM practitioners, researchers, and trade and industry, meet to discuss and prepare political actions to have CAM recognized throughout Europe (www.efcam.eu).

The research and political initiatives of these organizations are discussed in a number of journals, founded during the past thirty years. The most important among them are: *Advances in Mind–Body Medicine* (www.advancesjournal.com), *Integrative Medicine* (www.imjournal.com), *Journal of Complementary and Integrative Medicine* (www.degruyter.com/view/j/jcim), and *European Journal of Integrative Medicine* (www.europeanintegrativemedicinejrnl.com).

Reference

Harrington, A. (2008). *The Cure Within: A History of Mind–Body Medicine*. New York: Norton.

seems an appropriate term for these policies and collective activities, dedicated to the maintenance and improvement of public health (see Ottawa Charter for Health Promotion of the World Health Organization (www.who.int/healthpromotion/conferences/previous/ottawa/en/).

Social healthcare will have two basic components – health education and health policies – to be pursued simultaneously and in close coordination. The aim of health education will be to make people understand how their behavior and their environment affect their health, and to teach them how to cope with stress in their daily lives. Comprehensive programs of health education with this emphasis can be integrated into the school system and be given central importance. At the same time they can be accompanied by public health education through the media to counteract the effects of advertising of unhealthy products and lifestyles, which is now all-pervasive.

An important aim of health education will be to foster corporate responsibility. The business community needs to learn much more about the health hazards of its production processes and its products. It should develop and demonstrate concern about public health, become aware of the health costs generated by its activities, and formulate corporate policies accordingly.

Health policies, to be established by governments at various administrative levels, will consist of legislation to prevent health hazards from being created, and also of social policies that provide for people's basic needs, so as to minimize social stress. Measures needed to provide an environment that would encourage and make it possible for people to adopt healthy ways of living include the following:

- restrictions on all advertising of unhealthy products;
- "health taxes" on individuals and corporations who generate health hazards, to offset the medical costs that inevitably arise from these hazards;

- social policies to improve education, employment, civil rights, and economic levels of large numbers of impoverished people;
- development of nutritional policies that provide incentives for industry to produce more nutritious foods, free of toxic chemicals.

Careful study of these suggested policies shows that any one of them, ultimately, requires a different economic system if it is to be successful. There is no way to avoid the conclusion that the present global economy itself has become a fundamental threat to our health. Moreover, it is also increasingly evident that social and ecological health – the health of our planet – are inextricably intertwined. These issues will be discussed in some detail in the last three chapters of our book.

15.3.2 Integrative therapy

In the systemic approach to therapy, the first and most important step will be to make patients aware, as fully as possible, of the nature and extent of their imbalance. This means that their problems will have to be put into the broad context from which they arose, and this will involve a careful examination of the multiple aspects of a particular illness by therapist and patient. Psychological counseling will play an important role in this process.

The main purpose of the first encounter between patient and general practitioner, apart from emergency measures, will be to explore with the patient the nature and meaning of the illness, and to find possibilities of changing the patterns in the patient's life that have led to it. To assess the relative contribution of biological, psychological, social, and environmental factors to the illness of a particular person is the essence of the science and art of general practice. It requires not only some basic knowledge of human biology, psychology, and social science but also experience, wisdom, compassion, and concern for the patient as a human being.

Health professionals administering primary care of this kind need not be medical doctors, nor experts in any of the scientific disciplines concerned, but they will have to be sensitive to the multiple influences affecting health and illness, and able to decide which of these are most relevant, best known, and most manageable in a particular case. If necessary, they will refer the patient to specialists in the relevant areas, but even when such special treatments are needed, the object of the therapy will still be the whole person.

The basic aim of any therapy will be to restore the patient's balance, and since the underlying model of health acknowledges the organism's innate tendency to heal itself, the therapist will try to intrude only minimally and help to create the environment most conducive to the healing. Such an approach to therapy will be multidimensional, involving treatments at several levels of the mind/body system, which will often require a multidisciplinary team effort. Healthcare of this kind will require many new skills in disciplines not previously associated with medicine and is likely to be intellectually richer, more stimulating, and more challenging than a medical practice that adheres exclusively to the biomedical model.

The reorganization of healthcare will also mean that hospitals will have to be transformed into more humane institutions, in agreement with the holistic orientation of the systems view of health. This will involve creating comfortable and therapeutic environments with good, nourishing food, family members included in patient care, and other such sensible improvements.

Drugs will be used mainly in emergency cases and as sparingly and as specifically as possible. This means that healthcare will need to be liberated from the current excessive influence of the pharmaceutical industry. Physicians and pharmacists will collaborate in selecting from the many thousands of pharmaceutical products only a limited number of drugs, which, according to "evidence-based" medicine," are fully adequate for effective medical care – see Cochrane Collaboration (www.cochrane.org).

These changes will be possible only with a thorough reorganization of medical education. To prepare medical students and other health professionals for the new systemic approach will require a considerable broadening of their scientific basis and much greater emphasis on the behavioral sciences and on human ecology. Such an educational program, dealing with various levels of individual and social health, will be based on the systems view of life and will study the human condition in health and illness within an ecological context. This will be a good foundation for more detailed medical studies and will provide all health professionals with a common language for their future collaboration in health teams.

15.4 Concluding remarks

The forces promoting the new systemic conception of health and healing work both inside and outside the medical system. While integrative medicine has established itself as a new scientific discipline, there has been a quiet health revolution among the general public, carried out by individuals and newly formed organizations dissatisfied with the existing systems of medical care.

These groups have embarked on an extensive exploration of alternative approaches, including promotion of healthy living habits, combined with recognition of personal responsibility for health and of the individual's potential for self-healing; a strong interest in traditional healing arts from various cultures that integrate physical and psychological approaches to health; and formation of integrative healthcare centers, many of them experimenting with a broad range of complementary therapies.

This health revolution goes hand in hand with a worldwide renaissance of sustainable, community-oriented agriculture, based on the recognition of the fundamental interdependence between a healthy soil, healthy individuals, and healthy communities. From an even broader perspective, the current health revolution can be seen as part of a global movement dedicated to creating a sustainable world for our children and for future generations – a vision that is fully consistent with the systems view of life. We shall discuss the structure, values, strategies, and challenges of this global movement in the last part of this book.

IV

Sustaining the web of life

16

The ecological dimension of life

Having discussed the systems view of life in its biological dimension (Chapters 7–11), its cognitive dimension (Chapters 12–13), and its social dimension (Chapter 14), and having proposed a conceptual framework that integrates these three dimensions (Section 14.3), we shall now, in the last part of our book, include the ecological dimension in our synthesis of the systemic conception of life.

In fact, we could have (and, perhaps, should have) begun this discussion with the ecological dimension, since it is well known that no individual organism can exist in isolation. Animals depend on the photosynthesis of plants for their energy needs; plants depend on the CO_2 produced by animals, as well as on the nitrogen fixed by microorganisms at their roots; and together plants, animals, and microorganisms regulate the entire biosphere and maintain the conditions conducive to life. As Harold Morowitz (1992, p. 54) reminds us,

> Sustained life is a property of an ecological system rather than a single organism or species. Traditional biology has tended to concentrate attention on individual organisms rather than on the biological continuum. The origin of life is thus looked for as a unique event in which an organism arises from the surrounding milieu. A more ecologically balanced point of view would examine the proto-ecological cycles and subsequent chemical systems that must have developed and flourished while objects resembling organisms appeared.

Indeed, according to Gaia theory (discussed in Section 8.3.3), the evolution of the first living organisms went hand in hand with the transformation of the planetary surface from an inorganic environment to the self-regulating biosphere. "In that sense," writes Morowitz (1992, p. 6), "life is a property of planets rather than of individual organisms."

On the other hand, ecology is a relatively new science, compared with the traditional disciplines of biology, psychology, and sociology. This is why we have chosen to introduce the ecological perspective after discussing the biological, cognitive, and social dimensions of life, following more or less the historical sequence in which the corresponding disciplines developed.

16.1 The science of ecology

Ecology, from the Greek *oikos* ("household"), is the study of the "Earth Household." More precisely, it is the scientific study of the relationships between the members of the Earth

Household – plants, animals, and microorganisms – and their natural environment, living and nonliving. The basic ecological unit is the ecosystem, defined as a community of different species in a particular area, interacting with its nonliving, or abiotic, environment (air, minerals, water, sunlight, etc.) and with its living, or biotic, environment (i.e., with other members of the community). The ecosystem, then, consists of a biotic community and its physical environment.

It is evident from these definitions that the science of ecology is inherently multidisciplinary. Since ecosystems interlink the living with the nonliving world, ecology must be grounded not only in biology, but also in geology, atmospheric chemistry, thermodynamics, and other branches of science. And when it comes to assessing the impacts of human activities on the biosphere, which is becoming ever more urgent, we have to add a whole new range of fields to ecology, including agriculture, economics, industrial design, and politics.

The importance of studying, within the general framework of ecology, the pervasive influence of human activities on ecosystems, as well as the reciprocal influence of their deterioration on human health and well-being, also makes it clear that ecology, today, is not only a rich and fascinating area of study but is also highly relevant to assessing, and hopefully influencing, the future fate of humanity. One of the great challenges of our time is to build and nurture sustainable communities, and to do so we can learn many lessons from ecosystems, because ecosystems are, in fact, communities of plants, animals, and microorganisms that have sustained life for billions of years. We shall come back to this critical role of ecology later on in this chapter (in Section 16.3).

16.1.1 Development of basic ecological concepts

The inherently multidisciplinary nature of ecology suggests that the systems view of life, which also is multidisciplinary, could provide an ideal framework and language for ecological studies. Indeed, as we showed in our historical review of the emergence of systems thinking (Section 4.1), the development of ecology and that of systemic thinking were closely connected from the very beginning. Ecology originated in the naturalist tradition of the late nineteenth century, and its first key concepts were refined in the 1920s and 1930s in a series of interdisciplinary dialogues with organismic biologists, Gestalt psychologists, and other early systems thinkers.

As we discussed in Section 4.1, the concepts formulated and explored in the foundational years of ecology included those of the biosphere, the ecosystem, the natural environment, and the ecological community. From the beginning of ecology, ecological communities were conceived as networks of organisms interlinked through feeding relations. This idea led not only to the concepts of food chains and food cycles, later to be expanded to the contemporary concept of food webs, but, eventually, also to the momentous recognition of the network as the basic pattern of organization of all living systems.

When the English zoologist Charles Elton (1927) introduced the concepts of food chains and food cycles in his pioneering book *Animal Ecology*, he brought a dynamic perspective into ecology that would become the principal theoretical framework for subsequent generations of ecologists – the flow of energy and matter from one organism to another. Elton also introduced the concept of an ecological niche, defining it as the role that an animal plays in a community in terms of what it eats and is eaten by; and he was one of the first ecologists to study populations – groups of individuals of the same species living in a particular area, or habitat, at a given time.

Evelyn Hutchinson, an American limnologist (fresh-water scientist) who was strongly influenced by Elton's ideas, developed niche theory further and made it popular among ecologists. Another limnologist and student of Hutchinson, Raymond Lindeman, coined the term "trophic" – from the Greek *trophe* ("food") – to discuss feeding relations. Lindeman introduced the notion of "trophic level" to describe the position an organism occupies in a food chain. The first trophic level is that of green plants, or primary producers. In the process of photosynthesis, plants convert solar energy into chemical energy and bind it in organic substances, while oxygen is released into the air to be taken up again by other plants, and by animals, in the process of respiration.

The second trophic level is that of herbivores, or primary consumers, who eat plants and pass on the energy of their organic matter up the food chain when they are eaten in turn by carnivores, the secondary consumers. Further levels of carnivores and omnivores (animals eating both plants and other animals) constitute successive trophic levels. Thus, nutrients are passed on through the food chain, while some energy is dissipated as heat through respiration and as waste through excretion. However, at each level the so-called decomposers (bacteria, fungi, insects, and others) break down dead or decaying organisms into basic nutrients, to be taken up once more by green plants. In doing so, the decomposers turn the food chain into a food cycle (see Figure 16.1), in which nutrients and other basic elements continually cycle through the ecosystem, while energy is dissipated at each stage. The only waste generated by the ecosystem as a whole is the heat energy of respiration, which is radiated into the atmosphere and is replenished continually from the sun through photosynthesis.

In addition to the dynamics of ecological cycles and flows, ecologists from the very beginning studied the directional structural changes of an ecosystem as a whole, known as ecological succession. One of the first to do so was the botanist Henry Cowles (1899), who showed in a pioneering study of sand dunes on Lake Michigan how these dunes are initially colonized by drought-resistant plants that give rise to shrubs, to be succeeded by small trees, and eventually by forests.

A decade later, the plant ecologist Frederic Clements (1916) proposed a theory of succession in which predictable patterns of vegetation development eventually lead to a stable "climax community," characteristic of a particular climate. Clements suggested that such ecological succession toward a climax community is analogous to the development of an organism to adulthood. This view was challenged as early as the 1920s by Henry Gleason (1926), who offered a more complex and less deterministic theory of succession in which chance factors play a much greater role. Over the subsequent decades, this debate led to extensive research on the dynamics of vegetation change in terrestrial ecosystems (see Chapin *et al.*, 2002).

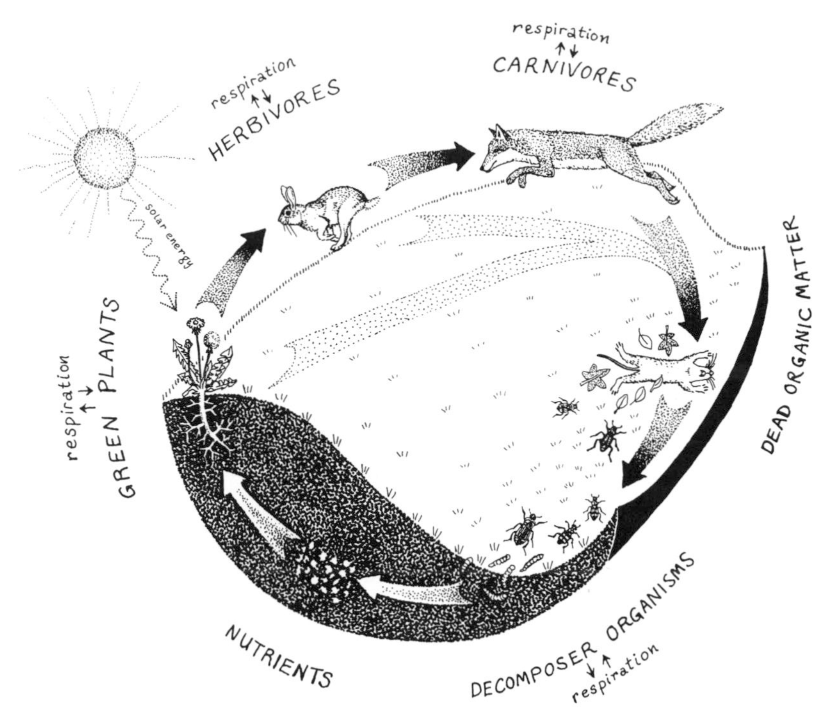

Figure 16.1 A typical food cycle (from Capra, 1996).

Ecosystems come in all sizes. They may be as small as a rotting log or as large as an ocean. A basic principle of ecology is the recognition that ecosystems, like all living systems, form multilevel structures of systems nesting within other systems. At the largest levels, regional communities of plants and animals that extend over millions of square kilometers are known as biomes. Among terrestrial ecosystems, ecologists have identified eight major biomes: tropical, temperate, and conifer forests; tropical savanna, temperate grassland, chaparral (shrubland), tundra, and desert. And finally, the largest ecological unit is the biosphere, the global sum of all ecosystems. According to Gaia theory (Section 8.3.3), the biosphere is tightly coupled to the Earth's rocks (lithosphere), oceans (hydrosphere), and atmosphere in such a way that together they form a self-regulating planetary system.

In the 1950s, the American ecologist Eugene Odum (1953) wrote the first ecology textbook, *Fundamentals of Ecology*, in which the basic principles and concepts of ecology were discussed for the first time in a clear and systematic exposition. Odum's text influenced a whole generation of ecologists. In addition to the lucid and detailed discussions of fundamental ecological concepts, the book also provides an overview of the main branches of ecology according to the habitats studied: fresh-water ecology, marine ecology, and terrestrial ecology.

Odum's textbook was coauthored in part with his brother, Howard Odum, and together the Odum brothers also published several studies of ecosystems in which they documented the movement of energy and materials through the system in a series of flow diagrams. Such "Odum flow diagrams" have since become standard practice in the analysis of ecosystems.

16.1.2 Branches of ecology

As ecologists developed and refined the basic concepts discussed in the previous section, they organized the various branches of their science accordingly. Thus, population ecology is concerned with the structure, spatial distribution, growth, and migrations of animal and plant populations; evolutionary ecology is the study of the natural selection and evolution of populations; and community ecology is concerned with species interactions, focusing especially on understanding the nature and consequences of biodiversity within ecological communities.

Recent advances in science and technology (e.g., aerial photography and satellite imagery), as well as increasing concern about human impacts on the environment, gave rise to several new branches of ecology. These include conservation ecology, concerned with the maintenance of biological diversity; human ecology, which studies the wide range of relationships between humans and the natural environment; and global ecology, concerned with ecological phenomena on a global scale.

Within our conceptual framework of the systems view of life, we shall discuss two areas of ecology in some detail. The first is ecosystem ecology, also known as systems ecology – the theoretical study of the structure and dynamics of ecosystems. The second is human ecology, which includes critical issues like sustainability and the global manifestations of climate change.

16.2 Systems ecology

Systems ecology, or ecosystem ecology, is concerned with the ecosystem as an integrated and interactive system of biological and physical components. Hence, this branch of ecology should reflect most explicitly the systems view of life we have been discussing throughout this book. In this section, we shall examine to what extent this is actually the case. To do so, we list below the basic systemic characteristics of biological life, as discussed in Chapters 7 and 8, so as to determine whether ecosystems satisfy those characteristics.

Characteristics of biological life

(1) A living system is materially and energetically open; it is a dissipative structure, operating far from equilibrium. There is a continual flow of energy and matter through the system.
(2) It is self-organizing, its structure being organized by the system's own internal rules.
(3) Its dynamics are nonlinear and may include the emergence of new order at critical points of instability.
(4) It is operationally closed – an autopoietic, bounded network.

(5) It is self-generating; each component helps to transform and replace other components, including those of its semipermeable boundary.
(6) Its interactions with the environment are cognitive – that is, determined by its own internal organization.

16.2.1 Ecosystems as dissipative structures

When we look at the literature of systems ecology, we can easily see that ecologists, so far, have concentrated almost exclusively on the first three of these characteristics. Ever since the pioneering studies of Howard Odum, the flows of energy and matter through the ecosystem – from the Sun, the atmosphere, and the soil to primary producers, consumers, and decomposers – have been the main subject of systems ecology. Ecosystem energetics are analyzed within the framework of thermodynamics in terms of nutrient pools and fluxes; food chains, food webs, and trophic levels; and decomposition and nutrient cycling (see, e.g., Chapin *et al.*, 2002; Smith and Smith, 2006).

In these analyses, ecologists realized early on that an ecosystem is materially and energetically open, its main energy source being the Sun, and its net waste the heat energy of respiration, radiated into the atmosphere (see Figure 16.1). In addition, the nonlinear nature of all trophic dynamics was appreciated with the recognition of the food web as a basic pattern of organization in ecosystems. Accordingly, systems ecologists soon began to pay attention to the phenomenon of self-organization.

As the molecular biologist and historian of science Evelyn Fox Keller (2005) points out, it is helpful to distinguish between different meanings of the term "self-organization" during different periods of modern science when discussing its application to ecosystems. To the cyberneticists in the 1940s, self-organization meant the spontaneous emergence of order in machines featuring feedback loops, which were conceived as models of living systems (see Section 5.3.5). This idea was used just a few years after its inception to analyze self-regulating and self-balancing mechanisms in ecosystems in a classic paper by Evelyn Hutchinson (1948), titled "Circular Causal Systems in Ecology." A few decades later, the cybernetic concept of self-organization was explored extensively in the flow diagrams introduced by Howard Odum.

With the advent of complexity theory in the 1980s, the original cybernetic conception of self-organization changed. Self-organization now came to mean the spontaneous emergence of new order in complex systems governed by nonlinear dynamics (see Section 8.3). However, as we have emphasized, feedback (both self-balancing and self-amplifying) is still an important feature of these processes of emergence in dynamic systems.

In this more recent conception of self-organization, ecosystems are understood as dissipative structures operating far from equilibrium, and the forms of new order are represented mathematically by attractors emerging at bifurcation points. Due to its complexity and mathematical sophistication, this conceptual framework is much more challenging, and ecologists have begun only recently to use it in their analyses of ecosystems (see, e.g., Kay, 2000).

16.2.2 Are ecosystems autopoietic?

Nevertheless, our first three characteristics of biological life have all been applied successfully by systems ecologists to studies of ecosystem structures and processes. Unfortunately, the same cannot be said about the other three characteristics relating to autopoiesis. In fact, the concept of autopoiesis is glaringly absent from the literature on systems ecology. For example, in a multiauthor book, titled *Theoretical Studies of Ecosystems: The Network Perspective* (Higashi and Burns, 1991), the editors state in their preface: "The diversity of viewpoints and approaches for theoretical study of ecosystems that are collected here share the perception of ecosystems as networks, and represent the state-of-the-art." But the question of whether these networks are self-generating, or autopoietic, is not discussed by any of the book's authors. Nor is it addressed in another multiauthor book, *Handbook of Ecosystem Theories and Management* (Jørgensen and Müller, 2000), published a decade later. Perhaps this should not be too surprising, since the theory of autopoiesis is still not broadly accepted in mainstream biology, even though it has been embraced enthusiastically in many biological "niches."

As we have mentioned, Maturana and Varela originally proposed that the concept of autopoiesis should be restricted to the description of cellular networks, and that the broader concept of "operational closure," which does not specify production processes, should be applied to all other living systems (see Section 14.3.3). Other authors distinguish between first-order (unicellular) and second-order (multicellular) autopoiesis, and in our discussion we suggested that biological life might be seen as a system of interlocked autopoietic systems (Section 7.3). We have also reviewed the very lively debate on autopoiesis in social systems (Section 14.3.3), which is in stark contrast to the almost total silence on the question of autopoiesis in ecosystems.

It may well be that the pathways and processes in ecological networks are not yet known in sufficient detail to decide whether these networks can be described as autopoietic. However, it would certainly be as interesting to engage in discussions on autopoiesis with ecologists as it has been with social scientists.

To begin with, we can say that a function of all components in a food web is to transform other components within the same network. As plants take up inorganic matter from their environment to produce organic compounds, and as these compounds are passed on through the ecosystem to serve as food for the production of more complex structures, the entire network regulates itself through multiple feedback loops. Individual components of the food web continually die, to be decomposed and replaced by the network's own processes of transformation. Whether this is sufficient to define an ecosystem as autopoietic remains to be seen and will depend, among other things, on a clear understanding of the system's boundary.

The defining feature of an autopoietic system is that it continually recreates itself within a boundary of its own making (see Section 7.1). In a cell, for example, the membrane surrounding the cell is continually regenerated and maintained by internal cellular processes, and it contributes to these processes, in turn, by regulating the flow of nutrients from the cell's environment.

In ecosystems, the situation is less clear-cut. To begin with, there are several different boundaries: the atmosphere, the soil, the boundary between a small ecosystem nesting inside a larger one, and the boundaries between large-scale ecosystems or patches, as they are called in landscape ecology. How these different boundaries influence the functioning of the corresponding ecosystems, and in particular how they affect the flows of materials through them, is still poorly understood (see Cardenasso *et al.*, 2003).

According to Gaia theory, the atmosphere is tightly coupled to life on Earth, its gases being continually removed and replenished by living organisms. On the other hand, the atmosphere may be seen as "semipermeable," not unlike a cellular membrane, since it lets through certain frequencies of sunlight but absorbs others, while also protecting the biosphere from high-energy cosmic rays (see Capra, 1975). Hence, it may be argued that the atmosphere constitutes a boundary in the sense of autopoiesis. However, whether this notion can be applied to a particular ecosystem and the portion of atmosphere above it seems debatable; and whether a similar argument can be made for the soil between a terrestrial ecosystem and the Earth's crust is even less evident.

As far as the boundaries between patches in a landscape are concerned, the situation here is quite different from the membranes surrounding cells. In a multicellular organism each cell has its own membrane, and the cells are interlinked by so-called protein channels through which they exchange chemical and electrical signals. Adjacent ecosystems, by contrast, share a single boundary, which may have some characteristics in common with one or the other patch, or may be completely distinct. These boundaries may be wide or narrow, depending on the gradients of changing characteristics from one patch to the other (Cadenasso *et al.*, 2003).

For example, if a meadow adjacent to a forest is mown close to the trees, the boundary may be very narrow. But if parts of the meadow near the trees are left unmown, the transition between grasses and trees may be more gradual, and thus the boundary much wider, with shrubs and young trees growing at the interface. In both cases the boundary will influence the flows of materials, organisms, and energy between the two patches. However, it is far from clear how this regulation of flows can be associated with either one or the other ecosystem in the sense of autopoiesis.

We conclude from these considerations that the question of whether, and how exactly, the concept of autopoiesis applies to ecosystems, is still wide open and well worth in-depth discussions within the conceptual framework of systems ecology.

16.2.3 Autopoiesis and the Gaia system

When we shift our perception from ecosystems to the planet as a whole, we encounter a global network of processes of production and transformation, which has been described in some detail in the Gaia theory of James Lovelock and Lynn Margulis (see Section 8.3.3).

In fact, there seems to be more evidence for the autopoietic nature of the Gaia system than for ecosystems.

The Earth system operates on a very large scale in space and also involves very long time scales. It is thus not so easy to think of Gaia as being alive in a concrete manner. Is the whole planet alive, or just certain parts; and if the latter, which parts? To help us picture Gaia as a living system, Lovelock (1991) has suggested a redwood tree as an analogy. As the tree grows, there is only a thin layer of living cells (known as the cambium) around its perimeter, just beneath the bark. All the wood inside, more than 97% of the tree, is dead. Similarly, the Earth is covered with a thin layer of living organisms – the biosphere – reaching down into the ocean about 8–9.6 km (5–6 miles) and up into the atmosphere about the same distance. Thus the living part of Gaia is but a thin film around the globe.

Just as the bark of a tree protects the tree's thin layer of living tissue from damage, life on Earth is surrounded by the protective layer of the atmosphere, which shields us from ultraviolet light and other harmful influences and keeps the planet's temperature within a range appropriate for life to flourish. Neither the atmosphere above us nor the rocks below us are alive, but both have been shaped and transformed considerably by living organisms, just like the bark and the wood of the tree. Outer space and the Earth's interior are both part of Gaia's environment.

According to Gaia theory, the Earth's atmosphere is created, transformed, and maintained by the biosphere's metabolic processes. Bacteria play a crucial role in these processes, influencing the rate of chemical reactions and thus acting as the biological equivalent of enzymes in a cell. As we have mentioned, the atmosphere is semipermeable, like a cell membrane, and forms an integral part of the planetary network. For example, it created the protective greenhouse in which early life on the planet was able to unfold 3 billion years ago, even though the Sun was then 25% less luminous than it is now.

The Gaia system is clearly self-generating. The planetary metabolism converts inorganic substances into organic, living matter and back into soil, oceans, and air. All components of the Gaian network, including those of its atmospheric boundary, are produced by processes within the network. A key characteristic of Gaia is the complex interweaving of living and nonliving systems within a single web. This results in feedback loops, known to ecologists as biogeochemical cycles, of vastly differing scales. They may extend over hundreds of millions of years, while the organisms associated with them have very short life spans.

The CO_2 cycle is an impressive illustration of such a giant feedback loop (see Harding, 2009; Lovelock, 1991). The Earth's volcanoes have spewed out huge amounts of CO_2 for millions of years. Since CO_2 is one of the main greenhouse gases, Gaia needs to pump it out of the atmosphere, which otherwise would get too hot for life. Plants and animals recycle massive amounts of CO_2 and oxygen in the processes of photosynthesis, respiration, and decay. However, these exchanges are always in balance and do not affect the level of CO_2 in the atmosphere. According to Gaia theory, the excess of CO_2 in the atmosphere is removed and recycled by a vast feedback loop, which involves rock weathering as a key ingredient.

In the process of rock weathering, silicate rocks (granite and basalt) combine with rainwater and CO_2 to form various chemicals known as carbonates. The CO_2 is thus taken

Figure 16.2 Oceanic alga (coccolithophore) with chalk shell (from Capra, 1996).

out of the atmosphere and bound in liquid solutions. These are purely chemical processes that do not require the participation of life. However, Lovelock (1991) and others discovered that the presence of soil bacteria, fungi, lichens, and plants vastly increases the rate of rock weathering. In a sense, these organisms act as biological catalysts for the process of rock weathering.

The carbonates are then washed down into the ocean, where tiny algae, invisible to the naked eye, absorb them and use them to make exquisite shells of chalk (calcium carbonate). So the CO_2 that was in the atmosphere has now ended up in the shells of those minute algae (Figure 16.2). In addition, ocean algae also absorb CO_2 directly from the air.

When the algae die, their shells rain down to the ocean floor where they form massive sediments of chalk and limestone (both forms of calcium carbonate). Thanks to plate tectonics, these sediments gradually sink into the mantle of the Earth and melt, whereupon some of the CO_2 contained in the molten rocks is spewed out again by volcanoes and sent on another round in the great Gaian cycle.

The entire cycle – linking volcanoes to silicate rock weathering, to soil bacteria, to oceanic algae, to limestone sediments, and back to volcanoes – acts as a giant feedback loop, which contributes to the regulation of the Earth's temperature. As the Sun gets hotter, the growth of organisms on the rocks and in the soil is stimulated, which increases the rate of rock weathering. This in turn pumps more CO_2 out of the atmosphere and thus cools the planet. According to Lovelock and Margulis, similar feedback cycles – interlinking plants and rocks, animals and atmospheric gases, and microorganisms and the oceans – regulate the Earth's climate, the amount of oxygen in the atmosphere, and other important planetary conditions.

In the Gaia system, the components of the oceans, soil, and air, as well as all the organisms of the biosphere, are continually replaced by the planetary processes of production and transformation. It seems, therefore, that the case for Gaia being an autopoietic network is very strong. Indeed, Lynn Margulis, coauthor of the Gaia theory, asserted confidently: "There is little doubt that the planetary patina – including ourselves – is autopoietic" (Margulis and Sagan, 1986, p. 66).

The confidence of Margulis in the idea of a planetary autopoietic web stemmed from three decades of pioneering work in microbiology. To understand the complexity, diversity, and self-organizing capabilities of the Gaian network, an understanding of the microcosm – the nature, extension, metabolism, and evolution of microorganisms – is absolutely essential. Life on Earth began around 3.5 billion years ago, and for the first 2 billion years the living world consisted entirely of unicellular microorganisms (see Chapter 11). During the first billion years of evolution, bacteria – the most basic forms of life – covered the planet with an intricate web of metabolic processes and began to regulate the temperature and chemical composition of the atmosphere so that it became conducive to the evolution of higher forms of life.

Plants, animals, and humans are latecomers to the Earth, having emerged from the microcosm less than 1 billion years ago. And even today the visible living organisms function only because of their well-developed connections with the bacterial web of life. "Far from leaving microorganisms behind on an evolutionary 'ladder'," wrote Margulis and Sagan (1986, pp. 14, 21), "we are both surrounded by them and composed of them . . . [We have to] think of ourselves and our environment as an evolutionary mosaic of microcosmic life."

During life's long evolutionary history, over 99% of all species that ever existed have become extinct, but the planetary web of bacteria has survived, continuing to regulate the conditions for life on Earth as it has for the past 3 billion years. According to Margulis, the concept of a planetary autopoietic network is justified because all life is embedded in a self-organizing web of bacteria, involving elaborate networks of sensory and control systems which we are only beginning to recognize. Myriads of bacteria, living in the soil, the rocks, and the oceans, as well as inside all plants, animals, and humans, continually regulate life on Earth: "It is the growth, metabolism, and gas-exchanging properties of microbes . . . that form the complex physical and chemical feedback systems which modulate the biosphere in which we live" (Margulis and Sagan, 1986, p. 271). And since the systems view of life defines these self-organizing processes ultimately as cognitive, the microbial web of life must be considered a cognitive system.

16.3 Ecological sustainability

In the unfolding of life on Earth during the past 3.8 billion years, the emergence and extinction of species has not been a steady process but has produced several dramatic fluctuations – so-called mass extinctions in which more than 50% of animal species perished (see G.T. Miller, 2007). From fossil and geological evidence, paleontologists have identified five such mass extinctions during the past 500 million years (i.e., since the evolution of plants and land animals; see Figure 11.1), including the extinction of dinosaurs 65 million years ago.

Estimates of current extinction rates, due to deforestation and the destruction of other habitats, indicate that the Earth is now in the midst of a sixth mass extinction (Leaky and Lewin, 1995). The current extinction event, however, is unique both in its magnitude and its cause. Whereas all previous extinctions were caused by natural physical phenomena – volcanic explosions, glaciations, or the impact of an asteroid – the current mass extinction is caused, for the first time, by the activities of a single species: *Homo sapiens*. As a recent report of the Royal Society of London (Magurran and Dornelas, 2010) noted,

> There are very strong indications that the current rate of species extinctions far exceeds anything in the fossil record . . . Never before has a single species driven such profound changes to the habitats, composition and climate of the planet.

Because of this dire situation, which threatens the very survival of humanity (as we shall discuss in more detail in Chapter 17), the issue of sustaining life on Earth has moved to center stage in recent years. Concern with the environment is no longer one of many "single issues." It is the context of everything else – our lives, our businesses, our politics. The great challenge of our time is to build and nurture sustainable communities and societies. Hence, to convey a clear understanding of sustainability has become a critical role of ecology (see G.T. Miller, 2007; Steffen *et al.*, 2004).

16.3.1 Defining sustainability

The concept of sustainability was introduced in the early 1980s by Lester Brown, founder of the Worldwatch Institute (see Section 17.4) and one of the most authoritative environmental thinkers (Brown, 1981). A few years later, Brown, Flavin, and Postel defined a sustainable society as one that "satisfies its needs without jeopardizing the prospects of future generations" (Brown *et al.*, 1990). Around the same time, the report of the World Commission on Environment and Development (1987), known also as the "Brundtland Report," presented the notion of "sustainable development":

> Humankind has the ability to achieve sustainable development – to meet the needs of the present without compromising the ability of future generations to meet their own needs.

These definitions of sustainability are important moral exhortations. They remind us of our responsibility to pass on to our children and grandchildren a world with as many opportunities as the ones we inherited. However, they do not tell us anything about how to build a sustainable society. This is why there has been much confusion about the meaning of sustainability, even within the environmental movement. We should also note here that the notion of "sustainable development" is rather problematic, as we discuss in Section 17.2.2.

The key to an operational definition of ecological sustainability is the realization that we do not need to invent sustainable human communities from scratch but can model them after nature's ecosystems, which *are* sustainable communities of plants, animals, and microorganisms. Since the outstanding characteristic of the "Earth Household" is its inherent ability to sustain life, a sustainable human community is designed in such a manner that its ways of life, businesses, economy, physical structures, and technologies *do not interfere with nature's inherent ability to sustain life* (Capra, 2002). Sustainable communities evolve their patterns of living over time in continual interaction with other living systems, both human and nonhuman. Hence, sustainability does not mean that things do not change. It is a dynamic process of coevolution rather than a static state.

16.3.2 Ecological literacy

Our operational definition of sustainability – to design a human community in such a way that its activities do not interfere with nature's inherent ability to sustain life – implies that the first step in this endeavor must be to understand how nature sustains life. In other words, we need to understand the principles of organization that ecosystems have evolved to sustain the web of life. In recent years, this understanding has become known as ecological literacy, or "ecoliteracy" (Capra, 1993, 1996; Orr, 1992). Being ecoliterate means understanding the basic principles of ecology, or principles of sustainability, and living accordingly.

The systems view of life discussed in this book provides an appropriate framework for the conceptual link between ecological and human communities. Both are living systems exhibiting common principles of organization. They are networks that are operationally closed but open to continual flows of energy and resources; they are self-organizing, operate far from equilibrium, and evolve by means of their inherent creativity, resulting in the emergence of new structures and new forms of order.

Of course, there are many differences between ecosystems and human communities. There is no self-awareness in ecosystems, no language, no consciousness, and no culture (as we discussed in Chapter 12), and therefore no justice or democracy; but also no greed or dishonesty. We cannot learn anything about those human values and shortcomings from ecosystems. But what we *can* learn and must learn from them is how to live sustainably. During more than 3 billion years of evolution, the planet's ecosystems have organized themselves in subtle and complex ways so as to maximize their sustainability. This wisdom of nature is the essence of ecoliteracy.

Based on the systemic understanding of ecosystems outlined in Section 16.2, we can formulate a set of principles of organization that may be identified as basic principles of ecology, and use them as guidelines to build sustainable human communities.

The first of those principles is interdependence. All members of an ecological community are interconnected in a vast and intricate network of relationships, the web of life. They derive their essential properties and, in fact, their very existence from their relationships to other things. Interdependence – the mutual dependence of all life processes on one another – is the nature of all ecological relationships. The behavior of every living member of the ecosystem depends on the behavior of many others. The success of the whole community depends on the success of its individual members, while the success of each member depends on the success of the community as a whole.

Understanding ecological interdependence means understanding relationships. It requires the shifts of perception that are characteristic of systems thinking – from the parts to the whole, from objects to relationships, from quantities to qualities (see Section 4.3). A sustainable human community is aware of the multiple relationships among its members, as well as of the relationships between the community as a whole and its natural and social environment. Nourishing the community means nourishing all these relationships.

The fact that the basic pattern of life is a network means that the relationships among the members of an ecological community are nonlinear, involving multiple feedback loops. Linear chains of cause and effect exist very rarely in ecosystems. Thus a disturbance will not be limited to a single effect but is likely to spread out in ever-widening patterns. It may even be amplified by interdependent feedback loops, which may completely obscure the original source of the disturbance.

The cyclical nature of ecological processes is an important principle of ecology. The ecosystem's feedback loops are the pathways along which nutrients are continually recycled. Being open systems, all organisms in an ecosystem produce wastes, but what is waste for one species is food for another, so that the ecosystem as a whole remains without solid waste. Communities of organisms have evolved in this way over billions of years, continually using and recycling the same molecules of minerals, water, and air. The lesson for human communities here is obvious. A major clash between economics and ecology derives from the fact that nature is cyclical, whereas our industrial systems are linear. Our businesses take resources, transform them into products plus waste, and sell the products to consumers, who discard more waste when they have consumed the products. Sustainable patterns of production and consumption need to be cyclical, imitating the cyclical processes in nature. To achieve such cyclical patterns, we need to fundamentally redesign our businesses and our economy, as we shall discuss in more detail in Chapter 18.

Solar energy, transformed into chemical energy by the photosynthesis of green plants, is the primary source of energy driving the ecological cycles. The implications for maintaining sustainable human communities are again obvious. Solar energy in its many forms – sunlight for solar heating and photovoltaic electricity, wind and hydropower, biomass, etc. – is the only kind of energy that is renewable, economically efficient, and environmentally benign. By disregarding this ecological fact, our political and corporate leaders again and again endanger the health and well-being of millions around the world.

To describe solar energy as economically efficient assumes that the costs of energy production are counted honestly. This is not the case in most of today's market economies. The "free market" does not provide consumers with proper information, because the social and environmental costs of production are not part of current economic models (see Section 3.5). Corporate economists treat not only the air, water, and soil as free commodities but also the delicate web of social relations, which is severely affected by continuing economic expansion. Private profits are being made at public costs in the deterioration of the environment and the general quality of life, and at the expense of future generations. The marketplace simply gives us the wrong information. There is a lack of feedback, and basic ecological literacy tells us that such a system is not sustainable.

Partnership is an essential characteristic of sustainable communities. The cyclical exchanges of energy and resources in an ecosystem are sustained by pervasive cooperation. Indeed, ever since the creation of the first nucleated cells over 2 billion years ago, life on Earth has proceeded through ever more intricate arrangements of cooperation and coevolution (see Section 9.6.3). Partnership – the tendency to associate, establish links, live inside one another, and cooperate – is one of the hallmarks of life. In the memorable words of Margulis and Sagan (1986, p. 15): "Life did not take over the globe by combat, but by networking."

Here again we notice the basic tension between the challenge of ecological sustainability and the way in which our present societies are structured, the tension between economics and ecology. Economics emphasizes competition, expansion, and domination; ecology emphasizes cooperation, conservation, and partnership.

All the principles of ecology mentioned so far are closely interrelated. They are just different aspects of a single fundamental pattern of organization that has enabled nature to sustain life for billions of years. In a nutshell, nature sustains life by creating and nurturing communities. Sustainability is not an individual property but a property of an entire web of relationships. It always involves a whole community. This is the profound lesson we need to learn from nature. The way to sustain life is to build and nurture community. A sustainable human community interacts with other communities – human and nonhuman – in ways that enable them to live and develop according to their nature.

Once we have understood the basic pattern of organization that ecosystems have evolved to sustain themselves over time, we can ask more detailed questions. For example, what is the resilience of these ecological communities? How do they react to outside disturbances? These questions lead us to two further principles of ecology – flexibility and diversity – which enable ecosystems to survive disturbances and adapt to changing conditions.

The flexibility of an ecosystem is a consequence of its multiple feedback loops, which tend to bring the system back into balance whenever there is a deviation from the norm due to changing environmental conditions. For example, if an unusually warm summer results in increased growth of algae in a lake, some species of fish feeding on these algae may flourish and breed more, so that their numbers increase and they begin to deplete the algae. Once their major source of food is reduced, the fish will begin to die out. As the fish population drops, the algae will recover and expand again. In this way, the original disturbance generates a fluctuation around a feedback loop, which eventually brings the fish/algae system back into balance.

Disturbances of that kind happen all the time, because things in the environment change all the time, and thus the net effect is continual fluctuation. The variables we observe in an ecosystem – population densities, availability of nutrients, weather patterns, etc. – always fluctuate. This is how ecosystems maintain themselves in a flexible state, ready to adapt to changing conditions. The web of life is a flexible, ever-fluctuating network. The more variables are kept fluctuating, the more dynamic is the system, the greater is its flexibility, and the greater is its ability to adapt to changing conditions. As we discussed in our previous chapter, loss of flexibility always means loss of health.

All ecological fluctuations take place between tolerance limits. There is always the danger that the whole system will collapse when a fluctuation goes beyond those limits and the system can no longer compensate for it. The same is true of human communities. Lack of flexibility manifests itself as stress (see Section 15.2.2). In particular, stress will occur when one or more variables of the system are pushed to their extreme values, which induces increased rigidity throughout the system. Temporary stress is an essential aspect of life, but prolonged stress is harmful and destructive to the system. These considerations lead to the important realization that managing a social system – a company, a city, or an economy – means finding the *optimal* values for the system's variables. If one tries to maximize any single variable instead of optimizing it, this will invariably damage the system as a whole.

The diversity of an ecosystem is closely connected to the system's network structure. A diverse ecosystem will be resilient, because it contains many species with overlapping ecological functions that can partially replace one another. When a particular species is destroyed by a severe disturbance so that a link in the network is broken, a diverse community will be able to survive and reorganize itself, because other links in the network can at least partially fulfill the function of the destroyed species. In other words, the more complex the network is, the richer is its pattern of interconnections, and the more resilient it will be; and since the complexity of the network is a consequence of its biodiversity, a diverse ecological community is resilient.

In human communities, ethnic and cultural diversity may play the same role. Diversity means many different relationships, many different approaches to the same problem. A diverse community is a resilient community, capable of adapting to changing situations. However, diversity is a strategic advantage only if there is a truly interconnected community, sustained by a web of relationships. If the community is fragmented into isolated groups and individuals, diversity can easily become a source of prejudice and friction. But if the community is aware of the interdependence of all its members, diversity will enrich all the relationships and thus enrich the community as a whole, as well as each individual member.

16.3.3 Education for sustainable living

In the coming decades the survival of humanity will depend on our ecological literacy – our ability to understand the basic principles of ecology and to live accordingly. This means that ecoliteracy must become a critical skill for politicians, business leaders, and professionals in all spheres, and should be the most important part of education at all levels – from primary and secondary schools to colleges, universities, and the continuing education and training of professionals.

We need to teach our children, our students, and our corporate and political leaders, the fundamental facts of life – that one species' waste is another species' food; that matter cycles continually through the web of life; that the energy driving the ecological cycles flows from the sun; that diversity assures resilience; that life, from its beginning more than 3 billion years ago, did not take over the planet by combat but by networking.

This basic ecological knowledge, which is also ancient wisdom, is now being taught increasingly in schools, universities, and various centers of learning. In a previous chapter we profiled two such institutions – the Cortona Week and Schumacher College – where knowledge is pursued in an interdisciplinary way, and where the ecological and spiritual dimensions of education are explicitly emphasized (Section 13.6). In the following chapter we shall mention several other centers of learning, all of them part of the new global civil society, which share a systemic and ecological perspective. In this section we want to concentrate specifically on how ecoliteracy is being taught today in schools, colleges, and universities.

Schooling for sustainability

At the Center for Ecoliteracy (CEL) (www.ecoliteracy.org) in Berkeley, California, which was cofounded by one of us (F.C.), scientists and educators have developed a special pedagogy for teaching ecological literacy in primary and secondary schools (Stone, 2009; Stone and Barlow, 2005). Called "schooling for sustainability," it is a pedagogy to teach the basic principles of ecology and the skills that are necessary to build and nurture sustainable communities. Schooling for sustainability offers a systemic, participatory, and experiential approach. Since its inception almost twenty years ago, CEL has worked with schools in hundreds of cities on six continents to implement its pedagogy.

The systemic nature of ecoliteracy derives from the fact that ecology itself is essentially a science of relationships and, moreover, is inherently multidisciplinary, as we discussed at the outset of this chapter. Teaching ecology, therefore, requires a conceptual framework that is quite different from that of the conventional academic disciplines, and teachers notice that at all levels of teaching, from very small children to university students. From the beginning, the focus of the ecoliteracy curriculum is on relationships, patterns, and context.

Mapping relationships and studying patterns involves visualizing. This is the reason why, throughout our intellectual history, artists have contributed significantly to the advancement of science whenever the study of patterns was in the forefront. The two most famous examples, perhaps, are Leonardo da Vinci, the great genius of the Renaissance, whose whole scientific life was an exploration of patterns, and the German poet Goethe in the eighteenth century, who made significant contributions to biology through his study of patterns (see Introduction). For educators, this opens the door for integrating the arts into the school curriculum. There is hardly anything more effective than the arts – be they the visual arts or music and the other performing arts – for developing and refining a child's natural ability to recognize and express patterns. Thus, the arts can be a powerful tool for teaching systemic thinking.

In addition, the arts enhance the emotional dimension that is increasingly being recognized as an essential component of the learning process. Indeed, artistic expression is used extensively, especially with younger children, in the schools that have adopted CEL's pedagogy. One particular form is a poetry contest, held annually by the organization River of Words (www.riverofwords.org), in which children are encouraged to explore and express their understanding of nature through poetry while being in the natural world. As the organization's founder, Pamela Michael, illustrates with examples from several finalists and prize winners, the poetic testimonies of these children, aged between 8 and 17, are stunningly beautiful and deeply moving (Pamela Michael, "Helping Children Fall in Love with the Earth," in Stone and Barlow, 2005).

The fact that ecology is inherently multidisciplinary means that ecoliteracy cannot be taught as a single isolated discipline. Indeed, schooling for sustainability involves weaving the principles of ecology into the entire curriculum so that, ideally, ecoliteracy becomes the school's central focus. Experience in dozens of schools has shown that this is done best by getting students engaged both intellectually and emotionally in a concrete ecological project – growing a school garden, greening the campus through ecodesign, or cultivating collaboration and partnerships throughout the school community – and then designing the whole curriculum around this central project (see Stone, 2009).

Integrating the curriculum in this way will be possible only if there is pervasive collaboration among teachers, administrators, and parents, since it involves complex coordination of schedules, team planning, restructuring of teaching blocks, extended preparations over the summer, and so on. In other words, the conceptual relationships among the various disciplines can be made explicit in a systemic pedagogy only if there are corresponding human relationships among the teachers and school administrators. Ideally, the whole school is transformed into a learning community in which teachers, students, administrators, and parents are all interlinked in a network of relationships, working together to facilitate learning.

This highlights once more the central importance of community in schooling for sustainability. As we discussed in Section 16.3.2, the basic principles of sustainability can be viewed both as principles of ecology and as principles of community. Understanding community is essential for understanding sustainability, and building community in schools is essential for teaching ecoliteracy in the proper systemic and multidisciplinary way.

Describing CEL's pedagogy as participatory means that the teaching does not flow from the top but takes place in cyclical exchanges of ideas and information among teachers and students. The focus is on learning, and everyone in the community is a teacher and a learner. In addition, the experiential and emotional dimensions are important characteristics of schooling for sustainability. The intention is that students not only *understand* ecology but also *experience* it in nature – in a school garden, on a farm, or in a riverbed – and that they also experience community while they become ecologically literate. Otherwise, they could leave school and be first-rate theoretical ecologists but care very little about nature, about the Earth. Schooling for sustainability is dedicated to creating experiences that lead to an emotional relationship with the natural world.

When ecoliteracy is taught in this systemic, participatory, and experiential way, the content of the lessons is designed by the teachers and students. CEL provides a broad-based and integrative conceptual framework that can be used for selecting and designing a variety of specific units. The core of this framework consists of six "principles of ecology" – that is, six basic ecological concepts that represent a convenient summary of the principles of organization of all living systems.

In their joint explorations of ecoliteracy with numerous schools over many years, CEL's educators have found that growing a school garden and using it as a resource for cooking school meals is an ideal way of experiencing these basic ecological concepts. Metabolism – the intake, digestion, and transformation of food – is a central characteristic of life (see Section 7.4.3), and hence food is an ideal vehicle for teaching the principles of ecology. Gardening reconnects children to the fundamentals of food, and thus to the fundamentals of life, and it is also an ideal way to teach them about culture, health, and the environment.

In this scheme, the first three principles of ecology are networks, flows, and cycles. As we have discussed in this book, the systemic understanding of living networks is expressed in the theory of autopoiesis (see Chapter 7), and the systemic understanding of the flows of energy and matter through a living system is reflected in the theory of dissipative structures (see Chapter 8). These theories are too technical to be taught in schools. What can be taught, however, is that one of the key characteristics of living networks is the fact that all their nutrients are passed along in cycles. In an ecosystem, energy flows through the network, while water, oxygen, carbon, and all other nutrients move in the well-known ecological cycles. Similarly, the blood cycles through our body, and so does the air, the lymph fluid, and so on. Wherever we see life, we see networks; and wherever we see living networks, we observe cyclical flows. This is an essential part of the systemic understanding of life, and this is why the teaching of ecoliteracy emphasizes the web of life, the flows of energy, and the cycles of nature.

In the school garden, students learn how plants depend on sunlight, on water from the ground, and on CO_2 from the air for their photosynthesis, as well as depending on fungi and bacteria at their roots to absorb nitrogen, on pollinators to reproduce, and so on. All these are different strands in the web of life; all are interdependent.

In the garden, they learn about food cycles, and they integrate the natural food cycles into their cycles of planting, growing, harvesting, composting, and recycling. Through this practice, they also learn that the garden as a whole is embedded in larger systems that are again living networks with their own cycles. The food cycles intersect with these larger cycles – the water cycle, the cycle of the seasons, and so on – all of which are links in the planetary web of life.

In the school garden, they learn that the energy that drives the ecological cycles flows from the sun. This solar energy, together with water and CO_2, is transformed into chemical energy by the photosynthesis of green plants. And from there, the energy flows through the entire food web, as animals eat plants and are eaten by other animals, and the dead organic matter that is left over is decomposed by worms, bacteria, and fungi in the soil.

Once students understand these principles of networks, cycles, and flows, they are introduced to another three basic ecological concepts: nested systems, dynamic balance, and development. They learn that throughout nature we find living systems nesting within other living systems. Through gardening, they become aware how they themselves are part of the web of life, and over time the experience of ecology in nature gives them a sense of place. They become aware of how we are embedded in many nested systems – in an ecosystem, in a landscape with a particular flora and fauna, and in a particular social system and culture.

The principle of dynamic balance is slightly more complex. It means understanding that all ecological cycles act as feedback loops, so that the ecological community regulates and organizes itself, maintaining a state of dynamic balance characterized by continual fluctuations. In the garden, the students may observe aphids on certain plants and ladybugs eating them. The teacher might use this opportunity to explain that, if an unusually mild winter results in increased growth of aphids, some ladybugs feeding on them may flourish and breed more, so that their numbers increase and they begin to deplete the aphids. Once their major source of food is reduced, the ladybugs will begin to die out. As the ladybug population drops, the aphids will recover and expand again. Eventually, the aphid/ladybug system will balance itself out by means of continual fluctuations.

Finally, the students can easily observe that not all changes are cyclical, that there are also changes of development. Indeed, through gardening they experience growth and development on a daily basis. Moreover, the understanding of growth and development is essential not only for gardening, but also for education. While the children learn that their work in the school garden changes with the development and maturing of the plants, the teachers' methods of instruction and the entire discourse in the classroom changes with the development and maturing of the students.

For children, being in the garden is something magical. As one ecoliteracy teacher put it, "one of the most exciting things about the garden is that we are creating a magical childhood place for children who would not have such a place otherwise, who would not be in touch with the Earth and the things that grow. You can teach all you want, but being out there, growing and cooking and eating, that's an ecology that touches their heart."

Ecoliteracy in higher education

CEL does not work in higher education, concentrating its efforts on primary and secondary schools. However, a corresponding organization in Boston, called Second Nature (www.secondnature.org), promotes ecoliteracy at the university level. For twenty years, they have worked with more than 500 colleges and universities to help make the principles of sustainability fundamental to every aspect of higher education. Second Nature has launched a national movement in the USA, known as "Education for Sustainability (EFS)," and they maintain EFS networks at the state, regional, and national levels.

To promote education for sustainability at the university level, Second Nature advocates systemic thinking and interdisciplinary learning that includes the academic community as a whole. They work to overcome the current fragmentation in the academic world by demonstrating the interdependencies and interconnections among seemingly separate and competing social challenges – such as population, the economy, health, social justice, national security, and the environment. Like the educators at CEL, they teach that systems thinking is essential to understanding the complex, nonlinear systems that are characteristic of both society and the natural world.

In 2011, representatives from two dozen EFS member organizations published a report titled "EFS Blueprint" (www.secondnature.org/efsblueprint), in which they mapped out a strategy for systemic change in higher education to accelerate education for sustainability. In reviewing the current state of affairs at American colleges and universities, the report notes that on the vast majority of campuses progress is still spotty, with many campus initiatives operating in isolation from each other and lacking a systemic, integrative vision.

However, during the last five years the EFS movement has observed a number of encouraging indicators of progress. More than 113 new academic degree programs in sustainability have been established, and over 1,100 interdisciplinary degree programs in environmental education are now in existence; 70 new sustainability centers have opened on campuses; 905 campus buildings are LEED-certified (the recognized standard for "green architecture"), and more than 3,000 buildings are registered with LEED – that is, are in the process of being certified; over 500 campuses have reported their greenhouse gas emission inventories, and 330 have formulated climate action plans.

As to be expected, education for sustainability at colleges and universities involves more political activism than similar programs in primary and secondary schools. For example, a network of colleges and universities was created a few years ago to address the challenge of climate change by making institutional commitments to reduce greenhouse emissions from their campuses, and to promote research and educational efforts accordingly. Known as the American College & University Presidents' Climate Commitment (ACUPCC), the network now has close to 700 members. Its mission is to implement comprehensive plans in pursuit of climate neutrality, to educate students about sustainability, and to serve as a model for society at large.

Political activism on college and university campuses goes hand in hand with the global efforts of a worldwide network of nongovernmental organizations (NGOs). We shall review the principal projects and campaigns of this global network, together with the major problems it seeks to address, in our last chapter. Our review will show that a large number of organizations in this new global civil society explicitly embrace the systems view of life we have been discussing throughout this book.

16.4 Concluding remarks

In this chapter we have extended our synthesis of the systemic conception of life to the ecological dimension. We have noted that, today, the study of ecosystems and of our interactions with them is critical for the survival and well-being of humanity on the planet. Thus ecology not only is a rich and exciting field for theoretical studies, in which the systems view of life is essential, but is also of paramount practical importance.

Having discussed the theoretical aspects of ecology in this chapter, we shall now turn to its practical aspects – its technological, social, and political implications – in the last two chapters of our book.

17

Connecting the dots

Systems thinking and the state of the world

The great challenge of our time, as we mentioned in our previous chapter, is to build and nurture sustainable communities and societies, designed in such a way that our activities do not interfere with nature's inherent ability to sustain life. The first step in this endeavor is to understand the principles of organization that nature's ecosystems have evolved to sustain the web of life; we need to become, as it were, ecologically literate.

We also emphasized in the previous chapter that the basic principles of ecology – interdependence, the cyclical nature of ecological processes, flexibility, diversity, etc. – are basic systemic properties of all living systems. This is why the systemic understanding of life not only holds great intellectual fascination but is also tremendously important from a practical point of view. It is the cognitive foundation of our endeavor to move toward a sustainable future.

17.1 Interconnectedness of world problems

Once we become ecologically literate, once we understand the processes and patterns of relationships that enable ecosystems to sustain life, we will also understand the many ways in which our human civilization, especially since the Industrial Revolution, has ignored these ecological patterns and processes and has interfered with them. And we will realize that these interferences are the fundamental causes of many of our current world problems.

When we look at the state of the world today, what is most evident is the fact that the major problems of our time – energy, the environment, climate change, food security, financial security – cannot be understood in isolation. They are systemic problems, meaning that they are all interconnected and interdependent. One of the most detailed documentations of the fundamental interconnectedness of world problems is the recent book by Lester Brown, *Plan B* (Brown, 2008; see also Brown, 2009, 2011a). Lester Brown, founder of the Worldwatch Institute and, more recently, of the World Policy Institute, has been for many years one of the most authoritative environmental thinkers. In this book, he demonstrates with impeccable clarity how the vicious circle of demographic pressure and poverty leads to the depletion of resources – falling water tables, shrinking forests, collapsing fisheries, eroding soils, and so on – and how this resource depletion, exacerbated by climate change, produces failing states whose governments can no longer provide security for their citizens, some of whom in sheer desperation turn to terrorism.

As we mentioned in the Preface to this book, all these problems, ultimately, must be seen as just different facets of one single crisis, which is largely a crisis of perception. It derives from the fact that most people in our society, and especially our large social institutions, subscribe to the concepts of an outdated worldview, a perception of reality inadequate for dealing with our overpopulated, globally interconnected world.

Plan B is Lester Brown's "road map for saving civilization." It is the alternative to business as usual (Plan A), which leads to disaster. The main message of the book is that there *are* solutions to the major problems of our time – some of them even simple. But they require a radical shift in our perceptions, our thinking, our values. Since they are systemic problems, they require systemic solutions; and since the only viable solutions are those that are ecologically sustainable, they must incorporate the basic principles of ecology, or principles of sustainability.

To discuss the fundamental interconnectedness of today's global problems, we have designed a conceptual map, based on Brown's Plan B, which shows their complex interconnections (Figure 17.1). We shall begin with a brief overview of the entire conceptual map and will then, in the following sections, discuss the major problem areas in more detail before we turn to the corresponding systemic solutions.

The fundamental dilemma underlying the major problems of our time seems to be the illusion that unlimited growth is possible on a finite planet. This, in turn, reflects the clash between linear thinking and the nonlinear patterns in our biosphere – the ecological networks and cycles that constitute the web of life. This highly nonlinear global network contains countless feedback loops through which the planet balances and regulates itself. Our current economic system, by contrast, is fueled by materialism and greed that do not seem to recognize any limits.

There are actually three kinds of growth that have severe impacts on our natural environment and our well-being: economic growth, corporate growth, and population growth. The illusion of the viability of unlimited growth is maintained by economists who refuse to include the social and environmental costs of economic activities in their theories (see Section 3.5). Consequently, there are huge differences between market prices and true costs, as, for example, for fossil fuels. As Nicholas Stern, former chief economist of the World Bank, pointed out in his groundbreaking study on the costs of climate change, this amounts to a massive market failure (Stern, 2006).

Economic and corporate growth are pursued by global capitalism, the dominant economic system today. At the center of the global economy is a network of financial flows, which has been designed without any ethical framework. In fact, social inequality and social exclusion are inherent features of economic globalization, widening the gap between the rich and the poor and increasing world poverty.

In this new economy, capital works in real time, moving continually through global financial networks. These movements are facilitated by the "free-trade" rules, designed to support continuing corporate growth. Economic and corporate growth are pursued relentlessly by

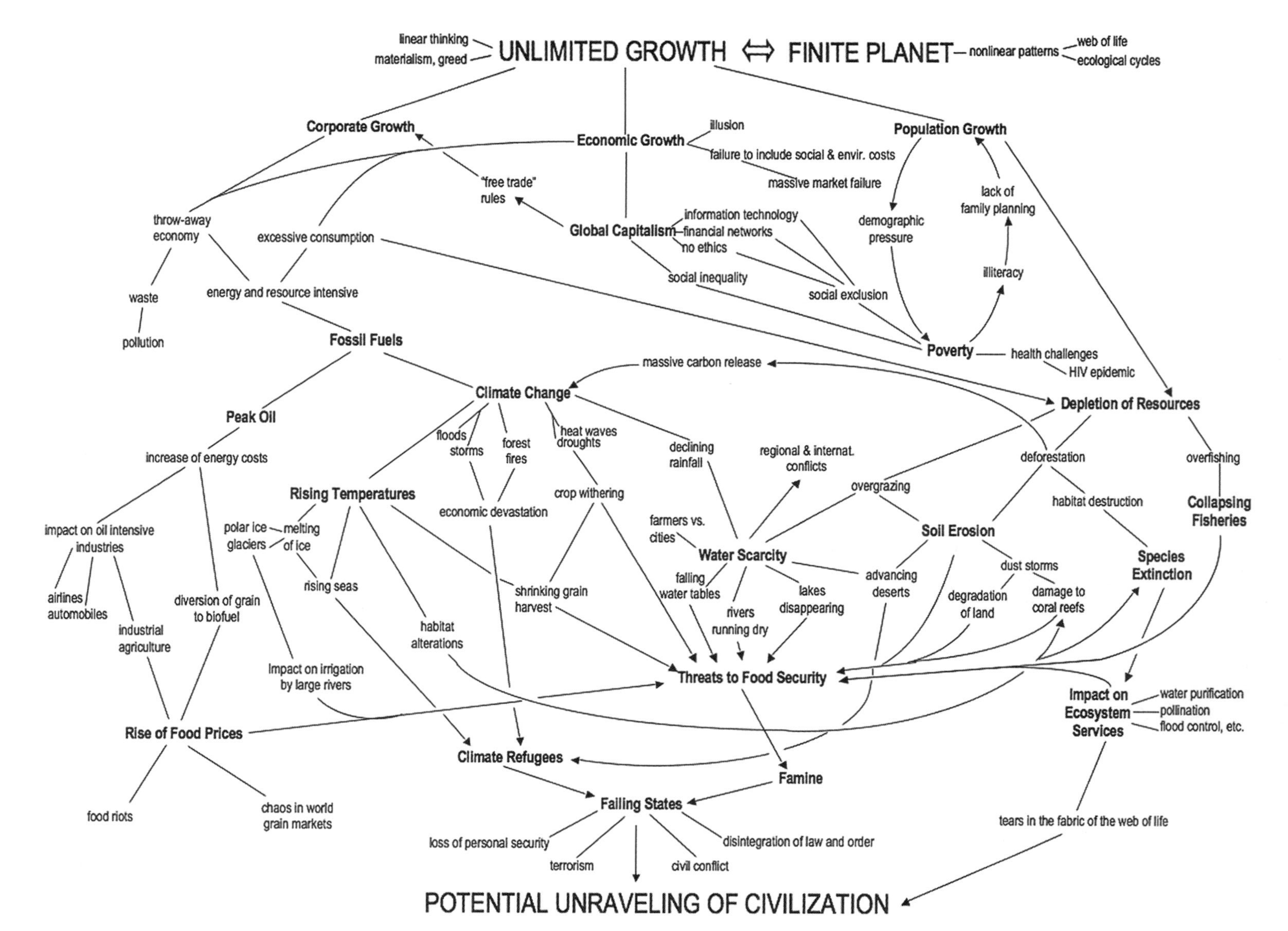

Figure 17.1 Interconnectedness of world problems (based on Brown, 2008).

promoting excessive consumption and a throw-away economy that is energy and resource intensive, generating waste and pollution, and depleting the Earth's natural resources.

Population growth and poverty form a vicious circle, or self-amplifying feedback loop. Rapid population growth reduces the available cropland and water supplies per person. The resulting poverty, often coupled with illiteracy, in turn increases the demographic pressure, as illiterate women typically have less access to family planning and thus have much larger families than literate women. The results of this mutual reinforcement of demographic pressure and poverty are the growing health challenges of the HIV epidemic and other infectious diseases, on the one hand, and further depletion of resources on the other hand.

Excessive consumption and waste in industrialized countries and rapid population growth in many developing countries combine to exert severe pressures on our natural resources, leading to overgrazing, deforestation, and overfishing. The results are well known – falling water tables, rivers running dry, lakes disappearing, shrinking forests, collapsing fisheries, eroding soils, grasslands turning into deserts – all of which are severe threats to food security.

Water scarcity has led to intense conflicts between farmers and cities, with farmers usually losing out, in addition to many political stresses in regional and international conflicts. Soil erosion results not only in the decline of soil fertility but also in an increasing number of dust storms that can travel for thousands of miles and cause further degradation of land, as well as damage to coral reefs.

Deforestation, especially in the tropical rainforests, results in the destruction of habitats of numerous plant and animal species, and consequently in their extinction. Indeed, we are now in the early stages of a massive species extinction which, for the first time in the history of the planet, is not caused by natural phenomena but by human behavior (see Section 16.3). As various forms of life disappear, so do the services they provide – water purification, pollination, flood control, etc. If the loss of these priceless ecosystem services continues, it could tear huge gaps in the fabric of the web of life.

All of these environmental problems are exacerbated by global climate change, caused by our energy-intensive and fossil-fuel-based technologies. This is aggravated by deforestation through the release of massive amounts of carbon into the atmosphere. Climate change manifests itself in increased floods, destructive storms, and forest fires, which cause economic devastation and give rise to large numbers of climate refugees. Other manifestations of climate change are severe heat waves and droughts that lead to crop withering, thus shrinking grain harvests and further threatening food security. In many regions of the world, the resulting decline of rainfall intensifies an already severe water scarcity.

Rising temperatures cause not only shrinking grain harvests but also the melting of ice – both glaciers and polar ice – and consequently the rise of sea levels. The shrinking of glaciers severely impacts the irrigation of rice and wheat fields by large rivers fed by those glaciers. These effects are additional huge threats to food security. Rising seas could potentially result in millions of climate refugees in the coming years. And finally, rising global temperatures alter many habitats and threaten the extinction of species living in them.

The excessive dependence on fossil fuels not only causes global warming but has also brought us close to "peak oil": after oil production reaches its peak, it will decrease worldwide, extraction of the remaining reserves will be more and more costly, and hence the price of oil will continue to rise. Most affected are the oil-intensive segments of the global economy, in particular the automobile industry, the airline industry, and industrial agriculture. Thus food prices are rising with rising oil prices, further threatening food security. There is now a serious risk that rising grain prices will lead to chaos in world grain markets and to food riots in low- and middle-income countries that import grain.

The search for alternative energy sources has recently led to increased production of ethanol and other biofuels, and since the fuel value of grain is higher on the markets than its food value, more and more grain is diverted from food to producing fuels. At the same time, the price of grain is moving up toward the oil-equivalent value.

The fact that the price of grain is now keyed to the price of oil is only possible because our global economic system has no ethical dimension. In such a system, as Brown (2008) points out, the question, "Should we use grain to fuel cars or to feed people?", has a clear answer. The market says, "Let's fuel the cars." This is even more perverse in view of the fact that the entire ethanol production in the USA could be replaced by raising average fuel efficiency by 20%, as could easily be done with the technologies available today.

Lester Brown's analysis makes it abundantly clear that virtually all our environmental problems are threats to our food security – water scarcity, soil erosion, collapsing fisheries, extreme climate events, and, most recently, the rise of food prices due to rising energy costs and increasing diversion of grains to biofuel. In addition, increased fuel consumption accelerates global warming, which results in crop losses in heat waves, and from the loss of glaciers that feed rivers essential to irrigation. When we think systemically and understand how all these processes are interrelated, we realize that the vehicles we drive, and other consumer choices we make, have a major impact on the food supply to large populations in other parts of the world.

As a result of these multiple threats to food security, world hunger is now on the rise again after a long steady decline. This worldwide famine and the large numbers of climate refugees have resulted in increasing numbers of failing states, characterized by the disintegration of law and order and the rise of civil conflict. The governments of these states can no longer provide security for their citizens. With increasing numbers of failing states, and ever increasing tears in the fabric of the web of life, caused by continuing species extinction, civilization itself could begin to unravel.

17.2 The illusion of perpetual growth

The obsession of politicians and economists with unlimited economic growth must be seen as one of the root causes, if not *the* root cause, of our global multifaceted crisis. As we discussed in Section 3.5.2, the goal of virtually all national economies is to achieve unlimited growth, even though the absurdity of such an enterprise on a finite planet should be obvious to all.

As the energy analyst and educator Richard Heinberg argues eloquently in his recent book, *The End of Growth*, the world is now colliding with three fundamental barriers to continuing economic expansion (Heinberg, 2011). The first is the depletion of important natural resources, and in particular the declining access to cheap and abundant fossil fuels, the situation also known as "peak oil." The second barrier is the proliferation of harmful environmental impacts arising from the extraction and use of resources, resulting in ever-increasing costs from both the impacts themselves and from efforts to avert them. The most catastrophic of these environmental impacts are the multiple global manifestations of climate change, which we shall discuss in more detail in Section 17.3.5.

The third barrier to perpetual growth is more abstract but no less severe: it is a financial barrier. As Heinberg explains in great detail, we have created a global monetary and financial system that requires growth. Economic activities are driven by money entering the economy through loans, and the interest on these loans can only be paid if the economy continues to grow. However, when economic growth collides with the two natural barriers of peak oil and climate collapse, financial systems built on expectations of perpetual growth are bound to fail, generating the multiple symptoms we have come to know only too well – massive unpaid debts, high rates of unemployment, chain reactions of defaults and bankruptcies, and so on.

Heinberg argues persuasively that resource depletion, environmental impacts, and systemic financial and monetary failures are now combining to make the resumption of conventional economic growth a near impossibility, and that policymakers who insist on pursuing the ideology of perpetual growth are really fleeing from reality.

The clear recognition of this dilemma is hampered by the fact that most economists use inadequate economic indicators. As we discussed in Section 3.5, countries measure their wealth in terms of their gross domestic product (GDP), a crude indicator resulting from adding up indiscriminately all economic activities associated with monetary values, while ignoring the many important nonmonetary aspects of the economy. The unlimited growth of GDP through the continuing accumulation of material goods is pursued relentlessly by virtually all economists and politicians, and is celebrated as the sign of a "healthy" economy. The problem that growth can also be harmful or pathological, like the growth of cancer, is rarely addressed; nor is the dilemma that unlimited material growth on a finite planet can only lead to disaster.

Indeed, it is evident from our graph in Figure 17.1 that undifferentiated economic growth is the root cause of our mountains of solid waste, our polluted cities, the depletion of our natural resources, and the energy crisis; and, because the continuing expansion of production is driven mainly by fossil fuels, it is also the root cause of the multiple disasters arising from peak oil and climate change.

This fatal dynamic is increasingly being recognized today. In a joint paper prepared for the Rio+20 Earth Summit in June 2012, the eighteen past winners of the Blue Planet Prize – the unofficial "Nobel Prize" for the environment – write: "The perpetual growth myth . . . promotes the impossible idea that indiscriminate economic growth is the cure for all the world's problems, while it is actually the disease that is at the root of our unsustainable global practices" (see Brundtland *et al.*, 2012, p. 41); the authors include Paul Ehrlich, James Hansen, Amory Lovins, James Lovelock, Karl-Henrik Robèrt, and Nicholas Stern).

17.2.1 From quantitative to qualitative growth

It seems, then, that our key challenge is to shift from an economic system based on the notion of unlimited growth to one that is both ecologically sustainable and socially just. From the perspective of the systems view of life, "no growth" cannot be the answer. Growth is a central characteristic of all life. A society, or economy, that does not grow will die sooner or later. Growth in nature, however, is not linear and unlimited. While certain parts of organisms, or ecosystems, grow, others decline, releasing and recycling their components, which become resources for new growth.

This kind of balanced, multifaceted growth is well known to biologists and ecologists. Capra and Henderson (2009) have proposed to call it "qualitative growth" to contrast it to the concept of quantitative growth used by today's economists. The recognition of the fallacy of the conventional concept of economic growth, the two authors suggest, is the first essential step in overcoming our economic crisis. In the words of social-change activist Frances Moore Lappé (2009):

> Since what we call 'growth' is largely waste, let's call it that! Let's call it an economics of waste and destruction. Let's define growth as that which enhances life – as generation and regeneration – and declare that what our planet most needs is more of it.

This notion of "growth which enhances life" is what is meant by qualitative growth – growth that enhances the quality of life. In living organisms, ecosystems, and societies, qualitative growth includes an increase of complexity, sophistication, and maturity. Unlimited quantitative growth on a finite planet is clearly unsustainable, but qualitative economic growth can be sustained if it involves a dynamic balance between growth, decline, and recycling, and if it also includes the inner growth of learning and maturing.

The focus on qualitative growth is fully consistent with the systems view of life. As we have emphasized several times in this book, this new science of life is essentially a science of qualities. This is relevant in particular to the understanding of ecological sustainability, since the basic principles of ecology – principles like interdependence or the cyclical nature of ecological processes – are expressed in terms of patterns of relationships, or qualities.

In fact, the new systemic conception of life makes it possible to formulate a scientific concept of quality. It seems that there are two different meanings of the term – one objective and the other subjective. In the objective sense, the qualities of a complex system refer to properties of the system that none of its parts exhibit. Quantities like mass or energy tell us about the properties of the parts, and their sum total is equal to the corresponding property of the whole – e.g., the total mass or energy. Qualities like stress or health, by contrast, cannot be expressed as the sum of properties of the parts. Qualities arise from processes and patterns of relationships among the parts. Hence, we cannot understand the nature of complex systems such as organisms, ecosystems, societies, and economies if we try to describe them in purely quantitative terms. Quantities can be measured; qualities need to be mapped (see Section 4.3).

With the recent emphasis on complexity, networks, and patterns of organization, the attention of scientists in the life sciences has begun to shift from quantities to qualities, and there has been a corresponding conceptual shift in mathematics. In fact, this began in physics during the 1960s with the strong emphasis on symmetry (see Section 8.4.3), which is a quality, and it intensified during the subsequent decades with the development of complexity theory, or nonlinear dynamics, which is a mathematics of patterns and relationships. The strange attractors of chaos theory and the fractals of fractal geometry are visual patterns representing the qualities of complex systems (see Sections 6.3 and 6.4).

In the human realm, the notion of quality always seems to include references to human experiences, which are subjective aspects. This should not be surprising. Since all qualities arise from processes and patterns of relationships, they will necessarily include subjective elements if these processes and relationships involve human beings.

For example, the quality of a person's health can be assessed in terms of objective factors, but it includes a subjective experience of well-being as a significant element (see Section 15.2). Similarly, the quality of a human relationship derives largely from subjective mutual experiences. To describe and explain the qualities of such subjective experiences within a scientific framework is known as the "hard problem" of consciousness studies, as we discussed in Chapter 12.

These considerations imply that, to properly assess the health of an economy, we need qualitative indicators of poverty, health, equity, education, social inclusion, and the state of the natural environment – none of which can be reduced to money coefficients or aggregated into a simple number. Indeed, several economic indicators of this kind have recently been proposed. They include the United Nations (UN) Human Development Index (HDI), launched in 1990, and the Calvert–Henderson Quality of Life Indicators, which assess twelve criteria and use monetary coefficients only where appropriate (see Capra and Henderson, 2009).

17.2.2 Growth and development

A few years after the concept of ecological sustainability was introduced by Lester Brown (1981), it was taken up by the UN in the so-called Brundtland Report (see Section 16.3). There, however, the term "sustainable" was coupled with "development," and this notion of "sustainable development" has been widely used ever since, unfortunately often without the ecological context that would give it its proper meaning.

The whole concept of "sustainable development" is rather problematic because, like "growth," "development" is used today in two quite different senses – one qualitative and the other quantitative. For biologists, development is a fundamental property of life. According to the theory of autopoiesis, a living organism continually responds to environmental influences with structural changes and over time will form its own, individual pathway of development (see Section 7.4). The fundamental dynamic underlying this process is the spontaneous emergence of order at critical points of instability. As we emphasized in our discussion of self-organization and emergence (Section 8.3), creativity – the generation of new forms – has been recognized in the systems view of life as a key property of all living systems. Life continually reaches out to create novelty.

This systemic understanding of development implies a sense of multifaceted unfolding; of living organisms, ecosystems, or human communities moving toward reaching their full potential. Most economists, by contrast, restrict the use of "development" to a single economic dimension, usually measured in terms of per capita GDP. The huge diversity of human existence is compressed into this linear, quantitative concept and then converted into monetary coefficients.

The economist Paul Ekins (1992) has summarized the concept of economic development, as it is presently played out on the world stage, in terms of three basic characteristics: it is economistic, Northern-oriented, and top-down. Development, Ekins points out, is a fairly recent economic concept. Before World War II, no one would have thought of development as an economic category. But from the second half of the twentieth century, it has been used almost exclusively in the narrow economistic sense. The entire world is arbitrarily categorized into "developed," "developing," and "less developed" countries, like in a football league, according to a single economic dimension. Development is measured only in terms of money and cash flows, ignoring all other forms of wealth – all ecological, social, and cultural assets.

Moreover, this economic "league table" is arranged according to Northern criteria. Countries considered as developed are those that have adopted the Northern industrial way of life. This makes development a Northern, profoundly monocultural concept: to be a developing country means to succeed in the aspiration of becoming more like the North. And finally, economic development is a top-down process. Decisions and control rest firmly in the hands of experts, managers of international capital, bureaucrats of state governments, and global financial institutions like the World Bank and the International Monetary Fund (IMF).

The linear, one-dimensional view of economic development, as used by most mainstream and corporate economists, and by most politicians, corresponds to the narrow quantitative concept of economic growth, while the biological and ecological sense of development corresponds to the notion of qualitative growth. In fact, the biological concept of development includes both quantitative and qualitative growth.

A developing organism, or ecosystem, grows according to its stage of development. Typically, a young organism will go through periods of rapid physical growth. In ecosystems, this early phase of rapid growth is known as a pioneer ecosystem, characterized by rapid expansion and colonization of the territory. The rapid growth is always followed by slower growth, by maturation, and ultimately by decline and decay, or, in ecosystems, by so-called succession (see Section 16.1). As living systems mature, their growth processes shift from quantitative to qualitative growth.

The distinction between the biological and the current economic sense of "development," and the association of qualitative economic growth with the former and purely quantitative growth with the latter, help to clarify the widely used but problematic concept of "sustainable development." If "development" is used in the current narrow economic sense, associated with the notion of unlimited quantitative growth, such economic development can never be sustainable, and the term "sustainable development" would thus be an oxymoron. If, however, the process of development is understood as more than a purely economic process, including social, ecological, cultural, and spiritual dimensions, and if it is associated with qualitative economic growth, then such a multidimensional systemic process can indeed be sustainable.

Such a broad, alternative view of development is advocated today by a number of scholars and activists, who see development as a creative process of increasing one's capabilities – as characteristic of all life – which requires, first and foremost, control over local resources (see Escobar, 1995; Esteva and Prakash, 1998; W. Sachs, 1992). In this view, development is not purely an economic process, but includes social, ecological, cultural, and ethical dimensions. It is a multidimensional and systemic process, in which the primary actors of development are the institutions of civil society – the NGOs based on kin, neighborhood, or common interests.

Because people are different and the places in which they live are different, we can expect development to produce cultural diversity of all kinds. The processes whereby it happens will be very different from the current global trading system. It will be based on the mobilization of local resources to satisfy local needs, and informed by the values of human dignity and ecological sustainability. Such truly sustainable development is based on the recognition that we are an inseparable part of the web of life, of human and nonhuman communities, and that enhancing the dignity and sustainability of any one of them will enhance all the others.

17.2.3 Qualifying economic growth

Let us now return to the central challenge of our economic and ecological crisis. How can we transform the global economy from a system striving for unlimited quantitative growth, which is manifestly unsustainable, to one that is ecologically sound and socially just? The concept of qualitative economic growth will be a crucial tool in this task. Instead of assessing the state of the economy in terms of the crude quantitative measure of GDP, we need to distinguish between "good" growth and "bad" growth and then increase the former at the expense of the latter, so that the natural and human resources tied up in wasteful and unsound production processes can be freed and recycled as resources for efficient and sustainable processes. A first step in this direction was the "Beyond GDP" conference in the European Parliament in November 2007, spearheaded by the European Commission together with the Club of Rome, the Organization for Economic Cooperation and Development (OECD), and the World Wildlife Fund (WWF) (see www.beyond-gdp.eu).

From the ecological point of view, the distinction between "good" and "bad" economic growth is obvious. Bad growth is growth of production processes and services that externalize social and environmental costs, are based on fossil fuels, involve toxic substances, deplete our natural resources, and degrade the Earth's ecosystems. Good growth is growth of more efficient production processes and services that fully internalize costs and involve renewable energies, zero emissions, continual recycling of natural resources, and restoration of the Earth's ecosystems.

Since their initial 2007 conference, the "Beyond GDP" partners have continued to work on a road map for qualifying economic growth. Many of the suggested reforms will involve shifts of perception from a product orientation to a service orientation and the "dematerializing" of our productive economies. For example, an automobile company should realize that it is not necessarily in the business of selling cars but rather in the business of providing mobility, which can also be achieved, among many other things, by producing more buses and trains and by redesigning our cities. Similarly, countries, and especially the USA, should realize that fighting climate change is today's most important and most urgent security issue. The US government should reduce the Pentagon's budget accordingly, while increasing funds for diplomacy, mitigating climate-related threats to global security, and building a new "green" economy.

17.2.4 From materialism to community

The shift from quantitative to qualitative growth will require deep changes not only at the social and economic levels, but also at the individual level. It will mean overcoming the pervasive cultural conditioning of materialism and turning from finding satisfaction in material consumption to finding it in human relationships and community. For most of us, this value shift is anything but easy, as we are bombarded daily with a stream of advertising messages assuring us that the accumulation of material goods is the royal road to happiness, the very purpose of our lives (see Dominguez and Robin, 1999).

The USA projects its tremendous power around the world to maintain optimal conditions for the perpetuation and expansion of production. The central goal of its vast empire – its overwhelming military might, impressive range of intelligence agencies, and dominant positions in science, technology, media, and entertainment – is not to expand its territory, nor to promote freedom and democracy, but to make sure that it has global access to natural resources and that markets around the world remain open to its products (Ramonet, 2000). Accordingly, political rhetoric in America moves swiftly from "freedom" to "free trade" and "free markets." The free flow of capital and goods is equated with the lofty ideal of human freedom, and material acquisition is portrayed as a basic human right, and increasingly even as an obligation.

This glorification of material consumption has deep ideological roots that go far beyond economics and politics. Its origins seem to lie in the universal association of manhood with material possessions in patriarchal cultures. The anthropologist David Gilmore studied images of manhood around the world – "male ideologies," as he puts it – and found striking cross-cultural similarities (Gilmore, 1990, p. 2). There is a recurring notion that "real manhood" is different from simple biological maleness, that it is something that has to be won. In most cultures, Gilmore shows, boys "must earn the right" to be called men. Although women, too, are judged by sexual standards that are often stringent, Gilmore notes that their very status as women is rarely questioned. Curiously, however, Gilmore does not mention the fact, widely discussed in feminist literature, that there is no need for women to prove their womanhood, because of their ability to give birth, which was perceived as an awesome, transformative power in pre-patriarchal cultures (see Rich, 1977).

In addition to well-known images of manliness like physical strength, toughness, and aggression, Gilmore found that in culture after culture, "real" men have traditionally been those who produce more than they consume. The author emphasizes that the ancient association of manhood with material production meant production on behalf of the community: "Again and again we find that 'real' men are those who give more than they take; they serve others. Real men are generous, even to a fault." (Gilmore, 1990, p. 15). Over time, there was a shift in this image, from production for the sake of others to material possession for the sake of one's self. Manhood was now measured in terms of ownership of valuable goods – land, cattle, or cash – and in terms of power over others, especially women and children. This image was reinforced by the universal association of virility with "bigness" – as measured in muscle strength, accomplishments, or number of possessions. In modern society, Gilmore points out, male "bigness" is measured increasingly by material wealth: "The Big Man in any industrial society is also the richest guy on the block, the most successful, the most competent . . . He has the most of what society needs or wants."

The association of manhood with the accumulation of possessions fits well with other values that are favored and rewarded in patriarchal culture – expansion, competition, and an "object-centered" consciousness. In traditional Chinese culture, these were called *yang* values and were associated with the masculine side of human nature. They were not seen as being intrinsically good or bad. However, according to Chinese wisdom, the *yang* values need to be balanced by their *yin*, or feminine, counterparts – expansion by conservation, competition by cooperation, and the focus on objects by a focus on relationships. As one of us (F.C.) has long argued, the movement toward such a balance is very consistent with the shift from mechanistic to systemic and ecological thinking that is characteristic of our time (Capra 1982, 1996).

The values of conservation, cooperation, and community are promoted today by many grass-roots movements and organizations in the new global civil society (see Section 17.4). Among them, the feminist movement and the ecology movement advocate the most profound value shifts, the former through a redefinition of gender relationships, the latter through a redefinition of the relationship between humans and nature. Both movements could contribute significantly to overcoming our culture's glorification of material consumption.

By challenging the patriarchal order and value system, the women's movement has introduced a new understanding of masculinity and personhood that does not need to associate manhood with material possessions. At the deepest level, feminist awareness is based on women's experiential knowledge that all life is connected, that our existence is always embedded in the cyclical processes of nature (see Spretnak, 1981). Feminist consciousness, accordingly, focuses on finding fulfillment in nurturing relationships rather than in the accumulation of material goods.

The ecology movement arrives at the same position from a different approach. Ecological literacy requires systemic thinking, which includes a shift of perspective from objects to relationships (see Section 4.3), and ecodesigners advocate a transition from an economy of goods to an economy of "service and flow," as we shall discuss in our last chapter. In such an economy, matter cycles continually, so that the net consumption of raw materials is drastically reduced.

Thus the rise of feminist awareness and the movement toward ecological sustainability could combine to bring about a profound change of thinking and values – from linear systems of resource extraction and accumulation of products and waste to cyclical flows of matter and energy; from the focus on objects and natural resources to a focus on services and human resources; from seeking happiness in material possessions to finding it in nurturing relationships. In the eloquent words of the biologist and environmental activist David Suzuki:

> Family, friends, community – these are the sources of the greatest love and joy we experience as humans. We visit family members, keep in touch with favourite teachers, share and exchange pleasantries with friends. We undertake difficult projects to help others, save frogs or protect a wilderness, and in the process discover extreme satisfaction. We find spiritual fulfillment in nature or by helping others. None of these pleasures requires us to consume things from the Earth, yet each is deeply fulfilling. These are complex pleasures, and they bring us much closer to real happiness than the simple ones, like a bottle of Coke or a new minivan.
>
> *(Suzuki and Dressel, 1999, pp. 263–4)*

These reflections testify to the important role of community in our task of creating a sustainable future. As we discussed in Section 16.3, building and nurturing community lies at the core of ecological sustainability, and community is also the most effective antidote against excessive material consumption. As Suzuki and Dressel point out in the passage quoted above, the happiness we derive from being an active member of a community – human or ecological – ultimately gives us a sense of spiritual fulfillment. indeed, the exhortation of a value shift from material consumption to human relationships and community can be found in the teachings of many spiritual traditions. The Dalai Lama (2000) expressed it beautifully:

> All sentient beings seek happiness. They all want to overcome pain and suffering at the sensory level. Human desires and goals come not only from sensory experience but also from pure imagination. This creates a mental level of pleasure and pain beyond the sensory. Material fulfillment – money, material goods, and so on – gives us satisfaction at the sensory level. But at the mental level, at the level of our imagination and desires, we need another kind of satisfaction which the physical level cannot provide. Material comfort alone is not an answer for humanity. I have met many people who live in great material comfort and yet are full of anxiety; and they tell me about their many problems.

The counter force to this mental disturbance is loving kindness. Human affection, caring, a sense of responsibility, and a sense of community – that is spirituality.

17.3 The networks of global capitalism

Let us now return to the impact of unrestricted, undifferentiated economic growth, and specifically to global capitalism, the dominant economic system today, which is the main driving force of economic and corporate growth. This new type of capitalism is profoundly different from the one formed during the Industrial Revolution (see Section 3.3) and from the Keynesian capitalism that was the dominant economic model for several decades after World War II (see Section 3.4).

The new capitalism, which emerged from the information technology revolution during the past three decades, is characterized by three fundamental features. Its core economic activities are global; the main sources of productivity and competitiveness are knowledge generation and information processing; and it is structured largely around networks of financial flows. This new global capitalism is also referred to as "the new economy," or simply as "globalization." Because of the global networks of informational and financial flows that form its very core, it is especially interesting to analyze the new global economy from a systemic perspective. To do so, we shall begin with a brief review of various aspects of globalization.

17.3.1 Understanding globalization

During the last decade of the twentieth century, a recognition grew among entrepreneurs, politicians, social scientists, community leaders, grass-roots activists, artists, cultural historians, and ordinary women and men from all walks of life that a new world was emerging – a world shaped by new technologies, new social structures, a new economy, and a new culture. "Globalization" became the term used to summarize the extraordinary changes and the seemingly irresistible momentum that were felt by millions of people.

Indeed, within a few years we became quite used to many facets of globalization. We rely on global communications networks for news, sports, and cultural events; we routinely use the Worldwide Web as a global information system; and through a variety of social networks we can enjoy staying in touch with friends around the world on a daily basis. Within the context of this section, however, we shall concentrate on the new global economy whose impact on our well-being has been much more problematic.

With the creation of the World Trade Organization (WTO) in the mid 1990s, economic globalization, characterized by "free trade," was hailed by corporate leaders and politicians as a new order that would benefit all nations, producing worldwide economic expansion whose wealth would "trickle down" to all. However, it soon became apparent to increasing numbers of environmentalists and grass-roots activists that the new economic rules established by the WTO were manifestly unsustainable and were producing a multitude of interconnected fatal consequences – social disintegration, a breakdown of democracy, more rapid and extensive deterioration of the environment, a series of financial crises, and increasing poverty and alienation.

In 1996, two books were published that provided the first systemic analyses of economic globalization. They were written in very different styles and their authors followed very different approaches, but their starting point was the same – the attempt to understand the profound changes brought about by the combination of extraordinary technological innovation and global corporate reach.

The Case Against the Global Economy is a collection of essays by more than forty grass-roots activists and community leaders, edited by Jerry Mander and Edward Goldsmith, and published by the Sierra Club, one of the oldest and most respected environmental organizations in the USA (Mander and Goldsmith, 1996). The authors of this book represent cultural traditions from many countries around the world. Most of them are well known among social-change activists. Their arguments are passionate, distilled from the experiences of their communities, and aimed at reshaping globalization according to different values and different visions. *The Rise of the Network Society* by Manuel Castells, professor of communication technology and society at the University of Southern California, is a brilliant analysis of the fundamental processes underlying economic globalization, published by Blackwell, one of the largest academic publishers (Castells, 1996).

During the years following the publication of these two books, some of the authors of *The Case Against the Global Economy* formed the International Forum on Globalization, a nonprofit organization that holds teach-ins on economic globalization in several countries. In 1999, these teach-ins provided the philosophical background for the worldwide coalition of grass-roots organizations that successfully blocked the meeting of the World Trade Organization in Seattle and made its opposition to the WTO's policies known to the world (see Section 17.4).

On the theoretical front, Manuel Castells published two further books to complete a series of three volumes on *The Information Age: Economy, Society and Culture* (Castells, 1997, 1998). This trilogy is a monumental work, encyclopedic in its rich documentation, which the sociologist Anthony Giddens (1996) has compared to Max Weber's *Economy and Society*, written almost a century earlier.

Castells' thesis is wide-ranging and illuminating. His central focus is on the revolutionary information and communication technologies that emerged during the last three decades of the twentieth century. As the Industrial Revolution gave rise to "industrial society," so the new information technology revolution is now giving rise to an "informational society." And since information technology has played a decisive role in the rise of networking as a new form of organization of human activity in business, politics, the media, and NGOs, Castells also calls the informational society the "network society" (see Section 14.4).

During the first decade of our new century, the attempts of scholars, politicians, and community leaders to understand the nature and consequences of globalization have continued and intensified (Cavanagh and Mander, 2004; Grewal, 2008; Hutton and Giddens, 2000). In the following pages, we shall synthesize the main ideas about economic globalization from the publications mentioned above. Applying our systemic perspective, we shall try to show how the rise of globalization has proceeded through a process that is characteristic of all human organizations – the interplay between designed and emergent structures (see Section 14.5).

17.3.2 The information technology revolution and the birth of global capitalism

The common characteristic of the multiple aspects of globalization is a global information and communications network based on revolutionary new technologies. The information technology revolution is the result of a complex dynamic of technological and human interactions, which produced synergistic effects in three major areas of electronics – computers, microelectronics, and telecommunications. The key innovations that created the radically new electronic environment of the 1990s all took place twenty years earlier, during the 1970s.

All these measures relied crucially on the new information and communication technologies, which made it possible to transfer funds between various segments of the economy and various countries almost instantly and to manage the enormous complexity brought about by rapid deregulation and new financial ingenuity. In the end, the information technology revolution helped to give birth to a new global economy – a rejuvenated, flexible, and greatly expanded capitalism.

17.3.3 Casino finance

In the new economy, capital works in real time, moving rapidly through global financial networks. From these networks it is invested in all kinds of economic activity, and most of what is extracted as profit is channeled back into the meta-network of financial flows. Sophisticated information and communication technologies enable financial capital to move rapidly from one option to another in a relentless global search for investment opportunities. Profit margins are generally much higher in the financial markets than in most direct investments; hence all flows of money ultimately converge in the global financial networks in search of higher gains.

The dual role of computers as tools for rapid processing of information and for sophisticated mathematical modeling has led to the virtual replacement of gold and paper money by ever more abstract financial products. These include "future options" (financial gains in the future, as anticipated by computer projections), "hedge funds" (high-risk investment funds that are used to buy and sell huge amounts of currencies within minutes to profit from tiny margins), and "derivatives" (packages of diverse funds, representing collections of actual or potential financial values). The end result of all these technological and financial innovations has been the transformation of the global economy into a giant, electronically operated casino. Here is how Castells (1996, pp. 434–5) describes the operations of these new financial markets, which have since become known as "casino finance":

The same capital is shuttled back and forth between economies in a matter of hours, minutes, and sometimes seconds. Favored by deregulation . . . and the opening of domestic financial markets, powerful computer programs and skillful financial analysts/computer wizards sitting at the global nodes of a selective telecommunications network play games, literally, with billions of dollars . . . These global gamblers are not obscure speculators, but major investment banks, pension funds, multinational corporations . . . and mutual funds organized precisely for the sake of financial manipulation.

At the existential human level, the most alarming feature of the new economy may be that it is shaped in very fundamental ways by machines. The "global market," strictly speaking, is not a market at all but a network of machines programmed according to a single value – moneymaking for the sake of making money – to the exclusion of all other values. In other words, the global economy has been designed in such a way that all ethical dimensions are excluded. However, the same electronic networks of financial and informational flows *could* have other values built into them. The critical issue is not technology, but politics and human values. And human values can change; they are not natural laws.

The process of economic globalization was purposefully designed by the leading capitalist countries (the "G7 nations"), the major transnational corporations, and by global financial institutions that were created for that purpose (see Mander, 2012). The most important of these financial institutions are the World Bank, the IMF, and the WTO. They are known collectively as the "Bretton Woods" institutions because they were established at a UN conference in Bretton Woods, New Hampshire, in 1944, in order to create an institutional framework for a coherent worldwide post-war economy.

The World Bank was originally created to finance the postwar reconstruction of Europe, and the IMF to ensure the stability of the international financial system. However, both institutions soon shifted their focus to promoting and enforcing a narrow model of economic development in the Third World, often with disastrous social and environmental consequences (see Sections 17.3.4 and 17.3.5). The ostensible role of the WTO is to regulate trade, prevent trade wars, and protect the interests of poor nations. In reality, the WTO implements and enforces globally the same agenda that the World Bank and the IMF have imposed on most of the developing world.

In spite of the stringent rules established by the Bretton Woods institutions, the process of economic globalization has been far from smooth. Once the global financial networks reached a certain level of complexity, their nonlinear interconnections generated rapid feedback loops that gave rise to many unsuspected emergent phenomena. The resulting new economy is so complex and turbulent that it defies analysis in conventional economic terms. Thus Anthony Giddens admitted in 2000, when he was the director of the prestigious London School of Economics, "The new capitalism that is one of the driving forces of globalization to some extent is a mystery. We don't fully know as yet just how it works" (in Hutton and Giddens, 2000, p. 10).

In the global casino, the financial flows do not follow any market logic. The markets are continually manipulated and transformed by computer-enacted investment strategies, subjective perceptions of influential analysts, political events in any part of the world, and – most significantly – unsuspected turbulences caused by the complex interactions of capital flows in this highly nonlinear system. These largely uncontrolled turbulences are as important in setting prices and market trends as are the traditional forces of supply and demand.

Global currency markets alone involve the daily exchange of over $3 trillion, and since these markets largely determine the value of any national currency, they contribute significantly to the inability of governments to control economic policy. As a result, the 1990s saw a series of severe financial crises, from Mexico (1994) to the Asian Pacific (1997), Russia (1998), and Brazil (1999). These crises made evident that the financial networks of the new economy are inherently unstable. They produce random patterns of informational turbulence that may destabilize any company, as well as entire countries or regions, regardless of their economic performances.

In the subsequent years, the global financial institutions were able to prevent further crises for almost a decade by making certain structural adjustments to the markets. At the same time, the global economy became ever more interconnected and its financial products ever more abstract; and in 2008 the system spun out of control again, and this time on a global scale.

The worldwide financial crisis and recession of 2008/9 was brought about by Wall Street bankers through a combination of greed, incompetence, and weaknesses inherent in the system (see Heinberg, 2011; Kroft, 2008). It began as a mortgage crisis, caused by the reckless marketing of risky "subprime" loans; then it slowly evolved into a credit crisis; and finally it became a full-blown global financial crisis.

During the mortgage crisis, the big Wall Street investment houses bought up millions of the least dependable mortgages, chopped them up into tiny bits and pieces, and repackaged them as exotic investment securities that hardly anyone could understand. For this repackaging they collected enormous fees. The complexity of these financial instruments lies at the heart of the credit crisis. Nobody knew exactly what they were made of, nor how they would behave.

These complex financial instruments were actually designed by mathematicians and physicists who used computer models to reconstitute the unreliable loans in ways that were supposed to eliminate most of the risks. But their models turned out to be wrong, because physicists and mathematicians are not experts in human behavior, and human behavior cannot be modeled mathematically.

The next phase of the credit crisis involved the creation of so-called credit default swaps (CDSs) that hid the risks until it was too late to do anything about them. CDSs are essentially side bets on the performance of the US mortgage markets and the solvency of some of the biggest financial institutions in the world; this is a form of legalized gambling that allows one to wager on financial outcomes without having to buy the stocks and mortgages.

With their massive sales of CDSs to mortgage investors, the investment banks created a huge, unregulated market that multiplied their losses. When homeowners began to default on their mortgages and Wall Street's high-risk securities also began to fail, the big investment houses and insurance companies who sold the CDSs had not set aside the money to pay all the insurance contracts they had written. This brought down three of the biggest firms on Wall Street (Bear Stearns, Lehmann Brothers, and AIG) and threatened all the others. At this point, the US government stepped in with a set of gigantic bailout packages for the largest Wall Street banks and insurance agencies, using American taxpayers' money to cover the losses of these private businesses, which were considered to be "too big to fail." Soon after receiving their loans and guarantees, the Wall Street banks continued to pay their top executives exorbitant bonuses.

In the new global capitalism, ethical considerations have been systematically excluded, whether they are about human rights, protecting the environment for future generations, or even the basic integrity of doing business honestly. The focus is now exclusively on making money, and the scale of personal rewards on Wall Street is so enormous at the top that greed has eclipsed all considerations of fairness and integrity.

In recent years, so many unethical practices have come to light that all trust and confidence have collapsed. In fact, during the credit crisis of 2008/9, this breakdown of trust was so complete that even the most reputable financial institutions would not lend to each other at all or, at best, just overnight. As a consequence, the money markets that banks organize between themselves completely froze. Credit – the very lifeblood of the economy – dried up, and panic ensued.

The most recent financial scandal, known as the LIBOR scandal, is also by far the largest, affecting payments on financial instruments worth trillions of dollars. It involved widespread rigging of key interest rates – used to determine the London Interbank Offered Rate (LIBOR) – for personal gain by traders at Barclays and several other major banks. In an exposé titled "The Rotten Heart of Finance," *The Economist* (July 7, 2012) noted that this huge global scandal "corrodes further what little remains of public trust in banks and those who run them".

During the last two decades, several authors have pointed out that our global economic system is still poorly understood and susceptible to severe destabilizing turbulences. But it seems that none of them could predict the scale of the system's collapse we recently witnessed on Wall Street. Nor do we know what may still be in store. Complexity theory tells us that, in highly nonlinear systems, small perturbations may give rise to dramatic and unpredictable changes. The current widespread anxiety about the global economy is based on such fears.

Indeed, it did not take long for the effects of the Wall Street crisis to reverberate throughout Europe, Asia, and the Middle East. The Eurozone countries and the UK experienced a dramatic slowing of growth; some Asian countries saw significant slowdowns, and by May 2009 the Arab world had lost an estimated $3 trillion due to the crisis – partly from a crash in oil prices. One year later, Greece faced a government debt crisis that threatened the economic integrity of the EU, with Ireland, the UK, Spain, and Portugal being the countries most at risk of losing their investors' faith (Heinberg, 2011). Today, the inherent instability of the global economy, due to its recurrent and largely uncontrolled turbulences, is evident to all.

It is interesting to apply the systemic understanding of life to the analysis of this phenomenon. The new economy consists of a global meta-network of complex technological and human interactions, involving multiple feedback loops operating far from equilibrium, which produce a never-ending variety of emergent phenomena. Its creativity, adaptability, and cognitive capabilities are certainly reminiscent of living networks. However, it does not display the stability that is also a key property of life. The information circuits of the global economy operate at such speed and use such a multitude of sources that they constantly react to a flurry of information, and thus the system as a whole is generating turbulences that are exceedingly difficult to control.

Living organisms and ecosystems, too, may become continually unstable, but if they do, they will eventually disappear because of natural selection, and only those systems that have stabilizing processes built into them will survive. In the human realm, corresponding regulatory mechanisms will have to be introduced into the global economy through appropriate financial policies, as we discussed in Section 17.2.3.

17.3.4 The social impact of economic globalization

In his trilogy on the Information Age, Manuel Castells provides a detailed analysis of the social and cultural impacts of global capitalism. He describes in particular how the new "network economy" has profoundly transformed the social relationships between capital and labor. In the global economy, money has become almost entirely independent of production and services by escaping into the virtual reality of electronic networks. Capital is global at its core, while labor, as a rule, is local. Thus, capital and labor increasingly exist in different spaces and times: the virtual space of financial flows and the real space of the local and regional places where people are employed; the instant time of electronic communications and the biological time of everyday life (Castells, 1996).

Economic power resides in the global financial networks, which determine the fate of most jobs, while labor remains locally constrained in the real world. Thus labor has become fragmented and disempowered. Many workers today, whether unionized or not, will not fight for higher wages or better working conditions out of fear that their jobs will be moved abroad.

As more and more companies restructure themselves as decentralized networks – networks of smaller units which, in turn, are linked to networks of suppliers and subcontractors – workers are employed increasingly through individual contracts, and thus labor is losing its collective identity and bargaining power. Indeed, in the new economy traditional working-class communities have all but disappeared.

Castells points out that it is important to distinguish between two kinds of labor. Unskilled, "generic" labor is not required to access information and knowledge beyond the ability to understand and execute orders. In the new economy, masses of generic workers move in and out of a variety of jobs. They may be replaced at any moment, either by machines or by generic labor in other parts of the world, depending on the fluctuations in the global financial networks.

"Self-educated" labor, by contrast, has the capacity to access higher levels of education, to process information, and to create knowledge. In an economy where information processing, innovation, and knowledge creation are the main sources of productivity, these self-educated workers are highly valued. Companies would like to maintain long-term, secure relationships with their core workers, so as to retain their loyalty and make sure that their tacit knowledge is passed on within the organization.

As incentive to stay on, such workers are increasingly offered stock options in addition to their basic salaries, which gives them a stake in the value created by the company. This has further undermined the traditional class solidarity of labor.

Economic inequality

Because of the fragmentation and individualization of labor and of the gradual dismantling of the welfare state under the pressures of economic globalization, the rise of global capitalism has been accompanied by rising social inequality and polarization (Castells, 1998). The gap between the rich and the poor has grown significantly, both internationally and within countries. According to the UN Human Development Report (UNDP), the difference in per capita income between the North and South tripled from $5,700 in 1960 to $15,000 in 1993. The richest 20% of the world's people now own 85% of its wealth, while the poorest 20% (who account for 80% of the total world population) own just 1.4% (UNDP, 1996). The assets of the three richest people in the world alone exceed the combined GNP of the least developed countries and their 600 million people (UNDP, 1999).

Within countries, the USA has by far the highest level of inequality among the advanced industrial countries. As the economist Joseph Stiglitz documents in great detail in his eye-opening book *The Price of Inequality*, America had been growing apart at an increasingly rapid rate already before the recent financial crisis (Stiglitz, 2012). During the last three decades, the bottom 90% have seen their income grow by only 15%, while that of the top 1% has grown by about 150%, and that of the top 0.1% by more than 300%.

By 2007, the year before the crisis, the income of the top 0.1% of US households was 220 times larger than the average income of the bottom 90%. Wealth was distributed even more unequally than income, with the wealthiest 1% owning more than a third of the nation's wealth.

The financial crisis exacerbated these inequalities in many ways. The wealthy lost more in stock market values, but those recovered relatively fast. In fact, the "recovery" since the recession overwhelmingly benefited the wealthiest Americans, the top 1% gaining 93% of the additional income created in the country in 2010, while almost one out of six Americans (and almost a quarter of American children) were left in poverty, and economic insecurity threatened many in the working and middle classes whose jobs, retirement incomes, and homes were all at risk.

CEOs, on the other hand, were remarkably successful in maintaining their high pay. By 2010, the ratio between their annual compensation and that of the typical worker was back to where it had been before the crisis – a shocking 243 to 1. This gap between CEO pay and that of the typical worker is far higher in America than in other countries. In Japan, for example, it is 16 to 1. Here is how Stiglitz (2012, p. 7) sums up the situation:

> The simple story of America is this: the rich are getting richer, the richest of the rich are getting still richer, the poor are becoming poorer and more numerous, and the middle class is being hollowed out. The incomes of the middle class are stagnating or falling, and the difference between them and the truly rich is increasing.

These rapidly increasing inequalities were brought about not only by the rise of global capitalism, but also – and perhaps even more – by specific government policies – tax cuts for the very rich, favorable tax rulings for keeping money offshore, deregulation, union busting, and favorable conditions for massive stock options for CEOs. The result has been a systematic transfer of wealth from the poor to the rich. As Stiglitz explains (p. 32), "there are two ways to become wealthy: to create wealth or to take wealth away from others," and in the USA much of the wealth at the top stems from wealth transfers instead of wealth creation.

Breakdown of democracy

Growing economic inequality is accompanied by growing imbalance of political power, as the super-rich – 400 billionaires with well over $1 trillion of net worth between them – have increasingly taken over the American political system. In a vicious circle (or self-amplifying feedback loop) between politics and economics, super-rich donors and big lobbies finance all the important electoral campaigns in exchange for favorable policies that further increase their wealth, leading to even larger campaign contributions, followed by yet more favorable financial policies and greater corporate influence on the political process (Sachs, 2011).

Indeed, the key policies of the first Obama administration were shaped significantly by corporate interests. Big Oil and Big Coal completely silenced the president and his secretary of energy (a physics Nobel Laureate) on the critical issue of climate change and prevented the US Senate from even discussing an appropriate climate and energy policy. The private health insurance industry forced the administration to exclude a national, "single-payer," healthcare system from all discussions of healthcare reform; and the military-industrial complex has steered American foreign policy to a remarkable extent (Sachs, 2011).

The Occupy Movement

By 2011, it had become abundantly clear that the global financial crisis and subsequent recession had perpetuated and even aggravated the economic inequalities created by global capitalism around the world. And suddenly, people all over the world began to rise up and forcefully expressed their outrage. The rapid spread of these popular and largely leaderless uprisings, as well as their mutual declarations of solidarity, were guided by the vast array of social media that have become critical political tools for communities and organizations in our age of globalization.

The uprisings began with a youth movement in Tunisia in December 2010, which spread to nearby Egypt and then to other countries in the Middle East, growing into huge social movements against the region's dictators that became known collectively as the Arab Spring. The protests in Cairo's Tahrir Square in January and February 2011, kept alive by hundreds of thousands for almost three weeks, provided the model of the occupation of public squares and buildings that inspired the concurrent occupation of the capitol of the US state of Wisconsin by tens of thousands protesting against antiunion legislation; the mass movement of *los indignados* ("the indignant," or "outraged") in Spain during the subsequent May and June; and finally the "Occupy Wall Street" movement, which began in September 2011 with the occupation of Zuccotti Park in New York's financial district and subsequently gave rise to the broader "Occupy Movement" in the USA and around the world (see van Gelder, 2011).

The specific grievances of these uprisings varied from country to country; those in the Middle East, in particular, were very different from the grievances in the West. However, as Joseph Stiglitz (2012) points out, there were some shared themes – a common understanding that in all those countries the economic and political systems had failed and were fundamentally unfair. Occupy Wall Street identified the essence of this feeling with the brilliant slogan "We are the 99%," which quickly caught on in the media and decisively shaped the American political dialogue. Soon government agencies, think tanks, and the media confirmed with numerous statistics that the super-rich, the "1%," indeed enjoyed high and unjustified levels of inequality; and even though the numbers of actual protesters were relatively small, two-thirds of the American public said that the Occupy Movement expressed their values and that they supported it.

During the winter of 2011–12, the visibility of the Occupy Movement diminished, partly because of the winter cold and of some internal fragmentation, but mostly because of concerted, often brutal, and seemingly nationally coordinated police repression (see La Botz *et al.*, 2012). At this time of writing, one year after the occupation of Zuccotti Park, the further evolution of the Occupy Movement is unclear. However, even at this stage of transition, the achievements of the movement are substantial. By highlighting the contrast between the 1% and the 99%, it has refocused the American political debate, made economic inequality a central theme in the 2012 presidential election, and brought about new political alliances. To quote Joseph Stiglitz once more, "The '99 percent' marks an attempt to forge a new coalition – a new sense of national identity, based not on the fiction of a universal middle class but on the reality of the economic divides within our economy and our society."

17.3.5 The ecological impact

According to the doctrine of economic globalization – known as "neoliberalism," or "the Washington consensus" – the free-trade agreements imposed by the WTO on its member countries (see Section 17.3.3) will increase global trade; this will create a global economic expansion; and global economic growth will decrease poverty, because its benefits will eventually "trickle down" to all. As political and corporate leaders like to say, the rising tide of the new economy will lift all boats.

The analysis by Manuel Castells (1996) shows clearly that this reasoning is fundamentally flawed. Global capitalism does not alleviate poverty and social exclusion; on the contrary, it exacerbates them. The Washington consensus has been blind to this effect because corporate economists have traditionally excluded the social costs of economic activity from their models (see Section 3.5). Similarly, most conventional economists have ignored the new economy's environmental cost – the increase and acceleration of global environmental destruction, which is as severe, if not more so, than its social impact.

As we discussed in Section 17.2, the central enterprise of current economic theory and practice – the striving for continuing, undifferentiated economic growth – is clearly unsustainable. Indeed, since the turn of this century it has become abundantly clear that our economic activities are harming the biosphere and human life in ways that may soon become irreversible (see Brown, 2008, 2009, 2011). In this precarious situation, it is paramount for humanity to systematically reduce our impact on the natural environment. As Al Gore (then a US senator) declared courageously in 1992, "We must make the rescue of the environment the central organizing principle for civilization" (Gore, 1992, p. 269).

Unfortunately, instead of following this admonition, global capitalism has significantly increased our harmful impact on the biosphere. In *The Case Against the Global Economy* (see Section 17.3.1), the late Edward Goldsmith, founding editor of the pioneering European environmental journal *The Ecologist*, gave a succinct summary of the environmental impact of economic globalization (Goldsmith, 1996). He illustrated how the increase of environmental destruction may be linked to increasing economic growth with the examples of South Korea and Taiwan. During the 1990s, both countries achieved stunning rates of growth and were held up as economic models for the Third World by the World Bank. At the same time, the resulting environmental damage was devastating.

In Taiwan, for example, agricultural and industrial poisons severely polluted nearly every major river. In some places, the water was not only devoid of fish and unfit to drink but also was actually combustible. The level of air pollution was twice that considered harmful in the USA; cancer rates had doubled since 1965, and the country had the world's highest incidence of hepatitis. In principle, Taiwan could have used its new wealth to clean up its environment. However, competitiveness in the global economy is so extreme that environmental regulations were eliminated rather than strengthened in order to lower the costs of industrial production.

Resource depletion and environmental destruction

One of the tenets of neoliberalism is that poor countries should concentrate on producing a few special goods for export in order to obtain foreign exchange, and should import most other commodities. This emphasis on export has led to the rapid depletion of the natural resources required to produce export crops in country after country – diversion of fresh water from vital rice paddies to prawn farms; a focus on water-intensive crops, such as sugar cane, that result in dried-up river beds; conversion of good agricultural land into cash-crop plantations; and forced migration of large numbers of farmers from their lands. All over the world there are countless examples of how economic globalization is worsening environmental destruction (Goldsmith, 1996).

The dismantling of local production in favor of exports and imports, which is the main thrust of the WTO's free-trade rules, dramatically increases the distance "from the farm to the table." In the USA, the average ounce of food now travels more than 8,000 km (5,000 miles) before being eaten (Weber and Matthews, 2008), putting enormous stress on the environment. New highways and airports cut through primary forests; new harbors destroy wetlands and coastal habitats; and the increased volume of transport further pollutes the air and causes frequent oil and chemical spills. Studies in Germany have shown, moreover, that the contribution of nonlocal food production to climate change is between six and twelve times higher than that of local production, due to increased CO_2 emissions (Shiva, 2000).

As the ecologist and agricultural activist Vandana Shiva (2000, p. 112) points out, the impact of climate instability and ozone depletion is born disproportionately by the South, where most regions depend on agriculture and where slight changes in climate can totally destroy rural livelihoods. In addition, many transnational corporations use the free-trade rules to relocate their resource-intensive and polluting industries in the South, thus further worsening environmental destruction. The net effect, in Shiva's words, is that "resources move from the poor to the rich, and pollution moves from the rich to the poor."

The destruction of the natural environment in Third World countries goes hand in hand with the dismantling of rural people's traditional, largely self-sufficient ways of life, as American television programs and transnational advertising agencies promote glittering images of modernity to billions of people all over the globe without mentioning that the lifestyle of endless material consumption is utterly unsustainable. Edward Goldsmith (1996) estimated that, if all Third World countries were to reach the consumption level of the USA by the year 2060, the annual environmental damage from the resulting economic activities would be 220 times what it is today, which is not even remotely conceivable.

Since moneymaking is the dominant value of global capitalism, its representatives seek to eliminate environmental regulations under the guise of "free trade" wherever they can, lest these regulations interfere with profits. Thus the new economy causes environmental destruction not only by increasing the impact of its operations on the world's ecosystems but also by eliminating national environmental laws in country after country. In other words, environmental destruction is not only a side effect but also an integral part of the design of global capitalism.

Climate change

Among all the environmental problems engendered by global capitalism, climate change is by far the most dangerous, threatening the very existence of life as we know it on our planet. Climate change science is now a well-established scientific field with a history of several decades (see McKibben, 2012a). Its findings are unequivocal. The emissions of greenhouse gases due to human activities are now dangerously heating the planet. Climate scientists estimate that, to avoid an irreversible catastrophe, we need to cut global CO_2 emissions by 80% by 2020 (see Box 17.1). This will require decisive global action – "mobilizing to save civilization," as Lester Brown (2008) puts it.

Box 17.1
The basics of climate change science

When sunlight warms the surface of the Earth, a large portion of the reflected thermal radiation is absorbed by greenhouse gases in the atmosphere. In the early history of the planet, this "greenhouse effect" created the protective envelope in which life was able to unfold, and the giant feedback loops of the Gaia system maintained the Earth's atmosphere at a stable temperature range, conducive to life (see Section 16.2.3).

Since the Industrial Revolution, however, human activities have generated excessive greenhouse gas emissions. Thus, excessive amounts of heat have been trapped by the greenhouse effect, resulting in the global warming of the Earth's atmosphere beyond safe levels. The principal sources of human-induced greenhouse gases are the burning of fossil fuels and deforestation (CO_2 emissions), and the management of livestock (emissions of methane).

Warmer air means that there is more energy and more moisture in the atmosphere, and this can lead to a wide variety of consequences – floods, tornados, and hurricanes, but also droughts, heat waves, and wildfires. Indeed, during 2011 and 2012 we saw a whole series of climate catastrophes that are consistent with the predicted effects of global warming. Record heat waves in Pakistan, central Russia, and western Europe, triggering wildfires and droughts; mega-floods in Australia, Brazil, Sri Lanka, the Philippines, and South Africa, causing deluges and massive landslides; and the most deadly and destructive tornado season ever seen in North America.

Climate change science is highly complex, and exact predictions are very difficult. However, climate scientists have recently been able to identify some general patterns that seem to be consistent with the currently observed climate catastrophes. The most alarming discovery has been that human emissions of greenhouse gases have caused the Arctic to warm about twice as fast as the rest of the Northern Hemisphere, due to two unique feedback loops in the Arctic climate system. The first has been known for a long time. As the ice melts, it exposes the darker ocean underneath, which absorbs heat rather than reflecting it back into space, thus accelerating the ice melt.

The second feedback loop also has to do with the reflectivity, or "albedo," of ice. Recent studies of Greenland's ice sheet have revealed that its albedo gradually decreases as the ice warms, even before the actual melting. As ice crystals warm, they lose their jagged edges, becoming rounder and reflecting less light. Satellite data have shown a steady darkening of Greenland's albedo from a July average of 74% in 2000 to less than 65% in 2011 (McKibben, 2012c). In addition, the ice sheet was darkened by soot from wildfires in Colorado and Siberia – themselves provoked by climate change. The net effect is a vicious circle similar to the one involved in the complete ice melt. Loss of albedo leads to greater absorption of heat, which in turn accelerates the melting.

An even more recent discovery concerns the surprising effects of the self-perpetuating Arctic ice melt on the climate of northern Europe and North America. The key phenomenon here is the north polar jet stream, a powerful air current that steers weather systems from west to east around the northern hemisphere. The current forms a boundary between adjacent air masses, separating cold and wet air to the north from warmer and drier air to the south. The path of the jet stream has a meandering shape with the meanders themselves propagating eastward.

In a recent paper, the climate scientists Jennifer Francis and Stephen Vavrus show how Arctic warming may be the cause of the extreme weather conditions now observed in formerly temperate zones (Francis and Vavrus, 2012). The authors explain that the rapid warming of the Arctic has two distinct effects on the polar jet stream: it deepens the meanders, and it slows down their eastward progression. The steep meanders can drag severe ice and snow from the Arctic deep into southern regions and, conversely, may push hot and dry weather from the southern regions far to the north; and because of the meanders' slower progression, these extreme weather conditions – droughts or floods, for example – may be much more persistent than they used to be.

The world authority on the climate crisis is the Intergovernmental Panel on Climate Change (IPCC), a body of over 2,500 leading climate scientists. After 20 years of detailed studies and four unanimous reports, the IPCC has stated that it is "unequivocal" that the climate is changing and will continue to change, and that human generation of greenhouse gases is responsible for most changes since the 1950s (www.ipcc.ch).

The IPCC reports that the Earth's temperature has risen by 0.6 °C since 1970, and that dangerous climate change is considered inevitable beyond a rise of 2 °C. Concentrations of CO_2 in the atmosphere have been measured for the past 650,000 years, using air bubbles in Arctic ice. At no point before the industrial era did the CO_2 concentration exceed 300 parts per million (ppm). Current CO_2 concentration is 390 ppm and is increasing by about 2 ppm every year. Climate scientists estimate that, to avoid runaway climate change, we need to reduce CO_2 concentrations to a safe level of 350 ppm. This means that we need to cut global CO_2 emissions by 80% by 2020.

It is worth noting that the climate catastrophes experienced around the world during the past two years occurred with a temperature increase of less than 1 °C and with CO_2 concentrations of less than 400 ppm. Climate scientists predict that without dramatic action, we will reach up to 6 °C and up to 550 ppm by the end of the century.

So far, such decisive mobilization has not occurred, and the main stumbling block in one international climate conference after the other has been the government of the USA. As we have mentioned, the American oil and coal lobbies have effectively blocked all attempts by the president and the senate to develop a coherent climate and energy policy.

More than that, the fossil-fuel lobbies who are opposing climate-change legislation have financed sophisticated campaigns aimed at actively misleading the public about the nature and severity of the climate crisis. These campaigns are modeled on the disinformation campaigns of the tobacco industry. Their purpose is to systematically create doubt and confusion about the overwhelming scientific consensus concerning the threat of global warming (Oreskes and Conway, 2010). As a consequence, the American public remains largely ignorant about this vital issue.

The concerted obstruction of the fossil-fuel lobby is typical of big corporations in the global capitalist system, who always tend to put profits before ethical considerations. But this fact is not enough to explain their fierce and resourceful resistance against climate change legislation. As Bill McKibben, one of the most eloquent writers on global warming, points out in a recent article, titled "Global Warming's Terrifying New Math," the predicament of the fossil-fuel industry is encapsulated in two important numbers (McKibben, 2012b).

The first number is 565 gigatons (billion metric tons). It is the amount of CO_2 we can still emit into the atmosphere by mid century while staying below a temperature increase of 2 °C (the limit beyond which climate change is likely to spin out of control). At the current rate of global annual emissions of about 32 gigatons, the limit of 565 gigatons would buy us only 17 years before global climate collapse.

If this number is frightening, the second number is even more scary, writes McKibben. It is the amount of carbon locked up in the proven oil and coal reserves of the fossil-fuel companies and oil producing states: 2,800 gigatons. This is the amount of fossil fuel these companies and states are currently planning to burn – and it is about five times more than the safe limit of 565 gigatons!

As McKibben explains, these 2,800 gigatons of carbon are technically still in the soil, but, economically, they are already above ground, since they are listed as assets on the balance sheets of their owners. They give the fossil-fuel companies their value, and they figure in the national budgets of oil-producing countries. If these companies were to admit that, in order to avoid total climate collapse, they can pump out no more than 20% of their reserves, their values would plummet. At today's market value, they would collectively lose $20 trillion in assets.

This explains why Big Oil and Big Coal fight so hard against any restrictions on carbon emissions, even to the extent of systematically denying the science of climate change. As McKibben (2012b) concludes, "You can have a healthy fossil-fuel balance sheet, or a relatively healthy planet – but now that we know the numbers, it looks like you can't have both."

17.4 The global civil society

During the last two decades, the new economy's social and ecological impacts have been discussed extensively by scholars and community leaders, as we have documented in the preceding pages (Section 17.3). Their analyses make it abundantly clear that global capitalism in its present form is unsustainable – socially, ecologically, and even financially – and needs to be fundamentally redesigned (Cavanagh and Mander, 2004; Hutton and Giddens, 2000; Mander, 2012; Mander and Goldsmith, 1996; Korten, 2001; Shiva, 2005).

17.4.1 Core values of human dignity and ecological sustainability

In any realistic discussion of how to redesign the global economy, it is useful to remember that the current form of economic globalization has been consciously designed and thus can be reshaped. As we have pointed out, the "global market" is really a network of machines programmed according to the fundamental principle that moneymaking should take precedence over any other human value (see Section 17.3.3). However, human values can change; they are not natural laws. The same electronic networks of financial flows *could* have other values built into them. The critical issue is not technology, but ethics and politics.

As we discussed in our chapter on science and spirituality, ethics refers to a standard of human conduct that flows from a sense of belonging (see Section 13.2.4). When we belong to a community, we behave accordingly. In the context of globalization, there are two relevant communities to which we all belong. We are all members of humanity, and we all belong to the global biosphere. As members of *oikos*, the "Earth Household," it behooves us to behave in such a way that we do not interfere with nature's inherent ability to sustain the web of life. As members of the human community, our behavior should reflect a respect of human dignity and basic human rights.

Since the systemic conception of life encompasses biological, cognitive, and social dimensions, human rights, from a systems point of view, should be respected in all three of these dimensions. The biological dimension includes the right to a healthy environment and to secure and healthy food. Human rights in the cognitive dimension include the right of access to education and knowledge, as well as the freedom of opinion and expression. In the social dimension, finally, there is a wide range of human rights – from social justice to the right of peaceful assembly, cultural integrity, and self-determination.

In order to combine respect for these human rights with the ethics of ecological sustainability, we need to remember that sustainability – in ecosystems as well as in human society – is not an individual property but a property of an entire web of relationships: it involves a whole community (see Section 16.3). A sustainable human community interacts with other communities – human and nonhuman – in ways that enable them to live and develop according to their nature. In the human realm, therefore, sustainability is fully consistent with the respect of cultural integrity, cultural diversity, and the basic right of communities to self-determination and self-organization.

17.4.2 The Seattle Coalition

The values of human dignity and ecological sustainability, as outlined above, form the ethical basis for reshaping globalization. At the turn of this century, an impressive global coalition of NGOs formed around these core values. The numbers of international NGOs increased dramatically over the past few decades, from several hundred in the 1960s to over 20,000 by the end of the twentieth century. During the 1990s, a computer-literate elite emerged within these international NGOs. They began to skillfully use new communication technologies, and especially the internet, to network with one another and mobilize their members.

This networking became especially intense as they prepared joint protest actions for the meeting of the WTO in Seattle in November, 1999. For many months, hundreds of NGOs interlinked electronically to coordinate their plans and to issue a flurry of pamphlets, position papers, press releases, and books in which they clearly articulated their opposition to the WTO's policies (Barker and Mander, 1999). This literature was virtually ignored by the WTO, but had a significant impact on public opinion. The NGOs' educational campaign culminated in a two-day teach-in in Seattle before the WTO meeting, organized by the International Forum on Globalization and attended by over 2,500 people from around the world (Hawken, 2000).

On November 30, 1999, around 50,000 people belonging to more than 700 organizations took part in a superbly coordinated, passionate, and almost entirely nonviolent protest that permanently changed the political landscape of globalization.

The Seattle police turned out in force to keep the protesters away from the convention center where the WTO meeting took place, but they were unprepared for the street actions of a massive, well-organized network totally committed to shutting down the WTO. Chaos ensued, hundreds of delegates were blocked off in the streets or confined to their hotels, and the opening ceremony had to be canceled.

The WTO meeting broke down not only because of these massive demonstrations but also – and perhaps even more so – because of the way the major powers within the WTO bullied the delegates from the South (Khor, 1999, 2000). After ignoring dozens of proposals from developing countries, the WTO leaders excluded the delegates representing these countries from critical behind-the-scenes "green room" meetings and then pressured them to sign a secretly negotiated agreement. Infuriated, many developing countries refused to do so, thereby joining the massive opposition to the WTO that was going on outside the convention center. Faced with the prospect of rejection by developing nations in the final session, the major powers preferred to let the Seattle meeting collapse without even attempting to issue a final declaration. Thus the Seattle meeting, which was meant to be a celebration of the WTO's solidification, instead became the symbol of worldwide resistance.

After Seattle, smaller but equally effective demonstrations took place at other international meetings, and by the end of 2000, over 700 organizations from 79 countries had joined what they now officially called the International Seattle Coalition. Naturally, there is a great diversity of interests in these NGOs, which range from labor organizations to human rights, women's rights, religious, environmental, and indigenous peoples' organizations. However, there is remarkable agreement among them about the core values of human dignity and ecological sustainability.

17.4.3 "Another world is possible!"

During the subsequent years, the Seattle Coalition, or "Global Justice Movement," as it came to be called later on, not only organized a series of very successful protests at various meetings of the WTO, the G7, and the G8 but also held several World Social Forum meetings, most of them in Brazil, with the official motto "Another world is possible!" (Sen and Waterman, 2009). At these meeetings, the NGOs proposed a whole set of alternative trade policies, including concrete and radical proposals for restructuring the global financial institutions, which would profoundly change the nature of globalization (see Section 18.1.1).

The Global Justice Movement exemplifies a new kind of political movement that is typical of our Information Age. Because of their skillful use of the internet and the new social media, the NGOs in the coalition are able to network with each other, share information, and mobilize their members with unprecedented speed. As a result, the new global NGOs have emerged as effective political actors who are independent of traditional national or international institutions. They constitute a new kind of global civil society.

Civil society is traditionally defined as a set of organizations and institutions – churches, political parties, unions, cooperatives, and various voluntary associations – that form an interface between the state and its citizens. The institutions of civil society represent the interests of the people and constitute the political channels that connect them to the state. With the rise of the global network society, the nation state and its traditional institutions have been losing power, while a new kind of civil society, organized around reshaping globalization, has gradually emerged (Castells, 1996). It does not define itself *vis-à-vis* the state, but is global in its scope and organization. It is embodied in powerful international NGOs – Oxfam, Amnesty International, Greenpeace, and others – as well as in coalitions of hundreds of smaller organizations, all of which have become social actors in a new political environment.

According to the political scientists Craig Warkentin and Karen Mingst (2000), the new civil society is characterized by a shift of focus from formal institutions to social and political relationships among its actors. These relationships are structured around two different kinds of networks. On the one hand, NGOs rely on local grass-roots organizations (i.e., on living human networks); on the other hand, they skillfully use the new global communication technologies (i.e., electronic networks). By creating this unique link between human and electronic networks, the global coalition of NGOs has reshaped the political landscape.

17.5 Concluding remarks

We hope to have demonstrated in this chapter that the major problems of our time are systemic problems – all interconnected and interdependent – and that, accordingly, they require systemic solutions. This is why the systems view of life that we have discussed throughout this book not only is fascinating intellectually but also has tremendous practical relevance.

In addition to the academic fields where the systems view of life is being developed, systemic thinking is practiced today in numerous research institutes and centers of learning established by the global civil society (see Chapter 18). In fact, there is a continual exchange of ideas between the two areas, as quite a few academic systems thinkers are also teaching in civil-society institutions and are engaged in various projects and campaigns that develop systemic solutions to the world's problems.

One of the hallmarks of a systemic solution is that it solves several problems simultaneously. Agriculture offers a good example. If we changed from our chemical, large-scale industrial agriculture to organic, community-oriented, sustainable farming, this would contribute significantly to solving three of our biggest problems. It would greatly reduce our energy dependence, because we are now using (in the USA) one-fifth of our fossil fuels to grow and process food. The healthy, organically grown food would have a huge positive effect on public health, because many chronic diseases – heart disease, stroke, diabetes, and about 40% of cancers – are linked to our diet. And finally, organic farming would contribute significantly to fighting climate change, because an organic soil is a carbon-rich soil, meaning that it draws CO_2 from the atmosphere and locks it up in organic matter.

Today, hundreds of systemic solutions are being developed all over the world to solve problems of the economy, environmental degradation, energy, climate change, food insecurity, and so on (see Brown, 2008, 2009). In the following, and last, chapter of this book, we shall highlight some of the most far-reaching and most promising of these solutions.

18

Systemic solutions

18.1 Changing the game

In the preceding pages we chronicled the rise of a global civil society, based on the core values of human dignity and ecological sustainability, which exemplifies a new political movement, independent of traditional national and international institutions and typical of our Information Age.

To place the political discourse within a systemic and ecological perspective, we observe that the global civil society relies on a network of scholars, research institutes, think tanks, and centers of learning that largely operate outside our leading academic institutions, business organizations, and government agencies. There are dozens of these institutions of research and learning in all parts of the world today (see Box 18.1 for a short list). They all have their own websites and are interlinked with one another and with the more activist-oriented NGOs, for whom they provide the necessary intellectual resources. Their common characteristic is that they pursue their research and teaching within an explicit framework of shared core values.

Most of these research institutes are communities of both scholars and activists who are engaged in a wide variety of projects and campaigns. Among them, there are four clusters that seem to be focal points for the largest and most active grass-roots coalitions. The first is the challenge of reshaping the governing rules and institutions of globalization; the second is the task of raising awareness of the climate crisis and catalyzing leadership for developing appropriate energy and climate policies; the third issue cluster is the opposition to genetically modified (GM) foods and the promotion of sustainable agriculture; and the fourth is ecodesign – a concerted effort to redesign our physical structures, cities, technologies, and industries so as to make them ecologically sustainable.

In this chapter, we shall review various aspects of these endeavors that offer systemic solutions to the major problems of our time. We shall begin in this section with a review of proposals for reshaping globalization, developed by a task force of leading international NGOs, and we shall then discuss the emergence around the world of new ownership structures as alternatives to the dominant corporate structures – a growing ownership revolution.

Box 18.1

Research institutes and centers of learning of the global civil society

We have arranged this short list of NGOs in three groups, according to three dimensions of the systems view of life: the cognitive, the social, and the ecological dimensions. In the cognitive dimension, we find organizations collecting data about the state of the world and about proposed systemic solutions, as well as organizations offering education at various levels. In the social dimension, we find organizations dealing with economics, international relations, global social justice, and sustainable business; and in the ecological dimension we find the issues of climate change, energy, water, biodiversity, food and agriculture, and ecodesign. For more extensive lists, see Edwards (2010); Hawken (2008).

I Cognitive dimension	
State of the world; systemic solutions	
Earth Policy Institute	www.earth-policy.org
Worldwatch Institute	www.worldwatch.org
Wuppertal Institute	www.wupperinst.org
Tellus Institute	www.tellus.org
Global Footprint Network	www.footprintnetwork.org
Redefining Progress	www.rprogress.org
Education	
Center for Ecoliteracy	www.ecoliteracy.org
Second Nature	www.secondnature.org
Schumacher College	www.schumachercollege.org.
Resurgence Magazine	www.resurgence.org
Bioneers	www.bioneers.org
II Social dimension	
Economics	
International Forum on Globalization	www.ifg.org
New Economics Foundation	www.neweconomics.org
Foundation on Economic Trends	www.foet.org
International relations	
Institute for Policy Studies	www.ips-dc.org
Global Trade Watch	www.citizen.org/trade
Global social justice	
Third World Network	www.twnside.org.sg
The Cultural Conservancy	www.nativeland.org
Grameen Foundation	www.grameenfoundation.org
Sustainable business	
Natural Capitalism Solutions	www.natcapsolutions.org
Social Venture Network	www.svn.org

Business for Social Responsibility	www.bsr.org
Ecotrust	www.ecotrust.org
III Ecological dimension	
Climate change	
The Climate Reality Project	www.climaterealityproject.org
350.org	www.350.org
Sierra Club	www.sierraclub.org
Greenpeace	www.greenpeace.org
Rainforest Action Network	www.ran.org
Transition Network	www.transitionnetwork.org
Energy, water	
Rocky Mountain Institute	www.rmi.org
Council of Canadians	www.canadians.org
Biodiversity	
Navdanya	www.navdanya.org
Green Belt Movement	www.greenbeltmovement.org
Food and agriculture	
Via Campesina	www.viacampesina.org
Slow Food	www.slowfood.org
Sociedad Cientifica LatinoAmericana de Agroecologia (SOCLA)	www.agroeco.org/socla
Ecodesign	
Zero Emissions Research Initiatives	www.zeri.org
Biomimicry Institute	www.biomimicryinstitute.org
World Green Building Council	www.worldgbc.org
Global Ecovillage Network	www.gen.ecovillage.org

18.1.1 Reshaping globalization

Even before the Seattle teach-in in November 1999, the leading NGOs in the Seattle Coalition had formed an "Alternatives Task Force" under the leadership of the International Forum on Globalization (IFG) to synthesize the key ideas about alternatives to the current form of economic globalization. In addition to IFG, the task force included the Institute for Policy Studies (USA), Global Trade Watch (USA), the Council of Canadians (Canada), Focus on the Global South (Thailand and Philippines), the Third World Network (Malaysia), and the Research Foundation for Science, Technology, and Ecology (India).

After more than two years of meetings, the task force put together a draft interim report, "Alternatives to Economic Globalization," and disseminated it within the global coalition of NGOs. For the next three years, the report was enriched and refined through dialogues and workshops with scholars and grass-roots activists around the world before being published in its final version (Cavanagh and Mander, 2004).

The IFG synthesis of alternatives to economic globalization contrasts the values and organizing principles underlying the neoliberal Washington consensus (see Section 17.3.5) with a set of alternative principles and values. These include a shift from governments serving corporations to governments serving people and communities; the creation of new rules and structures that favor the local and follow the principle of subsidiarity ("Whenever power can reside at the local level, it should reside there"); respect for cultural integrity and diversity; and a strong emphasis on food sovereignty (the right to healthy and safe food, produced locally and sustainably); as well as core labor, social, and other human rights.

The alternatives report makes it clear that the Seattle Coalition does not oppose global trade and investment, provided that they help build healthy, respected, and sustainable communities. However, it emphasizes that the recent practices of global capitalism have shown that we need a set of rules stating explicitly that certain goods and services should not be commodified, traded, patented, or subjected to trade agreements.

In addition to already existing rules of this kind, which concern endangered species and goods that are harmful to the environment or to public health and safety – toxic waste, nuclear technology, armaments, etc. – the new rules would also concern goods that belong to the "global commons" – that is, goods that are part of the fundamental building blocks of life or of humanity's common inheritance. Included in this are goods like bulk fresh water, which should not be traded but should be given away to those in need; seeds, plants, and animals that are traded in traditional farming communities but should not be patented for profit; and DNA sequences that should neither be patented nor traded.

The authors of the report acknowledge that these issues constitute perhaps the most difficult, but also the most important, part of the globalization debate. Their main concern is to stem the tide of a global trading system where everything is for sale, even our biological heritage, or access to seeds, food, air, and water – elements of life that were once considered sacred.

In addition to the discussions of alternative values and organizing principles, the IFG synthesis includes concrete and radical proposals for restructuring the "Bretton Woods" institutions (see Section 17.3.3). Most of the NGOs in the Seattle Coalition felt that reforming the WTO, World Bank, and IMF was not a viable strategy, because their structures, mandates, purposes, and operating processes are fundamentally at odds with the core values of human dignity and ecological sustainability. Instead, the NGOs proposed a four-part restructuring process: dismantling the Bretton Woods institutions, unifying global governance under a reformed UN system, strengthening certain existing UN organizations, and creating several new organizations within the UN that would fill the gap left by the Bretton Woods institutions.

The report points out that we now have two strikingly different sets of institutions of global governance: the Bretton Woods triad and the UN. The Bretton Woods institutions have been more effective in implementing well-defined agendas, but these have been largely destructive and have been imposed on humanity in coercive, undemocratic ways. Indeed, in the view of the Seattle Coalition, these institutions bear major responsibility for burdening Third World countries with unpayable foreign debts and for implementing a misguided concept of development that has had disastrous social and ecological consequences (see Sections 17.3.4 and 17.3.5). The UN, by contrast, has been less effective, but its mandate is much broader, its decision-making processes are more open and democratic, and its agendas give much greater weight to social and environmental priorities.

The NGOs argue that limiting the powers and mandates of the IMF, World Bank, and WTO would create space for a reformed UN to fulfill its intended functions. The main thrust of their proposals is to decentralize the power of global institutions in favor of a pluralistic system of regional and international organizations, each of which would be checked by other organizations, agreements, and regional groupings. It seems indeed that such a less structured and more fluid system of global governance would be more appropriate for today's world, in which political authority is increasingly shifting to regional and local levels, giving rise to new types of political organizations, which Manuel Castells (1998) has termed "network states."

18.1.2 Reforming the corporation

As we noted in our previous chapter (in Section 17.2), the pursuit of unlimited economic growth appears to be the root cause of our multifaceted global crisis. The obsession with perpetual economic growth, in turn, is driven by the relentless pursuit of corporate growth, which is built into the very structure and legal framework of the corporation. Hence, there is a broad consensus among social-change activists today that reshaping global capitalism will not be possible without fundamental changes in corporate structures, brought about by a thorough reform of corporate law (Cavanagh and Mander, 2004; Heinberg, 2011; Kelly, 2001; Korten, 2001; Mander, 2012).

To discuss this issue, it will be useful to first review some basic terminology. Whereas the terms "company" and "firm" are generally used as synonyms of "business organization," a corporation is understood as an incorporated (i.e., registered) legal entity, owned by shareholders and controlled by a board of directors that is appointed by the shareholders. At the very heart of the corporate structure is the legal mandate to maximize returns for the corporation's shareholders, even if it means sacrificing the well-being (and, indeed, the continuing employment) of its employees, the livelihood of local communities, or the protection of the natural environment. This stringent mandate, often referred to as "fiduciary duty," is seen as the sole aim of the corporation. Rising share prices are exalted as the very definition of corporate success, and directors who fail to maximize shareholder returns can be sued.

Within a corporation, individual managers may be very caring people, embracing the ideals of social justice and environmental responsibility. But the organizational structure of the corporation and the severe constraints of its financial statements force them to behave in certain ways regardless of their personal values. The shareholders, in turn, are generally not excessively greedy and do not demand ever-increasing wealth. Very often they are not even aware of what happens to their money, leaving the details to their investment advisers. Like the corporation's managers, they may have great personal integrity but are trapped in a corporate structure that is devoid of all ethics. In other words, the problem of relentless corporate growth lies in the design of the system – it is a systemic problem.

It is worth noting that the primacy of the interest of shareholders in corporate governance is a relatively recent invention. The first major corporations – the East India Company, Hudson's Bay Company, and others – were chartered in the seventeenth century by European nations to lead their colonialist ventures. Two centuries later, smaller corporations were chartered in the USA at the state level for purposes that served the public good, like building roads and bridges. They were closely regulated and were allowed to exist only for a limited time. By the late nineteenth century, after the Civil War, this system was overturned. Corporations became private, were granted a plethora of new rights (including shareholders' limited liability), and began to see the financial gains of their shareholders as their sole purpose.

Today, the understanding of fiduciary duty as a mandate to maximize shareholder return is deeply ingrained in our business culture. Surprisingly, however, it does not have a firm legal basis. As the legal scholar D. Gordon Smith (1998) explains, it developed in so-called common law – law formulated by judges through court decisions, rather than by statutes adopted through legislation. Only in recent years has the concept of fiduciary duty been written into incorporation statutes, and even there, Smith points out, directors are generally required to act "in the interest of the corporation." It was judges again, not legislators, who narrowly defined corporate interest to mean shareholder interest alone. Common law, however, can easily be overturned by legislation. Thus, from a legal point of view, it would be easy to expand the concept of fiduciary duty to include the well-being of employees, the local community, and other stakeholders.

To do so would be consistent with the systemic understanding that all human organizations have a dual nature. On the one hand, they are social institutions designed for specific purposes; on the other hand, they are communities of people interlinked through a variety of informal networks (see Section 14.5). Today's widespread belief that corporations exist primarily to maximize returns to their shareholders favors the wealth of the shareholders over the well-being of the corporations' communities, which are as important for the flourishing of the organization as a whole.

In the corporate view of economics, shareholders should be paid as much as possible and employees as little as possible. And yet, it is the employees who contribute to companies year after year, while shareholders contribute very little beyond their initial investment. In fact, even the boards of directors do not govern in significant ways, except for protecting shareholder returns. The employees, by contrast, keep the corporation running on a daily basis. But in corporate governance they are largely invisible and have no rights to the value they help create.

The central mission of corporations to maximize the returns of their shareholders while minimizing the income of their employees is one of the main causes of the rising social and economic inequality we discussed in Section 17.3.4. In her illuminating book, *The Divine Right of Capital*, the business journalist Marjorie Kelly (2001) argues that this form of wealth discrimination is rooted in the ancient aristocratic ideology that those who own property or wealth are superior, and that their rights were granted by divine authority. Indeed, wealth privilege is the hallmark of aristocracy.

In the twentieth century, aristocracy gave way to democracy in many countries around the world, but financial aristocracy remains embedded in the structure of the modern corporation. "This concentrated wealth," Kelly points out, "controls not only corporations but also government. Rule by the financial aristocracy is the reality of life in America today" (2001, p. 10).

Foremost among the privileges of nineteenth-century aristocrats was the right to continuing streams of income without any duty to engage in productive activities. Today, the same privilege is bestowed upon the corporation's shareholders and is denied to its employees. Another basic aristocratic privilege was the exemption from paying taxes. This privilege, too, lives on in today's financial aristocracy whose members consider it fair that they should pay capital gains taxes below the level of employment taxes paid by workers.

Identifying the corporation only with its legal and financial structures and not with its communities of employees has led to the notion that corporations are pieces of property owned by the shareholders, like feudal estates owned by the aristocracy. At the same time, paradoxically, laws made for humans have been applied increasingly to corporations. In the USA, "commercial speech" (i.e., advertising) is protected, with a few exceptions, by the First Amendment, which grants citizens the freedom of speech. More recently (January 21, 2010), this decision to allow corporations to exercise "free speech" was interpreted by the Supreme Court (in the case *Citizens United* v. *Federal Election Commission*) to imply their right to unlimited financial contributions to political campaigns, thus vastly increasing their already excessive dominance of the American political system. However, while corporations are given the rights of people, they do not assume the responsibilities of human individuals, being designed so that none of their executives can be held fully accountable for corporate activities.

The European monarchies crumbled and gave way to democracies when their sustaining myth of the divine right of kings lost its credibility, Similarly, Marjorie Kelly argues, today's financial aristocracy will crumble and give way to an economic democracy when its core myth of the primacy of shareholder interest loses its credibility. Hence, the most important strategy in the attempt to reform the corporation will be to expose the core myth that shareholder returns must be maximized at the expense of human and ecological communities.

Once this has been achieved, it will be possible to expand the legal concept of fiduciary responsibility to include the well-being of the corporation's employees, of local communities, and of future generations. In fact, this would mean reviving the traditional purpose of the corporation to serve the public good; and, as Kelly points out, it will not be necessary to abandon the belief in a market economy in order to change the corporate structure, just as it was not necessary to abandon the belief in God in order to change the monarchy.

18.1.3 Redesigning ownership

After publishing her book in 2001, Marjorie Kelly continued to explore how to reform corporations in her capacity as publisher of the journal *Business Ethics* and later on as a fellow of the Tellus Institute, a research center developing strategies for the transition to a sustainable and just civilization. While discussing strategies to change corporate structures, she and her colleagues at the Tellus Institute realized that the fundamental issue that defines corporations and capital markets today is ownership. "In a way that many of us rarely notice," Kelly (2012, p. 10) writes, "ownership is the underlying architecture of our economy. It's the foundation of our world . . . Questions about who owns the wealth-producing infrastructure of an economy, who controls it, whose interests it serves, are among the largest issues any society can face. Issues of who owns the sky in terms of carbon emission rights, who owns water, who owns development rights, are planetary in scope. The multiplying crises we face today are entwined at their root with the particular form of ownership that dominates our world – the publicly traded corporation."

With this realization, Kelly began to look at forms of ownership that do not involve corporations at all. She spent the next ten years traveling around the world, studying community-owned businesses that embody a wide variety of ownership designs. At the end of her journey, Kelly concluded that we are at the beginning of an emerging ownership revolution (see guest essay by Marjorie Kelly, p. 402). The new forms of ownership go beyond capitalism (private ownership) and beyond socialism (state ownership). They include a radically new option of private ownership for the common good.

In her new book, *Owning Our Future*, Kelly (2012) tells the story of her journey and analyzes the ownership designs she encountered. They include worker-owned businesses operating "green" laundries, installing solar panels, or producing food in urban greenhouses, as well as the largest department store chain in the UK, owned 100% by its employees; wind farms in Denmark operated by "wind guilds," created by small investors; community land trusts, in which individual families own their homes and a nonprofit community owns the land beneath the homes, thus prohibiting speculative real-estate holdings; an organic dairy company in Wisconsin owned by 1,700 farm families; marine fisheries with catch shares that have halted or reversed catastrophic declines in fish stocks; cooperatives and nonprofit organizations in Latin America forming a "solidarity economy" to protect communities and ecosystems; conservation easements covering tens of millions of acres, which allow land to be used and farmed while being protected from development; and countless community banks, credit unions, and other varieties of customer-owned banks, thriving amidst the financial crisis.

What all these ownership designs have in common is that they create and maintain conditions for the flourishing of human and ecological communities. They serve the needs of life by building into the very fabric of their organizational structures tendencies to be socially just and ecologically sustainable. Kelly calls this new kind of ownership "generative ownership," because it generates well-being and real, living wealth. She contrasts it with the "extractive ownership" of the conventional corporate ownership model, whose central

Guest essay

Living enterprise as the foundation of a generative economy

Marjorie Kelly

Tellus Institute, Boston, Massachusetts

"What kind of economy is consistent with living inside a living being?" This was a question posed under a leafy canopy, deep in the woods of southern England, not far from Schumacher College, where I'd come as a teacher. I stood listening with a group of students as resident ecologist Stephan Harding asked what for me would become a pivotal question – the only question there is, really, as we negotiate the turn from the industrial age into a new age of civilization.

I'd come to Schumacher to share my learning from four years as cofounder of Corporation 20/20 at Tellus Institute in Boston, where I'd helped lead hundreds of experts in business, law, government, labor, and civil society to explore a critical question: *How could corporations be redesigned to incorporate social and ecological aims as deeply as financial aims?* Over twenty years as co-founder and publisher of *Business Ethics* magazine, I'd seen how corporations and financial markets had come to be the dominant institutions of society, how their profit-maximizing operating system had become the operating system of the planet. That design lay at the root of many major ills facing our society. But Stephan's talk helped me understand why redesigning corporations did not quite hit the mark as the solution: *You don't start with the corporation and ask how to redesign it. You start with life, with human life and the life of the planet, and ask, how do we generate the conditions for life's flourishing?*

If you stand inside a large corporation and ask how to make a sustainable economy, the conversation has to fit itself into the frame of profit maximization. ("Here's how you can make more money through sustainability practices.") Asking corporations to change their fundamental frame is like asking a bear to change its DNA and become a swan.

A better place to start – as the founding generation of America did – is by articulating truths we hold to be self-evident. That's what Stephan did in the forest, saying simply: *"A thing is right when it enhances the stability and beauty of the total ecosystem. It is wrong when it damages it."* The sustainability of the larger system comes first. Everything else must fit itself within *that frame*.

From maximizing profits to sustaining life

Central to the mandate of profit maximization is the imperative to grow – and that growth imperative threatens the Earth. What keeps that mandate in overdrive is the Wall Street demand for rising profits and stock price. Corporations, and the capital markets where their ownership shares trade, are the internal combustion engine of the capitalist economy. These organizational systems have become the main driving force of ecological systems.

In the short run, profit-maximizing companies can help in a rapid transition to a greener economy. But that transition might represent a brief moment in time. If civilization and planetary ecosystems are still functioning well fifty years from now (not a small *if*), what about the next fifty years? And the next hundred or thousand years beyond that? What kind of economy will be suited for ongoing life inside the living Earth? Will it be an economy dominated by massive corporations intent on earnings growth? That doesn't seem likely. In the long view, the question turns itself about: *Can we sustain a low-growth or no-growth economy indefinitely* without *changing dominant ownership designs?*

That seems unlikely. Probably impossible. How do we make the turn? What are the alternatives to *extractive design*, that seeking of endless extraction of financial wealth? Can we design economic architectures that are self-organized around serving the needs of life?

After my sojourn in England, this question set me on a quest, and I was heartened to find alternatives emerging in unsung, disconnected experiments across the globe. I studied employee ownership, tribal ownership, municipal ownership, commons ownership, social enterprise, community land trusts, and other models. If industrial-age ownership represents a monoculture model, emerging designs are rich in biodiversity. Yet they embody a coherent school of design – a common form of organization that brings the living concerns of the human and ecological communities into the world of property rights and economic power. I call it a family of *generative ownership designs*, aimed at generating the conditions for life to thrive. Together, they potentially form the foundation for a generative economy, a living economy with a built-in tendency to be socially fair and ecologically sustainable.

In ownership design, five essential patterns work together to create either extractive or generative design: purpose, membership, governance, capital, and networks. Extractive ownership has a *financial purpose*: maximizing profits. Generative ownership has a *living purpose*: creating the conditions for life. While corporations today have absentee membership, with owners disconnected from the life of enterprise, generative ownership has rooted membership, with ownership held in human hands. While extractive ownership involves governance by markets, with control by capital markets on autopilot, generative designs have mission-controlled governance, with control by those focused on social mission. While extractive investments involve casino finance, alternative approaches involve stakeholder finance, where capital becomes a partner rather than a master. Instead of commodity networks, where goods are traded based solely on price, generative economic relations are supported by ethical networks, which offer collective support for social and ecological norms.

I saw the power of stakeholder finance in the wind guilds of Denmark, groups of small investors who joined together to fund wind farms. Those wind guilds jumpstarted the wind industry in Denmark, where one-fifth of the nation's electric power today comes from wind, more than any other nation.

I saw the power of living purpose and rooted membership in the community forests of Mexico, where control over forests has often been granted to indigenous tribal peoples – like the Zapotec Indians of Ixtlan de Juarez in southern Mexico. At Ixtlan, problems of deforestation and illegal logging have become relatively unknown. Community members have incentive to be stewards, because forest enterprises employ hundreds of people harvesting timber, making furniture, and caring for the forest. These are living forests, communities of trees and humans, where the purpose is to live well together.

I again saw the power of rooted membership – combined with mission-controlled governance – on Martha's Vineyard, off the coast of Massachusetts, where I visited the South Mountain Company, an employee-owned firm specializing in sustainable design and construction. It's a consciously post-growth company. After the crash of 2008, it opted to shrink, in the most humane way possible. It could make that choice because it was owned and governed not by absentee owners, but by employees.

On a larger scale, I saw mission-controlled governance in Denmark, where the major pharmaceutical company Novo Nordisk produces 40% of the world's insulin in Kalundborg. That town is home to a famed example of "industrial symbiosis," where waste from making insulin is used by farmers as food for pigs or for fertilizer. That ecological design – stable for decades – is possible because governance of this major, publicly traded company is also stable. The company is legally controlled by a foundation, intent on the living purpose of defeating diabetes.

What makes generative designs a single family are the living purposes at their core, and the beneficial outcomes they tend to generate. More research remains to be done, but there is evidence that these models tend to create broad benefits, and to remain resilient in crisis. We've seen this, for example, in the success of the state-owned Bank of North Dakota in the 2008 crisis, which led more than a dozen states to pursue similar models. We've seen it in the resilience and responsible behavior of credit unions, which tended not to create toxic mortgages, and required few bailouts. We've seen it in the fact that workers at firms with employee stock ownership plans enjoy 2.5 times the retirement assets of comparable employees at other firms. And we've seen it in the fact that the Basque region of Spain – home to the massive Mondragon cooperative – has recently seen substantially lower unemployment than the country as a whole.

To move from isolated examples toward a fully generative economy, we may need a global movement of citizens, investors, and businesses, both profit and nonprofit, working together to create a pincer strategy – one arm aimed at reforming existing large companies, and the other aimed at promoting generative alternatives. We may need different designs in different sectors; generative private ownership may be appropriate for producing goods and services, for example, while commons ownership is better suited for natural resources. Government might incentivize and ultimately require a phase-in of generative ownership. At some point society must redesign the operating system of major corporations; otherwise, alternative designs may remain marginal, or face absorption. Yet forcing all major corporations to change their core purpose may be the wrong place to begin. Advancing generative alternatives could be a more likely place to win early successes – laying the ground for bigger wins in the future.

Through working for generative designs, we can advance the knowledge needed to create a truly generative economy, one that might have a built-in tendency to be socially fair and ecologically sustainable – an economy that would at last be consistent with living inside a living being.

feature is maximum financial extraction. In fact, she points out that "our industrial-age civilization has been powered by twin processes of extraction: extracting fossil fuels from the Earth and extracting financial wealth from the economy" (Kelly, 2012, p. 11). Within the conceptual framework of the systems view of life, we can recognize extractive ownership as the driving force of unlimited quantitative growth, and thus as one of the root causes of our multifaceted crisis, while generative ownership is fully consistent with the concept of qualitative growth discussed in our previous chapter (Section 17.2).

According to Kelly, the family of generative ownership designs are forming together a nascent generative economy, which involves no less than redesigning the fundamental economic architecture at the level of organizational purpose and structure. She emphasizes that this is an emergent phenomenon, brought about by self-organizing communities to sustain and enhance the flourishing of life.

18.2 Energy and climate change

Energy, defined in physics as "the ability to do work," is an indispensable prerequisite for any activity in natural and technological systems. Indeed, the Greek root of the term, *energeia*, means "activity." Energy is a quantitative measure of the amount of actual or potential activity.

Modern industrial society depends crucially on a continuous supply of abundant energy for its food production and manufacturing processes, for the lighting and heating of our homes and cities, and for our worldwide networks of transportation and communication. The very fabric of modern civilization is sustained by massive flows of energy. If our global energy supplies were to dry up, or even be reduced substantially, our entire infrastructure would collapse.

This massive use of energy is a relatively recent phenomenon. During the centuries before the Industrial Revolution, wind, water, and muscle power (exerted by animals and humans) provided the energy to move carts and ships, operate mills, and drive all other machines. Sunlight, captured by grasses and food crops, was the ultimate source of muscle power; and solar energy, stored in firewood, provided the heat for cooking and for smelting bronze and iron ores.

In the eighteenth century, steam engines, fueled primarily by coal, propelled the Industrial Revolution. Since coal has about twice the energy density of wood, this new source of power dramatically increased the efficiency of the machines used for manufacturing, mining, and transportation; and these changes, in turn, introduced profound transformations of the social, economic, and cultural conditions of the time.

In the late eighteenth and early nineteenth centuries natural gas, first produced from coal, was used to light houses and city streets; and in the second half of the nineteenth century petroleum emerged as a powerful and convenient source of energy. In the twentieth century the liquid fuels refined from petroleum, with energy densities more than three times that of wood, became the principal means of powering modern industrial societies.

The steam engines of the Industrial Revolution not only introduced machine-based manufacturing and transportation; they also stimulated scientists and engineers to explore the nature of energy in greater depth. Among the basic concepts of mechanics, energy is the one that took longest to be identified and precisely formulated. It is much more abstract than the concepts of mass, force, or momentum, and has become one of the most important concepts of modern physics.

Its importance is due to the fact that the total energy in any physical process is always conserved. Energy can change into many different forms – gravitational, kinetic, heat, chemical, etc. – but the total amount of energy in a particular process, or set of processes, never changes. There is no known exception to the conservation of energy. It is one of the most fundamental and far-reaching laws of physics.

The famous German philosopher and mathematician Gottfried Wilhelm Leibniz, who was a contemporary of Isaac Newton, is usually credited with being the first to recognize the conservation of energy. He called it a "living force" (*vis viva*) and defined it as the product of the mass of an object and its velocity squared. This corresponds to the modern definition of kinetic energy, $E_{kin} = \frac{1}{2} mv^2$. The term "energy" in its modern sense was first used in the early nineteenth century, and scientists and philosophers argued for many years about whether energy was some kind of substance or merely a physical quantity. It was only with the formulation of thermodynamics in the mid nineteenth century that energy was defined as the capacity of doing work, and the conservation of the total amount of energy through multiple transformations in mechanical and thermodynamic processes was clearly formulated.

As we discussed in Section 1.2.3, the law of the conservation of energy is also known as the first law of thermodynamics. The second law of thermodynamics is the law of the dissipation of energy. While the total energy involved in a process is always constant, the amount of useful energy is diminishing, dissipating into heat, friction, and so on. This is important for determining the efficiency of various energy sources. Careful analyses have recently revealed massive inefficiency and waste in most current industrial design. Ecological designers today are confident that astonishing reductions in energy and materials up to 90 percent – called factor ten because it corresponds to a tenfold increase in resource efficiency – are possible in developed countries with existing technologies and without any decline in people's living standards (Hawken *et al.*, 1999).

18.2.1 Energy – a systemic issue

Since all industrial and economic processes are powered by continuous flows of energy, the illusory pursuit of perpetual economic growth (discussed in Section 17.2) generates ever- increasing demands for energy; and as our access to cheap and abundant fossil fuels reaches its peak and begins to decline, our economic crisis and our energy crisis become inextricably linked (Butler, Wuerthner, and Heinberg, 2012).

The declining supplies of fossil fuels, however, are only one of the many facets of our twin crises of energy and economic growth. In addition to easily accessible coal, oil, and natural gas, we are also depleting numerous other natural resources – from water, copper, and steel to rare minerals like scandium, terbium, or yttrium, which are used widely in the aerospace and automotive industries, as well as in the manufacture of solar panels, wind turbines, and energy-efficient light bulbs (see *The Guardian*, UK, January 27, 2012).

Even more ominous are the multiple environmental impacts of unlimited economic growth, powered by fossil fuels, resulting in deforestation, soil erosion, water scarcity, and massive species extinction. The most dramatic consequences, however, have been the diverse manifestations of climate change – record heat waves, mega-floods, and ever more violent storms – which threaten the existence of life as we know it on our planet (see Section 17.3.5). These climate catastrophes have made it imperative to drastically curb greenhouse gas emissions and to implement sustainable climate policies on a global scale.

A secondary harmful effect of "peak oil" is the increasing frequency and severity of accidents involving fossil-fuel extraction. As the conventional oil and gas reserves are being depleted, energy companies are engaged in ever more extreme extraction methods, often operating under conditions near the limits of their technical capabilities. The explorations of such "extreme energy" may occur in inhospitable regions – e.g., at ocean depths of up to 3,000 m, or in the Arctic, where operating costs and environmental risks are extremely high. Or they may involve unproven technologies, such as the extraction of natural gas from shale by means of hydrofracturing, or "fracking," which turns fresh water into toxic fluids, contaminates aquifers with carcinogens, and releases methane (a greenhouse gas far more powerful than CO_2) into the atmosphere. In such extreme situations it is only a matter of time before major accidents occur. Indeed, during the years 2010–12 we witnessed not only the devastating Deepwater Horizon catastrophe in the Gulf of Mexico but also numerous disproportionately high incidences of asthma, infertility, and cancer in communities living near fracking operations (Dachille, 2011; Steingraber, 2012).

These considerations make it evident that our energy crisis is a systemic crisis requiring systemic solutions (Nader, 2010). Several solutions of this kind, to be reviewed in the following sections, have been proposed by scholars and activists in recent years. Unfortunately, however, the systemic thinking necessary to design and implement them is still very rare among our corporate and political leaders. Instead of taking into account the interconnectedness of our major problems, their "solutions" tend to focus on a single issue, thereby simply shifting the problem to another part of the system – for example, by producing more energy at the expense of biodiversity, public health, or climate stability.

Most of the efforts of the energy corporations are directed toward the exploitation of marginal fossil fuels – deepwater oil, tar sands, shale oil, and shale gas – by means of ever more advanced technologies at ever greater risks, while completely ignoring the threats of global warming. In fact, the fossil-fuel industry puts tremendous pressure on politicians and engages in deceptive publicity campaigns to prevent the public from understanding the nature and severity of the climate crisis, as we discussed in Section 17.3.5.

18.2.2 False solutions: "clean coal"

Coal, the first fossil fuel used extensively in the history of the Industrial Age, is still the leading emitter of greenhouse gases, accounting for 40% of CO_2 emissions globally. The large coal deposits in the USA, Russia, China, and Australia are the biggest fossil-fuel sources of CO_2; and coal is also by far the dirtiest fuel. When it is mined by mountain-top removal, mining companies literally blow off the tops of mountains to extract the seams of coal lying underneath. In the process, millions of tons of rubble and toxic waste are dumped into the streams and valleys below the mining sites. When coal is burned, it produces not only carbon emissions but also many other pollutants, which are blamed in the USA for thousands of heart attacks and premature deaths every year. These pollutants include highly toxic mercury, which rains down into streams and rivers, eventually accumulating in the food chain. In addition, coal mining, processing, and burning produce vast amounts of combustion ash and other toxic wastes.

Several years ago, the coal industry began to promote the notion of "clean coal" in its marketing campaigns. It is evident, however, that coal, the dirtiest and most destructive fossil fuel, can never be clean. Indeed, "clean coal" is not an actual new type of coal, but is essentially an advertising slogan used to describe new technologies of carbon capture and storage (CCS). In most scenarios, this involves applying heat and pressure to gasify the coal before burning it, removing CO_2 from the emissions, and then storing it underground in abandoned gas or oil wells. These technologies address only a small part of the damage caused by the mining, processing, and burning of coal; moreover, CCS is still an experimental technology, subject to feasibility studies. There are serious doubts that it will be commercially viable on a meaningful scale before another 15–20 years, if at all, which excludes CCS as an effective means to fight climate change, given the urgency of the task. Far more promising is a growing grass-roots campaign calling for a moratorium on the construction of new coal-fired power plants with the ultimate goal of closing every coal plant in the USA and replacing the coal with renewable energy sources and energy conservation (see Section 18.2.1).

Nuclear power

In previous decades, there was great hope that nuclear power might be the ideal clean fuel to replace coal and oil, but it soon became apparent that nuclear technology carries enormous risks and costs, and today there is a growing sense that it is not a viable solution. These risks begin with the contamination of people and the environment with cancer-causing radioactive substances during every stage of the fuel cycle – the mining and enrichment of uranium, the operation and maintenance of the reactor, and the handling and storage or reprocessing of nuclear waste. In addition, there are the unavoidable emissions of radiation in nuclear accidents and even during routine operation of power plants, as well as the unsolved problems of how to safely decommission nuclear reactors and store radioactive waste.

Still, in 2001 the nuclear industry launched the idea of a "nuclear renaissance," which gained some traction amidst rising fossil fuel prices and growing concerns about climate change. Nuclear power, the industry's lobbies claimed, would be a viable alternative to fossil fuels. As former US Vice President Dick Cheney put it, "America's electricity is already being provided through the nuclear industry efficiently, safely, and with no discharge of greenhouse gases or emissions" (quoted in Caldicott, 2006, p. vii). From a systemic perspective, it is evident that no part of this statement is true.

In Box 18.2, we have summarized the basic facts about nuclear energy in terms of the following seven "inconvenient truths": the production of nuclear electricity creates significant greenhouse gases and pollution; uranium supplies are very limited; the construction times

Box 18.2

Seven inconvenient truths about nuclear power

(1) Nuclear energy creates significant greenhouse gases and pollution. When the entire fuel cycle is considered, a nuclear plant emits 27% of the CO_2 emitted from a coal plant. Within only one or two decades, it will produce as many emissions as conventional sources of energy, as the concentration of available uranium ore declines and uranium becomes more and more difficult to extract and refine.

(2) Global uranium supplies are finite. If the world's total electricity demand were met by nuclear energy today, the accessible uranium would last less than nine years.

(3) If nuclear power were to actually replace fossil fuels, this would require the construction of one nuclear reactor per week for the next fifty years. Considering the 8–10 years it takes to build a new reactor, such an enterprise is simply not viable.

(4) The nuclear industry has never taken responsibility for the massive amounts of lethal radioactive waste that it produces continually. In spite of the global consensus on the appropriateness of storage in geological sites, no nation in the world has yet opened such a site.

(5) Historically, as well as technically, nuclear power and nuclear weapons are inextricably linked. Nuclear plants are essentially bomb factories, perpetuating grave concerns about nuclear weapons proliferation and nuclear terrorism. Al Gore stated: "During the 8 years I worked in the White House, every nuclear weapons proliferation problem we faced was connected to a reactor program."

(6) "New generations" of reactors are not only decades too late but also exhibit all of nuclear energy's inherent economic, environmental, safety, and proliferation problems.

(7) Nuclear power requires massive infusions of government (i.e., taxpayer) subsidies, relying on universities and the weapons industry for its research and development, and being considered far too risky for private investors. According to *The Economist* (1998), "Not one [nuclear power plant] anywhere in the world makes commercial sense."

Sources: Caldicott (2006); Gore (2009); Lovins (2009a, 2009b, 2011)

of nuclear reactors are prohibitively long; the problem of nuclear waste storage remains unsolved; nuclear power and nuclear weapons are inextricably linked; "new generations" of reactors exhibit the same problems and are decades too late; and, because of all these problems, nuclear power is not viable commercially, being unable to survive without massive government subsidies.

Each of these seven facts, by itself, is a compelling argument against nuclear power, without even invoking its tremendous health and safety risks. Together, they make an overwhelming case for phasing out nuclear energy and replacing it with the cheaper, cleaner, faster, and safer solutions discussed in the following pages.

If these arguments are so compelling, why then does the nuclear industry still receive massive subsidies from governments in developed countries, while the nuclear option is eagerly pursued by many countries in the Third World? We believe that the attraction of nuclear power to many governments worldwide is twofold. The first is the sophistication and high prestige of nuclear physics, the scientific foundation of nuclear energy technology, combined with the belief – deeply ingrained in modern industrial civilization – that after so many scientific and technological triumphs, physicists and engineers surely will be able, eventually, to design and build safe nuclear plants. This belief, even if it were justified, addresses at most two of the arguments against nuclear power listed in Box 18.2. Hence, it does not stand up to a proper systemic analysis of the issue.

An even more powerful attraction of nuclear energy is the inextricable link between weapons and reactor technologies (Caldicott, 2006). The underlying physics and the required raw materials are the same for both. In the 1950s, President Eisenhower coined the memorable phrase "Atoms for Peace," but today it is clear that the so-called peaceful use of nuclear power cannot be separated from its military use – neither technologically nor politically. Political leaders around the world are well aware of this fundamental fact. Even when they profess, or sincerely intend, to develop their nuclear capabilities only for peaceful use, the allure of even the potential capability to build nuclear weapons – with all its associations of military might, secrecy, and prestige – is so strong that they seem to be incapable (with very few exceptions) of abandoning nuclear technology, even though it does not make any economic nor ecological sense.

In the long run, nuclear power is destined to be phased out because it has no business plan. Indeed, since 2005 all US reactors have been subsidized 100% and still could not raise any private capital. In the meantime, renewable energy sources are sweeping the global energy market. In 2010, renewables won $151 billion of private investments and added over 50 GW (the equivalent of 50 typical power plants), while nuclear energy got zero private investment and kept losing capacity (Lovins *et al.*, 2011). Nuclear electricity now costs 2–3 times as much as electricity from new wind power, and by the time any new reactors are built they will not be able to compete with solar energy either (Lovins, 2011). And yet, the nuclear industry keeps lobbying for ever more lavish subsidies, thereby blocking billions of dollars from being invested in sustainable energy sources.

The case against nuclear power gathered new, and perhaps decisive, momentum after the nuclear disaster in Fukushima, Japan, in 2011, in which three reactors experienced full meltdown together with a series of hydrogen explosions and the release of large amounts of cesium-137 and other radioactive materials into the atmosphere and ocean. In the wake of this catastrophe, assessed as a level-7 accident (the maximum level) on the International Nuclear Event Scale, Japan announced plans to end its reliance on nuclear power by 2040. Germany and Belgium are also abandoning nuclear power; Italy canceled a long-planned nuclear revival; and even France, the most pronuclear country in the world, decided to lower its dependence on nuclear power from 75% to 50% of total electricity demand (*The Observer*, UK, October 27, 2012). From a systemic perspective, it is to be hoped that these moves will accelerate the much-needed shift from nuclear energy and fossil fuels to the sustainable, systemic solutions of the energy and climate crisis, which already exist and to which we shall now turn.

18.2.3 Raising climate change awareness

As we have noted, the climate crisis and the underlying energy crisis are systemic problems requiring systemic solutions. Like our other major problems – environmental degradation, food security, species extinction, and so on – they can be traced back to the illusory pursuit of unlimited economic growth on our finite planet (see Section 17.1). To solve the climate crisis, we need to go beyond fossil fuels, and to be able to do so, we need to use energy much more efficiently. However, this will not be enough: we also need to address the root problems of excessive material consumption and waste, which are inherent in the ideology of perpetual growth.

This means that, in the long run, the systemic solutions of the climate crisis will be the same as the solutions of the economic crisis that we discussed in this and the previous chapter: qualifying economic growth, redefining development, finding inner fulfillment in community, changing the structure and legal framework of corporations, and designing new forms of generative ownership. In the short run, however, the most urgent task is to accelerate the transition toward a future without fossil fuels, so as to survive the threat of global climate collapse.

In the following sections we shall review several comprehensive proposals of systemic strategies for going beyond fossil fuels, which make it evident that we have the knowledge and the technologies to design fossil-fuel-free energy systems. Moreover, these alternative energy systems are also feasible economically. The argument that effective climate policies are too costly at a time of economic crisis was disproved decisively in a comprehensive report on the economics of climate change, commissioned by the British Treasury from Sir Nicholas Stern, former chief economist of the World Bank (Stern, 2006). Known as the Stern Review, this report represents the most thorough economic analysis of climate change undertaken so far. It turns the economic argument about global warming on its head. Whereas, previously, many politicians and corporate economists insisted that curbing greenhouse gas emissions would be "bad for the economy," the Stern Review says the exact opposite. It states emphatically that the world has to act now or face devastating economic consequences.

Based on detailed ecological and economic modeling out to the end of this century, the report concludes that, if we do not act now to halt global warming, we will be faced with an economic downturn even more devastating than the current crisis. On the other hand, the Stern Review estimates that stabilizing greenhouse gas emissions would be relatively cheap; it would cost no more than 1% of global GDP.

In other words, the transition toward a future without fossil fuels is feasible today, both technologically and economically. The greatest obstacle is the lack of political will, especially in the USA where the fossil-fuel industry vehemently opposes any change in the status quo, spending millions of dollars on powerful lobbies and sophisticated disinformation campaigns (see Section 17.3.5). Hence, the first step in the transition toward alternative energy sources must be to raise public awareness about climate change and to inspire political action. In this section, we shall review the efforts of several individuals and organizations engaged in this vital campaign.

The Climate Reality Project

One of the most effective and most tireless climate campaigners has been former Vice President Al Gore. Gore served sixteen years in the US Congress (eight of them in the Senate) and another eight years as Vice President in the Clinton administration. During all these years, beginning in 1976 as a freshman in Congress, he held countless hearings and public events to spread awareness about the climate crisis and to build public support for congressional action. After winning the popular vote for president in 2000, but losing the presidency by a Supreme Court decision following irregularities in the vote count in Florida, Gore spent six years traveling around the world with an impressive slide show, presenting a compelling case that most global warming is caused by the greenhouse gas emissions of modern industrial society, and that its consequences for the planet will become irreversible unless we take decisive action. He estimates that in those years, he showed his presentation more than a thousand times.

In 2006, Gore wrote a bestselling book and an award-winning documentary film, both based on his slide show, titled *An Inconvenient Truth* (Gore, 2006). Both the book and the film have had a major impact on international public opinion about climate change, and in 2007 Gore was awarded the Nobel Peace Prize for his efforts, sharing it with the Intergovernmental Panel on Climate Change (IPCC). Following the film, Gore founded the nonprofit organization the Climate Project, later renamed the Climate Reality Project (www.climaterealityproject.org), and personally trained 1,200 volunteers to deliver his famous slide show and spread the message worldwide. By 2010, the number of volunteer presenters had increased to 3,500 worldwide, and they had delivered more than 70,000 presentations to a combined audience of over 7 million people.

350.org

One of the most prolific and articulate writers on climate change is the American activist Bill McKibben, whom we have already introduced in our previous chapter (McKibben, 1989, 2007, 2010, 2012a, 2012b, 2012c). In 2007, McKibben organized a nation-wide campaign, Step It Up 2007, to demand action on global warming by the US Congress. It involved 1,400 rallies at famous sites across the USA and had significant influence on the 2008 presidential campaigns of Barack Obama and Hillary Clinton. Two years later, he founded the international environmental organization 350.org (www.350.org), inaugurating it with 5,200 simultaneous demonstrations in 181 countries – the largest global coordinated rally ever.

The name "350.org" derives from the atmospheric CO_2 concentration of 350 ppm, which, according to the IPCC, is the safe upper limit (see Box 17.1). The organization describes itself as "a global movement to solve the climate crisis," and it attempts to do so by means of online campaigns, grass-roots organizing, and mass public actions. The most important of these public actions have been several mass protests against the Keystone XL pipeline; a global information campaign titled "connect the dots," aimed at drawing attention to the links between climate change and recent extreme weather conditions; and, most recently, a national movement to divest funds from fossil-fuel companies.

In 2011, the proposed Keystone XL pipeline became the signature issue of the American environmental movement. The pipeline would transport crude oil from tar sands in Alberta, Canada, to the Gulf of Mexico so that oil can be exported from the continent. The environmental risks involved in the extraction, processing, and transporting of the oil are extremely high. At its source, in the tar sands of Alberta, the mining of the oil-rich mixture, technically known as bitumen, has already destroyed vast areas of boreal forest. Bitumen is an extremely viscous form of petroleum that must be processed before it can flow in a pipeline; this requires enormous amounts of water and also involves various solvents that pollute the surrounding air and water. The resulting toxic sludge would then flow over a distance of 2,736 km (1,700 miles) with intermittent, and unavoidable, spills that would threaten some of the most sensitive land in North America, including the Ogallala aquifer, the fresh-water source vital to the Great Plains.

Oil from tar sands, moreover, is not only the world's dirtiest fuel but also the richest in carbon content. CO_2 emissions from tar sands have been estimated to be 20% higher than from average crude oil. This is why James Hansen, the leading authority on climate science, has stated that fully exploiting the Canadian tar sands would mean "game over" for the climate. This dire conclusion inspired thousands of activists, including Hansen himself, to join the protests organized by 350.org. At the first of these mass protests, in November 2011, an estimated 10,000 protesters surrounded the White House, demanding that President Obama deny the permit for the Keystone pipeline to run through the USA. In response to this and other subsequent mass protests, Obama postponed the final decision. At the time of writing, in August 2013, the decision is still pending.

The latest campaign of 350.org, named "Fossil Free," is a movement to divest stocks, bonds, or investment funds from fossil-fuel companies in order to overcome their resistance to responsible climate policies. This project is modeled after the campaign to divest from the Apartheid regime in South Africa. In the 1980s, large student demonstrations successfully pressured the boards of 155 US campuses – including some of the most famous in the country – to divest their stocks from companies doing business in South Africa. They were followed by divestments of funds held by American cities, states, and pension funds, all of which contributed significantly to the fall of the Apartheid government.

The activists of 350.org argue that, just as the investments in South Africa under Apartheid were unethical, so are the investments in fossil-fuel companies today, because they seriously endanger the well-being of humanity. The Fossil Free campaign is asking college and university presidents and boards to immediately freeze any new investment in fossil-fuel companies, and to divest from direct ownership and any commingled funds that include fossil-fuel public equities and corporate bonds within five years.

The demands of the campaign to 200 publicly traded fossil-fuel companies are derived from the basic facts of climate-change science (see Box 17.1): they need to stop exploring for new hydrocarbons; they need to stop lobbying in Washington and state capitols across the country to preserve their special tax breaks; and, most importantly, they need to pledge to keep 80% of their current reserves underground forever. The Fossil Free campaign began in November 2012 and spread like wildfire to over 190 US college campuses in just over a month. After the first semester of the campaign, two small colleges, Unity College and Hampshire College, had divested their endowments from fossil-fuel companies, and over a dozen colleges and universities had begun serious discussions of the issue with their students (see www.gofossilfree.org).

"Beyond Coal"

There is broad consensus among climate-change scientists and activists today that our efforts to stabilize the climate will be won or lost with coal. Coal is not only the world's largest source of carbon emissions but also the dirtiest fuel, as we discussed in Section 18.2.2. In fact, coal-fired power plants are not even viable economically if their health and environmental costs are counted. A recent study, published in the *American Economic Review*, the country's most prestigious economics journal, concluded that the economic damage caused by air pollutants from burning coal exceeds the value of the electricity produced (Muller *et al.*, 2011).

On the other hand, James Hansen estimates that we still have a chance to bring the atmospheric CO_2 concentration back down to the safe level of 350 ppm if we cut off coal, the largest source of CO_2. Hansen is therefore advocating an immediate moratorium on new coal-fired power plants, followed by a phase-out of existing coal plants over the next few decades (Hansen, 2012).

This is exactly the goal of a new and very successful campaign launched by the Sierra Club, one of the oldest and most influential environmental organizations in the USA. Known as "Beyond Coal" (www.sierraclub.org/coal), the campaign aims to replace dirty coal with clean energy by mobilizing grass-roots activists in local communities. The first goal is to prevent the issuing of permits and the construction of new coal-fired power plants; the second goal is to close the 492 existing coal plants; and the third goal is to replace the retired plants with clean energy sources (conservation and renewable energy).

The Beyond Coal campaign is supported by several other environmental organizations, including Friends of the Earth, Greenpeace, 350.org, and the Rainforest Action Network; and it also has strong public support. A national opinion poll found that coal is the preferred source of electricity for only 3% of Americans (Opinion Research Corporation, 2007). With all that support, the campaign has been very successful. At the time of writing (in August 2013), 153 proposed coal plants have been taken off the board, and 149 of the existing plants are scheduled for closure with specific retirement dates.

In July 2011, Beyond Coal got a major boost when Mayor Bloomberg of New York, one of the most successful businessmen of his generation, gave the Sierra Club a grant of

$50 million to support the campaign. This was important not only because it allowed the campaign to dramatically expand its activities. The fact that one of the country's best-known and wealthiest entrepreneurs strongly supports the phasing out of coal has great symbolic significance and will have many ripple effects. In fact, Lester Brown believes that it may well be a tipping point in the struggle to stabilize the climate. "As the United States closes its coal-fired power plants, it sends a message to the world," Brown (2011b) writes. "With Michael Bloomberg's grant bolstering the Sierra Club's well-organized program to phase out coal, we can now imagine a coal-free United States on the horizon."

If coal and eventually also oil and natural gas need to be phased out to stabilize the climate, and if nuclear power is not an alternative option, how can we build energy sources that are clean, efficient, abundant, and renewable? In the following sections we shall review three different strategies for designing such energy systems, each of which is documented in a book titled with the name of the strategy:

(1) *Plan B* by Lester Brown and the Earth Policy Institute (Brown, 2008, 2009);
(2) *Reinventing Fire* by Amory Lovins and Rocky Mountain Institute (Lovins *et al.*, 2011);
(3) *The Third Industrial Revolution* by Jeremy Rifkin and the Foundation on Economic Trends (Rifkin, 2011).

Parts of these three strategies overlap one another, while other parts are complementary and mutually reinforcing. Together they present compelling evidence that the transition to a fossil-fuel-free future is possible and affordable today with existing technologies.

18.2.4 Plan B

Plan B is Lester Brown's "road map to save civilization" (Brown, 2008). It is his alternative to business as usual, which will lead to disaster. Its scope is much broader than energy and climate change, but stabilizing the climate is a principal component of Brown's scenario. Plan B proposes several simultaneous actions involving systemic solutions that reinforce one another, thus generating synergistic effects. All of the actions proposed are based on existing technologies and illustrated with successful examples from countries around the world. Brown also provides a budget for each of his proposals, based on estimates derived from numerous authoritative studies. Since its first publication in 2008, Brown has updated his proposals in two further books (Brown, 2009, 2011a), and he also regularly posts updates on specific issues on the website of his Earth Policy Institute (www.earth-policy.org).

Eradicating poverty and stabilizing population

The three main components of Plan B are eradicating poverty and stabilizing population, stabilizing climate, and restoring the Earth's ecosystems. For the first component, eradicating poverty and stabilizing population (two closely interdependent goals), Brown proposes to fill several current funding gaps centering on education and health. The programs to be funded include universal primary education, eradication of adult illiteracy, school lunch programs (one of the most effective ways to get children to school), universal basic healthcare, assistance to pregnant women and preschool children, and reproductive health and family planning. The proposals make it clear that we now have the technologies and financial resources to achieve the twin goals of eradicating poverty and stabilizing population.

Brown also notes that today, we are well on our way to stabilizing population. There are now 46 countries in the world with, essentially, zero population growth. They include a large part of Asia (China, South Korea, and Japan), as well as countries in Western and Eastern Europe. North America, too, is moving in the right direction, and Latin America is doing surprisingly well. The two big regions we need to concentrate on, according to Brown, are the Indian subcontinent (population 1.6 billion) and Sub-Saharan Africa (population 860 million). "What we need to do there," Brown explains, "is to eradicate poverty, which we have the resources to do now, and make sure that women everywhere have access to reproductive healthcare and family planning services" (Brown, 2012a).

Stabilizing climate

The second component of Plan B, stabilizing climate, urges us to massively reduce carbon emissions, and do it quickly. The goal of Plan B, in accordance with the strong advice by the IPCC (see Box 17.1), is to cut CO_2 emissions by 80% by 2020. This will stabilize atmospheric CO_2 around 350 ppm and will help keep future temperature rise to a minimum.

The proposed reduction of CO_2 emissions is to be achieved, on the one hand, by expanding the Earth's forest cover and, on the other hand, by raising energy efficiency and developing renewable sources of energy. The first priority in stabilizing climate is to replace all coal- and oil-fired electricity generation with renewable resources. As we discussed in our previous section, this is well within our reach. Options excluded from Plan B, because they are not economically viable, are "clean coal" and nuclear power.

As Brown points out, there is an enormous potential for increasing energy efficiency in our throw-away economy. Raising resource productivity by a factor of 10 (and thus dramatically increasing energy efficiency) is possible today with existing technologies in a wide range of industries. These include the petrochemical, steel, cement, and building industries. Switching to more energy-efficient appliances – e.g., compact fluorescent light bulbs (CFLs) – and restructuring transport systems can lead to further massive energy savings. As old-fashioned incandescent light bulbs are being phased out by 2014 in the USA (in compliance with the Energy Independence and Security Act of 2007), and being replaced by CFLs and LEDs, electricity use for lighting will drop by up to 80%. The measures proposed in Plan B will more than offset projected growth in energy use between now and 2020.

The centerpiece of the new energy economy proposed in Plan B is wind, combined with plug-in hybrid cars that will run largely on wind energy. A second important element is thermal energy, largely from rooftop solar heaters. Photovoltaics, solar electric and thermal power plants, geothermal energy, energy from biomass, and hydropower are the other components in Plan B's diverse energy budget.

Harnessing the wind

Lester Brown's projections, based on current trends, clearly show that we now have the technologies to build a new energy economy from renewable sources. Wind energy, in particular, has emerged as the most promising source, growing spectacularly beyond even the most optimistic recent projections (Brown, 2012b). The total capacity for the world's wind farms, now generating power in about 80 countries, led by China and the USA, is near 240,000 MW (the equivalent of 240 power plants). Over the past decade, wind-generating capacity grew at nearly 30% per year worldwide, and higher growth rates are expected for this decade.

Wind energy displays a combination of attractive features that cannot be matched by any other energy source. It is abundant, widely distributed, can be developed quickly, scales up easily, uses no water and no fuel, and can never be depleted. In the USA, three wind-rich states – North Dakota, Kansas, and Texas – have enough harnessable wind energy to satisfy all national electricity needs.

Unlike fossil-fuel plants, wind farms do not require water for cooling. As wind farms replace coal and gas plants, massive amounts of water will be freed for irrigation – another systemic solution! And while it may take a decade or more to build a nuclear power plant, a typical wind farm can be built in a year.

An obvious issue with wind is its variability. However, as wind farms multiply and are interconnected through transmission grids, this becomes less of a problem. Since no two wind farms have the same wind profile, each farm added to the grid will reduce overall variability. A study at Stanford University has shown that, as the number of interconnected wind farms increases, the whole grid becomes more and more similar to a single farm with steady wind speed, and thus steady deliverable wind power (Archer and Jacobson, 2007).

In densely populated areas, there is often local opposition to wind turbines on aesthetic grounds. However, one should not compare a landscape dotted with wind turbines with a pristine landscape, but rather with landscapes disfigured by oil rigs, mountain-top removal, or the mining of tar sands. Wind turbines take up only 1% of the land covered by a wind farm, and most of the space under and between the turbines can be used – e.g., for agriculture. In previous centuries, windmills were often part of a country's culture, and many of these traditional windmills are regarded as works of art today. Maybe this should be a challenge for our contemporary artists to design wind turbines that are efficient and also beautiful.

In the spacious ranching and farming regions of the USA, there is no opposition to wind turbines. On the contrary, they are extremely popular in the Midwest and the Great Plains for economic reasons. With no investment on their part, ranchers and farmers can earn several thousand dollars a year for each wind turbine on their land; and since these turbines take up only 1% of the land, the farmers can harvest electricity while also producing cattle, wheat, or corn. In fact, Lester Brown (2012b) predicts that in years to come, wind royalties will dwarf the ranchers' earnings from cattle sales.

Plan B calls for developing 3 million MW of wind-generating capacity worldwide, enough to meet 40% of the world's electricity needs. This will require building 300,000 wind turbines per year over the next decade. It seems daunting until we realize that automakers are producing 70 million cars and trucks a year worldwide. Most of the wind turbines could even be produced in currently idling automobile assembly plants, creating thousands of high-paying, long-term jobs.

If current trends continue, the development of the wind-generating capacity proposed in Plan B should be a very achievable goal. It will go a long way toward building an energy system that is not climate-disruptive, does not pollute the air, and can last as long as the Sun itself.

Restoring the Earth

The third major component of Plan B, restoring the Earth's ecosystems, proposes to launch a vast international effort to protect and restore forests, conserve and rebuild soils, regenerate fisheries, protect animal and plant diversity, and plant millions of trees to sequester carbon. Protection of the Earth's forests includes, among many other conservation practices, paper recycling to reduce the quantity of wood used to make paper in the North, and replacing inefficient firewood cookers with alternative devices to reduce firewood use in the South. In addition, Plan B includes numerous tree-planting projects around the world, both for flood reduction and soil conservation, and for carbon sequestration. Moreover, it proposes to end net deforestation worldwide.

The restoration of fisheries involves the establishment of a worldwide network of marine reserves, covering 30% of the ocean surface. The protection of biological diversity involves the creation and maintenance of plant and animal reserves. In the book, Brown cites numerous success stories, which show that we *can* restore the Earth. Brown acknowledges that it will not be cheap: in his estimate, the additional annual funding needed to restore the Earth will be $113 billion. But he is quick to add that the world cannot afford to not make these investments (Brown, 2008, p. 174).

The three main components of Plan B – eradicating poverty and stabilizing population, stabilizing climate, and restoring the Earth's ecosystems – are interlinked in multiple ways with two further goals. The first, called "feeding 8 billion well," involves raising cropland productivity with new agricultural techniques like intercropping, raising irrigation productivity, and producing animal protein more efficiently by moving down the food chain. Brown points out that all these actions (to be discussed in more detail in Section 18.3.3 below) also help to reduce greenhouse gas emissions.

The other supplementary goal concerns urbanization, the second demographic trend today after population growth. As our cities continue to grow, an inherent conflict between the automobile and the city is becoming ever more apparent. This has given rise to a new urbanism, also known as the "ecocity" movement, which aims to redesign our cities so that they become community-friendly and ecologically sustainable (see Section 18.4.1).

The proposals for urban design in Brown's Plan B, under the headline "designing cities for people," involve the creation of parks, bicycle paths, and car-free zones; the planting of trees; and the restructuring of public transportation. They also include integrating the city into local ecosystems (with rooftop gardens, urban agriculture, etc.), reducing urban water use (e.g., with the help of composting toilets and recycling systems), and upgrading squatter settlements. All these measures improve public health, reduce air pollution and carbon emissions, and transform the quality of urban life.

Reorienting national budgets

The proposals in Plan B include several suggestions for restructuring national economies. The key to such economic restructuring will be the creation of an honest market, in which the market prices reflect actual costs (see Section 17.1). To do so we need to restructure the tax system by reducing taxes on work and raising them on environmentally destructive activities. This restructuring is known as "tax shifting" because it would be revenue neutral for the government. Taxes would be added to existing products, forms of energy, services, and materials, so that their prices would better reflect their true costs, while equal amounts would be subtracted from income and payroll taxes.

To be successful, tax shifting needs to be a gradual, long-term process in order to give new technologies and consumption patterns sufficient time to adapt, and it needs to be implemented predictably in order to encourage industrial innovation. Such a long-term, incremental shift of taxation will gradually drive wasteful, harmful technologies and consumption patterns out of the market.

The most efficient means for such tax shifting is a carbon tax. Paid by the primary producers – the oil and coal companies – it would permeate the entire fossil-fuel economy. Plan B proposes a worldwide carbon tax of $240 per metric ton, to be phased in at the rate of $20 per year for 12 years. In addition, Plan B proposes to phase in a gasoline tax of 40¢ per gallon per year for the next 12 years, and offsetting it with a reduction in income taxes. This would raise the US gasoline tax to the $4–5 level prevailing today in Europe and Japan, still short of the indirect costs of $12 per gallon.

Another urgent budget measure is the removal of hidden subsidies, also known as "perverse subsidies" (Myers and Kent, 2009). Today, the governments of the industrial world use $700 billion of their taxpayers' money each year to subsidize unsustainable and harmful industries and corporate practices. Examples include the billions of dollars paid by Germany to subsidize the extremely harmful coal-burning plants of the Ruhr Valley; the huge subsidies the US government gives to its automobile industry, which was on corporate welfare during most of the twentieth century; and the subsidies given to agriculture by the OECD, totaling $300 billion per year, which is paid to farmers *not* to grow food although millions in the world go hungry; as well as the millions of dollars the US government offers to tobacco farmers to grow a crop that causes disease and death.

All of these are perverse subsidies indeed. They are powerful forms of corporate welfare that send distorted signals to the markets. While they support inequity and environmental degradation, the corresponding life-enhancing and sustainable enterprises are portrayed by the same governments as being "uneconomical." It is high time to eliminate these immoral forms of government support, and shift fossil-fuel energy subsidies to the development of renewables, reduce airline subsidies, and shift subsidies from road construction to rail construction, from the fishing industry to the creation of marine parks, and so on.

A detailed study of Brown's proposals and of their costs shows not only that we have the knowledge and the technologies to carry them out; we also have the necessary money. The annual expenditures needed to meet the social goals of Plan B (universal primary education, eradication of illiteracy, universal basic healthcare, etc.) are $77 billion. The energy restructuring proposed for stabilizing climate involves no public costs; and the annual costs of restoring the Earth's ecosystems (planting trees, protecting top soil, stabilizing water tables, restoring fisheries, etc.) are $113 billion. The grand total is $190 billion.

This looks like a lot of money until we realize that this budget is only one-third of the US military budget, or one-sixth of the global military budget (2006 figures). As Brown points out, the financial resources for Plan B would easily be available if our political leaders realized that this, in a sense, is the new defense budget, the one that addresses the most serious threats to our security.

18.2.5 "Reinventing Fire"

The second systemic strategy for going beyond fossil fuels, titled evocatively "Reinventing Fire," was developed by the physicist and ecodesigner Amory Lovins and his colleagues at Rocky Mountain Institute (RMI). Some of their proposals cover the same ground as Lester Brown in his Plan B (and, of course, Brown is familiar with Lovins' work), but the basic approach of the RMI team is quite different. Their focus is narrower than Brown's – "to create a clear and practical vision of a fossil-fuel free future for the United States, backed up by quantitative analysis, and to map a pathway to achieve that future" (Lovins, 2009c) – and their approach is essentially a design approach, but a new kind of systemic and ecoliterate design.

Their key strategy is to redesign energy systems in such a way that their efficiency is increased multifold, with massive savings that will be so attractive to business people that business will become the driving force of the entire process. "Reinventing Fire," Lovins (2009c) explains, is "a business-led transition from oil, coal, and ultimately gas, to efficiency and renewables . . . [It] encompasses diverse activities, all aimed at eliminating fossil-fuel use and shifting toward radical energy efficiency and abundant renewable energy."

Amory Lovins, one of the world's leading and best-known experts on energy, has worked on issues of energy efficiency and ecological sustainability for over forty years. During those four decades he pioneered several innovative ideas that were at first met with suspicion by the energy industry but have since been embraced and implemented worldwide. They include his early advocacy in the 1970s of a "soft energy path," based on efficient energy use and a diversity of renewable energy sources (Lovins, 1977; see also Capra, 1982); the concept of "negawatts" as units of saved energy, which he began to promote in the 1990s to express the fact that saving fuel through energy conservation and greater efficiency generally costs less than buying fuel; and the hypercar, a revolutionary design concept of an ultralight and aerodynamic car with a hybrid drive, pioneered by RMI in the 1990s.

In 1982, Amory and Hunter Lovins founded Rocky Mountain Institute (RMI), an independent research institute focused on the "efficient and restorative use of resources." RMI calls itself a "think-and-do-tank" because its researchers design innovative solutions and also form consulting teams engaged in various initiatives to help companies implement these solutions. In 2004, RMI published *Winning the Oil Endgame*, the first detailed strategy for ending America's oil dependency (Lovins *et al.*, 2004). "Reinventing Fire" is a revised and updated version of this earlier effort, based on new technologies developed in the intervening seven years and on the recent spectacular rise of wind power and other renewable energy sources. Whereas the previous book was received cautiously by the business world, Lovins is confident that now, after the global financial crash and the recent frightening climate-change findings, business will be much more receptive to his ideas. "There is a strong business case," Lovins (2009c) writes, "to get the United States and the world completely off fossil fuels."

The book, *Reinventing Fire* (Lovins *et al.*, 2011), is meticulously researched and is supported by hundreds of credible references and a vast set of data. It is easy to read, however, since most of the data are presented in well-designed charts, often in the form of colorful graphics, and are interspersed with anecdotal success stories and numerous real-world examples. The entire book, as well as individual sections, has been peer-reviewed by over eighty experts in various fields, as listed in the Acknowledgments. It is an impressive distillation of many years of research by the RMI team and of numerous outside studies.

The book begins with a visionary, and even poetic, paragraph describing a future without fossil fuels:

> Imagine fuel without fear. No climate change. No oil spills, dead coal miners, dirty air, devastated lands, lost wildlife. No energy poverty. No oil-fed wars, tyrannies, or terrorists. Nothing to run out. Nothing to cut off. Nothing to worry about. Just energy abundance, benign and affordable, for all, for ever.
>
> *(Lovins* et al.*, 2011, p. xi)*

The authors then summarize the main features of the scenario that could lead to such an ideal future. It is a road map toward a US economy in the year 2050 without any oil, coal, or nuclear power, and with one-third less natural gas, with a rate of growth of 2.6% (not their preference, they add, but the rate given in official projections). They estimate that the transition to such a fossil-fuel-free economy will cost *\$5 trillion less* than business as usual (without even counting the costs of carbon emissions and other hidden costs), and they emphasize that it could be accomplished with existing technologies and without any acts of Congress, led simply by entrepreneurial businesses. This transition to a fossil-free future would reduce carbon emissions by 82–86%; it would considerably strengthen the country's national security, create plenty of jobs, and improve public health – clearly a systemic solution *par excellence*.

Integrative design

If the vision of "Reinventing Fire" seems unrealistic to most of today's economists, engineers, and politicians, this is because it is an utterly systemic vision that cannot be understood with the linear thinking that is still prevalent in these professions. At the very core of the RMI strategy lies a systemic, nonlinear approach to design, which Amory Lovins calls "integrative design." It means designing a modern industrial structure – a car, factory, or office building – as a whole system, identifying a multitude of relationships between various components, or subsystems, and then optimizing the entire system for multiple benefits rather than optimizing individual components for single benefits. Integrative design is systems thinking in action. It involves collaborative teams of architects, designers, and engineers from the very beginning of the design process, who look for interrelationships and synergies between their areas of expertise, and who are able to allow for the spontaneous emergence of new ideas and solutions (see Section 14.5.4). With this systemic approach, radical energy efficiencies, far beyond anything in conventional engineering, can be achieved at very low cost through design solutions that have snowball effects. For example, making a car out of ultralight and ultrastrong materials, like carbon-fiber composites, generates a cascade of secondary effects, many of which result in further weight reductions and energy savings. Similar cascading energy savings can be achieved in the integrative designs of buildings and industrial processes, as we discuss below; and since it is generally cheaper to save fuel costs than to buy fuel, the entire process of integrative design can be very lucrative for business. This allows Amory Lovins to claim that climate protection (by eliminating fossil fuels) is not costly but profitable.

A further characteristic of RMI's systemic approach is that it integrates the redesign of the four most energy-intensive sectors of the US economy – transportation, buildings, industry, and electricity (see Box 18.3). For example, the energy problems of automobiles and the electricity grid are easier to solve together than separately; energy savings in motors that run industrial pumps and fans can also be applied to commercial buildings, and so on. We shall now review each of these four sectors in some detail.

Transportation: running cars without oil

Transportation uses over 70% of US oil (see Box 18.3), most of it for running cars. Hence, RMI's road map toward a future without fossil fuels begins with making cars run without oil or natural gas – a strategy pioneered by Amory Lovins in the 1990s with his hypercar concept and refined over the last twenty years.

The redesign of automobiles combines three innovations. The cars are ultralight, weighing two or three times less than steel cars; they display high aerodynamic efficiency, moving along the road several times more easily than standard cars; and they are propelled by a hybrid-electric drive, which combines an electric motor with fuel that produces the electricity for the motor on board. These innovations strongly reinforce each other and, together, lead to radically increased fuel efficiency, to the point where the transition to fully electric cars becomes affordable, thus eliminating fossil fuels.

Box 18.3
Energy use in the USA

Figure 18.1 shows a simplified graph of energy sources and end uses as percentages of total energy production. Percentages are rounded to integer numbers, and energy flows of less than 1% are not shown. Because of this rounding, totals may not equal the sum of their components.

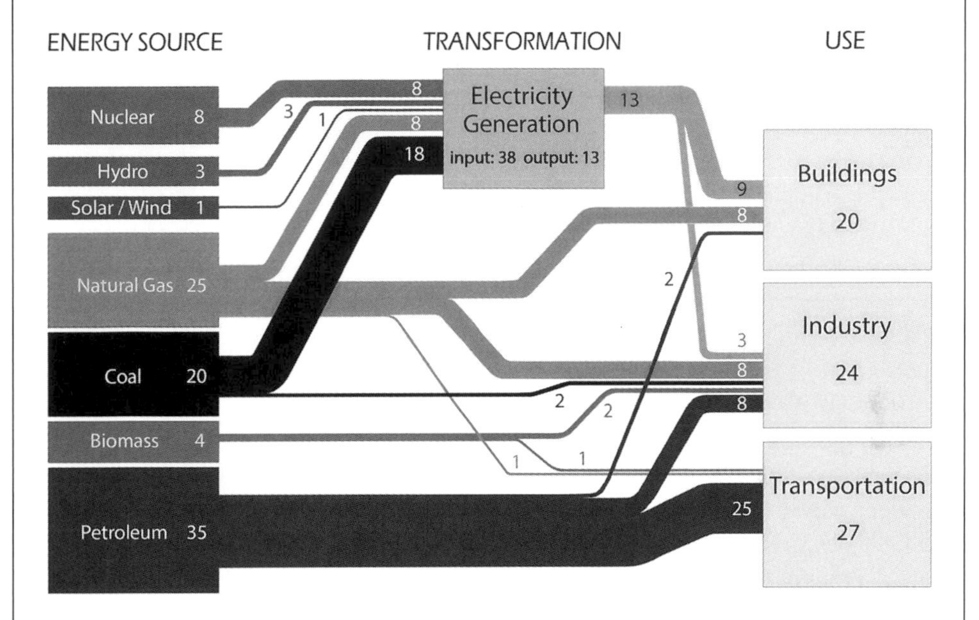

Figure 18.1 US energy use in 2011 (adapted from Lawrence Livermore Laboratory Report, October 2012).

Fossil fuels provide 80% of US energy, renewables provide 8%, and nuclear power provides 8%. About half the electricity is made from coal. Almost three-quarters of the oil fuels transportation; three-quarters of the electricity powers buildings; and the rest of both runs factories.

The graph also shows that electricity production from fossil fuels is very inefficient. Two-thirds of the energy is wasted, mostly in the form of waste heat. Renewables, by contrast, generally produce no waste heat and require no cooling water.

Hybrid cars can use gasoline or a variety of cleaner options, including advanced biofuels made from organic waste without displacing any cropland. The cleanest, most efficient, and most elegant way to power a hybrid car is to use hydrogen in a fuel cell. Hydrogen, the universe's lightest and most abundant element, is commonly used as rocket fuel. A fuel cell is an electrochemical device that combines hydrogen with oxygen to produce electricity and water – and nothing else! The entire operation is silent and reliable, and does not generate any pollution or waste. This makes hydrogen the ultimate clean fuel.

Fuel cells were invented in the nineteenth century, but until recently were not produced commercially (except for the US space program), because they were bulky and uneconomical. This situation changed radically during the last decade when several technological breakthroughs made it possible to create compact and highly efficient units, ideal for powering ultralight cars.

Hydrogen exists in abundance but must be separated from water (H_2O) or natural gas (CH_4) before it can be used as a fuel. This is not technically difficult, but requires a special infrastructure, which nobody in the fossil-fuel economy was interested in developing. However, recent studies by General Motors and by independent experts found that implementing a hydrogen infrastructure would cost less than sustaining the equivalent oil-fueling capacity (Lovins, 2003).

At present, natural gas is the most common source of hydrogen, but separation from water with the help of renewable energy sources (especially wind power) will be the most economical – and cleanest – method in the long run. When that happens, we will have created a truly sustainable system of energy generation. As in nature's ecosystems, the energy we need for our mobility will be supplied by the Sun, either directly as electricity or as hydrogen to be used in fuel cells.

The energy strategies developed for cars also apply to trucks and airplanes. During the last seven years, RMI has helped companies with large fleets of heavy trucks save up to half their fuels costs with better logistics and design, switching from diesel to natural gas with the long-term goal of using biodiesel and, eventually, hydrogen fuel cells. Similar savings can be achieved in airplanes, for which liquid hydrogen has been established as a feasible fuel by the US Air Force and by major airplane makers.

The automotive revolution envisaged by Amory Lovins and his colleagues is now well under way, not only in Europe but also in other parts of the world. In China, for example, energy efficiency is now the top strategic goal for national development. China strongly favors electric vehicles, and the first two carbon-fiber electric cars entered production in Germany in 2013.

Even in the USA, the fuel efficiency of cars is increasing very fast, partly due to the new standards introduced by President Obama. When he bailed out the automotive industry in Detroit during the financial crisis of 2008–10, car makers had to make a commitment to double the fuel efficiency of their new cars by 2020, which is now resulting in substantial fuel savings. Lovins points out that this process could be accelerated with temporary "feebates" – that is, rebates for efficient, new automobiles paid for by fees on inefficient ones. Such programs have been implemented very successfully in Europe where they have tripled the speed of improving automotive efficiency.

In the RMI scenario, ultralight cars, trucks, and airplanes will make it possible to switch fuels to eliminate all oil. Cars can use any mixture of hydrogen fuel cells, electricity, or advanced biofuels. Trucks and planes can realistically use hydrogen or advanced biofuels, and the trucks could use natural gas as a transition fuel; but no vehicles will need oil. Lovins estimates that saving or displacing barrels of oil at $25, rather than buying them for over $100, will add up to $4 trillion of net savings over the next 40 years, without counting any of the hidden costs of oil.

As the automotive revolution progresses, customers will buy the new ultralight, safe, pollution-free, silent, and super-efficient models not just because they want to save energy and protect the environment but simply because these models will be better cars. People will switch to them just as they switched from mechanical typewriters to computers and from vinyl records to CDs. Eventually, the only steel cars with combustion engines on the road will be a small number of vintage Jaguars, Porsches, Alfa Romeos, and other classic sports cars.

Since the automobile industry is the world's largest, followed by the related oil industry, the automotive revolution will have a profound impact on industrial production as a whole. The dramatic shifts from steel to carbon fibers and from gasoline to hydrogen will ultimately replace today's steel, petroleum, and related industries with radically different types of environmentally benign and sustainable production processes.

From the systemic perspective, however, the automotive revolution alone will not solve the multiple health, social, and environmental problems caused by the excessive use of cars. Only fundamental changes in our patterns of production and consumption and in the design of our cities, including efficient systems of public transportation, will accomplish that. Fortunately, these changes are also under way. The "ecocity" movement, which we have already mentioned, tries to counteract the urban sprawl and its high automobile dependence that have become so typical of our modern cities by redesigning them so that they become ecologically healthy (see Section 18.4.1).

Moreover, after a century of growth, the US automobile fleet has started to shrink, peaking in 2008 and then declining very slowly but steadily. As Lester Brown (2012a) has observed, one reason for this hopeful trend is a cultural shift among young people who are no longer part of a car culture in the way previous generations were. They tend to live in cities, using public transportation and bicycles, or scooters, rather than cars. Their status symbols are not heavy, noisy, and fast driving machines but sophisticated electronic devices for local and global communication. We believe that, ultimately, this is one more sign of the fundamental shift of metaphors from the machine to the network that is characteristic of our age.

Redesigning electric systems

Since the long-term goal of the automotive revolution is a transition to fully electric cars, it is not surprising that the redesign of the transportation system and that of the electric system are closely interlinked. Indeed, Amory Lovins affirms that automobile and electricity problems are easier to solve together than separately. The central theme of RMI's "Reinventing Fire" strategy – that switching fuels is easier after achieving radical energy efficiency – applies also to how we make electricity. Today, about three-quarters of US electricity powers buildings, and the rest runs factories (see Box 18.3). The RMI team is working on saving electricity in both of these sectors.

It turns out that energy productivity in commercial buildings can be tripled, or even quadrupled, by applying integrative design to carry out what RMI calls a "deep retrofit." In 2010 the RMI team designed such a retrofit for the iconic Empire State Building, which is now saving over 40% of the building's energy use. This involved replacing 6,500 windows onsite (in a temporary window factory on an empty floor) with "super windows" that pass light but reflect heat, as well as installing better lighting systems and office equipment. This cut the maximum cooling load by a third. Renovating and reducing cooling systems, rather than adding bigger ones, then saved $17 million of capital cost, helping to pay for the other improvements and reducing the payback time to just three years. RMI is now planning to design similar retrofits for at least 500 buildings within five years. The ultimate goal is to retrofit the entire US commercial building stock by 2050, with average energy savings of at least 50%.

The other major sector using electricity is industry, and here, too, substantial savings of energy and costs can be made by increasing the efficiency of industrial processes. By applying the principles of integrative design to industry, RMI was able to obtain energy reductions of 30–60% with a few years' payback on retrofits, and of 40–90% in new factories with generally lower capital costs. Such large energy savings are possible because, until recently, integrative design for radical energy efficiency was not part of industrial design. It was not discussed in engineering textbooks, and hence most industrial processes were designed in ways that are surprisingly inefficient.

Pumps are an important example to illustrate this situation, since 60% of the world's electricity runs motors and half of that runs pumps and fans. Amory Lovins tells how the redesign of a standard industrial pumping loop saved at least 86% of electric energy, simply by replacing long, thin, crooked pipes by short, fat, straight pipes (Lovins, 2012). Naturally, this also reduced the size of the pumps, motors, and electrical systems, and this more than paid for the new pipes by decreasing capital costs.

Large amounts of energy are wasted today not only in badly designed industrial processes but also in the generation and transmission of electricity. In the USA, an amazing two-thirds of primary energy is wasted in electricity production, mostly in the form of waste heat (see Box 18.3). As Lovins (2009c) points out, "Our power plants discard as waste heat more energy than Japan uses. We should either use that wasted energy (as Europe profitably does) or design it out."

The last step in the RMI strategy – after achieving radical energy efficiency – is to switch from fossil fuels to renewable sources of electricity such as solar and wind power, hydrogen fuel cells, and advanced biofuels. As we discussed in Section 18.2.4, renewable energy – chiefly wind and solar power – is now growing so fast that, combined with efficiency, it can easily replace coal-fired and nuclear power plants. Moreover, closing coal plants will also reduce oil use considerably, since over 40% of freight-train diesel fuel is used to transport coal. Renewable power is no longer a niche market. For each year during 2008–12, half of the world's new generating capacity has been renewable. China has become the world leader, followed by Germany, which like the United States has more solar jobs than steel jobs.

Transforming the electricity grid

A third element of "Reinventing Fire" – in addition to radical energy efficiency and renewable energy sources – is the integration of modern information and communication technology with the electric grid. The result, known as the "smart grid," is an electric system that uses a multitude of smart chips and instant communication to interlink and coordinate countless small generators whose lower costs, lead times, and financial risks make the system far superior to a centralized grid. The US electric grid, in particular, has become overcentralized and brittle in recent years, and thus very vulnerable to cascading and, potentially, paralyzing blackouts caused by natural disasters or terrorist attacks. These risks are greatly reduced when distributed renewables – rooftop solar photovoltaics, small-scale wind turbines, small hydropower plants, and many others – are organized into local microgrids, which normally interconnect but can also stand alone in emergency situations.

RMI has developed special simulations of electricity grids which show that partially or wholly renewable smart grids can deliver highly reliable power when the renewable sources are forecast, integrated, and diversified with regard to both type and location (Lovins, 2012). This is true for both large continental areas, like the USA or Europe, and also for smaller areas embedded in a larger grid; and this is how Europe is now shifting to renewable electricity – 36% of Denmark's electricity and 45% of Portugal's are now renewable-powered.

Smart grids can also be designed to send price signals to customers via "smart meters" and related infrastructure, enabling them to adapt their electricity usage to save money if they so desire, thus improving the efficiency and economy of the entire grid. The most surprising capability of smart grids, however, appears when plug-in, hybrid-electric cars are integrated into them. This is what RMI calls the "smart garage" (see Burns, 2008). The basic idea is that utilities would sell cheap electricity to the car's owner at night (when there is usually more wind power), who would then sell it back for a higher price to the smart grid during peak hours at daytime, and then drive home from work on gasoline or hydrogen. This would create revenue for the car owners and would also help the utilities smooth out variations in the load by using the storage capacity of cars as a power reserve. Amory Lovins estimates that, ultimately, this reserve electric capacity on wheels could far exceed the capacity now in power stations.

A worldwide transition

As we have mentioned, the focus of "Reinventing Fire" is narrower than that of Lester Brown's Plan B (Section 18.2.4). The RMI strategy does not address social issues like poverty and population growth, nor does it attempt to restore the Earth's ecosystems by expanding its forest cover or protecting its biodiversity. Some of these goals, however, will be supported by the transition from fossil fuels to renewable energy sources; and we should also note that many of the additional proposals of Plan B – protecting and restoring forests, conserving and rebuilding soils, creating parks and planting trees in our cities, etc. – will further reduce the carbon content of the atmosphere. Thus the two strategies of Plan B and Reinventing Fire, both based on deep systemic thinking, support and complement one another in many ways.

Another difference between the two approaches that Lester Brown emphasizes is the necessity of tax shifting and of removing hidden subsidies, whereas Amory Lovins and his colleagues – frustrated by the persistent gridlock in the US Congress – propose to "end-run" such ineffective and corrupt political institutions by emphasizing smart business strategy over public policy. Lovins affirms that the policy innovations suggested in Reinventing Fire – e.g., a "feebate" system for new cars to accelerate automotive fuel efficiency – can all be implemented by federal administrative actions or by policies at the state level. Business, however, will often be reluctant to adopt the radically new strategies designed by RMI without supporting government incentives. Hence, the transition to a future without fossil fuels would be vastly accelerated by collaborative efforts between business and government.

The RMI road map was designed for the USA but is applicable worldwide. Indeed, as we have seen, many parts of it are already being implemented in various parts of the world. What we are witnessing here is nothing less than the beginning of a profound technological, economic, and cultural transformation around the world. As Lovins (2012) puts it,

> Fire made us human; fossil fuels made us modern. But now we need a new fire that makes us safe, secure, healthy, and durable... [This is] not just a once-in-a-civilization business opportunity, but one of the most profound transitions in the history of our species. We humans are inventing a new fire, not dug from below but flowing from above, not scarce but bountiful, not local but everywhere, not transient but permanent, not costly but free.

18.2.6 The Third Industrial Revolution

The third systemic strategy for going beyond fossil fuels has been developed and promoted by the economist and activist Jeremy Rifkin. Its main elements – renewable energy sources, hydrogen as the principal fuel and energy-storage system, electric plug-in vehicles, and smart grids – are also part of Amory Lovins' Reinventing Fire. But in contrast to RMI's business-centered approach, Rifkin and his team at the Foundation for Economic Trends (a research institute examining new trends in science and technology and their economic and environmental impacts) work mainly with government agencies, helping them to develop master plans for the transition to a future hydrogen economy run by distributed power.

At the core of Rifkin's scenario lies the idea that combining renewable energies with internet technologies will allow individuals to generate their own green energy in their homes, offices, and factories, and to share it with one another across widely distributed smart electricity grids, just as people now create their own stories and images and share them on the internet across widely distributed social media.

Rifkin argues that throughout human history great economic revolutions occurred when new communication technologies converged with new energy systems. In the nineteenth

Box 18.4

The five pillars of the Third Industrial Revolution (from Rifkin, 2011)

The five pillars of the Third Industrial Revolution are as follows:

(1) shifting to renewable energy (photovoltaic electricity, wind power, small distributed hydro power, bioenergy from waste, and geothermal energy);
(2) transforming the building stock into mini-power plants to collect renewable energies on site (from the sun on the roof, the wind coming up the external walls, the sewage flowing out of the house, and the geothermal heat underneath the building);
(3) deploying hydrogen and other storage technologies in every building and throughout the infrastructure to store intermittent energies (e.g., creating solar electricity on the roof when the sun is shining, using it to power the building, splitting hydrogen from water with the surplus electricity, sequestering it in storage systems, and then transforming it back into electricity by a fuel cell when the sun does not shine);
(4) using internet technology to transform the power grid of every continent into an energy-sharing "intergrid" that acts just like the internet (millions of buildings generating small amounts of energy locally, selling surplus back to the grid, and sharing electricity with their neighbors);
(5) transforming the automobile fleet to electric plug-in and fuel-cell vehicles, powered by electricity and hydrogen from buildings converted into mini-power plants; and, moreover, using these vehicles to buy and sell electricity on the smart intergrid.

century, the introduction of steam power into print technology greatly increased the speed and efficiency of printing. Thus vast quantities of newspapers, magazines, and books could be produced, encouraging mass literacy for the first time in history. Together with the introduction of public schooling in Europe and America, this created a print-literate workforce, capable of organizing the complex operations of the steam-powered factories and rail systems of the "First Industrial Revolution," as Rifkin calls it.

In the twentieth century, liquid fuels refined from petroleum became the principal source of energy, while electronic communications (the telephone, radio, and later television) came to be the new media to manage and market a second Industrial Revolution – the oil economy, its consumer culture, and the age of the automobile. In the early twenty-first century, Rifkin claims, we are now at the cusp of another convergence of new communication technologies and energy regimes. The fusion of internet communication and renewable energies will usher in a Third Industrial Revolution, and with it a new economic narrative for the transition to a future without fossil fuels.

In his book of the same title, Jeremy Rifkin lays out his vision of the Third Industrial Revolution – a renewable energy regime, loaded by buildings, partially stored in the form of hydrogen, distributed via smart grids, and connected to plug-in, zero-emission vehicles (Rifkin, 2011). The entire system is interactive and seamlessly integrated. To build the necessary infrastructure, Rifkin outlines a strategy comprising five pillars (see Box 18.4).

He strongly emphasizes the interdependence of these five pillars and the critical need to integrate them at every stage of their development. For example, when the contributions of renewable energy sources to the electricity grid exceed 15%, the grid must be digitized and made smart to handle the intermittent nature of these new energy sources; a hydrogen infrastructure must be developed for energy storage; and buildings must be retrofitted to harness renewable energies on site, send the surpluses back to the grid, or use them to power electric vehicles.

Rifkin's scenario does not include increasing energy efficiency, which is the cornerstone of RMI's design-oriented road map. However, the two scenarios seem to be quite complementary. While Amory Lovins and the RMI team are working with business leaders and designers on radical energy efficiency. Rifkin and his colleagues work with governments at various levels to develop master plans for building smart energy grids and other infrastructure components of the future hydrogen economy.

For over a decade, Jeremy Rifkin has promoted his scenario of the Third Industrial Revolution tirelessly and with considerable success at professional conferences and in strategy meetings with business and political leaders. It is strange, but, unfortunately, not surprising, that, even though he lives in Washington and lectures at a prestigious American business school, Rifkin's ideas have found a far more enthusiastic reception in Europe where he now works most of the time.

Rifkin has been an adviser to the EU for the past ten years. In this capacity, he advised several European heads of state on issues related to the economy, climate change, and energy security during their presidencies of the European Council or the European Commission. He is the principal architect of the EU's long-term economic sustainability plan, which is based on the five-pillar strategy of the Third Industrial Revolution. In 2007, the plan was formally endorsed by the European Parliament as the EU's long-term economic vision and road map. It is now being implemented by various agencies within the European Commission as well as in the twenty-eight member states.

In 2008, Rifkin created a global economic development network, known as the Global CEO Roundtable, comprising 100 business leaders in renewable energy, construction, architecture, IT, transportation, and related fields. This business network is now collaborating with cities, regions, and national governments to design master plans for developing the infrastructures for an industrial future without fossil fuels.

Today, the EU, the biggest economy in the world, is virtually alone among the leading economic powers in asking deep questions about the survival of humanity on Earth. The strategies it has begun to implement not only testify to the emergence of a "moral compass," in Václav Havel's famous phrase (see Section 13.7), but also promise to create hundreds of thousands of new businesses and hundreds of millions of new jobs. Rifkin hopes that this will inspire the other world powers to follow the EU's leadership, and he sees signs that this is already happening in Asia, Latin America, and other parts of the world. He also argues that the Third Industrial Revolution is particularly relevant to the poorer countries in the developing world:

> We need to keep in mind that 40 percent of the human race still lives on two dollars a day or less, in dire poverty, and the vast majority have no electricity. Without access to electricity they remain "powerless," literally and figuratively. The single most important factor in raising hundreds of millions of people out of poverty is having reliable and affordable access to green electricity. All other economic development is impossible in its absence. The democratization of energy and universal access to electricity is the indispensable starting point for improving the lives of the poorest populations of the world.
>
> *(Rifkin, 2011, p. 63)*

Like Amory Lovins, Rifkin is fully aware that going beyond fossil fuels will amount to a profound technological, economic, and cultural transformation. He emphasizes different aspects of this transformation, however. Whereas Lovins extols the virtues of the "new fire" that humanity is now inventing, Rifkin highlights the revolutionary ways in which this new fire will be shared in smart distributed grids that interconnect our homes, offices, factories, and cars – all of which will silently produce, share, and use clean renewable energy.

Rifkin observes that the sharing of electricity within widely distributed energy networks will usher in a democratization of energy, which will fundamentally reorder human relationships in business, government, and civic life, just as the sharing of information in social networks has changed education, business, and politics. Some of these reflections are strikingly similar to those of the sociologist Manuel Castells when he describes the rise of the "network society" (see Section 14.4). Rifkin also addresses the issue of power (in its sociological, as well as in its technological sense) in these energy networks. He speaks of the transition from hierarchical to distributed, "lateral," and collaborative power much in the way that we discussed power in social networks as empowerment of others (Section 14.4.3).

The three scenarios we have discussed in this section – Lester Brown's Plan B, Amory Lovins' Reinventing Fire, and Jeremy Rifkin's Third Industrial Revolution – are all informed by ecological awareness and systemic thinking, and all three are mutually supportive and complementary. Lovins and Rifkin mainly deal with technological systems, but both take into account how these nonliving systems are embedded in living social and ecological systems, and how they all interact with and affect one another. Lester Brown's Plan B also deals with ecosystems directly, as in proposals involving forestry, plant and animal reserves, and agriculture.

In the following section, we shall turn to some of these ecological issues in more detail; we shall look specifically at agriculture where new ecological techniques are now being developed, and are being integrated into a systemic approach that promises to contribute significantly to solving several of the world's urgent problems.

18.3 Agroecology – the best chance to feed the world

From a systemic and ecological perspective, our global food crisis (see Section 17.1) is in many ways similar to our energy crisis, and both are, of course, interconnected. On the one hand, food is produced industrially by a system of agriculture that is highly centralized, energy-intensive, and fossil-fuel-based, creating health hazards for farm workers and consumers, and that is unable to cope with increasing climate disasters. On the other hand, a variety of agricultural techniques – often based on traditional practices – are now emerging around the world in which healthy, organic food is grown in decentralized, community-oriented, energy-efficient, and sustainable ways.

The ecologically oriented farming techniques are known variously as "organic farming," "permaculture," or "sustainable agriculture." In recent years, the term "agroecology" has increasingly been used as a unifying term, referring to both the scientific basis and the practice of an agriculture based on ecological principles. Among the leading authorities in this field, the writings of the Chilean agronomist Miguel Altieri and the Indian physicist and environmental activist Vandana Shiva have been the main inspiration for this section (Altieri, 1995; Shiva, 1993). Following Altieri, Shiva, and other agroecologists, we shall first summarize the nature and problems of conventional industrial agriculture, and will then contrast them with the principles and practices of agroecology.

18.3.1 The unsustainable nature of industrial agriculture

Industrial agriculture originated in the 1960s when petrochemical companies introduced new methods of intense chemical farming. For the farmers the immediate effect was a spectacular improvement in agricultural production, and the new era of chemical farming was hailed as the "Green Revolution." But a few decades later, the dark side of chemical agriculture became painfully evident.

It is well known today that the Green Revolution has helped neither farmers, nor the land, nor the consumers. The massive use of chemical fertilizers and pesticides changed the whole fabric of agriculture and farming, as the agrochemical industry persuaded farmers that they could make more money by planting large fields with a single highly profitable crop and by controlling weeds and pests with chemicals. This practice of single-crop monoculture entailed high risks of large acreages being destroyed by a single pest, and it also seriously affected the health of farm workers and people living in agricultural areas.

With the new chemicals, farming became mechanized and energy-intensive, favoring large corporate farmers with sufficient capital, and forcing most of the traditional single-family farmers to abandon their land. All over the world, large numbers of people left rural areas and joined the masses of urban unemployed as victims of the Green Revolution.

The long-term effects of excessive chemical farming have been disastrous for the health of the soil and for human health, for our social relations, and for the natural environment. As the same crops were planted and fertilized synthetically year after year, the balance of the ecological processes in the soil was disrupted; the amount of organic matter diminished, and with it the soil's ability to retain moisture. The resulting changes in soil texture entailed a multitude of interrelated harmful consequences – loss of humus, dry and sterile soil, wind and water erosion, and so on.

The ecological imbalance caused by monocultures and excessive use of chemicals also resulted in enormous increases in pests and crop diseases, which farmers countered by spraying ever larger doses of pesticides in vicious cycles of depletion and destruction. The hazards for human health increased accordingly as more and more toxic chemicals seeped through the soil, contaminated the water table, and showed up in our food.

In recent years, the disastrous effects of climate change have revealed another set of severe limitations of industrial agriculture. As Miguel Altieri and his colleagues at SOCLA – the *Sociedad Cientifica Latinoamericana de Agroecologia*, founded by Altieri – point out in a recent report (Altieri *et al.*, 2012), the Green Revolution was launched under the assumption that abundant water and cheap energy from fossil fuels would always be available, and that the climate would be stable. Neither of these assumptions is valid today. The key ingredients of industrial agriculture – agrochemicals, as well as fuel-based mechanization and irrigation – are derived entirely from dwindling and ever more expensive fossil fuels, water tables are falling, and increasingly frequent and violent climate catastrophes wreak havoc with the genetically homogeneous monocultures that now cover 80% of global arable land. Moreover, the energy-intensive practices of industrial agriculture contribute about 25–30% of global greenhouse gas emissions, further accelerating climate change. These considerations make it evident that, like energy generation from fossil fuels, a system of agriculture that is totally dependent on fossil fuels cannot be sustained in the long run.

18.3.2 Biotechnology in agriculture

In the 1990s, the agrochemical companies attempted to generate a new wave of technological optimism with the application of genetic engineering to agriculture, claiming that the health problems of chemical farming, as well as the problem of world hunger, could all be solved with the help of genetically modified organisms (GMOs). The biotech ads portrayed a brave new world in which nature would be brought under control. Plants would be genetically engineered commodities, tailored to the customers' needs. New crop varieties would be drought tolerant and resistant to insects and weeds. Fruits would not rot or bruise. Agriculture would no longer be dependent on chemicals and hence would no longer damage the environment. Food would be better and safer than ever before, and world hunger would disappear. Environmentalists and social justice advocates felt a strong sense of *déjà vu* when reading or hearing such optimistic but utterly naive projections of the future. They remembered vividly that very similar language had been used by the same agrochemical corporations when they promoted the Green Revolution several decades ago. In the words of the biologist David Ehrenfeld (1997),

> Like high-input agriculture, genetic engineering is often justified as a humane technology, one that feeds more people with better food. Nothing could be further from the truth. With very few exceptions, the whole point of genetic engineering is to increase the sales of chemicals and bio-engineered products to dependent farmers.

The simple truth is that most innovations in food biotechnology have been profit-driven rather than need-driven. For example, soybeans were engineered by Monsanto to be resistant specifically to the company's herbicide Roundup so as to increase the sales of that product.

Monsanto also produced cotton seeds containing an insecticide gene in order to boost seed sales. Technologies like these increase farmers' dependence on products that are patented and protected by "intellectual property rights," which make the age-old farming practices of reproducing, storing, and sharing seeds illegal. Moreover, the biotech companies charge "technology fees" in addition to the seed price, or force farmers to pay inflated prices for seed-herbicide packages (Altieri and Rosset, 1999).

Hazards of genetic engineering

Today, crops containing GMOs, also known as "transgenic" crops, already occupy about 12% of all arable land. There are many indications that this will exacerbate the problems of conventional industrial agriculture, while giving rise to a host of new problems caused by the hazards of genetic engineering.

As we mentioned in Section 9.4, the evocative term "genetic engineering" suggests to the general public that the transfer of genes between species to create new transgenic organisms is an exact, well-understood mechanical procedure. Indeed, it is usually presented as such in the popular press. In the words of the biologist Craig Holdrege (1996, pp. 116–17),

> We hear of genes being *cut* or *spliced* by enzymes, and of new DNA combinations being *manufactured* and *inserted* into the cell. The cell incorporates the DNA into its *machinery*, which begins to *read information* that is *encoded* in the new DNA. This *information* is then *expressed* in the *manufacture* of corresponding proteins that have a particular function in the organism. And so, as if resulting from such precisely determinate procedures, the transgenic organism takes on new traits.

This language is, of course, derived from the paradigm that views living organisms as machines; but since they are not, the reality of genetic engineering is far more complex and hazardous. To begin with, it is important to understand that geneticists cannot insert foreign genes directly into a cell because of natural interspecies barriers and other protective mechanisms that break down or inactivate foreign DNA. To circumvent these obstacles, scientists splice the foreign genes first into viruses, or into virus-like elements that are routinely used by bacteria to trade genes. These so-called "gene-transfer vectors" are then used to insert foreign genes into the selected recipient cells where the vectors, together with the genes spliced into them, embed themselves in the cell's DNA. If all the steps in this highly complex sequence work as planned, which is extremely rare, the result is a new transgenic organism.

The use of vectors to insert genes from the donor organism into the recipient organism is one of the main reasons why the process of genetic engineering is inherently hazardous. To overcome various natural barriers, geneticists construct a wide variety of aggressive infectious vectors that can easily recombine with existing disease-causing viruses to generate new virulent strains. In her eye-opening book, *Genetic Engineering: Dream or Nightmare?*, the geneticist Mae-Wan Ho (1998) speculated that the emergence of a host of new viruses and antibiotic resistances over the past decades may well be connected with the large-scale commercialization of genetic engineering during the same period.

At the current state of the art, geneticists cannot control what happens in the organism. They can insert a gene into the nucleus of a cell with the help of a specific gene-transfer vector, but they can never know whether the cell will incorporate it into its DNA, or where the new gene will be located, or what effects this will have on the organism. Thus, genetic engineering proceeds by trial and error in a way that is extremely wasteful. The average success rate of genetic experiments is only about 1%, because the living background of the host organism, which determines the outcome of the experiment, remains largely inaccessible to the engineering mentality that underlies our current biotechnologies.

Over the last two decades, there have been numerous studies of the risks of current biotechnologies in agriculture (see, e.g., Altieri, 2000; Altieri and Rosset, 1999; Shiva, 2000; Tokar, 2001). The studies revealed that most of those risks are a direct consequence of our poor understanding of genetic function. We have only recently come to realize that all biological processes involving genes are regulated by the cellular networks in which genomes are embedded, and that patterns of genetic activity change continually in response to changes in the cellular environment. Biologists are only just beginning to shift their attention from genetic structures to metabolic networks, and they still know very little about the complex dynamics of these networks (see Section 9.6.2).

We also know that all plants are embedded in complex ecosystems, both above the ground and in the soil, in which inorganic and organic matter moves in continual cycles. Again, we know very little about these ecological cycles and networks – partly because for many decades the dominance of genetic determinism resulted in a severe distortion of biological research, with most of the funding going into molecular biology and very little into ecology.

Since the cells and regulatory networks of plants are relatively simpler than those of animals, it is much easier for geneticists to insert foreign genes into plants. The problem is that, once the foreign gene is in the plant's DNA and the resulting transgenic crop has been planted, it becomes part of an entire ecosystem. The scientists working for biotech companies tend to know very little about the ensuing biological processes, and even less about the ecological consequences of their actions.

The most widespread use of plant biotechnology has been to develop herbicide-tolerant crops in order to boost the sales of particular herbicides. There is a strong likelihood that these transgenic plants will cross-pollinate with wild relatives in their surroundings, thus creating herbicide-resistant "superweeds." Evidence indicates that such gene flows between transgenic crops and wild relatives are already occurring (Altieri, 2000). Another serious problem is the risk of cross-pollination between transgenic crops and organically grown crops in nearby fields, which jeopardizes the organic farmers' important need to have their produce certified as truly organic.

Since one of the main objectives of plant biotechnology so far has been to increase the sales of chemicals, many of its ecological hazards are similar to those created by chemical agriculture. The tendency to create broad international markets for a single product generates vast monocultures that reduce biodiversity, thus diminishing food security and increasing vulnerability to plant diseases, insect pests, and weeds. These problems are especially acute in developing countries, where traditional systems of diverse crops and foods are being replaced by monocultures that push countless species to extinction and create new health problems for rural populations (Shiva, 1993).

Agribusiness and world hunger

The applications of genetic engineering to agriculture have aroused widespread resistance among the general public, which has grown into a worldwide political movement (see Robbins, 2001). Even though they may not understand the complexities of genetic engineering, most people around the world have a very basic, existential relationship to food and become suspicious when they hear about new food technologies being developed in secret by powerful corporations who try to sell their products without any health warnings, labels, or even discussion.

One of the main arguments of biotechnology proponents, used again and again to counter the widespread opposition to GMOs, is that transgenic crops are crucial to feed the world. Conventional food production, they maintain, will not keep pace with the growing world population. In the 1990s, Monsanto proclaimed: "Worrying about starving future generations won't feed them. Food biotechnology will." As Altieri and Rosset (1999) point out, this argument is based on two erroneous assumptions. The first is that world hunger is caused by a global shortage of food; the second, that genetic engineering is the only way to increase food production.

Development agencies have known for a long time that there is no direct relationship between the prevalence of hunger and a country's population density or growth. In their classic study, *World Hunger: Twelve Myths*, the development specialist Frances Moore Lappé and her colleagues at the Institute for Food and Development Policy gave a detailed account of world food production that surprised many readers. They showed that during the last three decades of the twentieth century, increases in global food production outstripped world population growth by 16%. During that time, food supplies kept ahead of population growth in every region except Africa. Indeed, many countries in which hunger was rampant exported more agricultural goods than they imported (Lappé *et al.*, 1998).

These statistics clearly show that the argument that biotechnology is needed to feed the world is highly disingenuous. The root causes of hunger around the world are unrelated to food production. They are poverty, inequality, and lack of access to food and land. In fact, in the Third World 78% of all malnourished children under 5 live in countries with food surpluses (Mulder-Sibanda *et al.*, 2002). People go hungry because the means to produce and distribute food are controlled by the rich and powerful: world hunger is not a technical but a political problem. If its root causes are not addressed, hunger will persist regardless of which technologies are used. Moreover, roughly one-third of all food produced for human consumption is wasted globally, most of it by consumers in Europe and North America. This amounts to 1.3 billion metric tons a year, enough to feed the entire African continent (Gustavsson *et al.*, 2011).

Biotechnology, of course, could have a place in agriculture in the future if it were used judiciously in conjunction with appropriate social and political measures, and if it could help produce better food without any harmful side effects. Unfortunately, the genetic technologies that are currently being developed and marketed do not fulfill these conditions at all.

Concentration of global food production

Recent experimental trials, cited by Altieri and Rosset (1999), have shown that GM seeds do not increase crop yields significantly. Moreover, there are strong indications that the widespread use of GM crops will not only fail to solve the problem of hunger but may also perpetuate and even exacerbate it. If transgenic seeds continue to be developed and promoted exclusively by private corporations, poor farmers will not be able to afford them, and if the biotech industry continues to protect its products by means of patents that prevent farmers from storing and trading seeds, the poor will become further dependent and marginalized. According to a Christian Aid Report by the economist Andrew Simms (1999), "GM crops are . . . creating classic preconditions for hunger and famine. Ownership of resources concentrated in too few hands – inherent in farming based on patented proprietary products – and a food supply based on too few varieties of crops widely planted are the worst option for food security."

For over a decade, an unprecedented concentration of ownership and control over food production has been under way through a series of massive mergers and because of the tight control afforded by genetic technologies. As Andrew Simms documented in the Christian Aid Report cited above, the ten leading agrochemical companies control 85% of the global food industry, and the top five control virtually the entire market for GM seeds. The goal of these corporate giants is to create a single world agricultural system in which they would be able to control all stages of food production and manipulate both food supplies and prices.

In their attempts to patent, exploit, and monopolize all aspects of biotechnology, the leading agrochemical corporations have bought up seed and biotech companies and have restyled themselves as "life sciences corporations." The traditional boundaries between pharmaceutical, agrochemical, and biotechnology industries are rapidly disappearing as corporations merge to form giant conglomerates under the life sciences banner. Thus Ciba-Geigy merged with Sandoz to become Novartis, Hoechst and Rhone Poulenc became Aventis, and Monsanto now owns and controls several large seed companies.

What all these "life sciences corporations" have in common is a narrow understanding of life, based on the erroneous belief that nature can be subjected to human control. This ignores the self-generating and self-organizing dynamic that is the very essence of life (see Chapter 7) and instead redefines living organisms as machines that can be managed from outside, patented, and sold as industrial resources. Thus life itself has become the ultimate commodity (see guest essay by Vandana Shiva on p. 438; see also Shiva, 1997, 2005).

18.3.3 Agroecology: a sustainable alternative

Over the last two decades, agroecological organic farming has greatly expanded around the world, and numerous studies have shown that it is a viable and sustainable alternative to industrial agriculture.

Guest essay

Seeds of life

Vandana Shiva

Navdanya, Dehradun, India

Seed – *bija* in Sanskrit, *shido* in Japanese, *zhangzi* in Chinese, *seme* in Italian, *semilla* in Spanish, *semence* in French, *Same* in German – is the self-urge of life to express herself, in her diverse expressions, her abundance, her permanent renewal and rejuvenation. All life begins in seed. Seed is not just the source of life; it is the very foundation of our being. For millions of years, seed has evolved freely, to give us the diversity and richness of life on the planet. For thousands of years, farmers, especially women, have evolved and bred seed freely in partnership with each other and with nature to further increase the diversity of that which nature gave us and adapt it to the needs of different cultures. Biodiversity and cultural diversity have mutually shaped one another. We have diversity of seeds because of the coevolution and co-creation by nature and farmers over 10,000 years.

Seed is the embodiment of millions of years of nature's evolution, and thousands of years of farmers' evolution and breeding. And it holds the potential of millions of years of future evolution. Seeds are therefore the repository of millennia of biological and cultural evolution. They hold the memory of the past and potential of the future.

Seed does not only hold the memory of time, evolution, and history. It also holds the memory of space, of the interactions within the web of life, of the pollinators such as bees and butterflies to whom the flowers of the seed gave their pollen and who then fertilized the plant so it could reproduce and renew itself. Seed is also the gift of millions of soil organisms which nourish the seeds and plants and are nourished by the organic matter the plants produce.

Seeds are the first link in the food chain and the repository of life's future evolution. As such, it is our inherent duty and responsibility to protect them and to pass them on to future generations. The growing of seed and the free exchange of seed among farmers has been the basis of maintaining biodiversity and of our food security.

Not all seeds are the same. There are varieties bred by farmers, which are also called indigenous varieties, native seeds, or heritage seeds. These seeds are fertilized through open pollination – that is, by birds, insects, and other natural pollinators – and are renewable. They can therefore be saved. But seed saving is seen as a problem by the agrochemical industry, which started out as a war industry and is now transforming itself into the biotechnology and so called life-science industries. These industries have transformed seed from being a self-organized renewable resource into a nonrenewable commodity to be bought every year.

Industrial agriculture goes hand in hand with industrial breeding, which has used different technological tools to consolidate control over the seed – from so-called high-yielding variety (HYV) seeds, to hybrids, genetically engineered seeds, and "terminator seeds," which are deliberately made sterile by killing the embryo. The tools might change, but the quest to control life and society does not.

The chemical industry is now bringing us GMOs. In genetically engineered seeds, toxic genes from bacteria have been introduced to plants that produce our food. Besides the risks of introducing genes from unrelated organisms, and thus scrambling the tree of life, GMOs go hand in hand with patents. Nonsustainability in agriculture, as in all aspects of life, has its roots in transforming what is renewable into a nonrenewable commodity.

The Latin root of the word "resource" is *resurgere* ("to rise again"). In the ancient meaning of the term, a natural resource, like all of life, is inherently self-renewing. This profound understanding of life is denied by the new "life-science corporations" when they prevent life's self-renewal in order to turn natural resources into profitable raw materials for industry.

Open-pollinated seed renews itself. Farmers have always saved seeds from their harvest to grow the next crop. And while saving seed, they select and breed – for taste, quality, diversity, and resistance to pests, diseases, drought, and floods. A seed renews itself over time as it grows into a crop, from which come new seeds, multiplied manifold. Unlike open-source and open-pollinated seeds, GMOs and hybrids are not renewable. They must be bought every year.

A reductionist, mechanistic science and a legal framework for privatizing seed, and knowledge of the seed, reinforce each other to destroy diversity, deny farmers innovation and breeding, enclose the biological and intellectual commons, create seed monopolies, and transform seed from a self-renewing resource to a patented commodity. This has led to the erosion of biodiversity in agriculture. What I have called the "monoculture of the mind" cuts across all generations of technologies to control the seed.

> While farmers breed for diversity, corporations breed for uniformity.
>
> While farmers breed for resilience, corporations breed for vulnerability.
>
> While farmers breed for taste, quality, and nutrition, industry breeds for industrial processing and long-distance transport in a globalized food system.

Monocultures of industrial crops and monocultures of industrial junk food reinforce each other, wasting the land, wasting food, and wasting our health. The privileging of uniformity over diversity, of quantity over quality of nutrition, has degraded our diets and displaced the rich biodiversity of our food and crops. It is based on the creation of a false boundary, which excludes both nature's and the farmers' intelligence and creativity. It has created a legal boundary to disenfranchise farmers of their seed freedom and seed sovereignty, and impose unjust seed laws to establish corporate monopoly on seed.

Multinational gene giants want to control the food system by controlling the seed.The only reason corporations genetically engineer seeds and crops is to claim patents on seeds, and hence collect royalties from life's renewal and farmers' creative work. Worse, by privileging uniformity and criminalizing diversity through a reductionist science of seed breeding, the rich diversity of our crops and the intimate links between biodiversity and cultural diversity, between seed and soil, seed and food, and, ultimately, between seed and freedom for all species in the web of life, are being broken.

Without this freedom to save, protect, and share renewable seeds, we will have neither bread nor freedom. It is to protect this fundamental freedom of life in its richness and diversity that I started Navdanya (www.navdanya.org) and the Global Citizens' Campaign for Seed Freedom (www.seedfreedom.in).

Reference

Shiva, V., R. Shroff, and C. Lockhart, eds. (2012). *Seed Freedom: A Global Citizens' Report*. Dehradun, India: Navdanya.

Basic agroecological principles

When farmers grow crops organically, they use technologies based on ecological knowledge rather than chemistry or genetic engineering to increase yields, control pests, and build soil fertility. They plant a variety of crops, rotating them so that insects that are attracted to one crop will disappear with the next. They know that it is unwise to eradicate pests completely, because this would also eliminate the natural predators that keep pests in balance in a healthy ecosystem. Instead of chemical fertilizers, these farmers enrich their fields with manure and tilled-in crop residue, thus returning organic matter to the soil to re-enter the biological cycle.

Organic farming is sustainable because it embodies ecological principles that have been tested by evolution for billions of years (see Section 16.3.2). Organic farmers know that a fertile soil is a living soil containing billions of living organisms in every cubic centimeter. It is a complex ecosystem in which the substances that are essential to life move in cycles from plants to animals, to manure, to soil bacteria, and back to plants. Solar energy is the natural fuel that drives these ecological cycles, and living organisms of all sizes are necessary to sustain the whole system and keep it in balance (Gliessman, 1998). Soil bacteria carry out various chemical transformations, such as the process of nitrogen fixation that makes atmospheric nitrogen accessible to plants. Deep-rooted weeds bring minerals to the soil surface where crops can make use of them. Earthworms break up the soil and loosen its texture; and all these activities are interdependent, combining to provide the nourishment that sustains life on Earth.

A key principle of agroecology is the diversification of farming systems. Mixtures of crop varieties are grown through intercropping (growing two or more crops in proximity), agroforestry (combining trees and shrubs with crops), and other techniques. Livestock is integrated into farms to support the ecosystems above the ground and in the soil. All these practices are labor-intensive and community-oriented, reducing poverty and social exclusion. In the words of Altieri (2000), "Agroecology raises agricultural productivity in economically viable, environmentally benign, and socially uplifting ways."

Resilience to climate extremes

Of critical importance for the future of agriculture is the observation that resilience to extreme climate events is closely linked to agricultural biodiversity, which is a key characteristic of agroecology. In recent years, several surveys conducted after major climate disasters – e.g., Hurricane Mitch in Central America (1998) and Hurricane Ike in Cuba (2008) – have shown that farms using agroecological practices suffered less damage than neighboring, conventionally farmed monocultures (studies cited in Altieri *et al.*, 2012). Other studies showed that diversified farming systems are able to adapt to and resist the effects of severe droughts, exhibiting greater yield stability and smaller decline of productivity than monocultures (cited in Altieri *et al.*, 2012). When soil is farmed organically, moreover, its carbon content increases, and thus organic farming contributes to reducing the CO_2 content of the atmosphere. In other words, agroecology not only is more resistant to global warming than industrial agriculture; it also helps to stabilize the climate, whereas industrial agriculture exacerbates climate change.

A renaissance in organic farming

The agroecological practices we have discussed are deeply rooted in the traditions of small-scale peasant farmers and are nourished by complex indigenous knowledge (Koohafkan and Altieri, 2010). In fact, even today family farms and indigenous people on 350 million small farms account for no less than half the global agricultural output for domestic consumption. Since the 1980s, thousands of projects have been launched by NGOs, farmers' organizations, and some university research centers, collaborating with peasant farmers around the world to apply general agroecological principles to local needs and circumstances, and improve yields while conserving natural resources and biodiversity (Altieri, 2004).

The current renaissance in organic farming is a worldwide phenomenon. Farmers in almost all countries of the world now produce organic food commercially. The total area being farmed sustainably is estimated at more than 30 million hectares, and the global market for organic food has grown to over $50 billion a year.

There is now abundant evidence that agroecology is a sound ecological alternative to the chemical and genetic technologies of industrial agriculture. The first global assessment of agroecologically based projects and initiatives in the developing world was conducted by the agroecologist Jules Pretty and his colleagues in 2003. They documented clear increases in food production over some 29 million hectares, with nearly 9 million households benefiting from increased food diversity and security (Pretty *et al.*, 2003). A re-examination of the data in 2010, extending the survey to 37 million hectares, showed that the average crop yield increase was 79%.

In the last two decades, the realization of the contribution of peasant agriculture and of agroecology to food security has gained worldwide attention. Two major international reports (De Schutter, 2011; IAASTD, 2009) state that, in order to feed 9 billion people in 2050, we urgently need to adopt the most efficient farming systems, and they recommend a fundamental shift toward agroecology as a way to boost food production. Based on broad consultations with scientists and extensive literature reviews, both reports contend that small-scale farmers can double food production within 10 years in critical regions by using agroecological methods already available (see also Godfray *et al.*, 2010).

To conclude this section, we would like to return to our starting point – the similarity and interconnectedness of our global food and energy crises – and re-emphasize that, in the long run, neither of them can be solved without addressing the root problems of excessive material consumption and waste, which are inherent in the ideology of perpetual growth. A recent report by Redefining Progress, a public policy institute focusing on sustainability in Oakland, California, eloquently makes this case:

Unfortunately, no form of agriculture – conventional or sustainable – can feed the world if we bank on continuous expansion of human demands. Feeding an ever-increasing population with its ever-increasing consumption habits cannot last, even with the most sustainable practices. Yet sustainable agriculture is the *best* chance we have to feed the world.

(Deumling et al., *2003)*

18.4 Designing for life

The systemic solutions discussed in this chapter all contribute to the ultimate goal of creating a sustainable future, and hence they all are informed by basic ecological awareness. The individuals and communities who design and implement these solutions are ecologically literate: they have realized that, in order to create and maintain sustainable societies, we need to honor, respect, and cooperate with nature; and that we can learn valuable lessons from nature's ecosystems – communities of plants, animals, and microorganisms that have sustained life for billions of years (see Section 16.3).

Such an approach to design, inspired by nature, is known as ecological design, or ecodesign (McDonough and Braungart, 2002; Orr, 2002; Van der Ryn and Cowan, 1996). In this section, we shall discuss the basic principles of ecodesign and shall illustrate them with numerous examples, many of which we have already mentioned in the preceding pages. All of them apply basic ecological knowledge to the fundamental redesign of our physical structures, cities, technologies, industries, and social institutions, so as to bridge the current gap between human design and the ecologically sustainable systems of nature.

From an ecological perspective, design consists of shaping flows of energy and materials for human purposes. "Ecological design," writes the environmental educator and philosopher David Orr (2002, p. 27), "is the careful meshing of human purposes with the larger patterns and flows of the natural world, and the study of these patterns and flows to inform human action." Thus, ecodesign principles reflect the principles of organization that nature has evolved to sustain the web of life (see Section 16.3.2).

To practice design in such a context requires a fundamental shift in our attitude toward nature, from finding out, as the naturalist and science writer Janine Benyus (1997, p. 2) puts it, "not . . . what we can *extract* from nature, but . . . what we can *learn* from her." Such a new attitude – learning from nature and cooperating with her, rather than attempting to control her; adapting our needs to the patterns and processes of the web of life, rather than the other way around – is a profound change indeed. It means no less than changing the principal motivation of design from designing for profits and market share to designing for life.

When we speak of the marvelous "design" of a butterfly's wings or a spider's silk thread, we need to remember that our language is metaphorical. Strictly speaking, these natural structures have not been designed but have emerged from the self-organizing processes that are inherent in all living systems (see Section 14.5.4). However, this does not change the fact that, from the perspective of sustainability, nature's "designs" and "technologies" are far superior to human science and technology. They were created and have been continually refined over billions of years of evolution, during which the inhabitants of the

Earth flourished and diversified without ever using up their "natural capital" – the planet's resources and ecosystem services on which the well-being of all living creatures depends.

18.4.1 An ecodesign revolution

Over the last two decades, there has been a dramatic rise in ecologically oriented design practices and projects, all of which are now well documented (see Hawken, Lovins, and Lovins, 1999, for comprehensive overall documentation, and the websites listed in Box 18.1 for up-to-date information on a wide variety of ecodesign projects). Many of the systemic solutions discussed in this chapter can be seen as ecodesign solutions in the broadest sense – from generative ownership designs that sustain and enhance the flourishing of life (Section 18.1.3), to farming practices that embody basic ecological principles (Section 18.3.3). In this section we shall review some examples from two broad areas of ecodesign: the incorporation of nutrient and waste recycling into industrial design, and the application of ecodesign principles to the built environment, in buildings as well as cities.

Ecological clustering of industries

The cyclical nature of ecological processes, sometimes expressed in the compact formulation, "Waste equals food," is one of the key principles of ecology. For human communities, it means that all products and materials manufactured by industry, as well as the wastes generated in the manufacturing processes, must eventually provide nourishment for something new (Hawken, 1993; McDonough and Braungart, 1998). A sustainable business organization would be embedded in an "ecology of organizations," in which the waste of any one organization would be a resource for another. In such a sustainable industrial system, the total outflow of each organization – its products *and* wastes – would be perceived and treated as resources cycling through the system.

Such ecological clusters of industries have actually been initiated in many parts of the world by an organization called Zero Emissions Research and Initiatives (ZERI), founded by the business entrepreneur Gunter Pauli in the early 1990s. Pauli introduced the notion of industrial clustering by promoting the principle of zero emissions and making it the very core of the ZERI concept. Zero emissions means zero waste. Inspired by nature, ZERI strives to eliminate the very idea of waste.

To appreciate how radical an approach this is, we need to realize that our current businesses throw away most of the resources they take from nature. For example, when we extract cellulose from wood to make paper, we cut down forests but use only 20–25% of the trees, discarding the remaining 75–80% as waste. Beer breweries extract only 8% of the nutrients from barley or rice for fermentation; palm oil is a mere 4% of the palm tree's biomass; and coffee beans are 3.7% of the coffee bush. Pauli's starting point was to recognize that the organic waste that is thrown away or burnt by one industry contains an abundance of precious resources for other industries. ZERI helps industries to organize themselves into ecological clusters, so that the waste of one can be sold as a resource to another, for the benefit of both (Pauli, 1998).

The principle of zero emissions ultimately implies zero material consumption. Like nature's ecosystems, a sustainable human community would use energy that flows from the Sun but would not consume any material goods without recycling them after use. In other words, it would not use any new materials. Moreover, zero emissions also means no pollution. ZERI's ecological clusters are designed to operate in an environment free of toxic wastes and pollution. Thus, the ecodesign principle "waste equals food" points to the ultimate solution for some of our major environmental problems.

ZERI now operates over sixty project centers on five continents in very diverse climates and cultural settings (see www.zeri.org). The clusters around Colombian coffee farms are good illustrations of the basic ZERI method. These farms are in crisis because of the dramatic drop in the price of coffee beans on the world market. Meanwhile, the farmers use only 3.7% of the coffee bush, returning most of the waste to the environment as landfill and pollution – smoke, waste water, and caffeine-contaminated compost. ZERI put this waste to work. Research showed that coffee biomass can be used profitably to cultivate tropical mushrooms, feed livestock, compost organic fertilizer, and generate energy.

The resulting ZERI cluster is pictured in Figure 18.2. To put it in greatly simplified terms, when the coffee beans are harvested, the remains of the coffee plant are used to grow shiitake mushrooms (a high-priced delicacy); the remains of the mushrooms (rich in protein) feed earthworms, cattle, and pigs; earthworms feed chickens; cattle and pig manure produces biogas and sludge; the sludge fertilizes the coffee farm and surrounding vegetable gardens, while the energy from the biogas is used in the process of mushroom farming.

The clustering of these productive systems generates several revenue streams in addition to the original coffee beans – from poultry, mushrooms, vegetables, beef, and pork – while creating jobs in the local community. The results are beneficial both to the environment and the community, there are no high investments, and there is no need for the coffee farmers to give up their traditional livelihood.Similar agricultural clusters, with beer breweries as their center instead of coffee farms, are operating in Africa, Europe, Japan, and other parts of the world. Other clusters have aquatic components; for example, a cluster in southern Brazil includes the farming of highly nutritious spirulina algae in the irrigation channels of rice fields (which otherwise are used only once a year). The spirulina is used as special enrichment in a "ginger cookie" program in rural schools to fight widespread malnutrition. This generates additional revenue for the rice farmers while responding to a pressing social need.

Technologies in the typical ZERI clusters are small-scale and local. The places of production are usually close to those of consumption, eliminating or radically reducing transportation costs. No single production unit tries to maximize its output, because this would only unbalance the system. Instead, the goal is to optimize the production processes of each component, while maximizing the productivity and ecological sustainability of the whole.

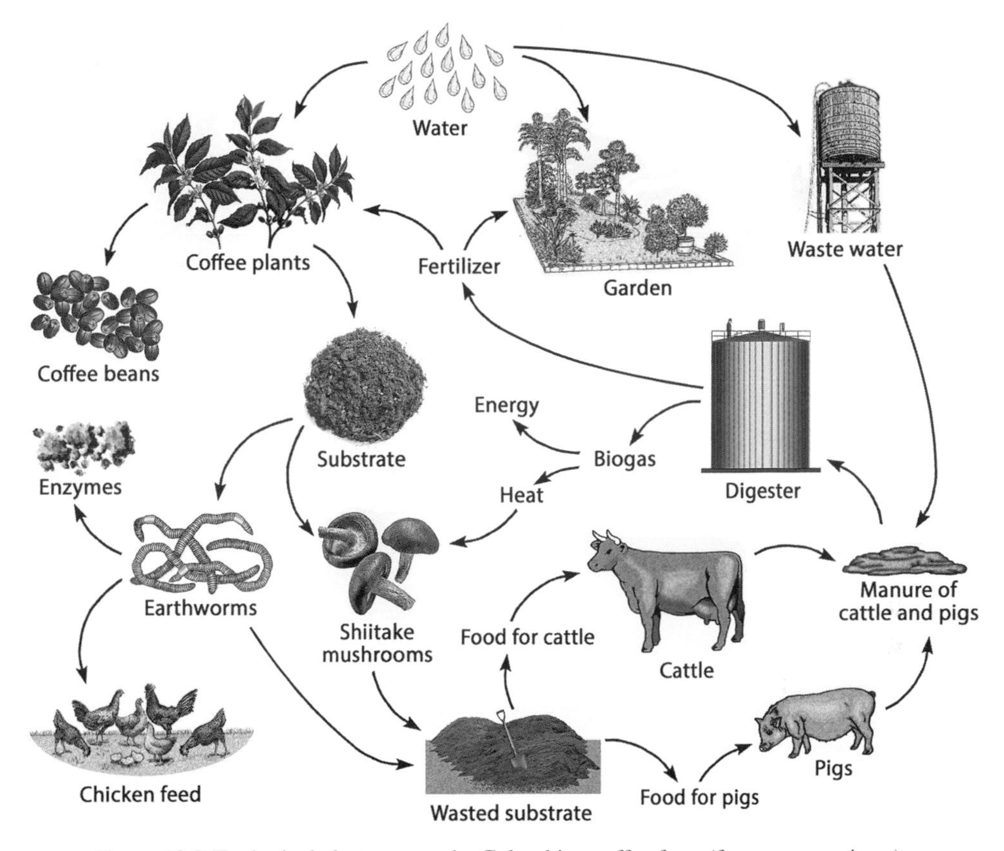

Figure 18.2 Ecological cluster around a Colombian coffee farm (from www.zeri.org).

An economy of service and flow

Most of the ZERI clusters involve organic resources and wastes. To build sustainable industrial societies, however, the ecodesign principle "waste equals food" and the resulting cycling of matter must extend beyond organic products. This concept has been best articulated by the ecodesigners Michael Braungart in Germany and William McDonough in the USA (McDonough and Braungart, 1998).

Braungart and McDonough speak of two kinds of metabolisms – a biological metabolism and a "technical metabolism." Matter that cycles in the biological metabolism is biodegradable and becomes food for other living organisms. Materials that are not biodegradable are regarded as "technical nutrients," which continually circulate within industrial cycles that constitute the technical metabolism. In order for these two metabolisms to remain healthy, great care must be taken to keep them distinct and separate, so that they do not contaminate each another. Things that are part of the biological metabolism – agricultural products, clothing, cosmetics, etc. – should not contain persistent toxic substances. Things that go into the technical metabolism – machines, physical structures, etc. – should be kept well apart from the biological metabolism.

In this vision of a sustainable industrial society, all products, materials, and wastes will be either biological or technical nutrients. Biological nutrients will be designed to re-enter ecological cycles to be consumed by microorganisms and other creatures in the soil. In addition to organic waste from our food, most packaging (which makes up about half the volume of our solid-waste stream) should be composed of biological nutrients. With today's technologies, it is quite feasible to produce packaging that can be tossed into the compost bin to biodegrade. As McDonough and Braungart (1998) point out, "There is no need for shampoo bottles, toothpaste tubes, yogurt cartons, juice containers, and other packaging to last decades (or even centuries) longer than what came inside them."

Technical nutrients will be designed to go back into technical cycles. Braungart and McDonough emphasize that the reuse of technical nutrients in industrial cycles is distinct from conventional recycling, because it maintains the high quality of the materials, rather than "downcycling" them into flowerpots or park benches. Technical metabolisms equivalent to the ZERI clusters have not yet been established, but there is definitely a trend to do so. The USA is not a world leader in recycling, but more than half of its steel is now produced from scrap. The new steel mini-mills do not need to be located near mines; they are located near the cities that produce the waste and consume the raw materials, saving considerable transportation costs.

Many other ecodesign technologies for the repeated use of technical nutrients are on the horizon. For example, it is now possible to create special types of ink that can be removed from paper in a hot-water bath without damaging the paper fibers. This chemical innovation would allow complete separation of paper and ink so that both can be reused. The paper would last 10–13 times longer than conventionally recycled paper fibers. If this technique were universally adopted, it could reduce the use of forest pulp by 90%, in addition to reducing the amounts of toxic ink residues that now end up in landfills (Hawken *et al.*, 1999).

If the concept of technical cycles were fully implemented, it would lead to a fundamental restructuring of economic relationships. After all, what we want from a technical product is not a sense of ownership but the service the product provides. We want entertainment from our DVD players, mobility from our car, cold drinks from our refrigerator, and so on.

From the perspective of ecodesign, it makes no sense to own these products and to throw them away at the end of their useful lives. It makes much more sense to buy their *services* – that is, to lease or rent them. Ownership would be retained by the manufacturer, and when one had finished using a product, or wanted to upgrade to a newer version, the manufacturer would take the old product back, break it down into its basic components – the "technical nutrients" – and use them in the assembly of new products, or sell them to other businesses. The resulting economy would no longer be based on the ownership of goods but would be an economy of service and flow. Industrial raw materials and technical components would continually cycle between manufacturers and users, as they would between different industries.

This shift from a product-oriented economy to a service-and-flow economy is no longer pure theory. One of the world's largest carpet manufacturers, Interface, based in Atlanta, has begun the transition from selling carpets to leasing carpeting services (Anderson, 1998). Similar innovations have been undertaken in the photocopying industry by Canon, in Japan, and in the automotive industry by Fiat, in Italy.

Green architecture

One area where ecodesign has led to a wide range of impressive innovations is the design of buildings. A well-designed commercial structure will display a physical shape and orientation that takes the greatest advantage of the Sun and wind, optimizing passive solar heating and cooling. That alone will usually save about one-third of the building's energy use. Proper orientation, combined with other passive-solar design features, also provides glare-free natural light throughout the structure, which usually provides sufficient lighting during daytime. Modern electric lighting systems can produce pleasant and accurate colors and eliminate all flicker, hum, and glare. Typical energy savings from such lighting are 80–90%, and this usually pays for the installation of the lighting systems within a year.

Perhaps even more impressive are the dramatic improvements in insulation and temperature regulation created by "superwindows," which keep people warm in winter and cool in summer without additional heating or cooling. Superwindows are covered with several invisible coatings that let through light but reflect heat, in addition to having double panes, the space between filled with heavy gas that blocks the flow of heat and noise. Buildings equipped with superwindows have shown that complete comfort can be maintained without any heating or cooling equipment, even with outdoor conditions ranging from severe cold to extreme heat.

Finally, ecodesigned buildings not only *save* energy by letting in natural light and keeping out the weather; they can even *produce* energy. Photovoltaic electricity can now be generated from wall panels, roofing shingles, and other structural elements that look and work like ordinary building materials but produce electricity whenever there is sunlight, even if it comes through clouds. A building with such photovoltaic materials as roofs and windows can produce more daytime electricity than it uses. Indeed, that is what millions of solar-powered homes around the world do every day.

These are just some of the most important recent innovations in the ecodesign of buildings. They are not confined to new buildings; they can also be implemented by retrofitting old structures (as we discussed in Section 18.2.5). The savings in energy and materials created by these design innovations are dramatic, and the buildings are also more comfortable and healthier to live and work in. As ecodesign innovations continue to accumulate, buildings will come ever closer to the vision of McDonough and Braungart (1998): "Imagine . . . a building as a kind of tree. It would purify air, accrue solar income, produce more energy than it consumes, create shade and habitat, enrich soil, and change with the seasons."

Ecocities

Similar considerations apply to urban design. The urban and suburban sprawl that characterizes most modern cities, especially in North America, has created very high automobile dependence with a minimal role for public transport, cycling, or walking. The consequences: high consumption of gasoline and correspondingly high levels of smog, severe stress due to traffic congestion, and loss of street life, community, and public safety.

The past three decades have seen the emergence of an international "ecocity" movement, which tries to counteract urban sprawl by using ecodesign principles to redesign our cities so that they become ecologically healthy. The city of Curitiba in Brazil pioneered this approach, and dozens of cities around the world have followed (see Register, 2001; Register and Peeks, 1997). By carefully analyzing transport and land-use patterns, the urban planners Peter Newman and Jeff Kenworthy found that energy use depends critically on city density (Newman and Kenworthy, 1998). As the city becomes denser, the use of public transport and the amount of walking and cycling increase, while the use of cars decreases.

Historic city centers with high density and mixed land use which have been reconverted into the car-free environments they were originally meant to be now exist in most European cities. Other cities have created modern car-free environments that encourage walking and cycling. These newly designed neighborhoods, known as "urban villages," display high-density structures combined with ample common green spaces. The application of ecodesign principles has brought these areas multiple benefits – significant energy savings and a healthy, safe, and community-friendly environment with drastically reduced levels of pollution.

The loss of community in our modern cities is the central concern of the landscape architect and urban designer Randolph Hester. He observes that nurturing a sense of community is not a goal in conventional urban design and that, as a consequence, most neighborhoods have lost the ability to engage in deliberation and cooperation. In addition, he notes that most urban design practices separate us from the natural environment and render us ecologically illiterate. To overcome these twin problems, Hester proposes to use urban design to transform our inhabited landscapes into ecologically sustainable, community-oriented, and joyful places (Hester, 2006). At the core of Hester's approach is his concept of "ecological democracy" – the attempt to foster direct citizen participation in ecologically oriented urban design.

18.4.2 Biomimicry – nature as model and mentor

To conclude our discussion of the principles and practices of ecological design, we shall now turn to a recently developed ecodesign branch, known as biomimicry, which is concerned with nature-inspired designs of specific structures and processes. The term "biomimicry" – from the Greek *bios* ("life") and *mimesis* ("imitation") – means "imitation of life." It was coined by the naturalist and science writer Janine Benyus in a book of the same title (Benyus, 1997). Her starting point was the basic idea underlying all ecodesign: that many of our design problems have been solved by living organisms and ecological communities during billions of years of evolutionary tinkering in elegant, efficient, and ecologically sustainable ways, and that we can learn valuable lessons from this evolutionary wisdom of nature.

Specifically, Benyus engaged in dialogues with scientists and engineers who were trying to understand how nature had developed specific structures and "technologies" that were far superior to our human designs. How do mussels produce glue that sticks to anything in water? How do spiders spin a silk thread that, ounce for ounce, is five times stronger than steel? How do abalone grow a shell that is twice as tough as our high-tech ceramics? How do these creatures manufacture their "miracle materials" in water, at room temperature, silently, and without any toxic byproducts?

Benyus realized that finding the answers to these questions and using them to develop bio-inspired technologies could provide fascinating research programs for scientists and engineers for decades to come. In fact, she soon found out that these programs have already begun. In her book, *Biomimicry: Innovation Inspired by Nature*, she takes us on a fascinating journey to numerous laboratories and field stations where interdisciplinary teams of scientists and engineers analyze the detailed chemistry and molecular structures of nature's most complex materials to use them as models for new human designs.

For example, scientists at the University of Washington have studied the molecular structure and assembly process of the smooth inner coating of the abalone shell, which shows delicate swirling color patterns and is hard as nails. They were able to mimic the assembly process at ambient temperatures and create a hard, transparent material that could be an ideal coating for the windshields of ultralight electric cars. German researchers have mimicked the bumpy, self-cleaning microsurface of the lotus leaf to produce a paint that will do the same for buildings. Marine biologists and biochemists have spent many years analyzing the unique chemistry used by blue mussels to secrete an adhesive that bonds under water. They are now exploring potential medical applications that would allow surgeons to create bonds between ligaments and tissues in a fluid environment. Physicists have teamed up with biochemists in several laboratories to examine the complex structures and processes of photosynthesis, hoping to eventually mimic them in new kinds of solar cells.

Benyus emphasizes that the deepest lessons of biomimicry lie in the exquisite ways in which organisms have adapted to their environments and to each other. To learn these lessons, she suggests that we should value nature as model, measure, and mentor. Taking nature as our model, we would ask, "How would nature do it?"; using her as our measure means asking, "What would nature not do?"; and respecting her as a mentor, we would ask, "Why does it work, and how does it work in detail?"

New biotechnologies

The basic idea of biomimicry, and of ecodesign more generally, is not new. Throughout human history, men and women have observed nature to find out how they could adapt her inventions for human use. An outstanding early pioneer of this practice is Leonardo da

Vinci, the great genius of the Renaissance. In the designs of his flying machines, Leonardo tried to imitate the flight of birds so closely that he almost gives the impression of wanting to become a bird. He called his flying machine *uccello* ("bird"), and when he drew its mechanical wings, he mimicked the anatomical structure of a bird's wing so accurately that it is often hard to tell the difference.

When Leonardo designed villas and palaces, he paid special attention to the movements of people and goods through the buildings, applying the metaphor of metabolic processes to his architectural designs. He applied the same principles to his designs of cities, viewing a city as a kind of organism in which people, material goods, food, water, and waste need to flow with ease for the city to be healthy.

Instead of trying to dominate nature, as Francis Bacon would advocate in the seventeenth century, Leonardo's intent was to learn from her as much as possible. He was always aware, and said so explicitly in his famous Notebooks, that nature's ingenuity was far superior to human design. In Leonardo's work as a designer and engineer, there are numerous examples of how he used natural processes as models for human design, and worked with nature rather than trying to dominate her, all of which show clearly that Leonardo worked in the spirit that the ecodesign movement and the practitioners of biomimicry are advocating today (see Capra, 2007).

What distinguishes the modern practitioners of biomimicry from their historical forerunners is the fact that they analyze and try to imitate biological structures and processes at the microlevel of biochemistry and molecular biology, sometimes even at the nanolevel of individual atoms and molecules. At those levels it becomes evident that there is a critical difference between human manufacturing processes that are noisy and energy-intensive, and often generate toxic wastes, and living organisms producing superior materials silently, at room temperature, and without toxic wastes. Plants, animals, and microorganisms produce their seemingly miraculous feats with the help of a wide variety of proteins, which until recently played no role in human technologies.

However, modern genetics has now given us the tools to create new kinds of biotechnologies in the service of biomimicry, the imitation of life. These would be biotechnologies of a radically different kind, motivated by the desire to learn from nature rather than control her, using nature as a mentor rather than merely as a source of raw materials. The development of such new biotechnologies would be a tremendous intellectual challenge. It would not involve modifying living organisms genetically but instead would use the techniques of genetic engineering to understand nature's subtle "designs" and use them as models for new human technologies, producing the appropriate proteins with the help of enzymes supplied by living organisms.

Recent advances

Since the publication of her pioneering book in 1997, Janine Benyus has been promoting biomimicry in countless lectures and seminars around the world, and she has also founded several organizations with that mission. In 1989, she co-founded the Biomimicry Guild together with the biologist Dayna Baumeister with the goal of bringing biologists to the design table. The guild provides biological counseling for companies, governments, and universities.

In 2005, Benyus founded the Biomimicry Institute, an educational organization offering courses in a range of educational settings, from schools to universities, as well as workshops for scientists, designers, and engineers. In 2010, Benyus and her colleagues integrated the two organizations into a hybrid organization (business and nonprofit), which they called "Biomimicry 3.8" to honor the 3.8 billion years during which nature has evolved brilliant designs and strategies.

In the last few years, the biomimicry movement has made great strides, which are discussed in some detail in a fascinating book by Jay Harman (2013), himself an inventor of a wide range of propellers, mixers, pumps, and turbines, inspired by nature's flow forms and working in ways that use less energy and are quieter than conventional devices. Harman titled his book *The Shark's Paintbrush* in reference to a special paint, developed by German researchers, that reduces the aerodynamic drag on the surfaces of ships and aircraft by mimicking the rough, yet slippery skin of sharks.

As Harman documents in this book, the US Patent office received over 900 patent applications in 2009 containing the words "biomimicry," "bio-inspired," or similar terms. In 2010, over 1,500 scientific papers related to biomimicry were published, and bio-inspired products have generated billions of dollars in annual sales. The Fermanian Business and Economic Institute, a business school in San Diego, California, estimates in a 2010 report that by 2025 biomimicry will inform 15% of all chemical manufacturing and waste management, as well as 10% of all architecture, engineering, and textile production. The institute has created a special index to measure activities related to biomimicry (number of scholarly papers, patents, grants, etc.), which they call the "Da Vinci Index" in honor of biomimicry's first great pioneer.

18.5 Concluding remarks

The ecodesign technologies and projects reviewed in the preceding sections all incorporate basic principles of ecology and therefore have some key characteristics in common. They tend to be small-scale projects with plenty of diversity, as well as energy efficient, nonpolluting, and community oriented. Moreover, they tend to be labor intensive, creating plenty of jobs; and the same is true for the new ownership designs discussed in Section 18.1.3. Indeed, the potential of creating local jobs through investment in green technologies, restoration of ecosystems, and redesigning of our infrastructure is enormous.

All these projects and initiatives are informed by systemic thinking, which is widespread in today's global civil society. As we have emphasized repeatedly, they address the fundamental interdependence of our global problems and also recognize the power of emergent solutions – from self-organizing communities creating new ownership models to teams of architects and engineers engaged in integrative design processes. At the deepest level, the systemic solutions reviewed in this chapter all embody an idea that has been a *leitmotiv* (or recurring theme) throughout our book: the fundamental change of metaphors from seeing the world as a machine to understanding it as a network.

The current ecodesign revolution, which is now well under way, provides compelling evidence that today the transition to a sustainable future is no longer either a technical or a conceptual problem. We have the knowledge and the technologies to build a sustainable world for our children and for future generations. What we need is political will and leadership. Such leadership is not limited to the political domain. In today's world, there are three centers of power: government, business, and civil society. All three of them (to varying degrees) need ecologically literate leaders, capable of thinking systemically. The collaboration between these three centers of power will be crucial for moving toward a sustainable future.

The question naturally arises: will this collaboration be realized in time, and will it act with enough urgency for human civilization to survive? As Lester Brown (2008, p. 5) puts it:

> We are in a race between tipping points in nature and our political systems. Can we phase out coal-fired power plants before the melting of the Greenland ice sheet becomes irreversible? Can we gather the political will to halt deforestation in the Amazon before its growing vulnerability to fire takes it to the point of no return? Can we help countries stabilize population before they become failing states?

To be sure, the transition to a sustainable future will not be easy. Gradual changes will not be enough to turn the tide; we also need some major breakthroughs. The task seems overwhelming but not impossible. From our new understanding of complex biological and social systems we have learned that meaningful disturbances can trigger multiple feedback processes that may rapidly lead to the emergence of a new order. Recent history has shown us some powerful examples of these dramatic transformations – from the fall of the Berlin Wall and the Velvet Revolution in Europe to the end of apartheid in South Africa.

On the other hand, complexity theory also tells us that these points of instability may lead to breakdowns rather than breakthroughs. So what can we hope for the future of humanity? In our opinion, the most inspiring answer to this existential question comes from one of the key figures in the recent dramatic social transformations, the great Czech playwright and statesman Václav Havel (1990, p. 181), who turns the question into a meditation on hope itself:

> The kind of hope that I often think about . . . I understand above all as a state of mind, not a state of the world. Either we have hope within us or we don't; it is a dimension of the soul, and it's not essentially dependent on some particular observation of the world or estimate of the situation . . . [Hope] is not the conviction that something will turn out well, but the certainty that something makes sense, regardless of how it turns out.

Bibliography

Abraham, R. and C. Shaw (1982–8). *Dynamics: The Geometry of Behavior*, vols. I–IV. Santa Cruz, CA: Aerial Press.

Aguilar, A.L.C., ed. (2009). *What Is Death?* Rome: Ateneo Pontificio Regina Apostolorum.

Akanuma, S., T. Kigawa, and S. Yojoyama (2002). Combinatorial mutagenesis to restrict amino acid usage in an enzyme to a reduced set. *Proceedings of the National Academy of Sciences of the United States of America*, **99**: 13549.

Alberts, B., D. Bray, J. Lewis, *et al.* (1989). *Molecular Biology of the Cell*, 2nd edn. New York: Garland Science.

Alexander, S. (1920). *Space, Time, and Deity*. London: Macmillan.

Altieri, M. (1995). *Agroecology: The Science of Sustainable Agriculture*. Boulder, CO: Westview Press.

(2000). The ecological impacts of transgenic crops on agroecosystem health. *Ecosystem Health*, **6**(1): 13–23.

(2004). Linking ecologists and traditional farmers in the search for sustainable agriculture. *Frontiers in Ecology and the Environment*, **2**: 35–42.

Altieri, M., C. Nicholls, F. Funes, and other members of SOCLA (Sociedad Cientifica Latinoamericana de Agroecologia) (2012). The scaling-up of agroecology: spreading the hope for food sovereignty and resiliency (www.agroeco.org/socla).

Altieri, M. and P. Rosset (1999). Ten reasons why biotechnology will not ensure food security, protect the environment and reduce poverty in the developing world. *Agbioforum*, **2**: 3–4.

Anderson, R. (1998). *Mid-Course Correction*. Atlanta, GA: Peregrinzilla Press.

Anella, F., C. Chiarabelli, D. De Lucrezia, and P.L. Luisi (2011). Stability studies on random folded RNAs ('never born RNAs'): implications for the RNA world. *Chemistry and Biodiversity*, **8**: 1422–33.

Anfinsen, C.B., R.R. Redfield, W.I. Choate, *et al.* (1954). Studies on the gross structure, cross-linkages, and terminal sequences in ribonuclease. *Journal of Biological Chemistry*, **207**(1): 201–10.

Arasse, D. (1998). *Leonardo da Vinci*. New York: Konecky & Konecky.

Archer, C. and M. Jacobson (2007). Supplying baseload power and reducing transmission requirements by interconnecting wind farms. *Journal of Applied Meteorology and Climatology*, **46**: 1701–17.

Ashby, R. (1952). *Design for a Brain*. New York: Wiley.

Atmanspacher, H. and R. Bishop, eds. (2002). *Between Chance and Choice: Interdisciplinary Perspectives on Determinism*. Charlottesville, VA: Imprint Academic.

Axelrod, R. (1984). *The Evolution of Cooperation*. New York: Basic Books.

Bachmann, P.A., P.L. Luisi, and J. Lang (1992). Autocatalytic self-replication of micelles as models for prebiotic structures. *Nature*, **357**: 57–9.

Bada, J.L. (1997). Meteoritics – extraterrestrial handedness? *Science*, **275**: 942–3.

Baert, P. (1998). *Social Theory in the Twentieth Century*. New York: New York University Press.

Bain, A. (1870). *Logic, Books II and III*. London: Longmans, Green & Co.

Barker, D. and J. Mander (1999). *Invisible Government*. Sausalito, CA: International Forum on Globalization.

Barrow, J.D. (2001). Cosmology, life and the anthropic principle. *Annals of the New York Academy of Sciences*, **950**: 139–53.

Barrow, J.D. and F.J. Tipler (1986). *The Anthropic Cosmological Principle*. Oxford University Press.

Bateson, G. (1972). *Steps to an Ecology of Mind*. New York: Ballantine.

(1979). *Mind and Nature*. New York: Dutton.

Bedau, M.A. (1997). Weak emergence. *Philosophical Perspectives: Mind, Causation and World*, **11**: 375–99.

Bedau, M.A. and P. Humphreys, eds. (2007). *Emergence: Contemporary Readings in Philosophy and Science*. London: MIT Press.

Beerel, A. (2009). *Leadership and Change Management*. London: Sage.

Behe, M. (1996). *Darwin's Black Box: The Biochemical Challenge to Evolution*. New York: Free Press.

Bell, J.S. (2004). *Speakable and Unspeakable in Quantum Mechanics: Collected Papers on Quantum Philosophy*. Cambridge University Press.

Benedetti, F. (2009). *Placebo Effects*. New York: Oxford University Press.

Ben Jacob, E., I. Becker, Y. Shapira, *et al.* (2004). Bacterial linguistic communication and social intelligence. *Trends in Microbiology*, **12**: 366–72.

Benner, S.A. (1993). Catalysis: design versus selection. *Science*, **261**: 1402–3.

Benner S.A., S. Hoshika, M. Sukeda, *et al.* (2008). Synthetic biology for improved personalized medicine. *Nucleic Acids Symposium Series*, **52**: 243–4.

Benner, S.A. and A.M. Sismour (2005). Synthetic biology. *Nature Reviews Genetics*, **6**: 533–43.

Benyus, J. (1997). *Biomimicry*. New York: Morrow.

Bertalanffy, L. von (1968). *General System Theory*. New York: Braziller.

Bertoloni-Meli, D. (2006). *Thinking with Objects: The Transformation of Mechanics in the Seventeenth Century*. Baltimore, MD: Johns Hopkins University Press.

Birdi, K.S. (1999). *Self-Assembly Monolayer Structures of Lipids and Macromolecules at Interfaces*. New York: Plenum Press.

Bitbol, M. and Luisi, P.L. (2004). Autopoiesis with or without cognition: defining life at its edge. *Journal of the Royal Society Interface*, **1**: 99–107.

(2011). Science and the self-referentiality of consciousness, in Penrose *et al.*, eds., *Consciousness and the Universe*.

Bohr, N. (1934). *Atomic Physics and the Description of Nature*. Cambridge University Press.

(1958). *Atomic Physics and Human Knowledge*. New York: Wiley.

Bolli, M., R. Micura, and A. Eschenmoser (1997). Pyranosyl-RNA: chiroselective self-assembly of base sequences by ligative oligomerization of tetranucleotide-2′,3′-cyclophosphates (with a commentary concerning the origin of biomolecular homochirality). *Chemistry & Biology*, **4**: 309–20.

Bonabeau, E., M. Dorigo, and G. Theralulaz (1999). *Swarm Intelligence: From Natural to Artificial Systems*. New York: Oxford University Press.

Bondi, G. and O. Rickards (2009). *Umani da sei milioni di anni*. Roma: Caroci.

Borysenko, J. (2007). *Minding the Body, Mending the Mind*. New York: De Capo.

Bourgine, P. and J. Stewart (2004). Autopoiesis and cognition. *Artificial Life*, **10**(3): 327.

Boyer, P. (2008). Religion: bound to believe. *Nature*, **455**: 1038–9.

Brasier, M.D., O.R. Green, A.P. Jephcoat, *et al.* (2002). Questioning the evidence for Earth's oldest fossils. *Nature*, **416**: 76–7.

Breaker, R.R. (2004). Natural and engineered nucleic acids as tools to explore biology. *Nature*, **432**: 838–45.

Broad, C.D. (1925). *The Mind and Its Place in Nature*. London: Routledge and Kegan Paul.

Brower, D. (1995). *Let the Mountains Talk, Let the Rivers Run*. New York: HarperCollins.

Brown, L. (1981). *Building a Sustainable Society*. New York: Norton.

(2008). *Plan B 3.0*. New York: Norton.

(2009). *Plan B 4.0*. New York: Norton.

(2011a). *World on the Edge*. New York: Norton.

(2011b). A fifty million dollar tipping point? (www.earth-policy.org), posted August 10.

(2012a). Interview by John Wiseman (www.postcarbonpathways.net.au), posted July 31.

(2012b). Building a wind-centered economy (www.earth-policy.org), posted October 31.

(1990). Brown, L., C. Flavin, and S. Postel. Picturing a sustainable society, in L. Brown *et al.*, eds., *State of the World*. New York: Norton, p. 173.

Brundtland, G.H., *et al.* (2012). *Environment and Development Challenges: The Imperative to Act*. New York: UNEP Report.

Burns, C. (2008). The smart garage. *RMI Solutions Journal*, July.

Butler, T., G. Wuerthner, and R. Heinberg, eds. (2012). *Energy*. Sausalito, CA: Watershed Media.

Cadenasso, M., S. Pickett, K. Weathers, and C. Jones (2003). A framework for a theory of ecological boundaries. *BioScience*, **53**(8): 750–75.

Caldicott, H. (2006). *Nuclear Power Is Not the Answer*. New York: New Press.

Cannon, W. (1932). *The Wisdom of the Body* (rev. edn., 1939; repr. 1963). New York: Norton.

Capra, F. (2010/1975). *The Tao of Physics*. Boston: Shambhala.

(1982). *The Turning Point*. New York: Simon & Schuster.

(1985). Bootstrap physics: a conversation with Geoffrey Chew, in *A Passion for Physics: Essays in Honor of Geoffrey Chew*. Singapore: World Scientific, pp. 247–86.

(1986). Wholeness and health. *Holistic Medicine*, **1**: 145–59.

(1988). *Uncommon Wisdom*. New York: Simon & Schuster.

ed. (1993). *Guide to Ecoliteracy*. Berkeley, CA: Center for Ecoliteracy.

(1996). *The Web of Life*. New York: Anchor/Doubleday.

(2002). *The Hidden Connections*. New York: Doubleday.

(2007). *The Science of Leonardo*. New York: Doubleday.

(2013). *Learning from Leonardo*. San Francisco, CA: Berrett-Koehler.

Capra, F. and H. Henderson (2009). Qualitative growth, in *Outside Insights*. London: Institute of Chartered Accountants in England and Wales. October; posted on www.fritjofcapra.net.

Capra, F. and D. Steindl-Rast with Thomas Matus (1991). *Belonging to the Universe*. San Francisco, CA: Harper.

Carr, B., ed. (2007). *Universe or Multiverse?* Cambridge University Press.

Carter, B. (1974). Large number coincidences and the anthropic principle in cosmology, in *IAU Symposium 63: Confrontation of Cosmological Theories with Observational Data*. Dordrecht: Reidel, pp. 291–8.

Castells, M. (1996). *The Information Age*. Vol. I: *The Rise of the Network Society*. Malden, MA: Blackwell.

(1997). *The Information Age*. Vol. II: *The Power of Identity*. Malden, MA: Blackwell.

(1998). *The Information Age*. Vol. III: *End of Millennium*. Malden, MA: Blackwell.

(2000). Materials for an exploratory theory of the network society. *British Journal of Sociology*, **51**(1): 5–24.

(2009). *Communication Power*. New York: Oxford University Press.

Cavanagh, J. and J. Mander, eds. (2004). *Alternatives to Economic Globalization*. San Francisco, MA: Berrett-Koehler.

Chalmers, D. (1995). Facing up to the problem of consciousness. *Journal of Consciousness Studies*, **2**(3): 200–19.

Chapin, S., P.A. Matson, and H.A. Mooney (2002). *Principles of Ecosystem Ecology*. New York: Springer.

Chauvet, J.-M., E.B. Deschamps, and C. Hillaire (1996). *Dawn of Art: The Chauvet Cave*. New York: Harry N. Abrams.

Chiarabelli, C., J.W. Vrijbloed, D. De Lucrezia, *et al.* (2006a). Investigation of *de novo* totally random biosequences. II. On the folding frequency in a totally random library of *de novo* proteins obtained by phage display. *Chemistry and Biodiversity*, **3**: 840–59.

Chiarabelli, C., J.W. Vrijbloed, R.M. Thomas, and P.L. Luisi (2006b). Investigation of *de novo* totally random biosequences. I. A general method for *in vitro* selection of folded domains from a random polypeptide library displayed on phage. *Chemistry and Biodiversity*, **3**: 827–39.

Clements, F. (1916). *Plant Succession*. Washington, DC: Carnegie Institution of Washington, publ. 242.

Coen, E. (1999). *The Art of Genes: How Organisms Make Themselves*. New York: Oxford University Press.

Coleman, P. (2007). Frontier at your fingertips. *Nature*, **446**: 379–85.

Cook, B. (1971). *The Beat Generation*. New York: Scribner.

Coveney, P. and R. Highfield (1990). *The Arrow of Time*. London: W.H. Allen.

Cowles, H. (1899). Ecological relations of the vegetation on the sand dunes of Lake Michigan. *Botanical Gazette*, **27**(3).

Crick, F. (1994). *The Astonishing Hypothesis: The Scientific Search for the Soul*. New York: Scribner.

Cronin, J.R. and S. Pizzarello (1997). Enantiomeric excesses in meteoritic amino acids. *Science*, **275**: 951–5.

Dachille, K. (2011). The impact of hydraulic fracturing on communities. Fact Sheet, Network for Public Health Law, Carey School of Law, University of Maryland.

Dalai Lama, H.H. (2000). The values of spirituality: address to forum, Prague (unpublished, quoted from notes by F.C.).

(2005). *The Universe in a Single Atom*. New York: Morgan Road.

Damasio, A. (1999). *The Feeling of What Happens*. New York: Harcourt.

Dantzig, T. (2005). *Number: The Language of Science*. New York: Pi Press.

Darwin, C. (1859). *On the Origin of Species by Means of Natural Selection, or the Preservation of Favoured Races in the Struggle for Life*. London: John Murray.

(1882). *The Descent of Man, and Selection in Relation to Sex*, 2nd edn. London: John Murray.

Davies, P. (1983). *God and the New Physics*. New York: Simon & Schuster.
(1992). *The Mind of God*. New York: Touchstone Books.
(2006). *The Goldilocks Enigma*. London: Allen Lane.
Davis, P. and D.H. Kenyon (1989). *Of Pandas and People: The Central Question of Biological Origins*. Richardson, TX: Foundation for Thought and Ethics.
(1993). *Of Pandas and People: The Central Question of Biological Origins*, 2nd edn. Richardson, TX: Foundation for Thought and Ethics.
Dawkins, R. (1976). *The Selfish Gene*. Oxford University Press.
(1986). *The Blind Watchmaker*. New York: Norton.
(2003). A Devil's Chaplain: Reflections on Hope, Lies, Science, and Love. New York: Mariner Books.
(2006). *The God Delusion*. New York: Bantam Books.
De Duve, C. (1991). *Blueprint for a Cell: The Nature and Origin of Life*. Burlington, NC: Neil Patterson.
(2002). *Life Evolving: Molecules, Mind, and Meaning*. New York: Oxford University Press.
De Geus, A. (1997). *The Living Company*. Watertown, MA: Harvard Business Press.
De Lucrezia, D., M. Franchi, C. Chiarabelli, *et al*. (2006a). Investigation of *de novo* totally random biosequences. III. RNA foster: a novel assay to investigate RNA folding structural properties. *Chemistry and Biodiversity*, **3**: 860–8.
(2006b). Investigation of *de novo* totally random biosequences. IV. folding properties of *de novo* totally random RNAs. *Chemistry and Biodiversity*, **3**: 869–77.
Dembski, W. (1999). *Intelligent Design: The Bridge Between Science and Theology*. Downers Grove, IL: InterVarsity Press.
Dennett, D. (1991). *Consciousness Explained*. New York: Little, Brown.
Descartes, R. (2006/1637). *Discourse on Method*. Translated with introduction and notes by I. MacLean. New York: Oxford University Press.
De Schutter, O. (2011). *Agroecology and the Right to Food*. Report to the UN Human Rights Council, A/HRC/16/49.
de Souza, T., P. Stano, and P.L. Luisi (2009). The minimal size of liposome-based model cells brings about a remarkably enhanced entrapment and protein synthesis. *European Journal of Chemical Biology*, **10**: 1056–63.
de Souza, T., F. Steiniger, P. Stano, A. Fahr, and P.L. Luisi (2011). Spontaneous crowding of ribosomes and proteins inside vesicles: a possible mechanism for the origin of cell metabolism. *European Journal of Chemical Biology*, **12**: 2325–30.
Deumling, D., M. Wackernagel, and C. Monfreda (2003). Eating up the Earth. *Agriculture Footprint Brief*, Redefining Progress, July.
Devall, B. and G. Sessions (1985). *Deep Ecology*. Salt Lake City, UT: Peregrine Smith.
De Waals, F. (2006). *Good Natured: The Origin of Right and Wrong*. Cambridge, MA: Harvard University Press.
Diamond, J. (1992). *The Third Chimpanzee: The Evolution and Future of the Human Animal*. New York: HarperCollins.
Doi, N., K. Kakukawa, Y. Oishi, and H. Yanagawa (2005). High solubility of random-sequence proteins consisting of five kinds of primitive amino acids. *Protein Engineering Design Selection*, **18**: 279.
Dominguez, J. and V. Robin (1999). *Your Money or Your Life*. New York: Penguin.
Dutton, D. (2009). *The Art Instinct*. New York: Bloomsbury Press.
Dyson, F. (1985). *Origins of Life*. Cambridge University Press.

Eberhart, R.C., Y. Shi, and J. Kennedy (2001). *Swarm Intelligence*. San Francisco, CA: Morgan Kaufmann.
Edelman, G. (1992). *Bright Air, Brilliant Fire*. New York: Basic Books.
Edelman, G. and G. Tononi (2000). *A Universe of Consciousness*. New York: Basic Books.
Edwards, A. (2010). *Thriving Beyond Sustainability*. Gabriola Island, Canada: New Society.
Ehrenfeld, D. (1997). A techno-pox upon the land. *Harper's Magazine*, October.
Ehrenfels, C.V. (1960/1890). Über gestaltqualitäten. Repr. in F. Weinhandl, ed., *Gestalthaftes Sehen*. Darmstadt: Wissenschaftliche Buchgesellschaft.
Einstein, A. (1931). Maxwell's influence on the development of the conception of physical reality, in *James Clerk Maxwell: A Commemoration Volume, 1831–1931*. Cambridge University Press.
(1949). *The World As I See It*. New York: Philosophical Library.
Ekins, P. (1992). Lecture at Schumacher College (unpublished).
Ekland, E.H., J.W. Szostak, and D.P. Bartel (1995). Structurally complex and highly active RNA ligases derived from random RNA sequences. *Science*, **269**: 364–70.
El-Naggar, M.Y., *et al.* (2010). Electrical transport along bacterial nanowires from *Shewanella oneidensis* MR-1. *Proceedings of the National Academy of Sciences of the United States of America*, **107**(42): 18127–31.
Ellis, G., U. Kirchner, and W.R. Stoeger (2004). Multiverses and physical cosmology. *Monthly Notices of the Royal Astronomical Society*, **347**(3): 921–36.
Ellul, J. (1964). *The Technological Society*. New York: Knopf.
Elton, C. (1927). *Animal Ecology*. London: Sidgwick & Jackson (repr. 2001 University of Chicago Press).
(1958). *Ecology of Invasions by Animals and Plants*. London: Chapman & Hall.
Eschenmoser, A. and M.V. Kisakürek (1996). Chemistry and the origin of life. *Helvetica Chimica Acta*, **79**: 1249–59.
Escobar, A. (1995). *Encountering Development*. Princeton University Press.
Esteva, G. and M.S. Prakash (1998). Beyond development, what? *Development in Practice*, **8**(3): 280–96.
Farre, L. and T. Oksala, eds. (1998). Emergency, complexity, hierarchy, organisation. Selected papers from the ECHO III Conference (Espoo, Finland). *Acta Polytechnica Scandinavica*, **91**.
Field, R.J. (1972). A reaction periodic in time and space. *Journal of Chemical Education*, **49**: 308–11.
Fischer, C. (1985). Studying technology and social life, in M. Castells (ed.), *High Technology, Space, and Society*. Beverly Hills, CA: Sage.
Fisher, R.A. (1930). *The Genetical Theory of Natural Selection*. Oxford: Clarendon Press.
Forster, A.C. and G.M. Church (2006). Towards synthesis of a minimal cell. Molecular Systems Biology, **2**(45).
Fouts, R. (1997). *Next of Kin*. New York: William Morrow.
Fox, W. (1990). *Toward a Transpersonal Ecology*. Boston: Shambhala.
Fraenkel-Conrat, H. and R.C. Williams (1955). Reconstitution of active tobacco mosaic virus from its inactive protein and nucleic acid components. *Proceedings of the National Academy of Sciences of the United States of America*, **41**: 690–8.
Francis, J. and S. Vavrus (2012). Evidence linking Arctic amplification to extreme weather in mid-latitudes. *Geophysical Research Letters*, **39**: L06801.
Funqua, C., M.R. Parsek, and E.P. Greenberg (2001). Regulation of gene expression by cell-to-cell communication: acyl-homoserine lactone quorum sensing. *Annual Review of Genetics*, **35**: 439–68.

Futuyma, D.J. (1998). *Evolutionary Biology*, 3rd edn. Sunderland, MA: Sinauer.
Galbraith, J.K. (1984). *The Anatomy of Power*. London: Hamish Hamilton.
Garcia, L. (1991). *The Fractal Explorer*. Santa Cruz, CA: Dynamic Press.
Gelder, S. van, ed. (2011). *This Changes Everything*. San Francisco, CA: Berrett-Koehler.
Giddens, A. (1991). *Modernity and Self-Identity: Self and Society in the Late Modern Age*. Cambridge: Polity Press.
(1996). Out of place. *Times Higher Education Supplement*, December 13.
Gilmore, D. (1990). *Manhood in the Making*. New Haven, CT: Yale University Press.
Girotto, V., T. Pievani, and G. Vallortigara (2008). *Nati per credere*. Torino: Codice.
Gleason, H. (1926). The individualistic concept of the plant association. *Bulletin of the Torrey Botanical Club*, **53**: 7–26.
Gliessman, S.R. (1998). *Agroecology: Ecological Processes in Sustainable Agriculture*. Ann Arbor, MI: Ann Arbor Press.
Godfray, C., *et al.* (2010). Food security: the challenge of feeding 9 billion people. *Science*, **327**: 812–18.
Goldsmith, E. (1996). Global trade and the environment, in Mander and Goldsmith, *The Case Against the Global Economy*.
Goodenough, U., and T.W. Deacon (2006). Emergence and religious naturalism, in *Oxford Handbook of Science and Religion*. Oxford University Press.
Gorby, Y.A., *et al.* (2006). Electrically conductive bacterial nanowires produced by *Shewanella oneidensis* strain MR-1. *Proceedings of the National Academy of Sciences of the United States of America*, **103**(30): 11358–63.
Gore, A. (1992). *Earth in the Balance*. New York: Houghton Mifflin.
(2006). *An Inconvenient Truth*. Emmaus, PA: Rodale.
(2009). *Our Choice*. Emmaus, PA: Rodale.
Gorelik, G. (1975). Principal ideas of Bogdanov's tektology: the universal science of organization. *General Systems*, **20**: 3–13.
Gorlero, M., R. Wieczorek, A. Katarzina, *et al.* (2009). Ser-His catalyses the formation of peptides and PNAs. *FEBS Letters*, **583**: 153–6.
Gould, S.J. (1980). *The Panda's Thumb*. New York: Norton.
(1989). *Wonderful Life*. New York: Norton.
(1991). *Bully for Brontosaurus*. New York: Norton.
(1999). *Rocks of Ages*. New York: Ballantine Books.
(2002). *The Structure of Evolutionary Theory*. Watertown, MA: Harvard University Press.
Gould, S.J. and N. Eldredge (1977). Punctuated equilibria: the tempo and mode of evolution reconsidered. *Paleobiology*, **3**: 115–51.
Green, E.D. and M.S. Guyer (2011). Charting a course for genomic medicine from base pairs to bedside. *Nature*, **479**: 204–13.
Green, R. and J.W. Szostak (1992). Selection of a ribozyme that functions as a superior template in a self-copying reaction. *Science*, **258**: 1910–15.
Green, R.E., *et al.* (2010). A draft sequence of the Neanderthal genome. *Science*, **328**(5979): 710–22.
Greene, B. (1999). *The Elegant Universe*. New York: Norton.
Grewal, D.S. (2008). *Network Power: The Social Dynamics of Globalization*. New Haven, CT: Yale University Press.
Gustavsson, J., C. Cederberg, U. Sonesson, R. Van Ottersijk, and A. Meybeck (2011). *Global Food Losses and Food Waste*. United Nations FAO Report.
Haeckel, E. (1866). *Generelle Morphologie der Organismen*. Berlin: Reimer.

Haig, D. (2004). The (dual) origin of epigenetics, in *Cold Spring Harbor Symposia on Quantitative Biology*, vol. 69. Cold Spring Harbor, NY: Cold Spring Harbor Laboratory Press.

Halweil, B. (2000). Organic farming thrives worldwide, in *Vital Signs 2000*. Worldwatch Institute. New York: Norton.

Hansen, J. (2012). Coal: the greatest threat to civilization, in Butler *et al.*, eds., *Energy*.

Harding, S. (2004). Food web complexity enhances ecological and climate stability in a Gaian ecosystem model, in S. H. Schneider, *et al.*, eds., *Scientists Debate Gaia: The Next Century*. Cambridge, MA: MIT Press.

(2006). *Animate Earth: Science, Intuition and Gaia*. Totnes, UK: Green Books.

(2009). *Animate Earth*, 2nd edn. Totnes, UK: Green Books.

Harman, J. (2013). *The Shark's Paintbrush*. New York: Doubleday.

Harrington, A. (1997). *The Placebo Effect*. Watertown, MA: Harvard University Press.

Hauser, M. (2007). *Moral Minds: How Nature Designed Our Universal Sense of Right and Wrong*. New York: Little, Brown.

Havel, V. (1990). *Disturbing the Peace*. London and Boston: Faber and Faber.

(1997). Address to Forum 2000 conference. Prague (www.vaclavhavel.cz).

Hawken, P. (1993). *The Ecology of Commerce*. New York: HarperCollins.

(2000). N30: WTO showdown. *Yes!*, March.

(2008). *Blessed Unrest*. London: Penguin.

Hawken, P., A. Lovins, and H. Lovins (1999). *Natural Capitalism*. New York: Little, Brown.

Hawking, S. (1988). *A Brief History of Time*. New York: Bantam Books.

Hawking, S. and L. Mlodinow (2010). *The Grand Design*. New York: Bantam Books.

Heilbroner, R. (1978). Inescapable Marx. *New York Review of Books*, June 29.

(1980). *The Worldly Philosophers*. New York: Simon and Schuster.

Heims, S. (1991). *The Cybernetics Group*. Cambridge, MA: MIT Press.

Heinberg, R. (2011). *The End of Growth*. Gabriola Island, Canada: New Society Publishers.

Heisenberg, W. (1958). *Physics and Philosophy*. New York: Harper Torchbooks.

(1969). *Der Teil und das Ganze*. München: Piper (English edition (1971), titled *Physics and Beyond*, New York: Harper & Row).

Held, D. (1990). *Introduction to Critical Theory*. Berkeley, CA: University of California Press.

Henderson, H. (1978). *Creating Alternative Futures*. New York: Putnam.

(1981). *The Politics of the Solar Age*. New York: Doubleday/Anchor.

Hester, R. (2006). *Design for Ecological Democracy*. Cambridge, MA: MIT Press.

Higashi, M. and T.P. Burns, eds. (1991). *Theoretical Studies of Ecosystems: The Network Perspective*. New York: Cambridge University Press.

Hilborn, R. (2000). *Chaos and Nonlinear Dynamics*, 2nd edn. New York: Oxford University Press.

Hirao, I. and A.D. Ellington (1995). Re-creating the RNA world. *Current Biology*, **5**: 1017–22.

Ho, M.-W. (1998). *Genetic Engineering: Dream or Nightmare?* Bath: Gateway Books.

Holdrege, C. (1996). *Genetics and the Manipulation of Life*. Hudson, NY: Lindisfarne Press.

Hubbell, S.J. (2006). Neutral theory and the evolution of ecological equivalence. *Ecology*, **87**: 1387–98.

Humphrey, N. (2006). *Seeing Red: A Study on Consciousness*. Watertown, MA: Harvard University Press.

Huntley, H.E. (1970). *The Divine Proportion*. New York: Dover.

Hutchison, C.A., III, S.N. Peterson, S.R. Gill, R.T. Cline, O. White, C.M. Fraser, *et al.* (1999). Global transposon mutagenesis and a minimal mycoplasma genome. *Science*, **286**: 2165–9

Hutchinson, G.E. (1948). Circular causal systems in ecology. *Annals of the New York Academy of Sciences*, **50**: 221–46.

Hutton, W. and A. Giddens, eds. (2000). *Global Capitalism*. New York: The New Press.

Huxley J. (1942). *Evolution: The Modern Synthesis*. London: Allen and Unwin.

(1956). Epigenetics. *Nature*, **177**: 807–9.

IAASTD (International Assessment of Agricultural Knowledge, Science, and Technology for Development) (2009). Agriculture at a crossroads. *IAASTD Global Report*. Washington, DC: Island Press.

Jacob, F. (1982). *The Possible and the Actual*. Seattle, WA: University of Washington Press.

Jantsch, E. (1980). *The Self-Organizing Universe*. New York: Pergamon.

Jeong, H., B. Tombor, R. Albert, Z.N. Oltval, and A.L. Barabási (2000). The large-scale organization of metabolic networks. *Nature*, **407**: 651–4.

Jimenez-Prieto, R., M. Silva, and D. Perez-Bendito (1998). Approaching the use of oscillating reactions for analytical monitoring. *Analyst*, **123**: 1R–8R.

Johnson, E.T and C. Schmidt-Dannert (2008). Light-energy conversion in engineered microorganisms. *Trends in Biotechnology*. **26**(12): 682–9.

Johnston, J. and F. Baylis (2004). What happened to gene therapy? A review of recent events. *Clinical Researcher*, **4**(1): 11–15.

Jørgensen, S.E. and F. Müller, eds. (2000). *Handbook of Ecosystem Theories and Management*. Boca Raton, FL: CRC Press, Lewis Publishers.

Joyce, G.F. and E. Orgel (1993). Prospects for understanding the origin of the RNA world, in *The RNA World*. Cold Spring Harbor, NY: Cold Spring Harbor Laboratory Press, pp. 1–25.

Kauffman, S. (2008). *Reinventing the Sacred*. New York: Basic Books.

Kay, J. (2000). Ecosystems as self-organizing holarchic open systems, in F. Muller, ed., *Handbook of Ecosystem Theories and Management*. Boca Raton, FL: CRC Press, Lewis Publishers.

Keller, E.F. (2000). *The Century of the Gene*. Cambridge, MA: Harvard University Press.

(2005). Ecosystems, organisms, and machines. *BioScience*, **55**(12): 1069–74.

Kelley, K., ed. (1988). *The Home Planet*. New York: Addison-Wesley.

Kelly, M. (2001). *The Divine Right of Capital*. San Francisco, CA: Berrett-Koehler.

(2012). *Owning Our Future*. San Francisco, CA: Berrett-Koehler.

Khor, M. (1999/2000). The revolt of developing nations. *Third World Resurgence*. December/January.

Kim, J. (1984). Concepts of supervenience. *Philosophy and Phenomenological Research*, **45**: 153–76.

Kimura, M. (1968). Evolutionary rate at the molecular level. *Nature*, **217**: 624–6.

(1983). *The Neutral Theory of Molecular Evolution*. Cambridge University Press.

Kobayashi, K. and Y. Kanaizuka (1977). Reassembly of living cells from dissociated components in *Bryopsis*. *Plant & Cell Physiology*, **18**: 1373–7.

Kondepudi, D.K., R. Kaufman, and N. Singh (1990). Chiral symmetry breaking in sodium chlorate crystallization. *Science*, **250**: 975–6.

Kondepudi, D.K. and I. Prigogine (1981). Sensitivity of non-equilibrium systems. *Physica A: Statistical Mechanics and Its Applications*, **107**: 1–24.

Koohafkan, P. and M. Altieri (2010). *Globally Important Agricultural Heritage Systems*. United Nations FAO Report. Rome.

Korten, D. (2001). *When Corporations Rule the World*. San Francisco, CA: Berrett-Koehler.
Kroft, S. (2008). The bet that blew up Wall Street. *CBS Sixty Minutes*, October 16.
Kropoktin, P. (1902). *Mutual Aid*. London: William Heinemann.
Kuhn, T. (1962). *The Structure of Scientific Revolutions*. University of Chicago Press.
La Botz, D., R. Brenner, and J. Jordan (2012). The significance of Occupy. *Solidarity* (www.solidarity-us.org).
Lakoff, G. and M. Johnson (1980). *Metaphors We Live By*. University of Chicago Press.
(1999). *Philosophy in the Flesh*. New York: Basic Books.
Lakoff, G. and R. Núñez (2000). *Where Mathematics Comes From*. New York: Basic Books.
Lander, E. (2011). Initial impact of the sequencing of the human genome. *Nature*, **470**: 187–97.
Lappé, F.M. (2009). Liberation ecology. *Resurgence* (UK), January/February.
Lappé, F.M., J. Collins, and P. Rosset (1998). *World Hunger: Twelve Myths*. New York: Grove Press.
Leaky, R. and R. Lewin (1995). *The Sixth Extinction*. New York: Doubleday.
Lee, S.K., H. Chou, T.S. Ham, T.S. Lee, and J.D. Keasling. (2008). Metabolic engineering of microorganisms for biofuels production: from bugs to synthetic biology to fuels. *Current Opinion in Biotechnology*, **19**(6): 556–63.
Lehman, N. and G.F. Joyce (1993). Evolution *in vitro*: analysis of a lineage of ribozymes. *Current Biology*, **3**: 723–34.
LeShan, L. (1969). Physicists and mystics: similarities in world view. *Journal of Transpersonal Psychology*, **1**: 1–20.
Levin, J. (2002). *How the Universe Got Its Spots*. Princeton University Press.
Lewontin, R.C. (1991). Gene, organism and environment, in D.S. Bendall, ed., *Evolution from Molecules to Men*. Cambridge University Press, pp. 273–85.
Lincoln, T.A. and G.F. Joyce (2009). Self-sustained replication of an RNA enzyme. *Science*, **323**: 1229–32.
Livio, M. (2002). *The Golden Ratio*. New York: Broadway Books.
Lovelock, J. (1972). Gaia as seen through the atmosphere. *Atmospheric Environment*, **6**: 579.
(1979). *Gaia*. Oxford University Press.
(1988). *The Ages of Gaia*. New York: Norton.
(1991). *Healing Gaia*. New York: Harmony Books.
Lovelock, J. and L. Margulis (1974). Biological modulation of the Earth's atmosphere. *Icarus*, **21**: 471–89.
Lovins, A. (1977). *Soft Energy Paths*. New York: Harper & Row.
(2003). Twenty hydrogen myths. *RMI Report,* updated February 2005.
(2009a). Nuclear nonsense. *RMI Paper #10*.
(2009b). New nuclear reactors, same old story. *RMI Solutions Journal*, Spring.
(2009c). Reinventing Fire. *RMI Solutions Journal*, Fall.
(2011). Learning from Japan's nuclear disaster. *RMI Outlet*, March 19.
(2012). A 40-year plan for energy. TED Talk, www.ted.com.
Lovins, A., *et al.* (2004). *Winning the Oil Endgame*. Snowmass, CO: Rocky Mountain Institute.
(2011). *Reinventing Fire*. White River Junction, VT: Chelsea Green.
Luhmann, K. (1984). *Soziale Systeme*. Berlin: Suhrkamp.
Luhmann, N. (1990). *Essays on Self-Reference*. New York: Columbia University Press.
Luisi, P.L. (1997). Self-reproduction of chemical structures and the question of the transition to life, in C.B. Cosmovici, S. Bowyer, and D. Werthimer, eds., *Astronomical*

and Biochemical Origins and the Search for Life in the Universe. Milan: Editrice Compositori, pp. 461–8.
(2002). Toward the engineering of minimal living cells. *Anatomical Record*, **268**: 208–14.
(2003). Contingency and determinism. Philosophical Transactions of the Royal Society of London. Series A, **361**: 1141–7.
(2006). *The Emergence of Life: From Chemical Origins to Synthetic Biology*. Cambridge University Press.
(2007). Chemical aspects of synthetic biology. *Chemistry and Biodiversity*, **4**: 603–21.
(2008). The two pillars of Buddhism: consciousness and ethics. *Journal of Consciousness Studies*, **15**: 84–107.
(2009). *Mind and Life: Discussions with the Dalai Lama on the Nature of Reality*. New York: Columbia University Press.
(2011). The synthetic approach in biology: epistemological notes for synthetic biology, in P.L. Luisi and C. Chiarabelli, eds., *Chemical Synthetic Biology*. Chichester: Wiley, pp. 343–62.
(2012). On the origin of metabolism. *Chemistry and Biodiversity*, **9**: 1–11.
Luisi, P.L., M. Allegretti, T. de Souza, F. Steineger, A. Fahr, and P. Stano (2010). Spontaneous protein crowding in liposomes: a new vista for the origin of cellular metabolism. Chembiochem: *A European Journal of Chemical Biology*, **11**: 1989–92.
Luisi, P.L., A. Lazcano, and F. Varela (1996). What is life? Defining life and the transition to life, in M. Rizzotti, ed., *Defining Life: the Central Problem in Theoretical Biology*. Padua: University of Padova, pp. 149–65.
Luisi, P.L. and Stano, P., eds. (2011). *The Minimal Cell*. Dordrecht: Springer.
Lukes, S., ed. (1986). *Power*. New York: New York University Press.
Lutz, A., L.L. Greischar, N. Rawlings, M. Ricard, and R.J. Davidson (2004). Long-term meditators self-induce high-amplitude gamma synchrony during mental practice. *Proceedings of the National Academy of Sciences of the United States of America*, **101**(46): 16369–73.
MacArthur, R.H. (1955). Fluctuations of animal populations and a measure of community stability. *Ecology*, **36**: 533–6.
Mader, S.S. (1996). *Biology*, 5th edn. Dubuque, IA: W.C. Brown.
Magurran, A. and M. Dornelas (2010). Biological diversity in a changing world. *Philosophical Transactions of the Royal Society of London. Series B*, **365**: 3593–7.
Malthus, T.R. (1798). *An Essay on the Principle of Population*. London: J. Johnson.
Mandelbrot, B. (1983). *The Fractal Geometry of Nature*. New York: Freeman.
Mander, J. (1991). *In the Absence of the Sacred*. San Francisco, CA: Sierra Club Books.
(2012). *The Capitalism Papers*. Berkeley, CA: Counterpoint.
Mander. J. and E. Goldsmith, eds. (1996). *The Case Against the Global Economy*. San Francisco, CA: Sierra Club Books.
Manolio, T.A., *et al*. (2009). Finding the missing heritability of complex diseases. *Nature*, **461**: 747–53.
Mansfield, V. (2008). *Tibetan Buddhism and Modern Physics*. West Conshohocken, PA: Templeton Press,
Margulis, L. (1970). *Origin of Eukaryotic Cells*. New Haven, CT: Yale University Press.
Margulis, L. and D. Sagan (1986). *Microcosmos*. New York: Summit.
(1995). *What Is Life?* New York: Simon & Schuster.
(2002). *Acquiring Genomes*. New York: Basic Books.
Maslow, A. (1964). *Religions, Values, and Peak-Experiences*. Columbus, OH: Ohio State University Press.

Mason, S.F. and G.E. Tranter (1983). The parity violating energy difference between enantiomeric molecules. *Molecular Physics*, **53**: 1091–1111.

Matthew, W.P., B. Gerland, and J.D. Sutherland (2009). Synthesis of activated pyrimidine ribonucleotides in prebiotically plausible conditions. *Nature*, **459**: 239–42

Maturana, H. (1980/1970). Biology of cognition, in Maturana and Varela, *Autopoiesis and Cognition*.

Maturana, H. and B. Poerkson (2004). *From Being to Doing*. Heidelberg: Carl-Auer.

Maturana, H. and F. Varela (1980/1972). Autopoiesis: the organization of the living (original title De maquinas y seres vivos), in Maturana and Varela, *Autopoiesis and Cognition*.

(1980). *Autopoiesis and Cognition*. Dordrecht: D. Reidel.

(1998). *The Tree of Knowledge*, rev, edn. Boston: Shambhala.

Mayr, E. (1942). *Systematics and the Origin of Species*. New York: Columbia University Press.

(2000). Darwin's influence on modern thought. *Scientific American*, July, 79–83.

McBride, J.M. and R.L. Carter (1991). Spontaneous resolution by stirred crystallization. *Angewandte Chemie (International Edition in English)*, **30**: 293–5.

McDonough, W. and M. Braungart (1998). The Next Industrial Revolution. *Atlantic Monthly*, October.

(2002). *Cradle to Cradle*. New York: North Point Press.

McKibben, B. (1989). *The End of Nature*. New York: Random House.

(2010). *Eaarth*. New York: Time Books/Henry Holt.

ed. (2012a). *The Global Warming Reader*. London: Penguin.

(2012b). Global warming's terrifying math. *Rolling Stone*, October 4.

(2012c). The Arctic ice crisis. *Rolling Stone*, August 16.

McKibben, B., *et al.* (2007). *Fight Global Warming Now*. New York: Henry Holt.

McLaughlin, B.P. (1992). The rise and fall of British emergentism, in A. Beckermann, H. Flohr, and J. Kim. eds., *Emergence or Reduction: Essays on the Prospects of Nonreductive Materialism*. Berlin: de Gruyter, pp. 49–3.

McMichael, A.J. (2001). *Human Frontiers, Environments, and Disease*. Cambridge University Press.

McQueen, D., I. Kickbusch, L. Potvin, *et al.* (2010). *Health and Modernity*. New York: Springer.

Merchant, C. (1980). *The Death of Nature*. New York: Harper & Row.

Micozzi, M. (2006). *Fundamentals of Complementary and Alternative Medicine*. St. Louis, MO: Saunders Elsevier.

Mill, J.S. (1872). *System of Logic*, 8th edn. London: Longmans, Green, Reader and Dyer.

Miller, G.T. (2007). *Living in the Environment*, 15th edn. Belmont, CA: Brooks/Cole.

Miller, M.B. and B.L. Basler (2001). Quorum sensing in bacteria. *Annual Review of Microbiology*, **55**: 165–99.

Miller, S.L. (1953). Production of amino acids under possible primitive Earth conditions. *Science*, **117**: 2351–61.

Mills, G.C. and D. Kenyon (1996). *The RNA world: a critique. Origins & Design*, **17**(1).

Mingers, J. (1992). The problems of social autopoiesis. *International Journal of General Systems*, **21**: 229–36.

(1995). *Self-Producing Systems*. New York: Plenum.

(1997). *Self-Producing Systems: Implications and Applications of Autopoiesis*. New York: Plenum.

Mofid, K. and S. Szeghi (2010). Economics in crisis: what do we tell the students?, Share the World's Resources (www.stwr.org).

Monod, J. (1971). *Chance and Necessity*. New York: Knopf.
Morgan, C.L. (1923). *Emergent Evolution*. London: Williams and Norgate.
Morgan, G. (1998). *Images of Organizations*. San Francisco, CA: Berrett-Koehler.
Morowitz, H. (1992). *Beginnings of Cellular Life*. New Haven, CT: Yale University Press.
Moyers, B., B. Flowers, and D. Grubin (1993). *Healing and the Mind*. New York: Doubleday.
Mulder-Sibanda, M., F.S. Sibanda-Mulder, L. D'Alois, and D. Verna (2002). Malnutrition in food surplus areas. *Food and Nutrition Bulletin*, **23**(3), 253–61.
Muller, N., R. Mendelsohn, and W. Nordhaus (2011). Environmental accounting for pollution in the United States economy. *American Economic Review*, **101**(5), 1649–75.
Myers, N. and J. Kent (2009). *Perverse Subsidies*. Washington, DC: Island Press.
Nader, L., ed. (2010). *The Energy Reader*. Malden, MA: Wiley-Blackwell.
Newman, P. and J. Kenworthy (1998). *Sustainability and Cities*. New York: Island Press.
Newton, I. (1952/1730). *Opticks: A Treatise of the Reflections, Refractions, Inflections & Colours of Light*, repr. New York: Dover, from 4th edn, with foreword by Albert Einstein and preface by I. Bernard Cohen.
(1999/1687). *The Principia: Mathematical Principles of Natural Philosophy*, trans. I. Bernard Cohen and A. Whitman assisted by J. Budenz; preceded by guide to Newton's *Principia* by I. Bernard Cohen. Berkeley, CA: University of California Press.
Nicolis, G. and I. Prigogine (1977). *Self-Organization in Non-equilibrium Systems*. New York: Wiley.
Noble, D. (2006). *The Music of Life*. Oxford University Press.
Noyes, R.M. (1989). Some models of chemical oscillators. *Journal of Chemical Education*, **66**: 190–1.
Ntarlagiannis, D., E.A. Atekwana, E.A. Hill, and Y. Gorby (2007). Microbial nanowires: is the subsurface "hardwired"? *Geophysical Research Letters*, **34**(17).
Oberholzer, T., M. Albrizio, and P.L. Luisi (1995). Polymerase chain reaction in liposomes. *Current Biology*, **2**: 677–82.
Odum, E. (1953). *Fundamentals of Ecology*. Philadelphia: Saunders.
Oparin, A.I. (1924). *Proishkhozhddenie Zhisni*. Moskowski Rabocii (1938) (trans. as *The Origin of Life*. New York: Dover, 1957).
Opinion Research Corporation (2007). *A Post-Fossil-Fuel America: Are Americans Ready to Make the Shift?* Princeton, NJ.
Oppenheimer, J.R. (1954). *Science and the Common Understanding*. New York: Oxford University Press.
Oreskes, N. and E. Conway (2010). *Merchants of Doubt*. New York: Bloomsbury.
Ornish, D. (1998). *Love and Survival*. New York: HarperCollins.
Orr, D. (1992). *Ecological Literacy*. Albany, NY: State University of New York Press.
(2002). *The Nature of Design*. New York: Oxford University Press.
Oyama, S., E. Paul, P.E. Griffiths, and R.D. Gray, eds. (2003). *Cycles of Contingency: Developmental Systems and Evolution*. Cambridge, MA: MIT Press.
Paley, W. (1802). Natural theology, or, Evidences of the existence and attributes of the Deity: collected from the appearances of nature. Philadelphia: H. Maxwell (12th edn., Charlottesville, VA: Ibis, 1986).
Pauli, G. (1998). *Upsizing*. Sheffield: Greenleaf.
Peitgen, H.-O. and P. Richter (1986). *The Beauty of Fractals*. New York: Springer.
Peitgen, H.-O., H. Jurgens, D. Saupe, and C. Zahlten (1990). *Fractals: An Animated Discussion*, VHS/Color/63 minutes. New York: Freeman.
Pelletier, K. (2000). *The Best Alternative Medicine*. New York: Simon and Schuster.

Penrose, R. (1994). *Shadows of the Mind: A Search for the Missing Science of Consciousness*. New York: Oxford University Press.

Penrose, R., S. Hameroff, and S. Kak, eds. (2011). *Conciousness and the Universe*, Cambridge, MA: Cosmology Science.

Pert, C. (1997). *Molecules of Emotion*. New York: Scribner.

Pert, C., H.E. Dreher, and M. Ruff (1998). The psychosomatic network. *Alternative Therapies in Health And Medicine*, **4**(4).

Petto, A.J. and R.L. Godfrey (2007). *Scientists Confront Intelligent Design and Creationism*. New York: Norton.

Petzinger, T. (1999). *The New Pioneers*. New York: Simon & Schuster.

Pieper, D.H. and W. Reineke (2000). Engineering bacteria for bioremediation. *Current Opinion in Biotechnology*, **11**(3): 262–70.

Pievani, T., in Charles Darwin, *L'origine delle specie. Abbozzo del 1842. Lettere 1844–1858. Comunicazione del 1858*. Torino: Einaudi, 2009.

(2011). *La vita inaspettata*. Milano: Cortina.

Pigliucci, M. and G.B. Müller (2010). *Evolution: The Extended Synthesis*. Cambridge, MA: MIT Press.

Polanyi, K. (1968). *Primitive, Archaic, and Modern Economics*. New York: Doubleday/Anchor.

Postman, N. (1992). *Technopoly*. New York: Knopf.

Pressman, E.K., I.M. Levin, and L.S. Sandakchiev (1973). Reassembly of an *Acetabularia mediterranea* cell from the nucleus, cytoplasm, and cell wall. *Protoplasma*, **76**: 34–41.

Pretty, J., J. Morrison, and R. Hine (2003). Reducing food poverty by increasing agricultural sustainability in the development countries. *Agriculture, Ecosystems and Environment*, **95**: 217–34.

Prigogine, I. (1980). *From Being to Becoming*. San Francisco, CA: Freeman.

(1989). The philosophy of instability. *Futures*, **21**(4): 396–400.

Prigogine, I. and P. Glansdorff (1971). *Thermodynamic Theory of Structure, Stability and Fluctuations*. New York: Wiley.

Prigogine, I. and I. Stengers (1984). *Order out of Chaos*. New York: Bantam.

Primas, H. (1998). Emergence in exact natural sciences. *Acta Politechnica Scandinavica*, **91**: 86–7.

Qiu, X., D.C. Rau, A.V. Parsegian, L.T. Fang, C.M. Knobler, and W.M. Gelbart (2011). Salt-dependent DNA–DNA spacings in intact bacteriophageλ reflect relative importance of DNA self-repulsion and bending energies. *Physical Review Letters*, **106**(2): 28102–11.

Quack, M. (2002). How important is parity violation for molecular and biomolecular chirality? *Angewandte Chemie*, **41**: 4618–30.

Quack, M. and J. Stohner (2003a). Combined multidimensional anharmonic and parity violating effects in CDBrClF. *Journal of Chemical Physics*, **119**: 11228–40.

(2003b). Molecular chirality and the fundamental symmetries of physics: influence of parity violation on rotovibrational frequencies and thermodynamic properties. *Chirality*, **15**: 375–6.

Rahula, W. (1967). *What the Buddha Taught*, 2nd enlarged edn. London: Gordon Fraser.

Raine, D. and P.L. Luisi (2012). Open questions on the origin of life (OQOL). *Origins of Life and Evolution of the Biosphere*, **42**: 379–83.

Ramonet, I. (2000). The control of pleasure. *Le Monde Diplomatique*, May.

Randall, J.H. (1976). *The Making of the Modern Mind*. New York: Columbia University Press.

Register, R. (2001). *Ecocities*. Berkeley, CA: Berkeley Hills Books.

Register, R. and B. Peeks, eds. (1997). *Village Wisdom / Future Cities*. Oakland, CA: Ecocity Builders.

Revonsuo, A. and M. Kamppinen, eds. (1994). *Consciousness in Philosophy and Cognitive Neuroscience*. Hillsdale, NJ: Lawrence Erlbaum.

Rich, A. (1977). *Of Woman Born*. New York: Norton.

Richardson, G.P. (1992). *Feedback Thought in Social Science and Systems Theory*. Philadelphia: University of Pennsylvania Press.

Rifkin, J. (2011). *The Third Industrial Revolution*. New York: Palgrave Macmillan.

Riggs, A.D., R.A. Martienssen, and V.E.A. Russo (1996). *Epigenetic Mechanisms of Gene Regulation*, Introduction. Cold Spring Harbor, NY: Cold Spring Harbor Laboratory Press.

Riste, T. and D. Sherrington, eds. (1996). *Physics of Biomaterials: Fluctuations, Self-Assembly and Evolution*. Nato Science Series: E, Applied Sciences. Dordrecht: Kluwer.

Robbins, J. (2001). *The Food Revolution*. Berkeley, CA: Conari Press.

Ross, N.W. (1966). *Three Ways of Asian Wisdom*. New York: Simon & Schuster.

Roszak, T. (1969). *The Making of a Counter Culture*. New York: Doubleday (1995 edn., Berkeley, CA: University of California Press).

Ruiz-Mirazo, K. and P.L. Luisi, eds. (2009). *Open Questions on the Origins of Life*. Special issue: *Origins of Life and Evolution of Biospheres*, **40**(4–5): 353–497.

Runion, G.E. (1972). *The Golden Section and Related Curiosa*. Glenview, IL: Scott, Foresman and Co.

(1990). *The Golden Section*. Palo Alto, CA: Dale Seymour Publications.

Russell, B. (1961). *History of Western Philosophy*. London: Allen & Unwin.

Sachs, J. (2011). *The Price of Civilization*. New York: Random House.

Sachs, W., ed. (1992). *The Development Dictionary*. London: Zed Books.

Schilp, P.A., ed. (1949). *Albert Einstein: Philosopher-Scientist*. Evanston, IL: Library of Living Philosophers.

Schilthuizen, M. and A. Davison (2005). The convoluted evolution of snail chirality. *Naturwissenschaften*, **92**(11): 504–15.

Schneider, S., J. Miller, E. Christ, and P. Boston, eds. (2004). *Scientists Debate Gaia: The Next Century*. Cambridge, MA: MIT Press.

Schopf, J.W. (1992). Paleobiology of the Archean, in J.W. Schopf and C. Klein, eds., *The Proterozoic Atmosphere: A Multidisciplinary Study*. Cambridge University Press, pp. 25–39.

(1993). Microfossils of the early archean apex chert: new evidence of the antiquity of life. *Science*, **260**: 640–6.

(2002). When did life begin?, in J.W. Schopf, ed., *Life's Origin: The Beginnings of Biological Evolution*. London: University of California Press, pp. 158–77.

Schröder, J. (1998). Emergence: non-deducibility or downward causation? *Philosophical Quarterly*, **48**: 434–52.

Schumacher, E.F. (1975). *Small Is Beautiful*. New York: Harper & Row.

Scruton, R. (2009). *Beauty*. New York: Oxford University Press.

Searle, J. (1984). *Minds, Brains, and Science*. Cambridge, MA: Harvard University Press.

(1995). The mystery of consciousness. *New York Review of Books*, November 2 and 16.

Sen, J. and P. Waterman, eds. (2009). *World Social Forum: Challenging Empires*. Montreal: Black Rose Books.

Senge, P. (1990). *The Fifth Discipline*. New York: Doubleday.

Shapiro, R. (1984). The improbability of prebiotic nucleic acid synthesis. *Origins of Life*, **14**: 565–70.

(1988). Prebiotic ribose synthesis: a critical analysis. *Origins of Life*, **18**: 71–85.

Shear, J. and R. Jevning (1999). Pure consciousness: scientific exploration of meditation techniques. *Journal of Consciousness Studies*, **6**(2–3), 189–209.

Shimizu, Y., A. Inoue, Y. Tomari, *et al.* (2001). Cell-free translation reconstituted with purified components. *Nature Biotechnology*, **19**: 751–5.

Shiner, E.K., K.P. Rumbaugh, and S.C. Williams (2005). Interkingdom signaling: deciphering the language of acyl homoserine lactones. *FEMS Microbiology Review*, **29**: 935–47.

Shiva, V. (1993). *Monocultures of the Mind: Biodiversity, Biotechnology and Agriculture*. New Delhi: Zed Press.

(1997). *Biopiracy*. Boston: South End Press.

(2000). The world on the edge, in Hutton and Giddens, *Global Capitalism*.

(2005). *Earth Democracy*. Cambridge, MA: South End Press.

ed. (2007). *Manifestos on the Future of Food and Seed*. Brooklyn, NY: South End Press.

Siderits, M., E. Thompson, and D. Zahavi (2011). *Self, No Self? Perspectives from Analytical, Phenomenological, and Indian Traditions*. New York: Oxford University Press.

Siegel, D. (2010). *Mindsight*. New York: Bantam.

Simms, A. (1999). *Selling Suicide*. London: Christian Aid.

Smith, D.G. (1998). The shareholder primacy norm. *Journal of Corporation Law*, **23**(2).

Smith, H.O., C.A. Hutchinson, III, C. Pfannkoch, and C.J. Venter (2003). Generating a synthetic genome by whole genome assembly: phiX174 bacteriophage from synthetic oligonucleotides. *Proceedings of the National Academy of Sciences of the United States of America*, **100**: 15440–5.

Smith, R.S. and B.H. Iglewski (2003). *P. aeruginosa* quorum sensing systems and virulence. *Current Opinion in Microbiology*, **6**: 56–60.

Smith, T.M. and R.L. Smith (2006). *Elements of Ecology*, 6th edn. San Francisco, CA: Pearson – Benjamin Cummings.

Smolin, L. (2004). Scientific alternatives to the anthropic principle. hep-th/0407213.

(2006). *The Trouble With Physics*. New York: Houghton Mifflin.

Snow, C.P. (1960). *The Two Cultures*. Cambridge University Press.

Sonea, S. and M. Panisset (1993). *A New Bacteriology*. Burlington, MA: Jones and Bartlett.

Spencer, H. (1891/1854). *Essays: Scientific, Political and Speculative*. Library Edition. London: Williams and Norgate, vol. II.

(1857). Progress: its law and cause. *Westminster Review*, **67**: 445–85.

Spencer, J. and J. Jacobs, eds. (1999). *Complementary/Alternative Medicine*. St. Louis, MO: Mosby.

Sperry, R.W. (1986). Discussions: macro- versus micro-determinism. *Philosophy of Science*, **53**: 265–70.

Spretnak, C., ed. (1981). *The Politics of Women's Spirituality*. New York: Anchor/Doubleday.

Stano, P., S. Bufali, C. Pisano, *et al.* (2004). Novel camptothecin analogue (gimatecan)-containing liposomes prepared by the ethanol injection method. *Journal of Liposome Research*, **14**: 87–109.

Stano, P. and P.L. Luisi, eds. (2007). Basic questions about the origins of life: proceedings of the Erice International School of Complexity. *Origins of Life and Evolution of Biospheres*, **37**: 303–7.

(2008). Self-reproduction of micelles, reverse micelles, and vesicles: compartments disclose a general transformation pattern, in A. Leitmannova Liu, ed., *Advances in Planar Lipid Bilayers and Liposomes*, vol. VII. Amsterdam: Elsevier Academic Press, pp. 221–63.

Steffen, W., A. Sanderson, P. Tyson, *et al.* (2004). *Global Change and the Earth System*. Berlin: Springer.

Steindl-Rast, D. (1990). Spirituality as common sense. *The Quest*, **3**(2).

Steingraber, S. (2012). The whole fracking enchilada, in Butler *et al.*, Energy.

Stern, N. (2006). *The Stern Review on the Economics of Climate Change*. London: HM Treasury.

Sternberg, E. (2000). *The Balance Within*. New York: Freeman.

Stewart, I. (2002). *Does God Play Dice?*, 2nd edn. Malden, MA: Blackwell.

(1998). *Life's Other Secret*. New York: Wiley.

(2011). *The Mathematics of Life*. New York: Basic Books.

Stiglitz, J. (2012). *The Price of Inequality*. New York: Norton.

Stone, M. (2009). *Smart by Nature: Schooling for Sustainability*. Berkeley, CA: Watershed Media.

Stone, M. and Z. Barlow, eds. (2005). *Ecological Literacy*. San Francisco, CA: Sierra Club Books.

Strogatz, S. (1994). *Nonlinear Dynamics and Chaos*. Cambridge, MA: Perseus.

(2001). Exploring complex networks. *Nature*, **410**: 268–76.

Strohman, R. (1997). The coming Kuhnian revolution in biology. *Nature Biotechnology*, **15**: 194–200.

Stryer, L. (1975). *Biochemistry*. New York: Freeman.

Suzuki, D.T. (1963). *Outlines of Mahayana Buddhism*. New York: Schocken Books.

Suzuki, D. and H. Dressel (1999). *From Naked Ape to Superspecies*. Toronto: Stoddart.

Suzuki, S. (1970). *Zen Mind, Beginner's Mind*. New York: Weatherhill.

Talbot, M. (1980). *Mysticism and the New Physics*. London: Routledge & Kegan Paul.

Taylor, F. (1911). *Principles of Scientific Management*. New York: Harper & Row.

Tegmark, M. (2003). Parallel universes.*Scientific American*, April 14.

Teuscher, C., ed. (2010). *Alan Turing: Life and Legacy of a Great Thinker*. New York: Springer.

Thompson, D. (1917). *On Growth and Form*. Cambridge University Press (abridged edn., ed. J.T. Bonner, Cambridge University Press, 1961).

Thompson, E. (2007). *Mind in Life*. Cambridge, MA: Belknap Press of Harvard University Press.

Thompson, E. and F.J. Varela (2001). Radical embodiment: neural dynamics and consciousness. *Trends in Cognitive Sciences*, **5**: 418–25.

Tokar, B., ed. (2001). *Redesigning Life?* New York: Zed.

Tomasello, M. (1999). *The Cultural Origins of Human Cognition*. Cambridge, MA: Harvard University Press.

Tononi, G. and G. Edelman (1998). Consciousness and complexity. *Science*, **282**: 1846–51.

Toyota, H., M. Hosokawa, I. Urabe, and T. Yomo (2008). Emergence of polyproline II-like structure at early stages of experimental evolution from random polypeptides. *Molecular Biology and Evolution*, **25**(6): 1113–19.

Tranter, G.E. (1985). The parity-violating energy difference between enantiomeric reactions. *Chemical Physics Letters*, **115**: 286–90.

Tucker, R., ed. (1972). *The Marx–Engels Reader*. New York: Norton.

Turing, A. (1950). Computing machinery and intelligence. *Mind: A Quarterly Review of Psychology and Philosophy*, **59**(236): 433–60.
United Nations Development Programme (UNDP) (1996). *Human Development Report*. New York: Oxford University Press.
(1999). *Human Development Report*. New York: Oxford University Press.
Van der Ryn, S. and S. Cowan (1996). *Ecological Design*. Washington, DC: Island Press.
Varela, F. (1995). Resonant cell assemblies. *Biological Research*, **28**: 81–95.
(1996). Neurophenomenology. *Journal of Consciousness Studies*, **3**(4): 330–49.
(1999). Present-time consciousness. *Journal of Consciousness Studies*, **6**(2–3): 111–40.
(2000). *El fenomeno de la vida*. Santiago, Chile: Dolmen.
Varela, F.J., H.R. Maturana, and R.B. Uribe (1974). Autopoiesis: the organization of living systems, its characterization and a model. *Biosystems*, **5**: 187–96.
Varela, F. and J. Shear (1999). First-person methodologies: what, why, how? *Journal of Consciousness Studies*, **6**(2–3): 1–14.
Varela, F. J., E. Thompson, and E. Rosch (1991). *The Embodied Mind*. Cambridge, MA: MIT Press.
Veomett, G., D.M. Prescott, J. Shay, and K.R. Porter (1974). Reconstruction of mammalian cells from nuclear and cytoplasmic components separated by treatment with cytochalasin B. *Proceedings of the National Academy of Sciences of the United States of America*, **71**: 1999–2002.
Vernadsky, V. (1986/1926). *The Biosphere*. Oracle, AZ: Synergetic Press.
Vrooman, J.R. (1970). *René Descartes*. New York: Putnam.
Waddington, C.H. (1939). *An Introduction to Modern Genetics*. New York: Macmillan.
(1953). The epigenotype. *Endeavour*, **1**: 18–20.
(1957). *The Strategy of the Genes*. London: George Allen & Unwin.
Waks, Z. and P.A. Silver (2009). Metabolic engineering of microorganisms for biofuels production. *Applied Environmental Microbiology*, **75**(7): 1867–75.
Walde, P., R. Wick, M. Fresta, A. Mangone, and P.L. Luisi (1994). Autopoietic self-reproduction of fatty acid vesicles. *Journal of the American Chemical Society*, **116**: 11649–54.
Ward, P. and D. Brownlee (2000). *Rare Earth*. New York: Copernicus.
Warkentin, C. and K. Mingst (2000). International institutions, the state, and global civil society in the age of the World Wide Web. *Global Governance*, **6**: 237–57.
Watson, A. and J. Lovelock (1983). Biological homeostasis of the global environment: the parable of Daisyworld. *Tellus*, **35B**: 284–9.
Watson, J. (1968). *The Double Helix: A Personal Acccount of the Discovery of the Structure of DNA*. New York: Atheneum.
Watson, S. (1995). *The Birth of the Beat Generation*. New York: Pantheon.
Watts, A. (1957). *The Way of Zen*. New York: Vintage Books.
Weatherall, D. (1998). How much has genetics helped? *Times Literary Supplement*, January 30.
Weber, C. and H. Scott Matthews (2008). Food miles and the relative climate impacts of food choices in the United States. *Environmental Science and Technology*, **42**: 3508–13.
Weber, M. (1976/1905). *The Protestant Ethic and the Spirit of Capitalism*. New York: Scribner.
Weil, A. (1995). *Spontaneous Healing*. New York: Knopf.
(2009). *Why Our Health Matters*. New York: Hudson.

Weinberg, S. (1987). Anthropic bound on the cosmological constant. *Physical Review Letters*, **59**(22): 2607–10.
Weiss, P. (1971). *Within the Gates of Science and Beyond*. New York: Hafner.
(1973). *The Science of Life*. Mount Kisco, NY: Futura.
Weissbuch, I., H. Zepik, G. Bolbach, *et al.* (2003). Homochiral oligopeptides by chiral amplification within two-dimensional crystalline self-assemblies at the air–water interface; relevance to biomolecular handedness. *Chemistry*, **9**(8): 1782–94.
Wenger, É. (1998). *Communities of Practice*. Cambridge University Press.
Westhof, E. and N. Hardy, eds. (2004). *Folding and Self-Assembly of Biological Macromolecules*. Hackensack, NJ: World Scientific Publishing.
Wheatley, M. (1999). *Leadership and the New Science*. San Francisco, CA: Berrett-Koehler.
Wheatley, M. and M. Kellner-Rogers (1998). Bringing life to organizational change. *Journal of Strategic Performance Measurement*, April/May.
Whitehead, A.N. (1929). *Process and Reality*. New York: Macmillan (repr. 1960).
Whitesides, G. and M. Boncheva (2002). Beyond molecules: self-assembly of mesoscopic and macroscopic components. *Proceedings of the National Academy of Sciences of the United States of America*, **99**: 4769–74.
Whitesides, G. and B. Grzybowski (2002). Self-assembly at all scales. *Science*, **295**: 2418–21.
Wiener, N. (1948). *Cybernetics*. Cambridge, MA: MIT Press (repr. 1961).
(1950). *The Human Use of Human Beings*. New York: Houghton Mifflin.
Wilder-Smith, A.E. (1968). *Man's Origin, Man's Destiny*. Wheaton, IL: Harold Shaw.
(1987). *The Scientific Alternative to Neo-Darwinian Evolutionary Theory*. Costa Mesa, CA: The Word for Today Publishers.
Williams, R. (1981). *Culture*. London: Fontana.
Wilson, D. and D.A. Reeder (1993). *Mammal Species of the World*, 2nd edn. Washington, DC: Smithsonian Institute Press.
Wilson, E.O. (1975). *Sociobiology: The New Synthesis*. Watertown, MA: Harvard University Press (25th anniversary edition published in 2000).
Wimsatt, W.C. (1972). Complexity and organization, in K.F. Schaffner and R.S. Cohen, eds., *Proceedings of the Philosophy of Science Association*. Boston Studies in the Philosophy of Science. Dordrecht: Reidel, pp. 67–86.
(1976). Reductionism, levels of organization, and the mind–body problem, in G. Globus, G. Maxwell, and I. Savodinik, eds., *Consciousness and the Brain*. New York: Plenum Press, pp. 205–66.
Windelband, W. (2001/1901). *A History of Philosophy*. Cresskill, NJ: Paper Tiger.
Winfree, A.T. (1984). The prehistory of the Belousov–Zhabotinsky oscillator. *Journal of Chemical Education*, **61**: 661–3.
Winner, L. (1977). *Autonomous Technology*. Cambridge, MA: MIT Press.
World Commission on Environment and Development (1987). *Our Common Future*. New York: Oxford University Press.
Wrangham, R. and D. Peterson (1996). *Demonic Males*. New York: Houghton Mifflin.
Zukav, G. (1979). *The Dancing Wu Li Masters*. New York: Morrow.

Index

Note: page numbers in *italics* refer to figures and tables; those in **bold** refer to boxes.